W0042833

Nonlinear Superconductive Electronics and Josephson Devices

Nonlinear Superconductive Electronics and Josephson Devices

Edited by

Giovanni Costabile

University of Salerno
Baronissi, Italy

Sergio Pagano

CNR-Institute of Cybernetics
Arco Felice (Naples), Italy

Niels Falsig Pedersen

The Technical University of Denmark
Lyngby, Denmark

and

Maurizio Russo

CNR-Institute of Cybernetics
Arco Felice (Naples), Italy

Springer Science+Business Media, LLC

ISBN 978-1-4613-6719-2   ISBN 978-1-4615-3852-3 (eBook)
DOI 10.1007/978-1-4615-3852-3

Proceedings of a NATO Advanced Research Workshop on Nonlinear
Superconductive Electronics and of the Second Workshop on Josephson Devices,
held September 3–7, 1990, in Capri, Italy

© 1991 by Springer Science+Business Media New York
Originally published by Plenum Press New York in 1991
Softcover reprint of the hardcover 1st edition 1991

All rights reserved

No part of this book may be reproduced, stored in a retrieval system, or transmitted
in any form or by any means, electronic, mechanical, photocopying, microfilming,
recording, or otherwise, without written permission from the Publisher

PREFACE

The on-going developments, and the recent achievements of the superconducting electronics (especially in the field of Josephson junctions and the inherent nonlinear dynamics) inspired us to organize a conference where different groups working on the subject could meet and discuss the latest results of their investigations.

This idea was realized as two joint workshops, the NATO Advanced Research Workshop on Superconducting Electronics with Prof. N.F. Pedersen as chairman, and the 2nd Workshop on Josephson Devices, with Profs. G. Costabile and M. Russo as chairmen, held in Capri, Italy on September 3-7, 1990.

The Workshops were very successful. About 70 scientists from 12 countries (Denmark, France, Germany, Greece, Italy, Japan, People's Republic of China, Sweden, United Kingdom, USA, USSR, and Venezuela) enjoyed the many interesting and mostly informal occasions for scientific exchanges as well as the very pleasant weather of Southern Italy.

We are very grateful to the Institutions which made possible the realization and the success of the conference with their financial support : NATO Science Committee through the NATO International Scientific Exchange Programmes, University of Salerno, and Istituto di Cibernetica of the Consiglio Nazionale delle Ricerche (C.N.R.) of Italy. The conference was held under the auspices of the Progetto Finalizzato "Superconductive and Cryogenic Technologies" of C.N.R.

Finally special thanks go to our Conference Secretary Anna Maria Mazzarella for carrying out most of the organizative work, and for her continuous "on stage" support in solving all the problems which inevitably arise in such occasions.

G. Costabile and M. Russo

INTRODUCTION

The development of superconducting electronics has seen a boosting in the last years. A number of commercial devices start to use superconducting components both in analogic and digital circuits. Moreover, the continuous improvements in the fabrication technique and the development of the full-niobium Josephson junction technology make possible the production of complex superconductive circuits, characterized by a high degree of reproducibility and stability. Furthermore, after the discovery of the new class of high-critical-temperature ceramic superconductors, new perspectives of exciting and exotic applications have arisen. At the same time many theoretical aspects, regarding both basic and applied physics, have been further developed.

This book covers many recent achievements in the field of superconductive electronics with special emphasis on devices based on the Josephson effect. In the following paragraphs a brief summary of the covered topics is given.

Since the discovery of the Josephson effect great effort was put in developing very fast superconducting digital circuitry. The impresive recent Japanese results are reported in the paper by S. Takada (32,000 junction 4 bit Josephson processor).

On the side of the analog applications, the recent developments in the field of the Josephson voltage standard are reported by J. Niemeyer et al. (14000 junctions 10 Volts chip), while D. Andreone et al. , and H.G. Meyer et al. have reported new effects such as biharmonic junction drive, and synchronous array operation respectively. Great progress has characterized also the field of SIS (Superconductor Insulator Superconductor junction) mixers. Such devices are unique for the realization of highly sensitive millimeter-wave receivers, featuring noise levels down to the quantum limit. This subject is extensively covered by R. Blundell et al., and D. Winkler et al.

The most known component of superconducting electronics is the SQUID (Superconducting Quantum Interferometer Device) featuring an extreme sensitivity level in the detection of magnetic flux. Very recent achievements in this field are reported by M.B. Ketchen, while results on recent investigations on SQUIDs based on the new high-critical-temperature superconductors can be found in the papers by V. Foglietti, and G.J. Cui et al.

Other applications of superconductive electronics are covered in the papers by J.B. Green (fast analog signal processing), A. Barone et al. (nuclear particle detection), Y. Zhang (flux-flow three terminal device), and A. Andreone et al. (rf properties of thin superconducting films).

On the side of the basic physics a number of topics are covered. Noise effects in Josephson junctions are reviewed in the papers by J.B. Hansen (general noise) and P. Silvestrini (quantum fluctuations). The exotic world of ultra-small Josephson junctions is extensively explored by T. Claeson et al., and S.A. Hattel et al. Quantum simulations of Josephson junctions are reported in the paper by A. Davidson et al.

A great deal of work has been devoted in the study of nonlinear effects in large Josephson junctions, like solitonic wave propagation, and chaos. An extensive review of results on the interaction between solitons and electromagnetic field is given by G.Filatrella et al., while more specific aspects are reported in the papers by M. Salerno et al. (boundary coupling), H.D. Jensen et al. (long junctions in microstrip resonators), M. Cirillo (coupled long junctions), and W. Kretch et al. (long junction arrays).

Other aspects of nonlinear dynamics in long Josephson junctions are matter of the papers by A. Ustinov (soliton interaction with spatial inhomogenities), T. Skitiotis et al. (Josephson junction coupling), T. Doderer et al. (Josephson junction spatial structures imaging), S. Pagano et al. (critical static junction configurations), T. Holst et al. (plasma frequency tuning), L.E. Guerrero et al. (chaos in long Josephson junctions), and P.L. Christiansen et al. (new sine-Gordon solutions).

The discovery of the high-critical-temperature superconductivity has triggered a large effort for the understanding of the mechanism of superconductivity in copper oxides. Some structural and intrinsic aspects of this subject are investigated in the papers by G.A. Ovsyannikov et al. (vortex propagation in granular superconductors), and M.P. Soerensen (nonlinear solutions of the BCS gap equation).

The number of subjects covered in this volume, and the depth reached in the understanding of the basic physics involved, indicates that superconductive electronics has reached a state where it can be considered a mature technology capable of useful applications. Nevertheless there are many interesting aspects which deserve further investigations, and will hopefully provide the ground for future applications.

CONTENTS

PROGRESS ON JOSEPHSON COMPUTER

Susumu TAKADA

Electrotechnical Laboratory

1-1-4 Umezono, Tsukuba, Ibaraki, 305 Japan

INTRODUCTION

Josephson tunnel junctions are expected to be one of the most promising superconducting electronic devices which can be operated at ultra-low cryogenic temperature. Josephson tunnel junction has great advantage of ultra-high switching speed of pico second with very little electric power consumption of micro watt.

Josephson junction integration process has been developed for digital computer using Nb and NbN superconducting materials. Ultra-thin insulating films for tunnel barrier were fabricated by oxidizing surface of aluminum metal for Nb junctions, and by sputtering deposition of MgO for NbN junctions. Recently, Josephson LSI circuits have been developed on the basis of Nb junction integration technology. In order to construct a Josephson computer, many technologies such as circuit design of both logic and memory, integration process, system design, and low temperature packaging are required. Up to date there are many reports on the individual technology developed.

Several kinds of Josephson processor experiments have been reported[1,2,3,4], in which high speed clock cycles of above 1GHz[1,2], and full operation of 4-bit processor function[3] were achieved. On the Josephson random access memory (RAM), 1-kbit RAM chips[5,6] using Henkels' cell[7] and variable-threshold cell[8] have been demonstrated, in which 40% and 100% access operations were confirmed. Capacitor-coupled cell and vortex-transition cell have been applied to the 4-kbit RAM chips[9,10].

To our knowledge, a Josephson computer, which is composed of the processor and the memory chips and which can executes a computer program, has not yet been reported after the work of Josephson cross-sectional computer model by the IBM group[11].

In the Electrotechnical Laboratory we have developed the key technology for Josephson computer under the national project named "High-speed computing system for scientific and technological uses" from fiscal 1981 to 1989 in Japan. A Josephson junction integration process based on Nb and NbN has been developed to fabricate reliable Josephson LSI chips in that program. A current injection logic gate named "4JL" (four Josephson junction logic)[12] has been extended to construct any logic circuits including combination logic and sequential logic[13]. A multi-phase power supply method[14] has been developed to achieve high speed operation and to solve the punch-through problem. A 1-kbit RAM chip has been developed using a new type of Josephson RAM cell named variable threshold memory cell[15]. Josephson ROM(read-only memory) chip has been developed by employing a plane-like two-junction dc SQUID, in which the 128-step program consisting of 10-bit instruction codes can be stored[16].

In order to demonstrate the feasibility of the Josephson computer system, the Josephson computer ETL-JC1 (Electrotechnical Laboratory Josephson computer no. 1) has been designed and fabricated, in which the stored program can be executed.

Nonlinear Superconductive Electronics and Josephson Devices
Edited by G. Costabile *et al.*, Plenum Press, New York, 1991

In this paper, the key technology to develop the ETL-JC1, the experimental results and discussion will be presented.

EPHSON LSI KEY TECHNOLOGY

sephson tunnel junctions based on Nb and NbN have been fabricated using the tri-layer tion sandwich method with a self-aligned junction isolation process.   Significant ovement in junction quality has been achieved by introducing an artificial barrier into the tions.   Amorphous silicon(a-Si) was first applied successfully to fabricate Nb/Nb tions by Kroger et. al.[17].   Gurvich et. al.[18] have demonstrated Nb/Al-oxide/Nb tions, which show low leakage current below the gap voltage and a steep slope at the gap age.   In that junction process, the Al surface deposited on Nb electrode is thermally ized to make Al oxide tunnel barrier.

This junction is one of the most promising junctions from the viewpoint of digital ications.   Morohashi et. al.[19] reported that Nb/Al-oxide/Nb junctions were fabricated g a sputtering system in a conventional vacuum chamber.   Nakagawa et. al.[20] found the deterioration of the I-V characteristics of small junctions with several microns was to the internal stress in the tri-layer of the junction.   Significant improvement in the I-V e was obtained by introducing Nb underlayers beneath the tri-layer of the junction. ire 1 shows the I-V characteristics of the Nb/Al-oxide/Nb junction fabricated by the

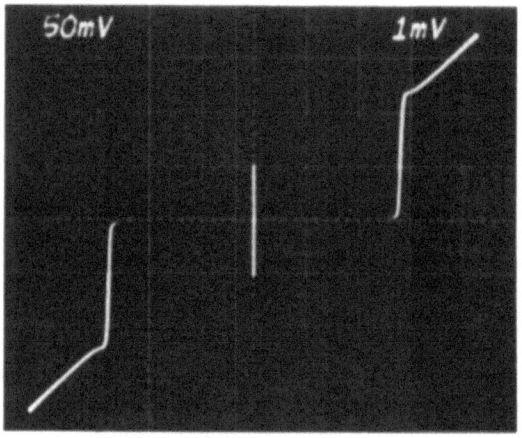

Fig. 1.   I-V characteristics of 3-μm Nb/AlOx/Nb junction
fabricated using the underlayer method.
Vertical: 50μA/div., horizontal: 1mV/div.

erlayer method.   This deterioration mechanism in the I-V characteristics has been stigated systematically by other authors[21,22].

here are many kinds of Josephson logic gates for high speed switching operations.   The ent injection logic gates have many advantages of high-current gain with wide-operating gin and small-gate size.   Much work has been done on the current injection gates using unction direct coupling gate (4JL[23]), (b)resistor coupling gate (JAWS[24], DCL[25], _[26], RCJL[27]) and (c)inductive coupling gate (CIL[28], MVTL[29]).

ι order to construct any logic circuits the 4JL gate has been extended to make a set of logic family.   Figure 2 shows schematic configurations of the 4JL gate family consisting of wo-input OR-, (b) two-input AND-, (c) AMP- and (d) timed INVERT-gates, and their shold curves[13].   Wide operating margin of more than ±35% is obtained in their gate. rder to suppress the backlash current after the switching in the next logic stage, the OR-s are used to drive the AND-gate, AMP-gate and INVERT-gate.   The AMP-gate is used btain a fan-out number of four in the logic circuit.

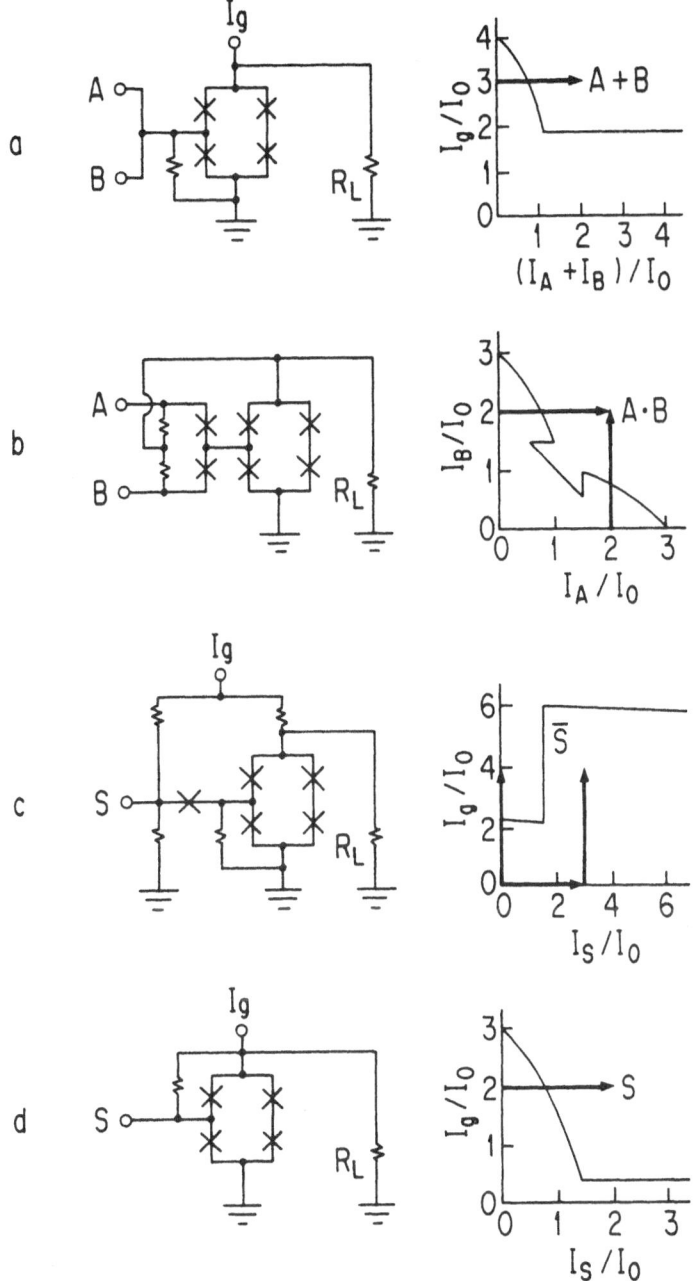

Fig. 2. A logic gate family of the 4JL gate, consisting of (a) OR-gate, (b) AND-gate, (c) AMP-gate, and (d) INVERT-gate with their threshold curves.

Various Josephson integrated circuits have been demonstrated in order to show high performance in digital applications using the 4JL-gate family. A 4-bit multiplier circuit consisting of 652 4JL gates was successfully demonstrated using NbN/NbN junctions with 2.5-μm design rules[30]. A multiplication time of 1ns was achieved in this circuit. The 2-bit Josephson ALU (arithmetic logic unit) has been demonstrated using a carry look-ahead algorithm, in which the 4JL gates were integrated using 3-μm Nb/Al-oxide/Nb junctions[31]. A logic delay time of 165ps was measured for A minus B arithmetic operation.

Data-latch function is quite important in the logic circuit, especially in the sequential logic circuits to store the data for a short time duration in which the alternative power supply reset Josephson junctions. The direct-coupled data-latch circuit[32] which consists of seven 4JL-gates and two single junctions was developed. The circuit configuration is shown in Fig.3. The latch circuit can generate the dual outputs of true (T) and complement (C), driving by a

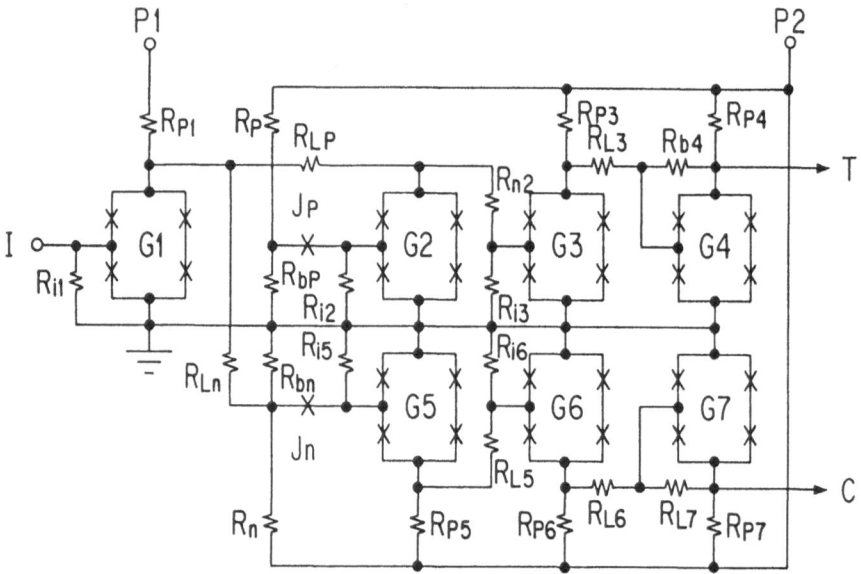

Fig. 3. A circuit diagram of the direct-coupled data-latch. The single Josephson junctions are presented by x.

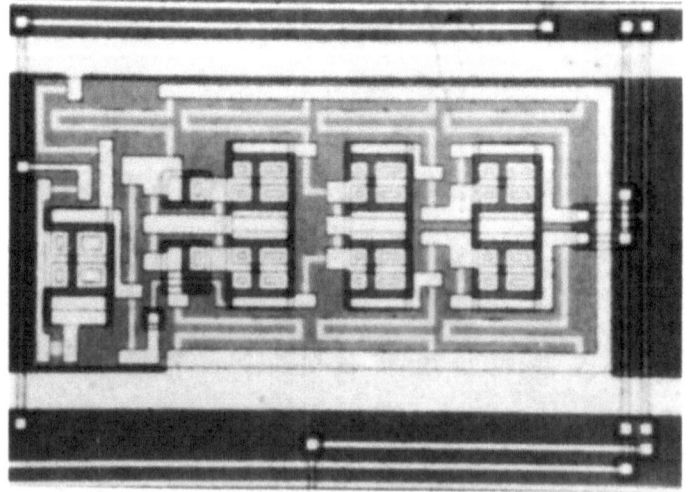

Fig. 4. A photograph of the direct-coupled data-latch circuit fabricated using 3-μm Nb/AlOx/Nb junctions.

two-phase power supply [32] of P1 and P2 . The single junctions of Jp and Jn in this figure are used to prevent the input signal from free-running for overlapping time of P1 and P2 power. Flip-flop function is carried out by making a feedback loop with a pair of the two data-latch circuits. This data-latch does not need any inductances for the data storage, so that the circuit can be reduced in size, resulting in high speed operation. Figure 4 shows the direct-coupled data-latch circuit fabricated using a 3-μm Nb/AlOx/Nb junction process. The occupied area is about 165μm x 300μm which is a few and ten times smaller than the conventional dc- and ac-data latches by the IBM[33,34]. The data latch circuit has been extended to construct sequential circuits such as a 4-bit shift-register[13] and a 4-bit digital counter[35].

A read-only memory (ROM) is very important to construct digital circuit system, especially in a computer system. This memory is used to store programs and to decode instruction codes in a computer. Some studies of a Josephson ROM unit have been reported[36, 37, 38]. In these studies a ZERO ROM cell was designed using either a two-junction dc SQUID in which control line keeps away from a normal position so as to reduce magnetic coupling, or a large junction instead of a two-junction dc SQUID so that it does not switch to the voltage state. A ONE ROM cell was designed using a two-junction dc SQUID without a damping resistor. All these ROM cells were consisting of two Josephson junctions and a bridge-like inductance. Since the superconducting loop of these ROM cells have no dumping resistor so that the operating margin of the ROM unit becomes small due to the resonance effect. In order to make the ROM chip for the ETL-JC1 a plane-like two-junction dc SQUID was employed[39]. The photomask pattern and the circuit diagram of (a) ONE and (b) ZERO ROM cells are shown in Fig.5. The damping resistor Rd was introduced to

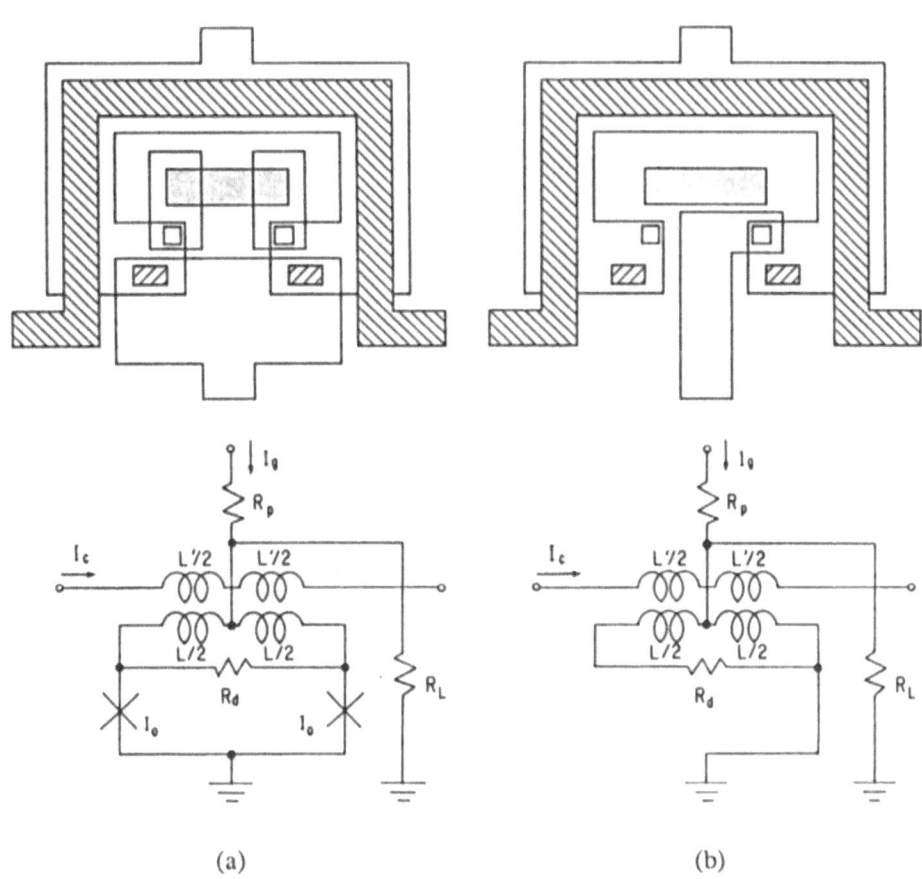

(a)   (b)

Fig. 5.   Photomask patterns and circuit diagrams of (a) ONE and (b) ZERO ROM cells consisting of a plane-like dc SQUID.

suppress resonance in the ONE ROM cell.   In the ZERO ROM cell a complete superconducting shunt was carried out using a top superconducting wiring layer.   This shunt makes no magnetic coupling since the junctions disconnected to the bit-line as shown in Fig.5(b).

Random access memory (RAM) is an essential element in ultra-high speed digital computer system.   Considerable efforts have been done in order to achieve high speed Josephson memory chips.   1-bit Josephson RAM [6] which consists of the variable threshold memory cell was developed for the Josephson computer ETL-JC1.   The circuit diagram of the RAM cell is shown in Fig.6(a).   The operating principle of this memory cell is based on an asymmetric dc SQUID.   This cell consists of a three-junction dc-SQUID writing gate and a single junction J2.   The information of the bit ONE is stored with a single quantum magnetic flux.   The cell has threshold curves which can distinguish the existence of the single quantum magnetic flux in the cell since the cell changes the threshold curve as shown in

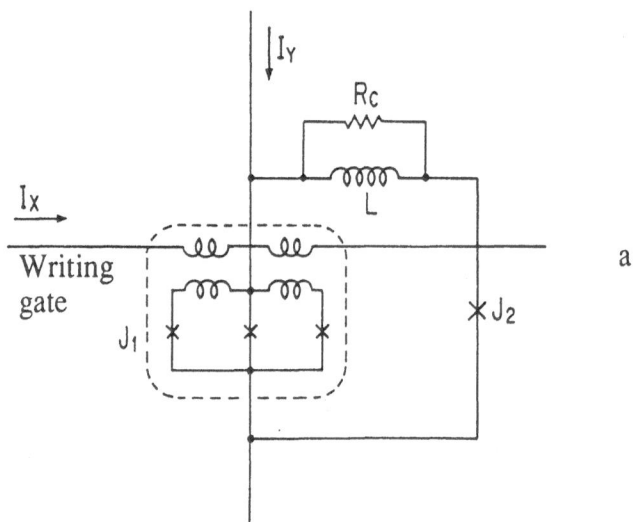

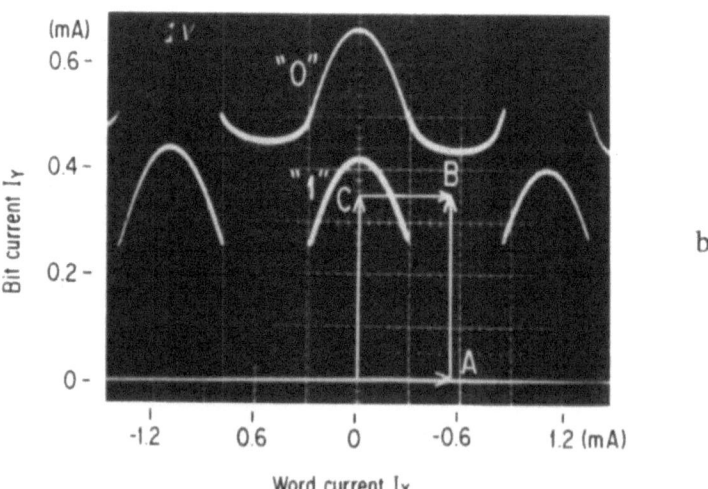

Fig. 6.   (a) circuit diagram of the variable threshold memory cell for RAM and
(b) threshold characteristics.

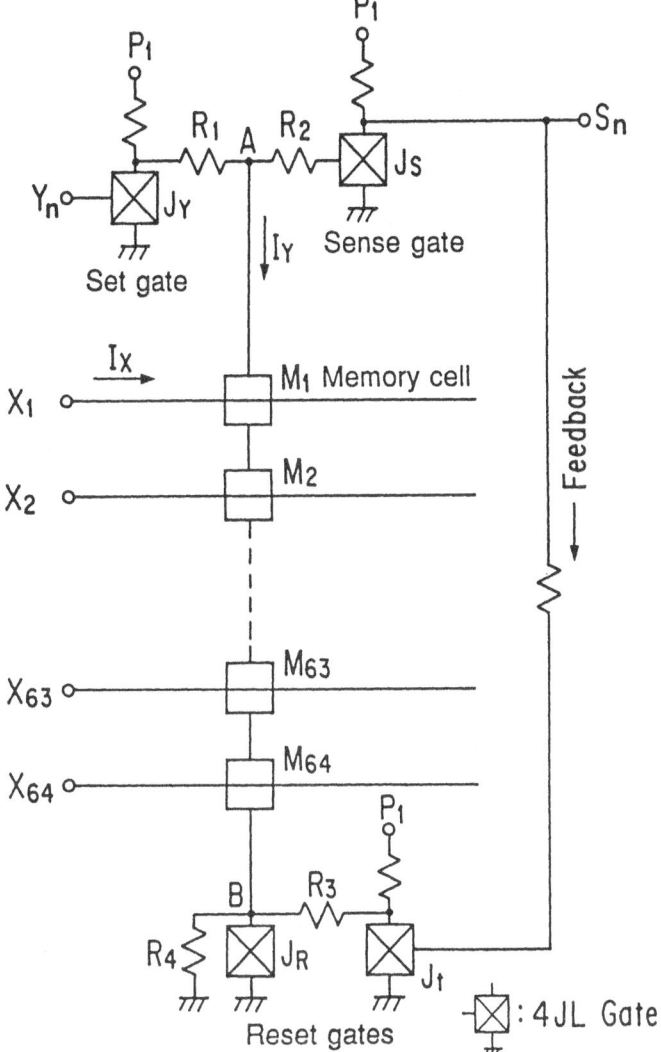

Fig. 7. Circuit to drive the RAM cells in the bit-line. The peripherally circuits
consists of the logic gates to set and reset the cells for NDRO.

Fig.6(b). The writing operation is carried out by driving the cell following the arrows of A
and B in the threshold curve. The read-out operation is carried out by the arrows of C and
B. The memory cells are connected and driven in series in the memory plane as shown in
Fig.7. It should be noted that the set and reset operation is performed by the logic gates
without any large superconducting closed loop. By this method, the drive current of the bit
line can be kept without current fluctuations due to the quantum magnetic flux in the large
superconducting loop, resulting in the wide operating margin in the RAM cells. By
introducing re-writing circuit of the feedback loop consisting of the logic gates, non
destructive read-out operation (NDRO) can be achieved.

Power system is very important to operate the Josephson LSI system, since the Josephson
junctions are essentially driven by the ac power in the latching mode operation. Figure 8
shows two kinds of power supply system of (a) conventional sinusoidal power supply and
(b) two-phase power supply. The sinusoidal power supply system has been investigated by

the IBM group[40]. The dc-latch[33] and ac-latch[34] using a large inductance storage loop were employed to store the data for a short period (Td) when the power supply goes to zero. In the two-phase power supply the dead time is eliminated and continuous logic operation can be performed. In the sinusoidal power supply some additional dead time is required to suppress the punch-through effect. The two-phase power supply can eliminate this time interval in the logic circuit. By this powering method the data latch circuit was significantly reduced in size without any inductances.

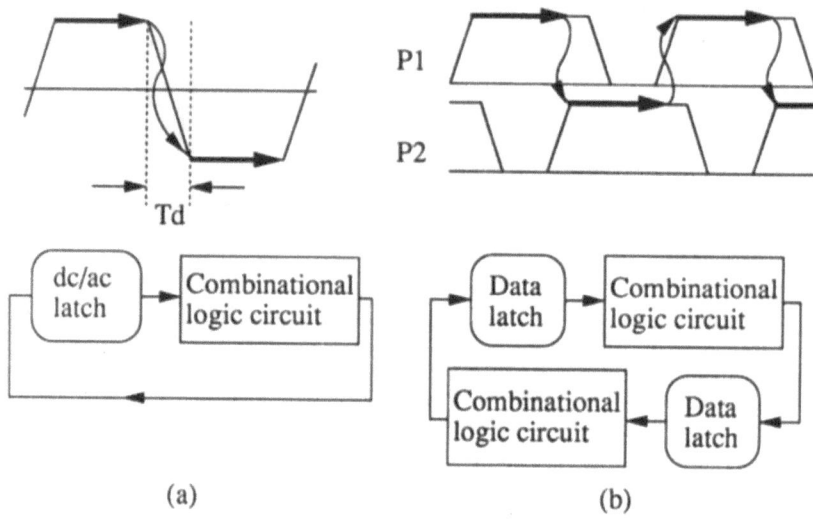

(a)   (b)

Fig. 8. Josephson power supply systems of (a) sinusoidal power supply and (b) two-phase power supply.

DESIGN OF THE ETL-JC1

The ETL-JC1 has been designed to demonstrate the feasibility of a Josephson computer system. Main feature of the ETL-JC1 is listed in Table I. A RISC (reduced instruction set computer) architecture was introduced to execute all instructions at every clock cycle. A Harvard architecture is employed, in which the data bus line is separated from the instruction bus line among the chips. The logic circuits are designed using dual-rail logic employing OR-gates and AND-gates. The logic function NOT is performed by using a direct-coupled data latch. All circuits are driven by the two-phase power supply. Arithmetic logic unit (ALU) is designed based on a ripple carry method for 4-bit data. Registers are prepared for the general purpose data storage and the accumulator. 7-bit registers are used to make two-level stack register for double-nesting subroutine call/return operation. 10-bit registers are used to hold 10-bit instruction codes in the instruction decoder. In order to store the 128-step program, the instruction ROM[39] was designed using the plane-like two-junction dc SQUID. 1-kbit RAM consists of 4-bit x 256-word memory plane using the variable threshold memory cell, in which access operation is carried out by 8-bit address data. 27 kinds of basic instruction codes to make a computer program is listed in Table II. The instruction codes include arithmetic logic function, conditional skip, write/read access for RAM, jumps, shifts, moves and subroutine call/return. These instructions are considered to be sufficient to make any computer program.

A block diagram of the ETL-JC1 is shown in Fig.9. The computer circuit is divided into four Josephson LSI chips, named RALU (register arithmetic logic unit), SQCU (sequence control unit), IROU (instruction read-only memory unit), and DRAU (data random access memory unit). These chips are connected each other as shown in the figure. The 128-step

Table I. Main feature of ETL-JC1

Architecture:	RISC (reduced instruction set computer) architecture, Harvard architecture
Arithmetic logic unit:	4-bit (ripple carry)
Register:	4-bit x 6 (ACC, MD, ML, ML', MH and MH'), 7-bit x 2 (two-level stack register), 10-bit x 2 (instruction)
Instruction ROM:	10-bit x 128-word, 7-bit address, 2-junction SQUID cell
Data RAM:	4-bit x 256-word, 8-bit address, Variable threshold memory cell
Functions:	16 arithmetic logic, 9 sequence control, 2 memory control
Logic gate:	Four-junction logic (4JL) gate family
Data latch:	Direct-coupled data-latch
Input/output:	10-bit instruction code, 2-bit external control/4-bit ACC, 7-bit program address
Power:	Tow-phase power supply, split power supply
Number of logic gates:	3658
Number of junctions:	22000 (Nb/AlOx/Nb tunnel junction)
Cycle time:	1 ns (by computer logic simulation)

Table II. 10-bit instruction codes for the ETL-JC1

MNEMONIC	MEANING	INSTRUCTION CODE MSB LSB	FUNCTION
JMP	Jump	0 0 0 A A A A A A A	address $\Rightarrow$ [PC]
CAL	Call subroutine	0 0 1 A A A A A A A	[S0] $\Rightarrow$ [S1] , [PC]+1 $\Rightarrow$ [S0] , address $\Rightarrow$ [PC]
RET	Return	0 1 0 0 x x x x x x	[S0] $\Rightarrow$ [PC] , [S1] $\Rightarrow$ [S0]
SZA	Skip if accumulator is zero	0 1 0 1 1 0 0 0 0 0	if [ACC]=0 $\Rightarrow$ SKIP and [PC]+2 $\Rightarrow$ [PC]
SNA	Skip if accumulator is not zero	0 1 0 1 0 1 0 0 0 0	if [ACC]≠0 $\Rightarrow$ SKIP and [PC]+2 $\Rightarrow$ [PC]
SPA	Skip if accumulator is positive	0 1 0 1 0 0 1 0 0 0	if [ACC]≥0 $\Rightarrow$ SKIP and [PC]+2 $\Rightarrow$ [PC]
SMA	Skip if accumulator is negative	0 1 0 1 0 0 0 1 0 0	if [ACC]<0 $\Rightarrow$ SKIP and [PC]+2 $\Rightarrow$ [PC]
SZC	Skip if carry flag is zero	0 1 0 1 0 0 0 0 1 0	if [C]=0 $\Rightarrow$ SKIP and [PC]+2 $\Rightarrow$ [PC]
SNC	Skip if carry flag is not zero	0 1 0 1 0 0 0 0 0 1	if [C]=1 $\Rightarrow$ SKIP and [PC]+2 $\Rightarrow$ [PC]
CMC	Complement carry flag	0 1 1 0 x x x x x x	complement [C] $\Rightarrow$ [C]
SWP	Swap register	0 1 1 1 0 0 x x x x	[MH] $\Leftrightarrow$ [MH'] , [ML] $\Leftrightarrow$ [ML']
RDM	Read memory	0 1 1 1 0 1 x x x x	$\Rightarrow$ READ , ([MH][ML]) $\Rightarrow$ [MD]
WTM	Write memory	0 1 1 1 1 0 x x x x	$\Rightarrow$ WRITE , [MD] $\Rightarrow$ ([MH][ML])
ADD	Addition	1 0 s s 0 0 0 0 d d	[ACC] plus [Rss] $\Rightarrow$ [Rdd]
AND	Logical AND	1 0 s s 0 0 0 1 d d	[ACC] • [Rss] $\Rightarrow$ [Rdd]
MOV	Move	1 0 s s 0 0 1 0 d d	[Rss] $\Rightarrow$ [Rdd]
OR	Logical OR	1 0 s s 0 0 1 1 d d	[ACC] + [Rss] $\Rightarrow$ [Rdd]
XOR	Exclusive OR	1 0 s s 0 1 0 0 d d	[ACC] $\oplus$ [Rss] $\Rightarrow$ [Rdd]
MVA	Move accumulator's data	1 0 x x 0 1 0 1 d d	[ACC] $\Rightarrow$ [Rdd]
CMA	Complement accumulator's data	1 0 x x 0 1 1 0 d d	complement [ACC] $\Rightarrow$ [Rdd]
SUB	Subtraction	1 0 s s 0 1 1 1 d d	[ACC] minus [Rss] $\Rightarrow$ [Rdd]
RTR	Rotate right	1 0 x x 1 0 0 0 d d	rotate right [ACC] $\Rightarrow$ [Rdd]
RTL	Rotate left	1 0 x x 1 1 0 0 d d	rotate left [ACC] $\Rightarrow$ [Rdd]
ADDI	Addition for immediate data	1 1 i i i i 0 0 d d	[ACC] plus i i i i $\Rightarrow$ [Rdd]
ANDI	Logical AND for immediate data	1 1 i i i i 0 1 d d	[ACC] • i i i i $\Rightarrow$ [Rdd]
MVI	Move immediate data	1 1 i i i i 1 0 d d	i i i i $\Rightarrow$ [Rdd]
ORI	Logical OR for immediate data	1 1 i i i i 1 1 d d	[ACC]+i i i i $\Rightarrow$ [Rdd]

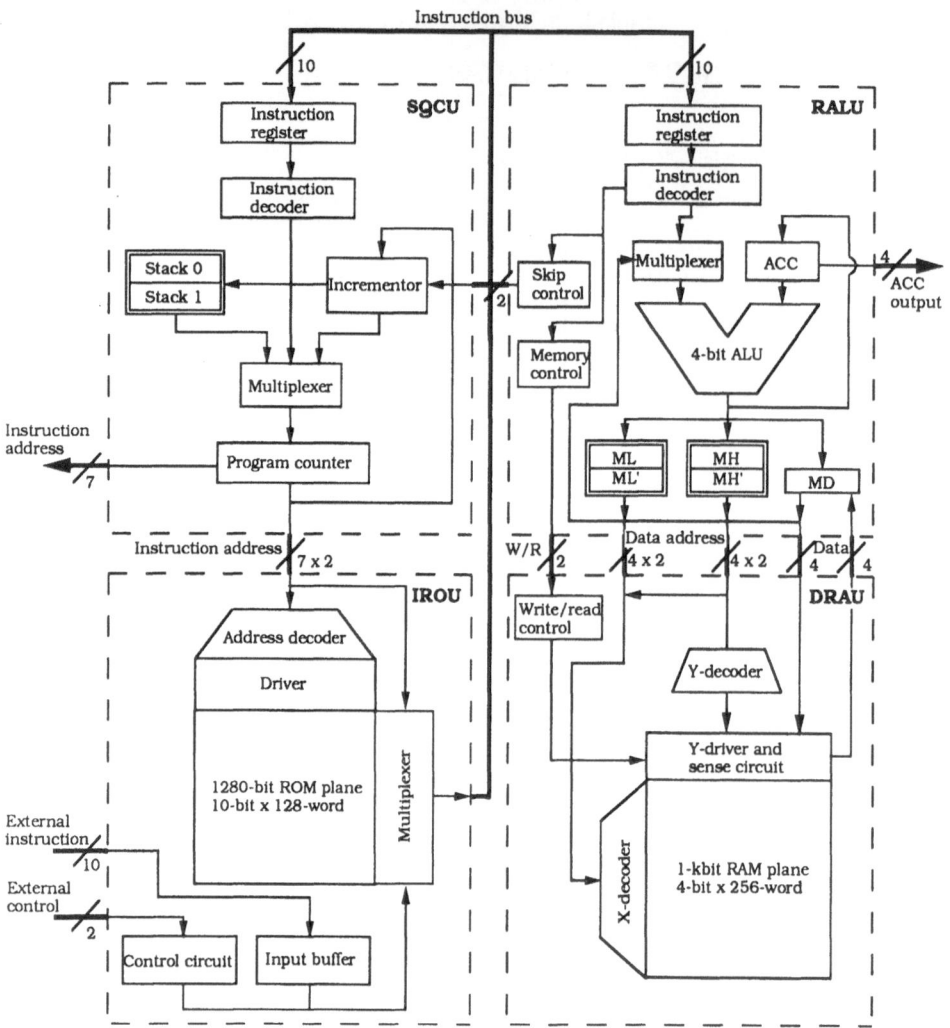

Fig. 9. A block diagram of the ETL-JC1.

computer program is installed in the 1,280-bit ROM plane on the IROU chip to operate the ETL-JC1. Every 10-bit instruction code of the program is fetched from the IROU chip through the 10-bit instruction bus line and is decoded in both RALU and SQCU chips. The RALU chip performs the arithmetic logic functions and the results are stored in the 4-bit ACC accumulator, MD, ML/ML' and MH/MH' registers. The skip control signal is generated in the RALU chip to control the program sequence in the SQCU chip. The SQCU chip controls the sequence of the computer program execution. Two 7-bit stack registers, stack1 and stack2 are used to store the return address of the subroutine call operation with a double nesting execution. The DRAU chip is only connected to the RALU chip to store the 4-bit data. The 4-bit data for the write/read operations for the 1-kbit RAM of the DRAU chip are stored in the MD register of the RALU chip. The 8-bit address data are divided into two 4-bit lower and higher addresses and are stored in ML and MH registers. Write and read modes of the DRAU chip are controlled by a 2-bit W/R signal, which is generated in the RALU chip.

In order to start the program execution, a 10-bit instruction "JMP" (jump to the program address) is loaded externally into the IROU chip. This instruction has the 7-bit data to indicate the initial program address.   To control the program address externally, the 2-bit control signal is used to switch the mode from internal to external in the IROU chip.   Once the initial 7-bit program address is assigned by the feeding "JMP" instruction externally, the program execution proceeds automatically.   This address is installed with the lower 7-bits in the 10-bit jump instruction code.   The experimental operation of the computer is verified by monitoring the 4-bit ACC output and the 7-bit program address output.

CAD SYSTEM FOR JOSEPHSON LSI PATTERN

Prior to making the Josephson LSI chips, all circuits of the ETL-JC1 were verified by the logic simulation using a CAD (computer aided design) system.   The logic circuits of the RALU and the SQCU chips were translated to the photomask pattern in the CAD system.   A standard-cell method was employed to make the layout of the logic gate cells and wiring among the logic gate cells.

Figure 10 shows the procedure for the simulation and the photomask pattern generation by the CAD system.   First, on the basis of the LSI circuit feature the LSI circuits are composed of a set of standard-cells, consisting of OR, AND, INVERT, LATCH, and BUFFER gates (see Ref. 41).   A set of the cell patterns of the standard-cells was designed using the pattern CAD system (Applicon AGS-860), taking account of the fabrication process.   The information is extracted from the cell pattern to indicate the cell size, the location of the via for input, output signals and powering, and wiring rule.   The layout cell model is defined in the logic CAD (Mentor Graphics IDEA) by using this information.   The logic cell model presents the electronic properties of the logic delay, logic function, fan-in, fan-out characteristics, latching operation for logic gates, by combining the standard semi-conductor logic gate model installed in the logic CAD.   Using this logic cell model all LSI circuits were designed and simulated to verify the full logic functions.   The longest logic path in the circuit was investigated by using this logic CAD.

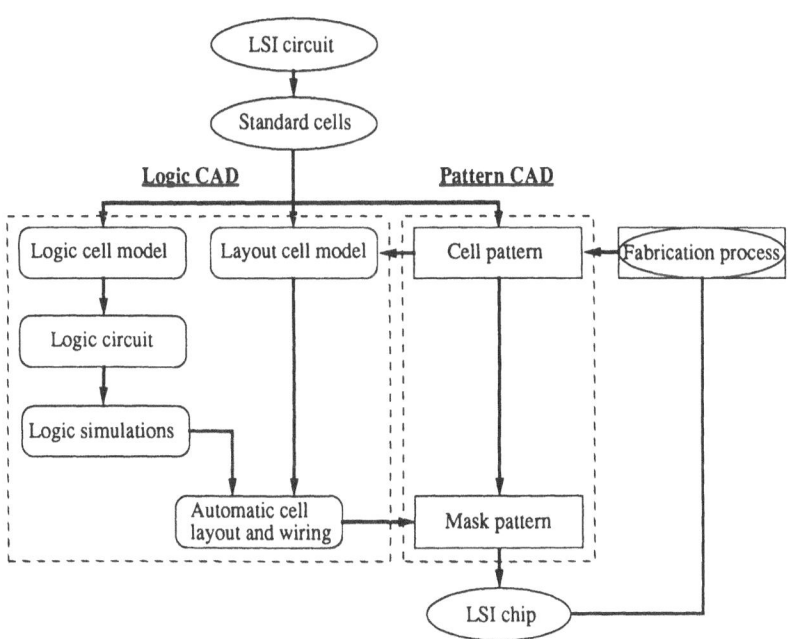

Fig.10. CAD system to design the Josephson LSI circuits and to make photomask pattern using a standard-cell method.

Figure 11 shows an example of the logic circuit presented by using the symbols of the logic cell models. The circuit is 1-bit register circuit used in the 7-bit stack register in the SQCU chip. Wiring ports denoted by alphabetic symbols in the figure are connected to other circuit in the LSI. In this diagram the number of fan-in and fan-out is checked by the logic CAD system.

The layout and wiring of the LSI circuit is carried out by combining the logic circuit and the layout cell model in the logic CAD shown in Fig.10. Finally, the photomask pattern of the LSI is generated to transfer the information of the logic CAD system to the pattern CAD system.

Figure 12 shows the CAD pattern of the RALU chip which is obtained following above

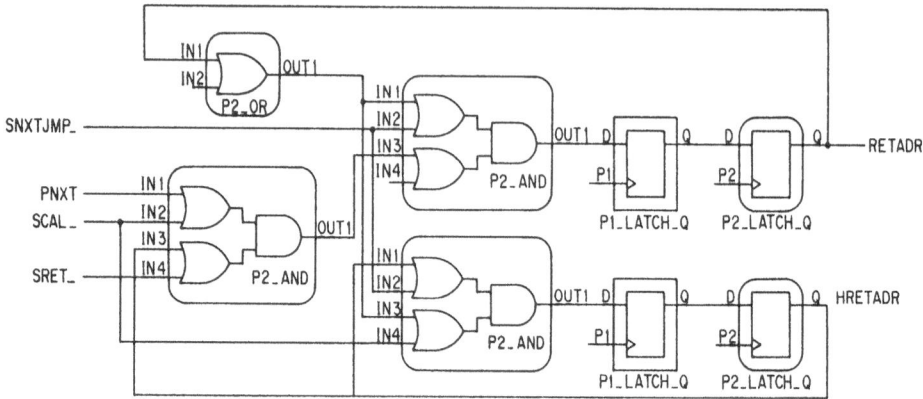

Fig.11. One-bit register circuit presented by logic cell model.

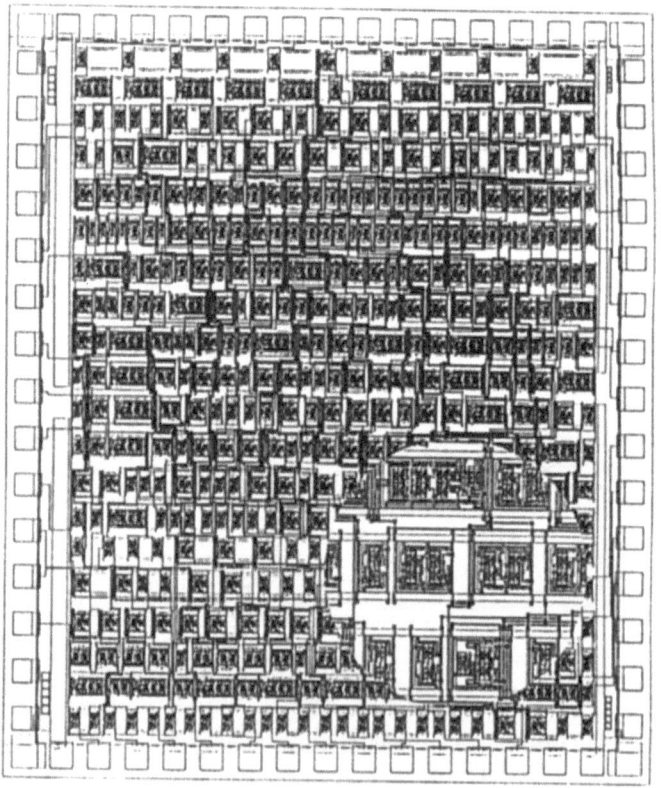

Fig.12. CAD pattern of the RALU chip, in which all layout of the logic gates and wiring were carried out.

procedure.   446 standard-gate cells which correspond to 1,273 logic gates are arranged in 20 lines.   The insertion figure shows an enlarged pattern of the standard-gate cells in the chip.

Table III shows a part of the 128-step program which was installed in the IROU chip for the ETL-JC1.   The source program was first written by the mnemonic expression using the computer VAX11/780.   This program was translated to the 10-bit instruction codes on the basis of Table II.   These instruction codes were transferred to the logic CAD system to make layout of ROM cells in the 1,280-bit ROM plane.   The program from the address of 0000011 to 0001011 is used to test the ACC, MD, ML and MH registers, using the MVI instruction which operates to move 4-bit data to the destination register immediately, MVA to move the data in ACC to the destination register, and MOV to move the data from the source register to the destination register.   The following program is used to test the operation for accessing the 4-bit data in the RAM from 0001100 to 0010100 and the arithmetic logic operations such as add, subtraction, logic AND, XOR and so on from 0010101 to 0101001. The 128-step program is fully presented in Ref.39.

Table III.   A part of the 128-step program for the ETL-JC1

Address	Instruction	Label	Mnemonic	Operand
0000000	0000000000	DUMMY1:	JMP	DUMMY1
0000001	0000000000		JMP	DUMMY1
0000010	0000000000		JMP	DUMMY1
0000011	1100011000	TEST1:	MVI	1, ACC
0000100	1000010101		MVA	MD
0000101	1100101000		MVI	2, ACC
0000110	1000010110		MVA	ML
0000111	1100111000		MVI	3, ACC
0001000	1000010111		MVA	MH
0001001	1001001000		MOV	MD, ACC
0001010	1010001000		MOV	ML, ACC
0001011	1011001000		MOV	MH, ACC
0001100	1101101001		MVI	6, MD
0001101	1001001010		MOV	MD, ML
0001110	1010001011		MOV	ML, MH
0001111	1011001000		MOV	MH, ACC
0010000	0111100000		WTM	
0010001	1111111001		MVI	F, MD
0010010	1001001000		MOV	MD, ACC
0010011	0111010000		RDM	
0010100	1001001000		MOV	MD, ACC
0010101	1001000010		ADD	MD, ML
0010110	1010001000		MOV	ML, ACC
0010111	1011000010		ADD	MH, ML
0011000	1010001000		MOV	ML, ACC
0011001	1011011100		SUB	MH, ACC
0011010	1001011100		SUB	MD, ACC
0011011	1010000100		AND	ML, ACC
0011100	1011010001		XOR	MH, MD
0011101	1001001000		MOV	MD, ACC
0011110	1000011001		CMA	MD
0011111	1001001000		MOV	MD, ACC
0100000	1011001111		OR	MH, MH
0100001	1011001000		MOV	MH, ACC
0100010	1101100100		ANDI	6, ACC
0100011	1110001111		ORI	8, MH
0100100	1011001000		MOV	MH, ACC
0100101	1100010010		ADDI	1, ML
0100110	1010001000		MOV	ML, ACC
0100111	1101010001		ADDI	5, MD
0101000	1001001000		MOV	MD, ACC
0101001	1000100000		RTR	ACC

FABRICATION

All LSI chips were fabricated using Nb/AlOx/Nb tunnel junctions with 3-μm design rules. Figure 13 shows a schematic cross section of the Josephson LSI chips. The Josephson junctions were made by using fabrication techniques of SNIP (self-aligned niobium isolation process)[42] with a Nb underlayer[20]. Josephson LSI chips were fabricated on 2-inch silicon substrates. The Nb superconducting ground plane was deposited by dc magnetron sputtering method. The ground plane insulation layer was composed of double insulating films of SiO and Nb_2O_5. The MgO thin film was used to protect the ground plane insulation layer during the reactive ion etching (RIE) process to make the Josephson junctions. The SiO evaporation film was used to make electric isolation among the superconducting layers. The Josephson critical current density was chosen to be 560 A/cm^2 in all LSI chips, which makes a maximum gate current of the typical 4JL gate to be 200 μA. The resistors were integrated using Pd thin film[43] which has a sheet resistance of 2 Ω. The two-level wiring layers were made by using Nb film and Pb-In alloy film. 8-level photomasks were used to fabricate RALU, SQCU and IROU chips, and 9 levels for DRAU chip. Details of the fabrication process are described in Ref. 44.

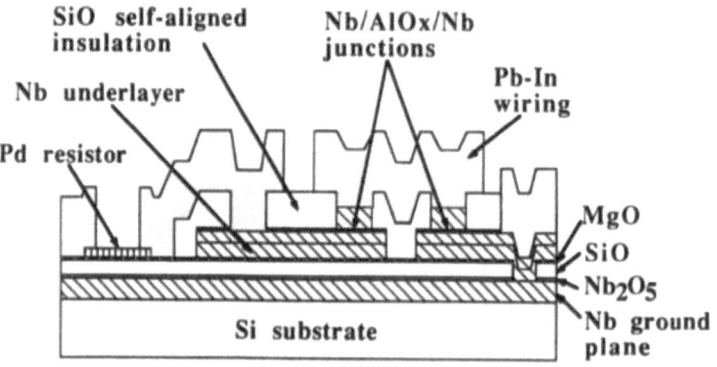

Fig.13. Schematic cross section of the Josephson LSI chips.

LSI CHIPS FOR ETL-JC1

Figure 14 shows (a) RALU , (b) SQCU, (c) IROU and (d) DRAU chips fabricated for the ETL-JC1. The RALU chip consists of 1,273 4JL gates. The size of the chip is 4.3mm x 5mm. Details of the chip are described in Ref. 41. The SQCU chip shown in Fig.14(b) was newly fabricated using Nb after converting the previous NbN circuit[30] to the Nb version using the CAD system. 593 4JL-gates were integrated on the chip. The size of the chip is 3.1mm x 3.2mm. This chip was driven by single port for the two-phase power supply. Full-sequence control functions were verified in the SQCU chip.

The IROU chip in Fig.14(c) has a 128-step program written by 10-bit instruction codes listed in Table II. The 1,280-bit ROM plane is composed of two-column x 64 lines. The chip size is 3.7mm x 5mm. An OR-type address decoder is implemented in this memory chip instead of the conventional AND-decoder in order to reduce the number of logic gates and delay time. All outputs of the decoder are inverted simultaneously by NOT-gates, which are driven by the two-phase power supply and an additional power named "split power supply"[39].

On the DRAU chip the 1-kbit Josephson RAM (random access memory) is integrated as shown in Fig.14(d). 256 words of 4-bit data can be written in and read out from the RAM chip. 100% memory cell access experiment was confirmed by this chip[6]. Size of the chip is 3.7mm x 3.7mm. The OR-type address decoder was implemented.

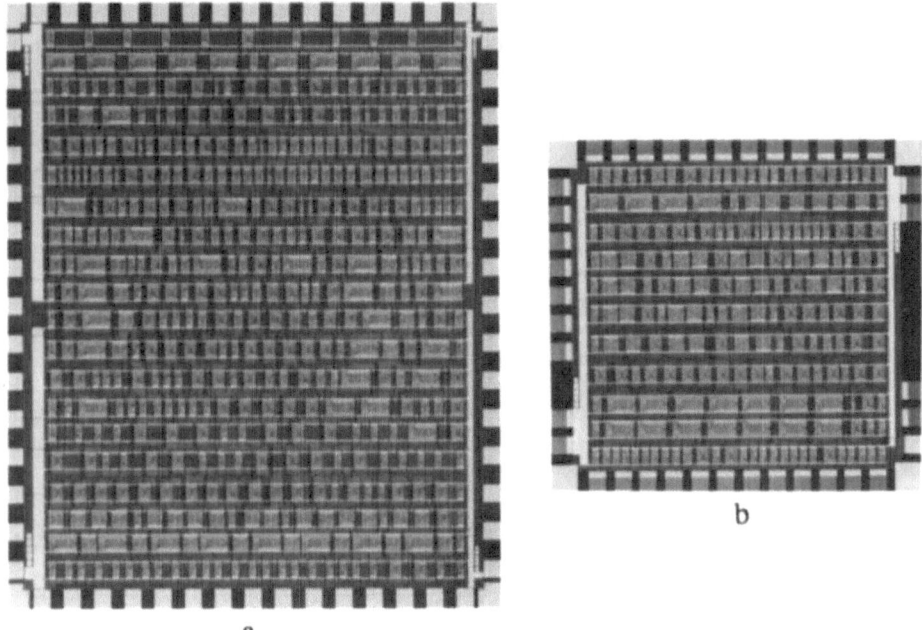

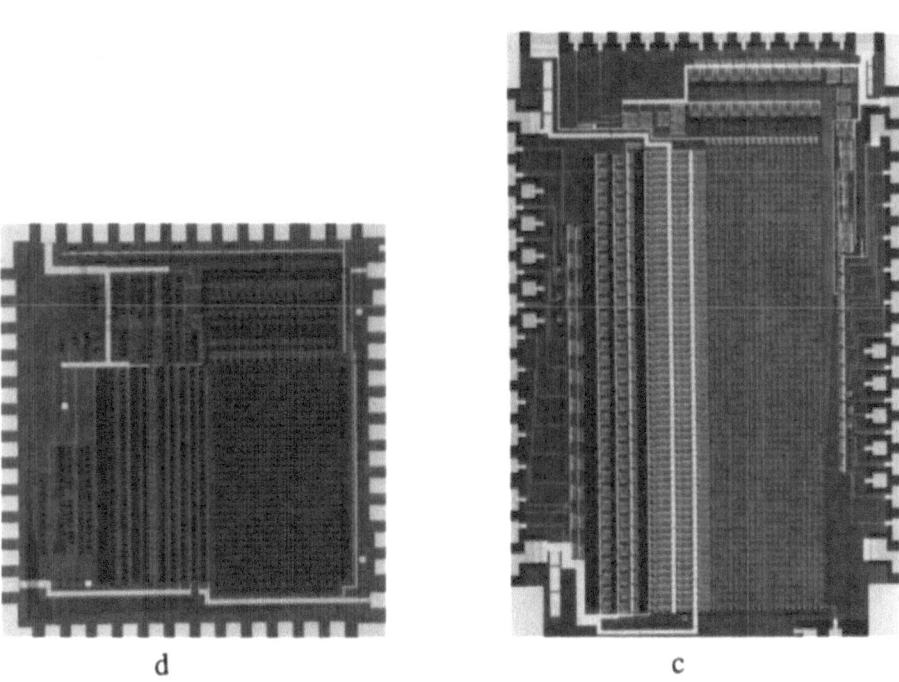

Fig.14. Photographs of the four LSI chips fabricated,(a) RALU chip,(b) SQCU chip, (c) IROU chip, and (d) DRAU chip ETL-JC1.

CHIP CONNECTION

The ETL-JC1 was constructed by connecting these four LSI chips each other on a nonmagnetic printed circuit board. Each chip was wire-bonded at the pads on the 24mm x 24mm chip carrier. Each chip carrier was mounted on the printed circuit board with a 68-pin socket. Figure 15 shows a photograph of the ETL-JC1. The printed circuit board is fixed on the end of a plastic pipe as shown at the bottom of the figure. The board size is about 80mm x 100mm. Chip-to-chip connection among four chips was carried out by using copper-metal strip lines on both front and back sides of the printed circuit board.. Coaxial cables with a characteristic impedance of 50 Ω were used to supply the power current to the chips. The output signals of the 4-bit ACC and the 7-bit program address were monitored by the coaxial cables at the printed circuit board as shown in the figure. To reduce the crosstalk among the signal lines and the power supply lines, the power supply lines were connected directly to the chips at the socket terminals with the coaxial cables at the back side of the printed circuit board (not seen in this figure).

In order to reduce an environmental static magnetic field to the chips, the liquid helium dewar was shielded by four-fold 1 mm thick permalloy metals with a size of 0.7 m diameter x 1.2 m height. Moreover, a double permalloy metal cylinder of 92 mm diameter x 135 mm long was placed in liquid helium to cover the printed circuit board. In this way the magnetic field was reduced to be less than a few micro gauss. All components of the chip carriers, sockets, screws, printed circuit board and so on were made of nonmagnetic materials.

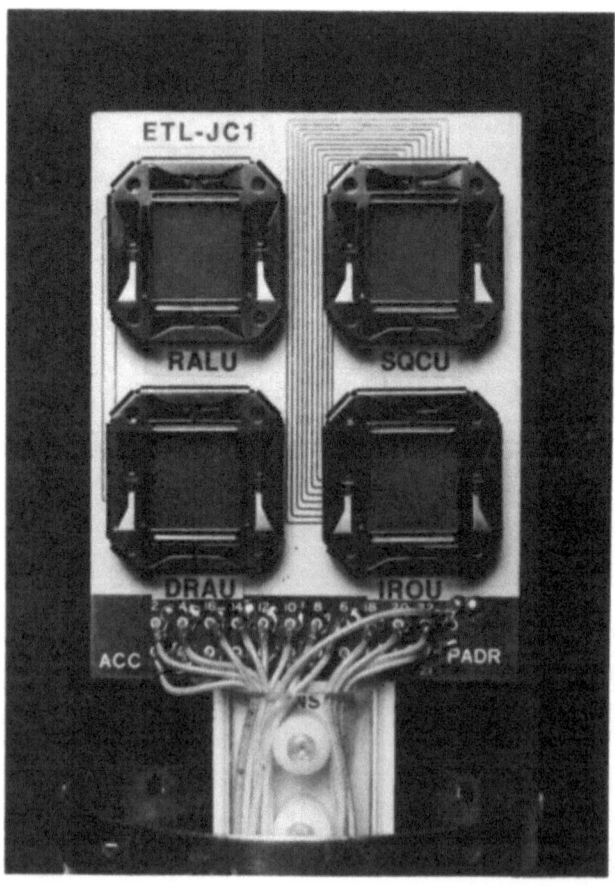

Fig.15. The ETL-JC1.

The experimental operation of the Josephson computer ETL-JC1 was carried out in liquid helium.   A two-phase power supply with a repetition frequency of a few kilohertz was used to drive all chips.   The split power was supplied only to the IROU chip to drive the ROM decoder circuit.

Figure 16 shows an example of the oscilloscope traces obtained by operating the ETL-JC1. The computer program from the address of 0000011 was executed corresponding to the program listed in Table III.   The upper four traces show the 4-bit data of the accumulator. The lower seven traces show the outputs of the 7-bit program counter in the SQCU chip which indicate the program address in the IROU chip.

At the beginning 10 cycles shown in Fig.16, the 4-bit ACC, MD, ML and MH registers are tested by storing a set of the 4-bit data.   In the next 9 cycles the write/read operations of the RAM chip are carried out.   Using the MD, ML, and MH registers, a set of the 4-bit data is written in the 1-kbit RAM of the DRAU chip and is read-out.   It should be noted that this operation is the first successful experiment which combines logic chips and memory chips. After these executions the arithmetic logic operations of add (ADD), subtraction (SUB), logic AND (AND), logic OR (OR), complement (CMA), exclusive OR (XOR), rotate right (RTR), and rotate left (RTL) for the 4-bit data are correctly carried out, corresponding to the program shown in Table III.   The 7-bit output of the program address in Fig.16 is incremented by +1 for the execution of each instruction, so that every instruction is sequentially executed according to the program sequence.

In this way all other program executions, which include the conditional skip, the subroutine call/return, the jump, and so on, were confirmed.   For all these program operations no failure appeared.

The operating margin was obtained to be more than ±10 % for the stable computer operation.   The total power dissipation was 6.2 mW in the computer which includes more than 22,000 single Josephson junctions and regulator junctions in the power lines.

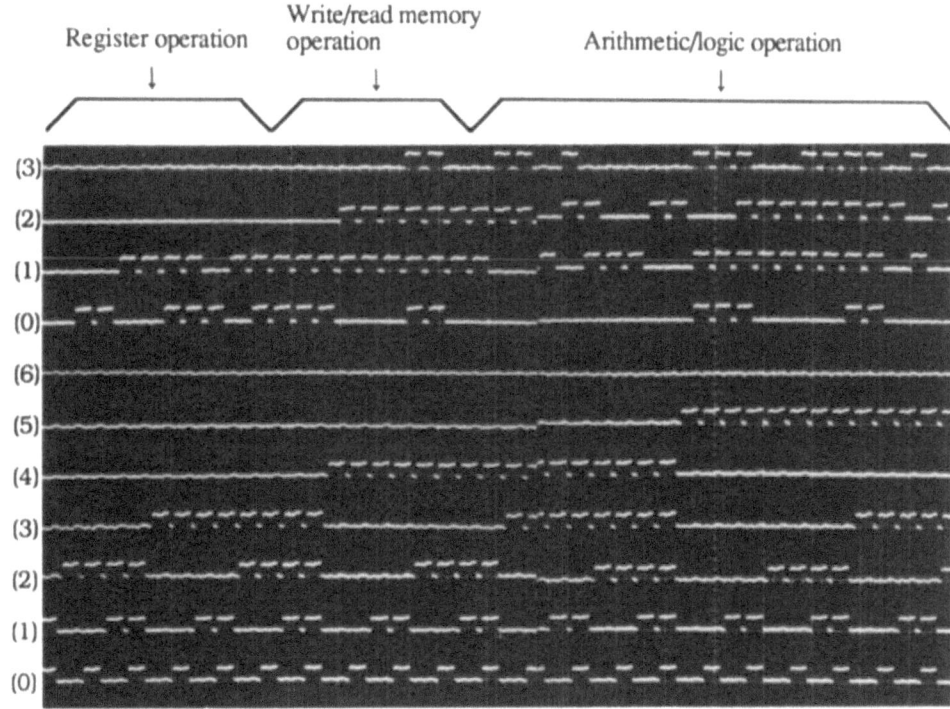

Fig.16. Experimental results of the computer operation.   Upper four traces are the 4-bit data in the ACC and lower seven traces corresponds to 7-bit program addresses.

DISCUSSION

The RALU chip, on which 1,273 4JL gates were integrated, has been fabricated in 1988 and tested in liquid helium to confirm the circuit operation. The exactly same RALU chip was used in the present ETL-JC1 experiment. It should be noted that the RALU chip was completely operated without any failure during the ETL-JC1 experiment. During this period the chip has been stored in atmosphere at room temperature and thermally cycled between room temperature and 4.2K more than ten times. Therefore, the Josephson LSI chips fabricated with the present process reveals high stability and reliability.

The magnetic flux trap in the superconducting films often suffer the logic operation of the LSI chips. A moat-guard method [45] was proposed to collect external magnetic flux in holes in the ground superconducting plane on the chips, resulting in the reduction of the magnetic fields at the logic gates effectively. In this experiment the moat was not employed. By shielding the external magnetic fields with permalloy magnetic shield cylinders the magnetic flux trap was reduced to operate the ETL-JC1 sufficiently.

The critical path which determines the clock cycles of the ETL-JC1 was carefully investigated in the CAD system and was found in the path which includes the skip control circuit in the RALU, the incrementor in the SQCU and the ROM plane in the IROU. Following this critical path the delay time in the RALU, SQCU and IROU was evaluated on the basis of experimental delay time of the switching gates. In the RALU chip the critical delay time which determines the clock cycles was 317ps. In the SQCU a delay time was 385ps. In the IROU chip a delay time of 390ps was measured. These delay times were obtained to operate the circuit at 75% nominal bias current condition. In the DRAU chip the access time was 520ps in the experiment [6]. These values mean that the ETL-JC1 can be operated at 1 gigahertz.

The chip connection technology will be required to transmit Josephson quick signals among the chips. A new circuit architecture such as multi-CPU(central processor unit) and a pipe-line method will be employed to the Josephson computer. It will be very important to develop the Josephson VLSI chips in the next research program.

SUMMARY

The first Josephson computer ETL-JC1 has been successfully demonstrated, in which the stored 128-step computer program which is written on the basis of RISC architecture was executed. The key technology to develop the ETL-JC1 has been summarized including logic and memory circuits driven by the two-phase power supply. The CAD system to develop the Josephson LSI chips based on Nb and NbN junctions has been described. The ETL-JC1 consisting of more than 22,000 Josephson junctions consumed an electric power of 6.2mW. 1 giga instructions per second operation can be expected in the Josephson computer with a single CPU by employing high speed power supply method.

ACKNOWLEDGEMENT

The author would like to thank the major contributors, H.Nakagawa, I.Kurosawa, M.Aoyagi, Y.Okada, S.Kosaka, and Y.Hamazaki for their collaboration to develop the Josephson technology for the ETL-JC1. He also expresses his thanks to Drs. H.Kashiwagi, K.Tamura, and T.Tsurushima in the Electrotechnical Laboratory for their encouragement throughout this work.

REFERENCES

1. S.Kotani, T.Imamura, and S.Hasuo, "A sub-ns Josephson 4-bit processor," in Extended Abstr. 1989 Int. Superconductivity Electronics Conf. (ISEC'89), Tokyo, pp.381-386, Jun. (1989).
2. Y.Hatano, S.Yano, H.Mori, H.Yamada, M.Hirano, and U.Kawabe, "A 4-bit Josephson data processor with dc output buffer," in Extended Abstr. 1989 Int. Superconductivity Electronics Conf. (ISEC'89), Tokyo, pp.375-380, Jun. (1989).

3. H.Nakagawa, S.Kosaka, I.Kurosawa, M.Aoyagi, Y.Hamazaki, Y.Okada, and S.Takada, "A Josephson 4-bit processor for a prototype computer," in Extended Abstr. 1989 Int. Superconductivity Electronics Conf. (ISEC'89), Tokyo, pp.387-390, Jun. (1989).

4. S.Kotani, A.Inoue, T.Imamura, and S.Hasuo, "A 1GOPS 8b Josephson digital signal processor," Digest of Technical Papers, 1990 IEEE Int. Solid-State Circuits Conference, pp.148-149, Feb. (1990).

5. S.Nagasawa, Y.Wada, M.Hidaka, H.Tsuge, I.Ishida, and S.Tahara, "570-ps 13-mW Josephson 1-kbit NDRO RAM," IEEE J. Solid-State Circuits, vol. 24, no. 5, pp. 1363-1371, Oct. (1989).

6. I.Kurosawa, H.Nakagawa, M.Aoyagi, S.Kosaka, and S.Takada, "A fully operational 1-kbit variable threshold Josephson RAM, " Digest of Technical Papers, 1990 VLSI Circuits Symposium, pp. 67-68, Jun. (1990).

7. W.H. Henkels, "Fundamental criteria for the design of high-performance Josephson nondestructive readout random access memory cells and experimental confirmation," J. Appl. Phys. vol. 50, no. 12, pp. 8143-8168, Dec. (1979).

8. I.Kurosawa, H.Nakagawa, S.Kosaka, and S.Takada, "Variable threshold Josephson memory fabricated with Nb/Al-oxide/Nb process," in Extended Abstr. 1987 Int. Superconductivity Electronics Conf. (ISEC'87), Tokyo, pp. 45-48, Aug. (1987).

9. H.Suzuki, N.Fujimaki, H.Tamura, T.Imamura, and S.Hasuo, "A 4k Josephson memory," IEEE Trans. Magn. vol. MAG-25, pp.783-788, Mar. (1989).

10. S.Tahara, I.Ishida, S.Nagasawa, M.Hidaka, N.Tsuge, and Y.Wada, "A 4-kbit Josephson nondestructive readout RAM operated at 580psec and 6.7mW," to be presented in 1990 Applied Superconductivity Conference, Snowmass, Sep. (1990).

11. M.B.Ketchen, D.J.Herrell, and C.J.Anderson, "Josephson cross-sectional model experiment," J. Appl. Phys. vol. 57, no.7, pp. 2550-2574, Apr. (1985).

12. S.Takada, S.Kosaka, and H.Hayakawa, "Current injection logic gate with four Josephson junctions," in Proc. 11th Conf. Solid State Devices, Tokyo 1979, also Jpn. J. Appl. Phys., vol. 19, suppl. 19-1, 1980, pp. 607-611, (1980).

13. S.Takada, I.Kurosawa, H.Nakagawa, M.Aoyagi, S.Kosaka, and F.Shinoki, "Josephson IC technology," in Superconductivity Electronics, K.Hara, ed., Ohmusha, LTD. & Prentice Holl, INC., Tokyo, (1987).

14. Y.Okada, Y.Hamazaki, E.Sogawa, H.Ohigashi, H.Nakagawa, and H.Hayakawa, "Josephson logic with multiple-phase pulsed power sources," IECE Tech. Rep. ED81-146, pp. 89-94, Mar. (1982), (in Japanese).

15. I.Kurosawa, H.Nakagawa, S.Kosaka,, M.Aoyagi, and S.Takada, "A 1-kbit Josephson memory using variable threshold cells," IEEE J. Solid-State Circuits, vol. 24, no. 4, pp. 1034-1040, Aug. (1989).

16. M.Aoyagi, H.Nakagawa, I.Kurosawa, Y.Okada, Y.Hamazaki, S.Kosaka, A.Shoji, and S.Takada, "A Josephson 10-bit instruction ROM unit for a prototype computer," in Extended Abstr. 1989 Int. Superconductivity Electronics Conf. (ISEC'89), Tokyo, pp.271-274, Jun. (1989).

17. H. Kroger, L. N. Smith, and D. W. Jillie, "Selective niobium anodization process for fabricating Josephson tunnel junctions," Appl. Phys. Lett. vol. 39, pp. 280-282, (1981).

18. M.Gurvitch, M. A. Washington, and H. A. Huggins, " High quality refractory Josephson tunnel junctions utilizing thin aluminum layers," Appl. Phys. Lett. vol. 42, no. 5, pp. 472-474, Mar. (1983).

19. S.Morohashi, F.Shinoki, A.Shoji, M.Aoyagi, and H.Hayakawa, "High-quality Nb/AlOx-Al/Nb Josephson junctions," Appl. Phys. Lett. vol. 46, pp. 1179-1181, June (1985).

20. H.Nakagawa, K.Nakaya, I.Kurosawa, S.Takada, and H.Hayakawa, "Nb/Al-oxide/Nb tunnel junctions for Josephson integrated circuits," Japan. J. Appl. Phys. vol. 25, no. 1, pp. L70-L72, Jan. (1986).

21. K.Kuroda and M.Yuda, "Niobium-stress influence on Nb/Al-oxide/Nb Josephson junctions," J. Appl. Phys. vol. 63, no. 7, pp. 2352-2357, Apr. (1988).

22. T.Imamura and S.Hasuo, "Effects of intrinsic stress on submicronmeter Nb/AlOx/Nb Josephson junctions," IEEE Trans. Magn. vol. MAG-25, no. 2, pp. 1119-1122, Mar. (1989).

23. H.Nakagawa, T.Odake, E.Sogawa, S.Takada, and H.Hayakawa, "Sub-10 ps logic operations in Josephson four-junction logic (4JL) gates," Jpn. J. Appl. Phys., vol. 22,

no. 5, pp. L297-L298, May. (1983).

24. T.A.Fulton, S.S.Pei, and L.N.Dunkleberger, "A simple high-performance current switched Josephson logic," Appl. Phys. Lett. vol 34, no. 10, pp. 709-711, May (1979).

25. T.Gheewala and A.Mukherjee, "Josephson direct coupled logic(DCL)," in IEDM Tech. Dig., pp. 482-484, Dec. (1979).

26. K.Hohkawa, M.Okada, and A.Ishida, "A novel current injection Josephson logic gate with high gain," Appl. Phys. Lett. vol. 39, no. 8, pp 653-655, (1981).

27. J.Sone, T.Yoshida, and H.Abe, "Resistor coupled Josephson logic," Appl. Phys. Lett. vol. 40, no. 8, pp. 741-743, (1982).

28. T.Gheewala, "Josephson logic circuits based on non-linear current injection in interferometer devices," Appl. Phys. Lett. vol. 33, no. 8, Oct. (1978).

29. N.Fujimaki, S.Kotani, S.Hasuo, and T.Yamaoka, "9 ps gate delay Josephson OR gate with modified variable threshold logic," Jpn. J. Appl. Phys., vol. 24, pp. L1-L2, (1985).

30. S.Kosaka, H.Nakagawa, H.Kawamura, Y.Okada, Y.Hamazaki, M.Aoyagi, I.Kurosawa, A.Shoji, and S.Takada, "Josephson address control unit IC for a 4-bit microcomputer prototype," IEEE Trans. Magn. vol. MAG-25, no. 2, pp. 789-794, Mar. (1989).

31. H.Nakagawa, I.Kurosawa, and S.Takada, "A Josephson 2-bit arithmetic logic unit," IEEE Trans. on Circuits and Systems, vol. CAS-34, no. 9, pp. 1123-1124, Sep. (1987).

32. H.Nakagawa, I.Kurosawa, S.Takada, and H.Hayakawa, "A Josephson counter circuit with two-phase power supply," in Extended Abstr. 17th Conf. Solid State Devices and Materials, Tokyo, pp. 751-752, (1986).

33. S. Dhong and T.Gheewala, "A Josephson dc powered Latch," Appl. Phys. Lett. vol. 38, no. 11, pp. 936-938, June (1981).

34. H.C.Jones and T.R.Gheewala, "AC powered Josephson latch circuits," IEEE J. Solid-State Circuits, vol. SC-17, no. 6, pp. 1201-1210, Dec. (1982).

35. H.Nakagawa, I.Kurosawa, S.Takada, and H.Hayakawa, "Josephson 4-bit digital counter circuit made by Nb/Al-oxide/Nb junctions," IEEE Trans. Magn. vol. MAG-23, no. 2, pp. 739-742, Mar. (1987).

36. H.Beha, "32K read-only-memory chip design in Josephson technology: A feasibility study," IEEE J. Solid-State Circuits, vol. SC-21, no. 2, pp. 353-361, Apr. (1986).

37. Y.Hatano, S.Yano, M.Hirano, Y.Tarutani, and U.Kawabe, "A subnonosecond Josephson data processor model," in Extended Abstr. 1987 Int. Superconductivity Electronics Conf. (ISEC'87), Tokyo, pp. 239-344, Aug. (1987).

38. I.Kurosawa, H.Nakagawa, A.Yagi, S.Takada, and H.Hayakawa, "A 2 Kbit-Josephson ROM," in IECE Nat. Conv. Rec., p. 117, Mar. (1984), (in Japanese).

39. M.Aoyagi, H.Nakagawa, I.Kurosawa, S.Kosaka, Y.Okada, Y.Hamazaki, and S.Takada,"A Josephson 10-b instruction 128-word ROM unit," IEEE J. Solid-State Circuits, vol.25, no. 4, pp.971-978, Aug. (1990).

40. P.C.Arnett and D.J.Herrell, "Regulated AC power for Josephson interferometer latching logic circuits," IEEE Trans. Magn. vol. MAG-15, no. 1, pp. 554-557, Jan. (1979).

41. H.Nakagawa, S.Kosaka, H.Kawamura, I.Kurosawa, M.Aoyagi, Y.Hamazaki, Y.Okada, and S.Takada, "A Josephson 4-bit RALU for a prototype computer," IEEE J. Solid-State Circuits, vol. 24, no. 4, pp. 1076-1084, Aug. (1989).

42. A.Shoji, F.Shinoki, S.Kosaka, M.Aoyagi, and H.Hayakawa, "New fabrication process for Josephson tunnel junctions with (niobium nitride, niobium) double-layered electrodes," Appl. Phys. Lett. vol. 41, no. 11, pp. 1097-1099, Dec (1982).

43. K.Kuroda, J.Nakano, M.Yuda, and M.Ueki, "3.0ps switching operation in all-Nb Josephson logic gates", Electron. Lett., vol.23, no. 4, pp. 163-165, Feb. (1987).

44. H.Nakagawa, I.Kurosawa, M.Aoyagi, and S.Takada, "Fabrication process for a Josephson computer ETL-JC1," to be presented in 1990 Applied Superconductivity Conference, Snowmass, Sep. (1990).

45. S. Bermon and T. Gheewala, "Moat-guarded Josephson SQUIDs," IEEE Trans. Magn. vol. MAG-19, no. 3, pp. 1160-1164, May (1983).

COOPERATIVE OPERATION OF LARGE JOSEPHSON SERIES ARRAYS
FABRICATED USING THE Nb/Al2O3/Nb-TECHNOLOGY *

J. Niemeyer, R. Pöpel

Phys-Techn. Bundesanstalt
Bundesalle 100,
D-3300 Braunschweig
FRG

VOLTAGE STANDARD

Introduction

A Josephson tunnel junction is an ideal frequency-to-voltage converter because an alternating supercurrent is generated across the tunnel barrier at finite dc voltages. The frequency of the supercurrent and the dc voltage are simply related by

$$V = h/2ef. \tag{1}$$

An external oscillator can be phase locked over a certain frequency range at its basic frequency f, and the harmonics nf_e (n = 0, 1, 2, 3, ...) to the tunnel junction oscillator, thus inducing a number of constant voltage steps in the dc characteristic of the junction[1]:

$$V_n = nh/2ef_e, \tag{2}$$

the precision of which are believed to be dependent only on the accuracy of the determination of the external microwave frequency f. At f = 70 GHz the first voltage step will appear at about 145 μA, provided the quotient $2e/h$ (e - electron charge, h - Planck's constant) to be fixed at 597,9 GHz/V [2]. A single junction generates up to seven steps - about 1 mV - in the special operation mode described below. Series arrays of tunnel junctions must therefore be used to increase the range of reference voltages without losing the fundamental exactness of a direct Josephson reference voltage by using resistive potentiometers. For a series array with m junctions the reference voltage range would span the range:

$$\pm V_m = \sum_{l=1}^{m} V_{l,n} \tag{3}$$

if $V_{l,n}$ is the contribution from the lth junction. Before designing the integrated circuit it has to be considered which mode of operation, the zero current step mode or the traditional constant voltage step mode, is most suitable for successfully operating a Josephson series array as a potentiometer.

Nonlinear Superconductive Electronics and Josephson Devices
Edited by G. Costabile *et al.*, Plenum Press, New York, 1991

21

In the simplest case, a series connection of only two junctions $(V_{1,n})$ and $(V_{2,n})$, Fig. 1 shows schematically the single junction characteristics with a small number of constant voltage steps and the resulting characteristics of the series connection $(V_{1,n} + V_{2,n})$. The normal state resistance of the two junctions and correspondingly the critical currents and the current widths of the constant voltage steps are assumed to differ by about 30%. The series array is biased by a single current source only. The incident microwave power ist adjusted to generate 7 zero current steps between $-V_{1,3}$ and $+ V_{1,3}$ in the characterisitc of the first junction and 5 steps between $-V_{2,2}$ and $+ V_{2,2}$ in the characteristic of the second junction (Fig. 1a). The series array provides 11 steps with the voltage difference between adjacent steps determined by the fundamental frequency of the external oscillator (n = 1 in eq.2. The maximum step number (11 in this example) between $-V_m$ and $+V_m$ (eq.3) is generated as the reference voltages of a single junction are multivalued for currents close to zero, allowing all 11 steps in the array characteristic to be induced by different combinations of the single junction steps. All steps can be biased by properly adjusting the dc current source (dashed line in Fig. 1a).

When the microwave power is increased, the current width of the constant voltage step usually decreases and the number of the steps increase. The characteristic (Fig. 1b). In this operation mode the reference voltage is single valued for to a particular current. The series array therefore produces a voltage difference between adjacent steps which tends to increase with the number of junctions. The step distance is generally doubled in the example of Fig. 1 because only two junctions are series connected. For n junctions the step distance would be exactly $(h/2e)$ mf, if all characteristics were identical. Occasional deviations to smaller step distances might occur, if a cerain parameter spread normally introduced by the fabrication process is present, but at the same time the parameter spread causes most of the steps to be very small (Fig. 1b).

Only by using indivdually biased junctions in a series array operated in the traditional step mode can the disadvantage of the strong influence of the junction parameter spread and the larger step distance be compensated [3], but the complicated bias procedure is difficult to apply for large series arrays (more than 20 junctions) which are needed for a potentiometer able to cover a large range of reference voltages. It might therefore be reasonable to use a series array of a large number of junctions operated in the traditional step mode - provided the fabrication process is extremely good - to construct a voltage standard with a single reference voltage output (e.g. 1 V). However, for a potentiometer with a continuous range of output reference voltages the use of a series array operated in the zero current step mode would be more advantageous [4], despite the fact that it introduces a delicate stability problem. As the constant voltage steps widely overlap and are multivalued for a particular bias current, a short interruption of the phaselock between the external and the Josephson oscillator would cause the bias to leave the originally selected constant voltage step and normally never returns to it. It is also difficult to select a particular constant voltage step (Fig.1a) especially if the number of junctions is very large. It is therefore advantageous to shunt the junction series array. This alters the overlapping step structure to a staircase-like characteristic by superimposing the shunt characteristic to the series array characteristic (Fig. 1c). A particular voltage step can now be selected by continuously decreasing or increasing the bias current. The range of stable constant voltage steps close to the desired reference voltage must be adjusted by optimizing the incident microwave power. At low rf power, the range of stable operation lies at low reference voltages and with increasing rf power it moves up to higher voltages. This series connection of a desired number of junctions properly designed must be integrated into a microwave coupling circuit allowing the microwave power to be uniformly distributed over all the junctions. To reach a homogeneous distribution of the microwave power in all junctions, the series array had to be arranged as the top line of a stripline [5].

A considerable improvement on facilitating more stable operation designed at PTB, was to divide the microwave stripline into several parallel paths [6,7,8]. This design was the basis of all further developments with more than 2000 junctions. Voltage steps between 0,1 V and 1,3 V could be kept stable for more than 5 hours and with stable current amplitudes of about 45 µA at 70 GHz (see Fig. 1b).These standards were used in precision voltage metrology for several years providing results of extremely high accuracy [9,10,11,12]. At the beginning of the standard development the circuits were fabricated in lead alloy technology.The main

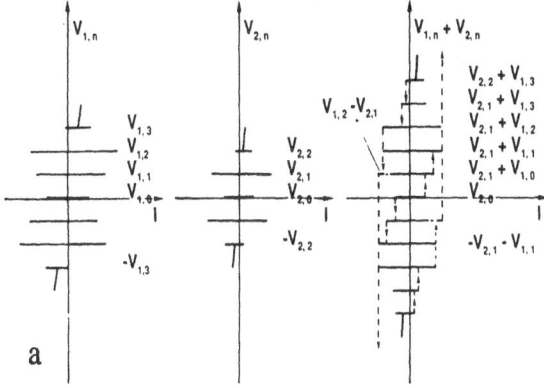

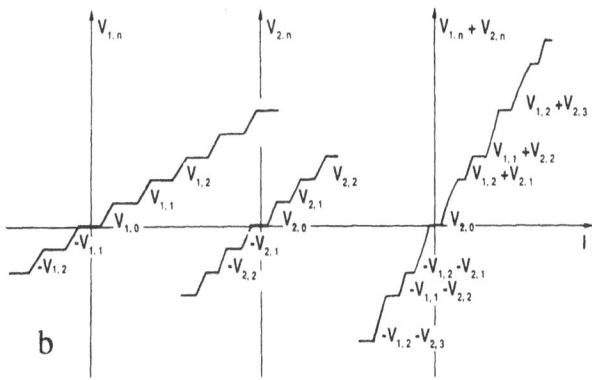

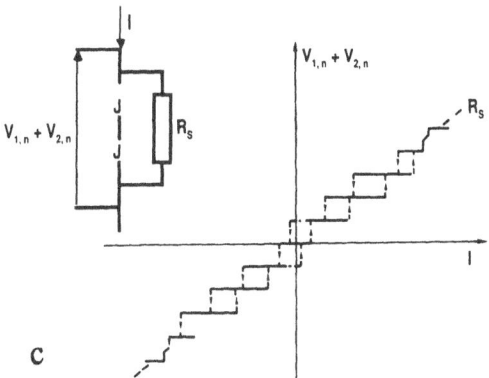

Figure 1. Schematic I-V characteristics of two single junctions (V_{1n}), (V_{2n}), and the series connection $(V_{1n} + V_{2n})$ of both under microwave radiation. All circuits are driven by a single current source (I). (a) In the zero current step mode.The reference voltages are multivalued. (The broken lines denote how a certain constant voltage step is selcted by adjusting the current source.) (b) In the traditional step mode. The reference voltages are single valued. (c) The *I-V* characteristic of the series connection of 2(a), but resistively shunted. (The steps can be selected by adjusting the current source continuously, broken line).

drawback of the lead alloy standards was the rather short durability caused by oxidation and diffusion processes, the mechanical instability of the lead alloy layers, and the delicate and inhomogeneous conditions on preparing the thin tunnel oxide layers with the help of the Greiner process[13]. To overcome these difficulties, two different techniques have been applied:

i) Successful devices with up to 19000 $Nb/Nb_2O_5/PbAu$ junctions have been fabricated providing reference voltages up to 10 V [14].

ii) The development described in this paper is concentrated on the design and operation of $Nb/Al_2O_3/Nb$ arrays[15,16].

The advantages of the latter systems are as follows: They are more durable than $Nb/Nb_2O_5/PbAu$ arrays, they can be fabricated with the smallest junction parameter spread over the complete wafer and in principle a $Nb/Al_2O_3/Nb$ junction can provide the largest constant voltage step amplitude[17]. Furthermore, the critical current density can be reduced after the fabrication by tempering the samples at temperatures between 200 ^{0}C and 250 ^{0}C in a defined way. The $Nb/Al_2O_3/Nb$ technique has therefore been used to produce new 1 volt chips with 2000 tunnel junctions and 10 volt chips with 20 160 junctions, respectively (Fig. 2). The $Nb/Al_2O_3/Nb$ tri-layer technique used to fabricate the arrays is described in detail elsewhere [18]

Design

The design of the new standard chips (Fig. 2) is simpler than that of the prior one described in[4]. The finline transforming the wave impedance of the E-band waveguide ($\approx$ 520 Ω) to the wave impedance Z_L of the stripline ($\approx$ 7 Ω) consists of exponential tapers. These tapers follow curves described by exp (- a_Tx) with a_T = 0.0006 μm^{-1} at the beginning and α_T = 0.0036 μm^{-1} at the end of the taper. The total length of the taper is 6.3 mm. The microwave power is split into 2^n (n = 3 ... 6) parts which are led via dc blocks to the same number of parallel paths containing the tunnel junctions. If the total number of junctions N is fixed, n determines the length of a single microwave path containing $N/2^n$ junctions connected in series. Each of the 2^n arrays is terminated by a matched lossy stripline to avoid microwave reflection from its open ends. Each stripline path is provided with the same amount of microwave power by the distribution circuit but is galvanically separated from this circuit by the dc blocks. After having passed through the distribution circuit, the microwave power reaches the tunnel junction array. In the case of 1 volt chips, the complete series array is split into 8 (n = 3, Fig. 2a) or 16 (n = 4, Fig. 2b) parallel stripline paths, respectively. According to[17], the junction area is determined to be 26 μm in length and 46 μm in width. Each junction stripline configuration in the designs of Fig.2 is terminated by a matched load to avoid the formation of standing waves by reflection of excess microwave power from the ends of the striplines. These loads are formed by the superconducting groundplane and the upper normal conducting strip conductors (Au or InAu). In addition, the microwave power is kept from the dc pads by low pass filters. The geometry of the 10 volt chip in Fig. 2c has been optimized to reduce the total circuit area to 10 mm x 27 mm and the total active junction area to only 7 mm x 13 mm. All geometrical lengths and widths are kept as small as possible, and the loads are placed at the end of each microwave path to increase the package density of the tunnel junctions. Also the single junction area is reduced to 22 μm in length and 46 μm in width. The junction array is divided into 64 paths (n = 6) so that only 315 junctions remain for one path. Each path is a stripline the upper layer of which is formed by a junction series array. The array striplines and the distribution striplines have a common groundplane (Nb 150 nm) and a dielectric layer (SiO 1-3 μm).

In all parts of the distribution circuit the stripline width w is much bigger than the dielectric height h, in which case the radiation at discontinuities and the edge effects of the high frequency field are negligible, and Z_L is given by the simple equation

a

b

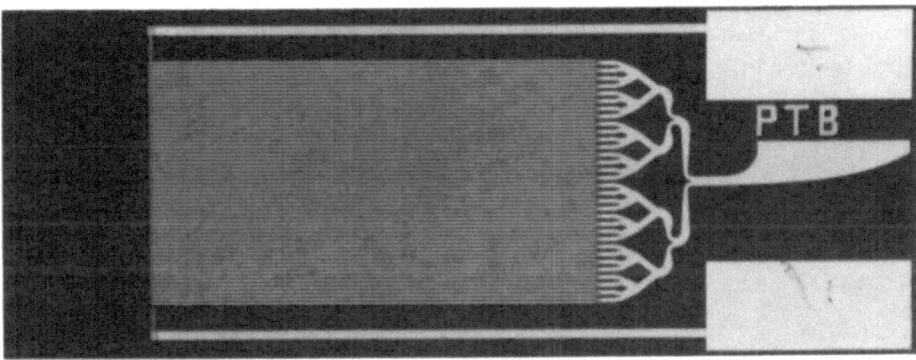

c

Figure 2. Three new voltage standard chips in Nb/Al$_2$O$_3$/Nb technique
a) 1 V standard with 2000 tunnel junctions split into 8 microwave paths,
chip size: 8 mm x 16 mm
b) 1 V standard split into 16 microwave paths, chip size: 10 mm x 16 mm
c) 10 V standard containing 20160 tunnel junctions, split into 64 microwave paths, chip size: 10 mm x 27 mm

$$Z_L = (\eta_0 h)/(\sqrt{\varepsilon_r} w), \quad (w/h > 20) \tag{4}$$

with ε_r the dielectric constant ($\varepsilon_r = 5.7$ for SiO, $\varepsilon_r = 3.5$ for SiO$_2$) and $\eta_0 = 120 \pi \Omega$ the wave impedance of the free space.. The microwave wavelength inside the stripline $\lambda_M = v_\phi/f$ is calculated from the phase velocity :

$$v_\phi = (c_0/\sqrt{\varepsilon_r})/(1/\sqrt{B}) \qquad \text{with}$$

$$B = 1+2 \, (\lambda_{Nb}/h_1) \coth (d_{Nb}/\lambda_{Nb}) \tag{5}$$

c_0 - speed of light,

λ_{Nb} - superconducting penetration depth of the Nb-films,

d_{Nb} - Nb-film thickness.

    This equation was derived by Swihart and Kautz [19, 20] for superconductors in the local limit. More exact calculations for λ_M in the anomalous limit reported later agree well with the results of Eq. (5). Measured values for the superconducting penetration depth can be taken from Henkels and Kircher [21].

    Fig. 3   shows a cross section of one stripline path. Most of the microwave power propagates through the thick SiO or SiO$_2$ film in the main stripline consisting of the ground plane (Nb) and the junction base electrodes (Nb) and only a small part passes the slots between the base electrodes to two junctions.

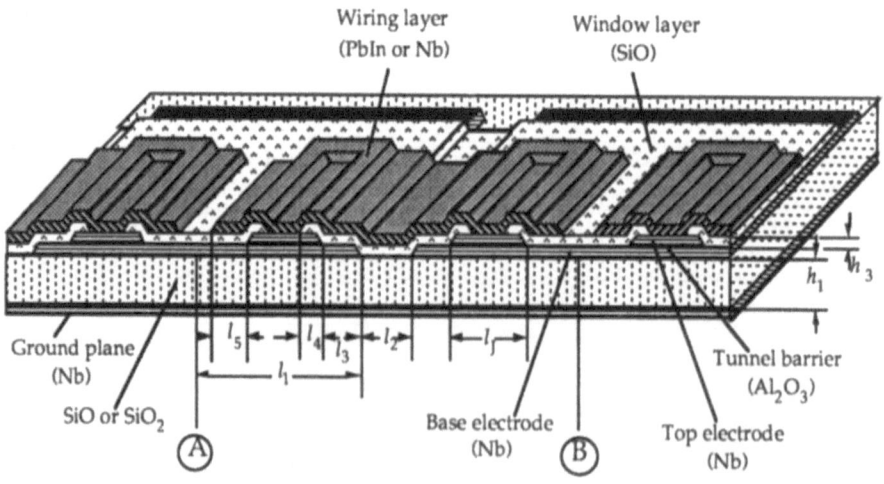

Figure 3. Cross section of one stripline path containing 4 junctions.

Thus from point A to point B a periodic sectipon of the stripline path consists of a connection of several thin film transmission lines (striplines) and every stripline is completely described by the characteristic impedance Z_L and the propagation constant $\gamma = \alpha + j\beta$ (α - attenuation constant, β - phase constant). These two quantities are usually related to the resistance R', the capacitance C' and the inductance L' normalized to the unit length [22] by

$$\alpha = \sqrt{\frac{1}{2}[-\omega^2 L'C' + \omega C'\sqrt{R'^2 + \omega^2 L'^2}]} \ ,$$

$$\beta = \sqrt{\frac{1}{2}[+\omega^2 L'C' + \omega C'\sqrt{R'^2 + \omega^2 L'^2}]} \ ,$$

(6)

$$Z_L = \sqrt{\frac{1}{\omega C'}[\omega L' - jR']} \quad \text{with}$$

$$R' = \frac{R_a + R_b}{w}, \quad C' = \varepsilon_0 \varepsilon_r \frac{w}{h} \text{ and } L' = \mu_0 \frac{h}{w} + \frac{X_a + X_b}{\omega w} \ ,$$

ω - circular frequency,

w - stripline width,

h - thickness of dielectric,

ε_0 - dielectric field constant,

ε_r - dielectric constant and

μ_0 - magnetic field constant.

Compared with the conductor losses, dielectric losses are negligible. This is confirmed by measurements [23]. The surface impedances $Z_{a,b} = R_{a,b} + jX_{a,b}$ of the two strip conductors - denoted by the indices a and b - can be obtained from an exact solution of the theory of the anomalous skin effect in superconductors of Mattis and Bardeen [24] which is described in anomalous skin effect in superconductors of Mattis and Bardeen [24] which is described in more detail in [25,26]. The results for the Nb films used (thickness $d_{Nb} = 150$ nm), the NbAl film ($d_{Al} = 18$ nm) and the PbIn film ($d_{PbIn} = 600$ nm) at 70 GHz and the temperature T = 4.2 K are listed in Table 1.

Table 1 Parameters of the array

$R_{Nb} = 4.38 \times 10^{-4}\ \Omega$	$l_1 = 38\ \mu m$	$w_1 = 70\ \mu m$	$h_1 = 3000$ nm
$X_{Nb} = 4.93 \times 10^{-2}\ \Omega$	$l_2 = 8\ \mu m$	$w_2 = 60\ \mu m$	$h_2 = 3300$ nm
$R_{NbAl} = 9.24 \times 10^{-3}\ \Omega$	$l_3 = 8\ \mu m$	$w_3 = 60\ \mu m$	$h_3 = 300$ nm
$X_{NbAl} = 5.54 \times 10^{-2}\ \Omega$	$l_4 = 2\ \mu m$	$w_4 = 60\ \mu m$	$h_4 = 300$ nm
$R_{PbIn} = 6.62 \times 10^{-4}\ \Omega$	$l_5 = 4\ \mu m$	$w_5 = 60\ \mu m$	$h_5 = 300$ nm
$X_{PbIn} = 3.76 \times 10^{-2}\ \Omega$	$l_J = 22\ \mu m$	$w_J = 46\ \mu m$	$h_J = 1.9$ nm

The complete electrical circuit of the periodic part of the stripline path in Fig. 3 is shown in Fig. 4. The stripline of length l_1 leads from point A to the beginning of the slot between the base electrodes and the line of length l_2 corresponds to the area where both dielectric layers overlap. The stripline of length l_3 leads from the slot to the junction input, and l_4 is the length of the shortened line starting from the junction input and formed by the top electrode, the window layer and the wiring layer.

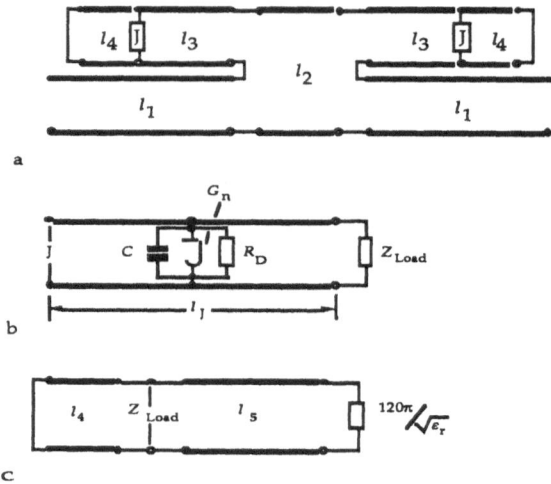

Figure 4. Electrical circuit of a periodic part of the stripline path.

It should be pointed out that this shortened stripline is not present in standard designs with Pb alloy junctions or $Nb/Nb_2O_5/PbInAu$ junctions. For the design using $Nb/Al_2O_3/Nb$, an additional layer - the top electrode - is needed. The dc series connection is formed by the wiring layer via windows in the second thinner SiO layer (window layer) which are smaller than the tunnel junction areas.

The junctions themselves are described by thin film striplines of length l_J including the junction capacitance C, the subgap resistance $R_D = 150 \, \Omega$ [5] and the junction rf admittance G_n = 2.1×10^{-3} S [27] (Fig. 4b).

On the other side the junctions are loaded to a parallel connection of a shortened line of length l_4 and a line of length l_5 terminated by the wave resistance of free space filled by the SiO dielectric with a dielectric constant $\varepsilon_r = 5.7$. All geometry parameters are listed in Table 1, too.

When the electrical circuit in Fig. 4 is calculated with the help of general equations of transmission line theory and matrix notation [28], two quantities are important: The total attenuation constant α_K of the stripline path and the attenuation P_w of the active power from the main stripline to one junction. The results for these quantities depending on the thickness h_3 of the window layer are given in Fig. 5 and Fig. 6 .

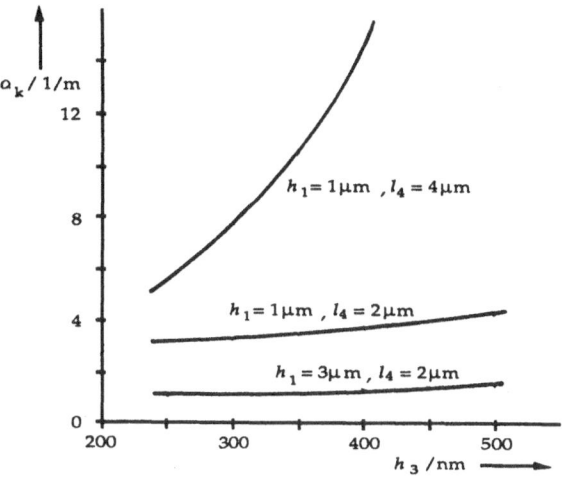

Figure 5. Total attenuation constant α_K of the stripline path depending on the thickness h_3 of the window layer, h_1 - thickness of the thick SiO or SiO_2 layer (Fig. 3), l_4 - length of the shortened line (Fig. 3).

In order to achieve a simpler adjustment of the window layer to the tunnel junction areas we started to make the SiO windows smaller by $l_4 = 4$ μm at each side. But according to the upper curve in Fig.5, this results in high values for the total attenuation constant α_K increasing with the thickness h_3 of the window layer. The upper curve in Fig.5 results from a resonance effect which occurs when the imaginary part of the imput impedance of the shortened line of length l_4 reaches the order of the imaginary part of the junction input impedance.

Decreasing l_4 from 4 μm to 2 μm avoids this resonance effect and increasing the thickness h_1 of the thick SiO or SiO_2 layer from 1 μm to 3 μm improves the attenuation constant α_K of the stripline path (lower curves in Fig.5).

Using the geometrical parameters of Table 1 we calculated α_K to 1.16 m^{-1} corresponding to 10.09 dB/m. One stripline path in the 10 V design containing 316 junctions has a length of 13.272 mm which than results in an attenuation of 0.134 dB, i.e. the last junction of the path receives about 3 % less power than the first one.

With the same geometrical parameters it can be deduced that one single junction receives $P_w = 46.95$ dB less microwave power than that propagating in the main stripline (Fig.6). To generate a 10 V step at the waveguide connection of the measuring set-up outside the He bath, a power of 18.7 dBm (74.1 mW) was needed. If we estimate the losses from this point to the beginning of one stripline path (waveguide attenuation and finline insertion loss: 10 dB, power splitting network: 0.08 dB, power splitting into 64 parts) the power at the path imput is about - 9.4 dBm (114 mW). Therefore with $P_w = 46.95$ dB, one junction receives an active power of only about 2.3 nW.

According to Fig. 6 P_w can be lowered, increasing the thickness of the window layer and decreasing the thickness of the thick SiO or SiO_2 layer. But this also increases the total attenuation constant aK, therefore the optimum design depends on that minimum power which the last junction of the stripline path in relation to the first one must receive to still produce a stable quantized voltage step.

The experimental confirmation of the resonance effect in Fig.5 can be seen in Fig.7. The dc current voltage characteristics of several 2000-junction arrays were observed irradiating the arrays with microwaves. Taking Nb/Al_2O_3/Nb arrays with $l_4 = 4$ mm, it was necessary

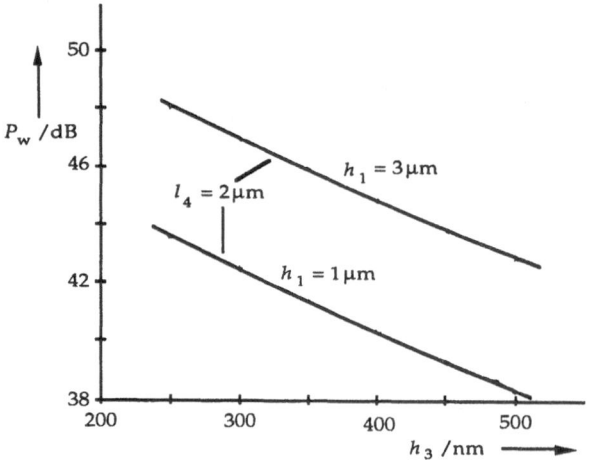

Figure 6. Attenuation P_w from the main stripline to one junction depending on the thickness h_3 of the window layer.

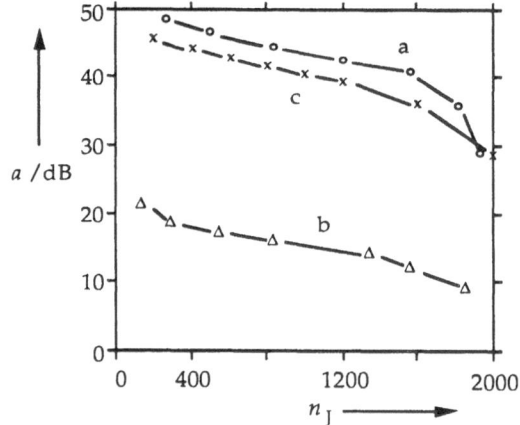

Figure 7. Measured relative microwave attenuation needed to suppress the critical currents of a 2000 junction array, n_J - number of junctions with suppressed critical currents, a. typical curve for a Pb-alloy array, b. typical curve for a Nb/Al$_2$O$_3$/Nb array with $l_4 = 4$ μm, c. typical curve for a Nb/Al$_2$O$_3$/Nb array with $l_4 = 2$ μm.

to increase the microwave power to about 10 dB on the attenuator to suppress all critical currents (curve b) and taking Nb/Al$_2$O$_3$/Nb arrays with l_4 = 2 mm about 20 dB less power was needed to suppress the critical currents (curve c). This was roughly the same power needed to suppress the critical currents of the other Pb-alloy arrays (curve a).

SYNCHRONOUS SWITCHING OF A LARGE NUMBER OF JOSEPHSON JUNCTIONS

Introduction

In the case of dc voltage standards, large Josephson junction series arrays integrated into a micro-stripline have proved to be a useful configuration for coupling a microwave signal homogeneously to a large number of junctions [17]. A section of such a stripline with a small number of junctions is shown in Fig.3. When the junctions are in the zero voltage state, (bias current I_b less than the smallest critical current of the array) the stripline is completely superconducting and should also propagate electromagnetic pulses - generated, for instance, by switching a Josephson junction to its normal conducting state - almost free of dispersion and with relatively small losses [20] ,at least at the stripline length discussed here.

A single Josephson junction induces very short current pulses in a load when it switches from the superconducting state to the normal state. In the case of highly hysteretic tunnel junctions (low damping), the switching time used for the junction of the array described here is mainly determined by the charging time of the junction capacitance C, and is estimated to be

$$\tau_s = CV_g/I_c \tag{7}$$

V_g is the gap voltage (see e.g. [29]). In this case, the junction is connected to a currrent source and effects such as "turn on delay" [30] are neglected.

If the series array on top of the stripline is connected to a current source, all the junctions are biased with the same current, and when the current is increased, the junction with the smallest critical current I_c first switches to its gap voltage. Compared with the normal junction switching time, the current sweep of the bias increases so slowly that that the current of the source might be considered constant on the switching time scale.

The rapid switch-on of the gap voltage of the first junction that switches induces a current pulse in the load represented by the stripline impedance which will be propagated along the line if the frequency spectrum of the pulse is within the margins of the stripline cut-off frequencies. The stripline described in Fig.1 acts as a "high-pass" [31] with a lower cut-off frequency of

$$f_1 = (1/2\pi)(Z_o l_p [\mu_o \, \varepsilon_o \varepsilon_r]^{1/2} C/2)^{-1/2}$$
$$\text{with } Z_o = (\mu_o/\varepsilon_o \varepsilon_r)^{1/2} \, h_1/w \tag{8}$$

Z_o is the characteristic impedance and l_p = 84 μm the periodic length of the stripline. ε_ρ = 5,7 denotes the dielectric constant of SiO. With C = 60 pF, h_1 = 3 μm and w = 70 μm (cf Fig.3) one obtains the f_1 = 12,5 GHz for cut-off frequency.

The propagation speed can be calculated from the phase velocity of a microwave passing the stripline (see eq. 5). λ_{Nb} is the penetration depth of Nb and may be assumed to be 35 nm in the circuits used here. As $\lambda_{Nb}/h_1 \gg 1$ and $d_{Nb}/\lambda_{Nb} > 1$ for the main stripline, v_ϕ may be estlmated from the simplified equation $v_\phi = c_o/\sqrt{\varepsilon_r}$. $d_{Nb} \sim$ 150 nm is the thickness of superconducting stripline electrodes. While travelling along the line the pulse passes the junctions on both sides of the junction that first switches to the normal state. A small amount of its rf power is coupled to each junction which then generates a current pulse $I_p(t)$. If I_b $+I_p(t)$ exceeds the critical current of a junction passed by the pulse, this junction will also switch to the normal state. The spread of the critical currents should therefore be as small as possible. The maximum number of junctions that can be switched depends on how far the pulse can travel before its power is attenuated to a level which is no longer sufficient to drive a junction to the normal state.

Experimental results

Fig.8 shows that in a very large series array of 2000 junctions, a certain number - about 400 - are switched directly to their normal state if the critical current $I_c \sim 3$ mA is large enough. The voltage level up to which the circuit switches (~1 V in this case) is indicated as a step in the critical current distribution, also showing that $I_p(t)$ reaches about 10 % to 20 % of the average critical current of the array.

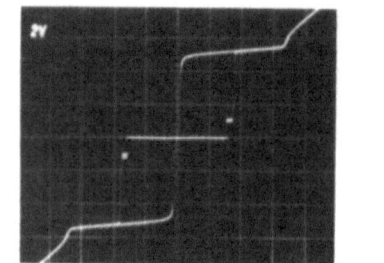
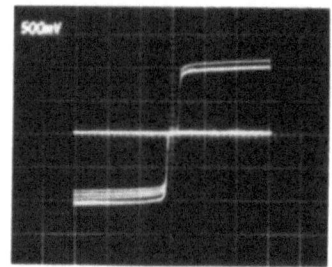

a   b

Figure 8. Dc characteristic of a Josephson series array of 2000 Nb/Al₂O₃/Nb junctions integrated into a stripline.
a) Horizontal: 2 mA/div, vertical: 2 V/div
b) Low voltage section of Fig. 2a. The sweep amplitude of the current bias is adjusted not to exceed the current step in the dc characteristic at about 1 V. The exposure time is long enough to show several cycles of the current sweep. The circuit directly switches to about 1 V.
Horizontal: 1 mA/div, vertical: 500 mV/div.

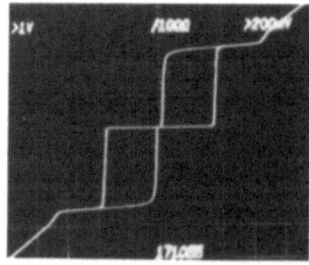
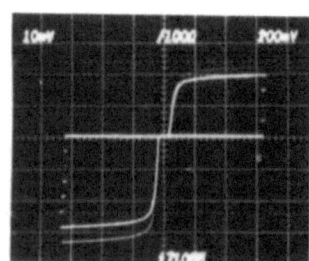

a   b

Figure 9. Dc characteristic of a Josephson series array of 1500 Nb/Al₂O₃/Nb junctions with the groundplane of the stripline omitted.
a) Horizontal: 5 mA/div, vertical: 1,5 V/div
b) Low voltage section of Fig. 3a.The single junction switching is visible. No synchronous switching of junctions occurs.
Horizontal: 2 mA/div, vertical: 10 mV/div.

32

If the groundplane of the circuit is omitted, so that a stripline configuration is no longer present, no switching is found (Fig.9). The number of junctions that are synchronously switched is proportional to the average critical current of the array. This does not depend on how the critical current has been reduced, for instance, by post-fabrication annealing at temperatures between 200 °C and 250 °C (Fig.10) or by applying an external magnetic field (Fig.11). If the spread of the critical currents is larger than that of the characteristic shown in Fig.8, for instance, the number of junctions synchronously switched decreases but is still proportional to the average critical current (cf the curve with the smaller slope in Fig.10).

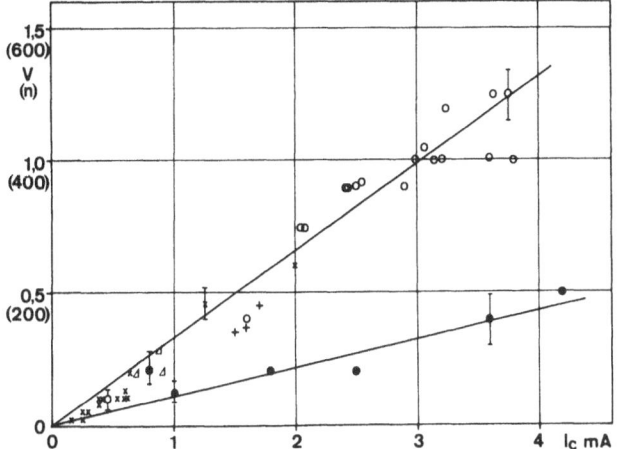

Figure 10. Dependence of the maximum junction number n synchronously switching to the normal state on the critical current of the circuit. n is equivalent to the sum gap voltage to which the circuit switches. The symbols denote different samples. The critical currents were reduced by post-fabrication annealing of the samples. The lower curve has been obtained from a sample with a relatively large current spread.

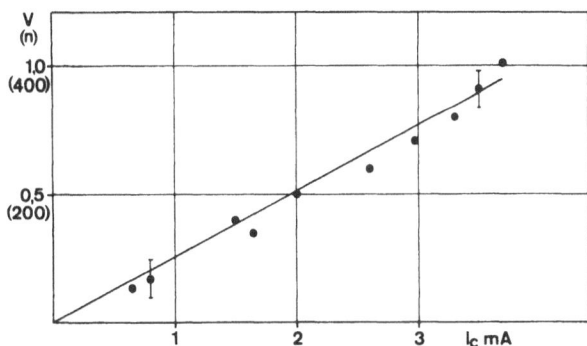

Figure 11. Dependence of the maximum junction number n synchronously switching to the normal state on the critical current of the circuit. The critical current is adjusted by applying a magnetic field to the circuit.

These results made it possible to design and fabricate a special circuit for synchronous switching. It consists of a series array with 250 junctions integrated into a microstripline as described above. Both ends of the line are teminated by a matched load. The average critical current was chosen to be large enough to allow all 250 junctions to switch synchronously. Since the stripline is short enough in this case and the critical current spread is small, the complete circuit switches directly to the normal state (Fig.12).

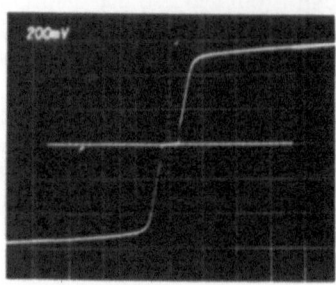

Figure 12. Dc characteristic of a series array of 250 Nb/Al$_2$O$_3$/Nb junctions with groundplane and both ends terminated by matched loads. All the junctions switch to the circuit's sum gap voltage of 560 mV. Horizontal: 1,5 mA/div, vertical: 200 mV/div.

If the junction at one end of the line is separately connected to a second current bias, it can be used to trigger the switch. The main current source biasing the complete array was swept with a relatively high repetition rate. The maximum current of the main bias was kept a few percent smaller than the lowest critical current of the array so that no switching occured from the zero state. A small additional bias current with a lower repetition rate was applied to the single junction causing this junction to switch to the normal state with the small repetition rate. By this means, the complete array was triggered to switch to the normal state (sum gap voltage of the circuit: 560 mV) with the repetition rate of the second current bias without increasing the current amplitude of the main current supply.The switching voltage of 560 mV is large enough to drive a semiconductor circuit. But even 1 V can easily be reached as Fig.8 shows

Switching time

Provided the model of the synchronous switching is correct, the total switching time τ may be estimated from the travelling time $\tau_t = L/v_\phi$ of the pulse through the stripline of the length L and the switching time of a single junction to: $\tau = \tau_t + \tau_s$. The circuit used to obtain the curve in Fig. 6 contains 250 junctions distributed over a stripline length $L = 10,5$ mm. With $v_\phi = 1,2 \cdot 10^8$ m/sec one obtains $\tau_t = 88$ ps. The single junction area amounts to 10^3 µm^2, which results in a junction capacitance of
$C = 60$ pF and together with $V_g = 2,5$ mV and $I_c = 5,4$ mA one obtains $\tau_s = 28$ ps for the switching time of a single junction. The total switching time is then about $\tau = 116$ ps. If the junction that triggers the switching is located in the center of the array, τ_t reduces to 44 ps which is not much larger than the switching time of a single junction. The midband frequency of the pulse is $f_m = 1/\tau_s = 36$ GHz, which is larger than the cut-off frequency f_l, so the stripline is correctly designed for the pulse propagation.

It should be noted that the series array used here has been designed for voltage standards and not for rapid switching. By using a smaller line width technique and larger critical current densities, it should be possible to reduce the travelling time of the pulse by at least a factor of five and the single junction switching time by a factor of two.

* This work has been carried out under contract with the Commission of the European Communities, contract No. 3108 (BCR)

REFERENCES

1. S. Shapiro, A. Janus and S. Holly, "Effect of Microwaves on Josephson Currents in Superconducting Tunneling", Rev. Phys. , 36: 223 (1964)

2. T. Quinn, "News from the BIPM: Representation of the Volt by Means of the Josephson Effect, Recommendation 1 (CI 1988)", Metrologia 26: 69 (1989)

3. T. Endo, M. Koyanagi and A. Nakamura, "High-Accuracy Josephson Potentiometer", IEEE Trans. Instr. Meas. IM-32:267 (1983)

4. M.T. Levinsen, R.Y. Chiao, M.J. Feldmann and B.A. Tucker, "An inverse ac Josephson effect voltage standard", Appl.Phys.Lett, 31: 776 (1977)

5. J. Niemeyer, J.H. Hinken and R.L. Kautz, "Microwave- induced constant voltage steps at one volt from a series array of Josephson junctions", Appl. Phys. Lett., 45: 478 (1984)

6. J. Niemeyer, E. Vollmer, J.H. Hinken and W. Meier, "Stable Constant Voltage Steps between 0.1 and 1.5 V from New Series Arrays of Josephson Tunnel Junctions",in :"SQUID 85", H.D.Hahlbohm, H. Lübbig (eds.), Walter de Gruyter, New York, 1163 (1985)

7. J. Niemeyer, L. Grimm, W. Meier, J.H. Hinken and E. Vollmer, "Stable Josephson reference voltages between 0.1 and 1.3 volts for high precision voltage standards",Appl. Phys. Lett 47: 1222 (1985)

8. C.A. Hamilton, R.L. Kautz, R.L. Steiner and F.L. Lloyd, "A practical Josephson voltage standard at one volt", IEEE Electron Device Lett., EDL-6: 623 (1985)

9. J. Niemeyer, J.H. Hinken, E. Vollmer, L. Grimm and W. Meier, "A New Josephson Voltage Standard with Tunnel-Junction Series Arrays", Metrologia 22: 213 (1986)

10. J. Niemeyer and L. Grimm, "Present Status of Josephson Voltage Metrology", Metrologia 25:135 (1988)

11. J. Niemeyer, L. Grimm, C.A. Hamilton and R.L. Steiner, "High precision measurement of a possible resistive slope of Josephson array voltage steps" IEEE EDL 7: 44 (1986)

12. R.L. Kautz and F.L. Lloyd, "Precision of series-array Josephson voltage standards", Appl. Phys. Lett. 51: 2043 (1987)

13. J.H. Greiner, "Oxidation of lead film by rf sputter etching in an oxygen plasma", J. Appl. Phys. 45:32 (1974)

14. C.A. Hamilton, F.L. Lloyd, K. Chieh and W.G. Goerke, "A 10 V Josephson voltage standard", IEEE Trans. Instrum. Meas., IM-38: 314 (1989)

15. J. Niemeyer, Y. Sakamoto, E. Vollmer, J.H. Hinken, A. Shoji, H. Nakagawa, S. Takada and S. Kosaka, "Nb/Al-oxide/Nb and NbN/MgO/NbN tunnel junctions in large series arrays for voltage standards", Jap. J. Appl. Phys., 25: L343 (1986)

16. Y. Sakamoto, T. Endo, Y. Murayama and T. Sakuraba, "Linearity of Quantized Voltages Generated by a Josephson-Junction Array", IEEE Trans Instr. Meas. 38:304 (1989)

17. J. Niemeyer, "Josephson series array potentiometer", Chap. 9 in: "Superconducting Quantum Electronics", V. Kose, Ed., Springer Verlag, Inc., New York (1989)

18. R. Pöpel, J. Niemeyer, R. Fromknecht, L.Grimm, W. Meier, "1 volt and 10 volt Series Array Josephson Voltage Standards in Nb/Al$_2$O$_3$/Nb Technique" to be published, J. appl. Phys. (1990)

19.   J.C. Swihart, "Field Solution for a Thin-Film Superconducting Strip Transmission Line", J. appl. Phys., 32: 461 (1961)

20. R.L. Kautz, "Picosecond pulses on superconducting striplines" J. appl. Phys. 49: 308 (1978)

21. W.H. Henkels and C.J. Kircher, "Penetration Depth Measurements On Type II Superconducting Films", IEEE MAG 13:63 (1977)

22. R.E. Matick, in "Transmission lines for digital and communication networks", Mc Graw-Hill, Inc., New York (1969)

23. R. Pöpel, "Auswertung der Mattis-Bardeen-Theorie und Messungen an supraleitenden Mikrostreifenleitungen", PhD thesis, University of Braunschweig (1986)
24. D.C. Mattis and J. Bardeen, "Theory of the anomalous skin effect in normal and superconducting metals", Phys. Rev., 111: 412 (1958)

25. R. Pöpel, "Exact solution of the Mattis Bardeen theory for thin superconducting films and bulk material", IEEE Trans. Magn., to be published (1991)

26. R. Pöpel, "Surface impedance and reflectivity of superconductors", J. Appl. Phys. 66: 5950 (1989)

27. T.F. Finnegan and S. Wahlsten, "Electromagnetic properties of small Josephson junctions coupled to microstrip resonators", Physica 94B: 219 (1978)

28. G. Matthaei, L. Young and E.M.T. Jones, in: "Microwave filters, impedance-matching networks, and coupling structures", Artech House Books, Inc., Dedham (1980)
29. K.K. Likharev, 129 pp, Gordon and Breach Science Publishers, New York 1986,.

30. D.G. McDonald, R.L. Peterson, C.A. Hamilton, R.E. Harris, R.L. Kautz, "Picosecond Applications of Josephson Junctions " IEEE Trans. on Electron. Devices ED-27: 1945 (1980)

31. J. Hinken, in: "Supraleiterelektronik" Springer Verlag, Berlin, 91, 1988

EXPERIMENTS ON JOSEPHSON JUNCTIONS DRIVEN BY A BI-HARMONIC RF SOURCE

D. Andreone, V. Lacquaniti, S. Maggi

Istituto Elettrotecnico Nazionale Galileo Ferraris
Torino, Italy

The amplitude of the steps at the reference voltages $V_n = n f(h/2e)$ in the I-V characteristic of a Josephson junction RF biased at a frequency f is an important parameter in designing devices for voltage standards. In fact the stability of the lock between the junction phase and the applied microwave field, which is a crucial problem for voltage standard application, is strongly dependent on the step amplitude[1]. It is well-known that in the simplified model of a voltage biased junction, where the spatial dependence of the phase is neglected, the expected amplitude ΔI_n of the nth step is a fraction of the critical current I_C given by the Bessel function of the first kind of integer order n, as in the equation[2]:

$$\Delta I_n = 2 I_C \, | \, \mathrm{J}_n\!\left(\frac{n V_{RF}}{V_n} \right) |$$

But due to the simplifications in that model many discrepancies have been found between experiments and the Bessel-like behaviour. Many causes have been identified. One example is phase bending in an extended junction because of the standing electromagnetic field in the barriers. The result of calculations in large area junctions that holds $\Delta I_n = 0$ when n is an odd integer and $\Delta I_n < |\mathrm{J}_n|$ for n even is well known[2,3]. Noise both in the junction and in the RF drive[4] and chaotic behaviour[1], when the driving frequency is near the plasma frequency, tend to reduce the step amplitude.

Recently[5] R. Monaco has demonstrated that a driving RF source rich in harmonics produces steps with maximum amplitude $\max(\Delta I_n / 2 I_C)$ greater than $\max |\mathrm{J}_n(n V_{RF}/V_n)|$. On the other hand one could guess that a suitable choice of the driving signal (frequency, amplitude, wave shape) can confer larger stability to phase-locked steps. Considering the case of a point-like junction voltage-biased by a bi-harmonic RF source (which could be reasonably imagined as a real instrument,) composed of both the signal of angular frequency ω and its second harmonic

Nonlinear Superconductive Electronics and Josephson Devices
Edited by G. Costabile *et al.*, Plenum Press, New York, 1991

37

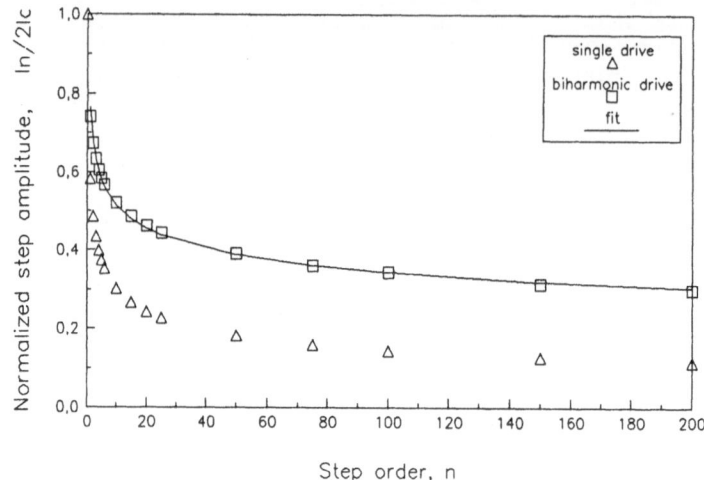

Fig. 1. - Expected maximum amplitude of the constant-voltage steps in a junction biased by a bi-harmonic drive (the two components are in phase, $\theta=0$) compared with the Shapiro steps obtained with a sinusoidal RF source (single drive). The fitted curve is represented by the equation $y = 0.766 \cdot n^{-0.174}$.

$$V_d(t) = V_{RF}[\cos\omega t + \rho \cos(2\omega t + \theta)]$$

one obtains the following expression for the amplitude of the nth step[5]

$$\Delta I_n = 2I_c \sum_{k=-\infty}^{\infty} J_k(a) J_{n+2k}\left(\frac{\rho a}{2}\right) \sin[(n+2k)\theta + \bar{\phi}]$$

where $a = 2eV_{RF}/\hbar\omega$. The step amplitudes depend both on the amplitudes of the two components of the driving signal and on their phases. In figure 1 the expected maximum amplitudes of the steps obtained working out this expression up to the order 200 when the two components are in phase ($\theta=0$) are shown. The enhancement of the steps with respect to the conventional Bessel-like behaviour obtained with a sinusoidal drive is of the order of 160 percent at $n \simeq 200$. Note that this maximum is obtained for decreasing ρ as n increases.

An experimental verification of this effect in real junctions could suggest the need for further calculations on more refined model of junctions but can give also some indications for the possible practical applications. For example in voltage standards larger step amplitudes could ease the design constraints in order to enhance the phase-lock stability.

The circuit diagram of the realized bi-harmonic drive is shown in figure 2. The fundamental component of the signal is generated in the 9-12 GHz range (X band) by means of a klystron. Part of the signal is used to drive a doubler in order to obtain the 2nd harmonic in the 18-24 GHz range (in

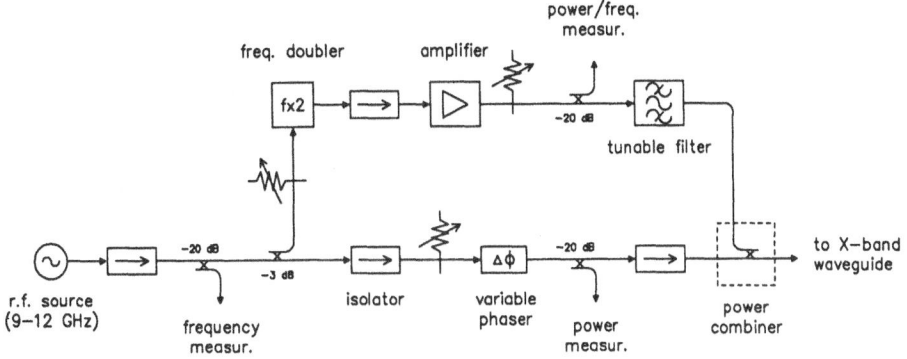

Fig. 2. - Schematic diagram of the RF circuit for generating the bi-harmonic signal $V_d(t) = V_{RF}[\cos(\omega t + \phi) + \rho \cos 2\omega t]$.

the K band). A high pass filter removes any trace of the fundamental in the output signal from the doubler. The 2nd harmonic is then amplified by means of a wide-band travelling-wave-tube amplifier while a tuneable filter reduces the band enabling a signal-to-noise ratio of 35 dB. Variable attenuators allow the regulation of the amplitudes and a phase shifter can change the phase of the fundamental component. Finally the two components are combined together and fed to the junction through either a 50-Ω microstrip line or a waveguide both for the X band, thus the second harmonic propagates out of mode. The available power at the Josephson device was 10 mW for both the components. The coupling factor was not estimated and also the ratio ρ of the components of the source remained undetermined because the coupling factor of the second harmonic may be different from the one of the fundamental.

We have experimented on two kinds of Josephson devices with different modes of operation with respect to RF biasing. In the devices of one kind the junctions are coupled to an external cavity and are driven at the resonant frequency of this cavity, in the devices of the other type the junctions are operated at their own resonant frequencies. In the former case the influence of the bi-harmonic drive is noticeable - the step amplitudes do increase - but the phenomenology is even more complicated than what we expect from the theory, because of the large area of the junctions and also because of the onset of chaos. In the latter case the influence of the bi-harmonic drive is still visible but apparently the step amplitudes do not increase beyond what one can obtain using a sinusoidal drive.

In figure 3 there is an example of how a bi-harmonic drive affects the I-V characteristics of an array of 20 junctions in series coupled to a microstrip resonator with a Q factor of the order of 1000. This type of device was developed for making a voltage standard at 100mV level and is described in Ref. 6. One can note in the curve 3(b) obtained with a sinusoidal drive that the steps are very small, compared to the amplitude (of the order of $0.15 \cdot I_c$) expected from the Bessel-like behaviour, and not well organized as collective steps of the 20 junctions because the sinusoidal drive has a frequency lower than the plasma frequency in the junctions and the critical currents in the array are widely spread. The addition of the 2nd harmonic increases the step amplitude at least three times, helping in organizing better collective steps in the array (curve 3a). The effect of the 2nd harmonic by itself was tested: it was strong enough to depress the critical current but not enough to generate steps. The effect of the 2nd harmonic was strongly reduced at frequencies slightly out of resonance. Rotating the phase of the fundamental component by 90°, the amplitudes of the steps vary from the maximum of the curve 3(a) to the minimum of the curve 3(b).

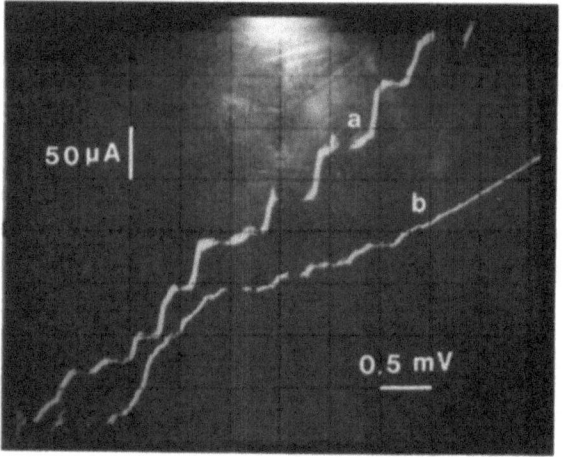

Fig. 3. - I-V characteristics of an array of 20 junctions in series (portions
centred at $V \simeq 37$mV and shifted vertically for convenience): the
lower curve **b** is obtained with a sinusoidal drive at 10 GHz, while
the upper curve **a** shows the effect when the 2nd harmonic at 20 GHz
is added. The step separation is 0.4mV corresponding to 20μV per
junction, as expected. The step order, at which each junction is
biased, is $n \simeq 90$. The Nb-Pb junctions have dimensions: 350μm x
50μm, critical currents: 2.5-4 mA, c.c. densities: 14-23 A/cm^2,
and plasma frequencies: 15-19 GHz.

The influence of changing the phase of the fundamental component can be seen in another example in figure 4 which shows the I-V characteristics of a single-junction device coupled to an external low-Q cavity (Q $\approx$100) taken in four different operating conditions. The steps obtained with the bi-harmonic drive and with the sinusoidal drive are compared graphically in figure 5. Also this junction does not follow the Bessel-like behaviour when biased with the sinusoidal drive because of the high impedance of the coupling circuit, of the extended dimensions and of the high plasma frequency which is comparable to the driving frequency. The two curves taken with the bi-harmonic drive with different settings for the phase of the fundamental component illustrate how the optimum amplitude of each step is obtained with proper values of the parameters ρ and θ: compare for example the steps n=5, n=-5 and n=10. The asymmetry observed between the steps with n positive and those with n < 0 is accounted for by the theory[5] but suggests some experimental complications in standard voltage application where voltage reversing is needed. The amplitude enhancements are of the expected order of magnitude taking into account the experimental errors.

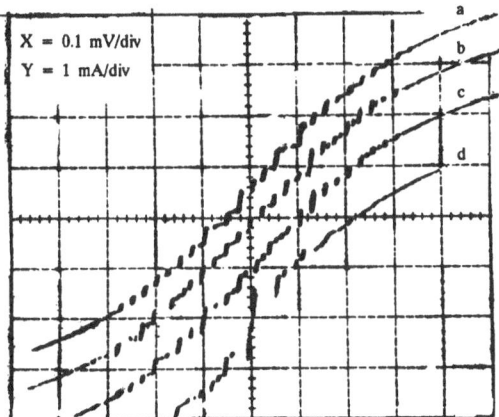

Fig. 4. - I-V characteristics, centred at V = 0 and shifted vertically, of a junction coupled to a low-Q cavity: curve **a** is obtained with a bi-harmonic drive (10.6GHz + 21.2GHz); curve **b** has the 10.6GHz phase rotated of 90°. Curve **c** is for the sinusoidal drive alone. Curve **d** shows the effect of the 2nd harmonic alone. The Nb-Pb junction has dimensions: 530μm x 200μm; critical current: 5 mA, c.c. density: 4.8 A/cm2, from which the plasma frequency: 9.2 GHz.

The last case examined is a junction operated at its own resonant frequency. In figure 6 there is a comparison between the steps obtained with the bi-harmonic drive at V $\simeq$ 2mV (step order n $\simeq$ 110) and those with a sinusoidal drive. The step amplitude is not as expected from Bessel-like behaviour: in fact $\Delta I_{110}/2I_c \simeq 0.05$ instead of 0.14. While the bi-harmonic drive modifies the step amplitude behaviour, it does not increase the step amplitudes significantly. Taken a particular step, the same maximum amplitude can be obtained either with the sinusoidal drive or with the bi-harmonic one provided proper settings of wave amplitudes and phases are used. Rotating the phase of the fundamental component the steps change from curve a to curve b. One cause of this behaviour worth exploring may be the current-biased model suggested by the mode of operation of this junction.

The experiments presented in this paper have illustrated the main features of Josephson junctions operated with a bi-harmonic drive. As the junctions used were not designed for this purpose, the RF induced steps deviate from the Bessel-like behaviour because the plasma frequency was in general of the order of the drive frequency, the dimensions were always large compared to the Josephson length, still the bi-harmonic drive enhances the step amplitudes. The only exception seems the case when the junction impedance is low compared to the coupling circuit.

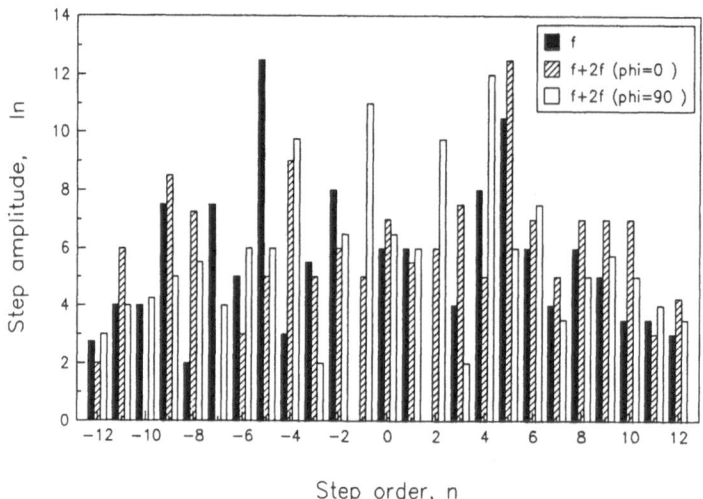

Fig. 5. - Graphical analysis of the steps amplitudes for curves **a**,**b** and **c** of fig. 4.

A bi-harmonic RF source may find a useful application as a drive of Josephson arrays for voltage standards operated at frequency as low as 10 GHz[6]. In fact the enhanced step amplitudes would ease the design parameters for optimum stability[1]. For example, to keep the plasma frequency lower than one third of the drive frequency a lower critical current should be used in comparison with the arrays operated at 70-90 GHz. On the other hand the number of steps per junction needed to reach the same voltage level is higher because of the lower step separation, of the order of 20 μV. All this comes up with low amplitude steps. However, if a bi-harmonic drive is considered, the amplitudes of the steps may be almost doubled already for $n \approx 10$; then the stability of the steps is expected 6 times better, as it increases exponentially with the step amplitude. This stimulates further experiments on properly designed Josephson junctions.

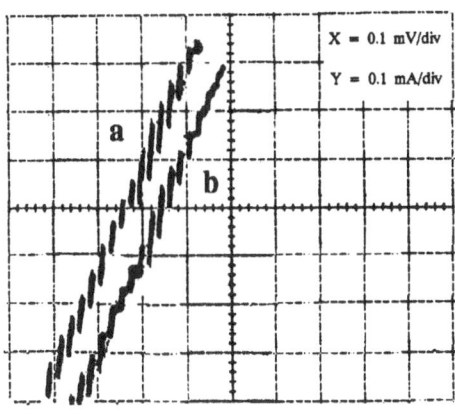

Fig. 6. - I-V characteristic of a junction RF biased at self-resonant frequency (portions centred at $V \simeq 2mV$ and shifted vertically): curve b is obtained with a sinusoidal drive at 9 GHz, while curve a shows the effect when the 2nd harmonic at 18 GHz is added. The step order is $n \simeq 110$ and are separated by 18.6 μV. The Nb-Nb junction has dimensions: $650\mu m$ x $50\mu m$; critical current: 1 mA; c.c. density: 3.3 A/cm^2; plasma frequency: 4.2 GHz.

ACKNOWLEDGMENTS

These experiments were originally suggested by R. Monaco, who provided also some samples. The microwave circuit has been set up with the instruments lent by CSELT of Turin. This work has been supported by the National Research Council, under the Progetto Finalizzato "Superconductive and Cryogenic Technologies".

REFERENCES

1. R.L. Kautz, C.A. Hamilton, and Frances L. Lloyd, Series-array Josephson voltage standards, IEEE Trans. Magn. MAG-23: 883 (1987).
2. A. Barone and G. Paternò, "Physics and applications of the Josephson effect", J. Wiley & S., New York (1982).
3. U. Klein, S. Schröder and J.H. Hinken, Calculation of the Shapiro step amplitude in two-dimensional Josephson tunnel junctions, in: "CPEM'90 Digest (Conference on Precision Electromagnetic Measurements)", G.R. Hanes, ed., Ottawa (1990).
4. G. P. Bava, D. Andreone, E. Bava, A. Godone, Influence of microwave phase noise on Josephson voltage steps, J. Appl. Phys. 53: 63256 (1982).
5. R. Monaco, Enhanced AC Josephson effect, J. Appl. Phys. 68: 679 (1990)
6. D. Andreone, V. Lacquaniti, G. Costabile, M. Lepore, S. Pagano and R. Monaco, Twenty-junction arrays for a Josephson voltage standard at 100mV level, in: "Stimulated effect in Josephson devices", M. Russo and G. Costabile, ed.s, World Scientific Publishing, Singapore (1990).

CO-OPERATIVE EXTERNAL SYNCHRONIZATION IN JOSEPHSON

SERIES ARRAYS FOR THE CRYOGENIC VOLTAGE STANDARD

Hans-Georg Meyer and Wolfram Krech

Friedrich-Schiller-Universität Jena
Institut für Festkörperphysik
Jena, Germany

1. INTRODUCTION

The modern cryogenic Josephson array voltage standard takes
advantage of two important principles. The first one is to use highly
hysteretic Josephson junctions with overlapping Shapiro steps near zero
current bias. This allows to obtain a stable quantized reference voltage
in the volt region without an individual dc biasing of each junction.
The second principle is to integrate the Josephson junctions into a
superconducting microwave transmission line. This proves as an effective
method to achieve a nearly uniform rf excitation of a very large number
of junctions essential for a maximum voltage output[2]. In the last few
years these two principles were successfully be used to realize Josephson
arrays and, finally, a quantized voltage level of about 12 V was
reached[3]. In the present paper we study in detail the formation of the
Josephson oscillations along the array.

2. DESCRIPTION OF THE VOLTAGE STANDARD ARRAY

The microwave chip for the voltage standard is designed in such a
way that the array itself acts as a microwave transmission line for the
rf bias but as a simple series connection for the dc bias. This enables
both, the necessary rf excitation of the junctions and the addition of
the single quantized voltage contributions. A simplified equivalent
circuit is shown in Fig. 1: The microwave power is introduced at the left
end of the stripline and that fraction reaching the other end is
dissipated by a load to get a traveling wave required for an uniform
junction excitation.

To describe the dynamics of the integrated junctions we have to use
a weak link theory. For simplicity, we restrict ourselves to the simple
Stewart-McCumber model although the Werthamer model is more appropriate
for the tunnel-type junctions applied in practical arrangements. (The
subsequent considerations can be easily transferred to the Werthamer
model[4]). In the frame of the Stewart-McCumber model the Josephson
junction is characterized by three parameters: the critical current I_c,

the normal-state resistance R, and the capacitance C; the current i_n

(n=1,2,...,N) passing the nth junction is given by

Nonlinear Superconductive Electronics and Josephson Devices
Edited by G. Costabile *et al.*, Plenum Press, New York, 1991

$$i_n = \beta_c \ddot{\phi}_n + \dot{\phi}_n + \sin \phi_n \, , \qquad (1)$$

where the current is normalized to the critical current; ϕ_n are the Josephson phase differences, the overdots denote a differentiation with respect to the normalized time $s = t \cdot 2eRI_c/\hbar$; the McCumber parameter β_c is given by $\beta_c = 2eR^2 I_c C/\hbar$. Due to the Josephson relation the normalized voltage can be obtained from the time derivative of the phase

$$v_n = \dot{\phi}_n \, . \qquad (2)$$

The microwave transmission line itself is described by the partial inductance L_s, the capacitance C_s and the resistance R_s where we have introduced normalized parameters in the following manner:

$$l_s = \frac{2eI_c L_s}{\hbar} \, , \qquad c_s = \frac{2eR^2 I_c C_s}{\hbar} \, , \qquad r_s = \frac{R_s}{R} \, .$$

The final load z_F as well as the auxiliary impedance $z_s = (1/r_s + 1/jl_s\omega)^{-1}$ are also normalized to R. The quantity r_s serves as a model parameter for the microwave attenuation in the superconducting transmission line; it can be calculated in the frame of the two-fluid model from the stripline parameters as the London penetration depth, the normal conductivity, and the geometry.

The bias current is composed of a ac and a dc component,

$$i = i_0 + i_1 \sin \omega s \, , \qquad (3)$$

where $\omega = \hbar f/2eRI_c$ is the normalized angular drive frequency.

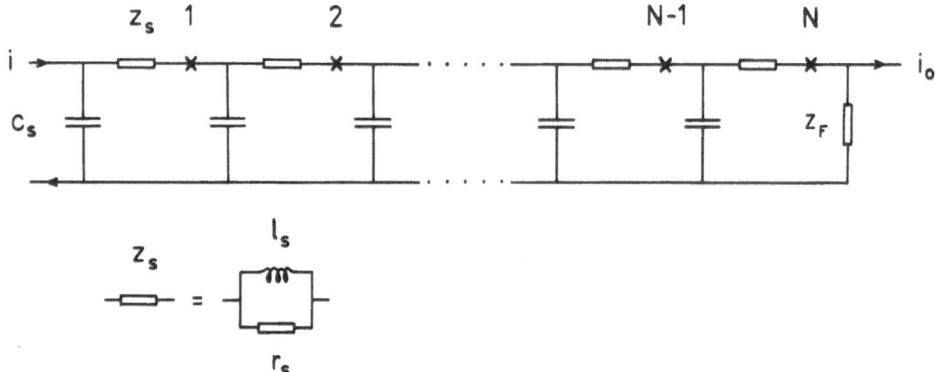

Fig. 1. Circuit diagram of a superconducting microwave transmission line with the integrated Josephson junctions marked by the crosses.

3. FORMATION OF THE JOSEPHSON OSCILLATIONS

About the Josephson oscillations in the array it must be supposed the following: Firstly, to get suitable Shapiro steps the oscillations must be synchronized with the external microwave drive and, secondly, for a good adaptation of the Josephson junctions to the transmission line the junctions must operate in a region with nearly harmonic junction currents and voltages. Therefore, we have to look for a solution of the type

$$\phi_n = \chi_n + m_n \omega s + j a_n e^{j\omega s} , \qquad (4)$$

where the more convenient complex notation is used; m_n is an integer indicating the synchronization state and a_n a complex amplitude responsible for the Josephson oscillations along the array. According to Eq. (2) the junction voltage is given by

$$v_n = m_n \omega - \omega\, a_n e^{j\omega s} ; \qquad (5)$$

since the mean voltage is quantized due to the external drive frequency ω the junction is trapped on a Shapiro step. Substituting the ansatz for the phase differences (4) into the Stewart-McCumber differential equation then we obtain the required harmonic current response

$$i_n = m_n \omega - J_{m_n} (A_n)\sin \psi_n - \omega\, z_J^{-1} a_n e^{j\omega s} \qquad (6)$$

if the oscillation amplitudes fulfill the condition

$$\left| \omega\, z_J^{-1} a_n \right| > 1 . \qquad (7)$$

Here, we have used the abbreviations

$$a_n = A_n e^{-j\chi_n^1} , \qquad \psi_n = \chi_n + m_n \chi_n^1$$

and

$$z_J = (1+j\, \beta_c \omega)^{-1}$$

as the junction passive impedance.

The complex amplitudes a_n controlling the junction current can be calculated from the equivalent circuit shown in Fig. 1 by the use of the conventional electrotechnical methods. As a result we find the following set of relations:

$$a_n = \frac{j z_J i_1 D_{N-n}}{\omega\, D_N} , \qquad (8a)$$

$$D_n = e^{jn\zeta} + \frac{u_F - e^{j\zeta}}{\sin \zeta} \sin n\zeta ; \qquad (8b)$$

$$\zeta = \cos^{-1}\left(1 - \frac{1}{2} lc_s \omega^2 + \frac{1}{2} jrc_s \omega \right) \quad , \tag{9}$$

$$u_F = 1 - lc_s \omega^2 + jc_s \omega \left(1 + z_F \right) \quad , \tag{10}$$

where the quantities

$$l = \frac{l_s}{1 + \dfrac{l_s^2 \omega^2}{r_s^2}} - \frac{\beta_c}{1 + \beta_c^2 \omega^2} \quad , \tag{11a}$$

$$r = \frac{\dfrac{l_s^2 \omega^2}{r_s + \dfrac{l_s^2 \omega^2}{r_s}}}{} + \frac{1}{1 + \beta_c^2 \omega^2} \quad . \tag{11b}$$

can be considered as the effective transmission line inductance and resistance in presence of the Josephson junctions; the first terms rise from the stripline, the second ones are the modifications due to the junctions.

Now, matching together Eqs. (6) and (8) we get an explicit expression for the junction current; its rf component

$$i_n^{ac} \sim a_n e^{j\omega s} \sim e^{j[\omega s + (N-n)\zeta]} + \frac{u_F - e^{j\zeta}}{\sin \zeta} \sin(N-n)\zeta \; e^{j\omega s} \tag{12}$$

shows that along the array the Josephson oscillations form a wave phenomena composed of a standing and a traveling wave; the dc component

$$i_n^{dc} = m_n \omega + J_{m_n}(|a_n|)\sin \psi_n \tag{13}$$

represents the Shapiro steps with the well-known Bessel function dependence of the step amplitude

$$H_n = | J_{m_n}(A_n) | \quad . \tag{14}$$

Here, it is important to note that in contrast to the single junction problem the step formation depends on the subscript n indicating the junction position in the array.

The standing wave component gives rise to a non-uniform rf excitation of the junctions in succession and, consequently, also to a non-uniform Shapiro step formation, cf. Eqs. (12) and (14). As a consequence, not treated here in detail, the maximum available reference voltage is reduced. Therefore, we have to favour a traveling wave situation which can be obtained due to the relation (12) by matching the load,

$$u_F = e^{j\zeta} \quad . \tag{15}$$

In this case, the Josephson oscillations form altogether a simple traveling wave with a propagation constant ζ

$$i_n^{ac} \sim e^{j(\omega s - n\zeta)} . \tag{16}$$

This situation can be compared with a finite chain of coupled oscillators excited at one side and transmitting the introduced power to the other side. In this way, a co-operative external synchronization of a large number of Josephson junctions is realized.

4. VOLTAGE STANDARD ARRAY DESIGN

The uniformity of the rf excitation is not only influenced by the standing wave but also by a finite attenuation of the Josephson wave (16). For a more detailed study of the wave propagation we take into account some practical aspects of the array design. The most important point of the junction design is to prevent a chaotic dynamics. For this end, the external drive frequency is usually chosen much greater than the plasma frequency $\beta_c^{-1/2}$, cf. Ref. 5,

$$\omega \, \beta_c^{1/2} \gg 1 \quad . \tag{17}$$

It is important to note that in the case of a large rf drive level $i_1 \gg 1$ a careful analytic analysis of the single junction instabilities leads to the following condition for chaos prevention :

$$\omega^2 \, \beta_c > 2(\, 2|J_m(A)| + |J_{m-1}(A)| + |J_{m+1}(A)| \,)$$

or

$$\omega \, \beta_c^{1/2} > 4.08 \cdot i_1^{-1/4} \quad . \tag{18}$$

This condition is much less restrictive than the relation (17) and allows the application of lower drive frequencies, for example, in the vicinity of the plasma frequency. Lower drive frequencies reduce the costs of the standard because of the much less expensive techniques for the frequency stabilization. However, the most arrays on which was reported use the above mentioned stronger condition (17). In addition, taking into account that the normalized drive frequencies applied in the present voltage standards are much smaller than unity then, the relations

$$\frac{1}{\beta_c \omega} \ll \omega \ll 1 \tag{19}$$

are fulfilled.

From the propagation constant (9) it can be concluded that we have a nearly ideal Josephson wave propagation and, in particular, a small attenuation for

$$r \ll l\omega \ll \frac{1}{c_s \omega} \, . \tag{20}$$

After all, these relations concern the transmission line design: The effective ohmic resistance of the microstripline must be much smaller than the inductive resistance and that one again much smaller than the capacitive one, cf. Eqs. (11a) and (11b).

The conditions (19) and (20) allow us to simplify considerably the expression for the propagation constant (9),

$$\zeta = \alpha - j\gamma \simeq \omega \, (lc_s)^{1/2} - \frac{1}{2} \, jr \, (c_s/l)^{1/2} \quad . \tag{21}$$

Here, the imaginary component γ corresponds to the Josephson wave attenuation which is of a particular interest. In the same manner we obtain from the matching condition (15) and Eq. (20) for the terminating impedance

$$z_F \simeq (1/c_s)^{1/2} - jl\omega \quad ; \tag{22}$$

apart from the small imaginary component the final load is approximately equal to the effective wave resistance of the transmission line.

Next, we have to inspect the consistency condition for the harmonic current response (7). In the traveling wave case we have

$$a_n = \frac{j \, z_J \, i_1}{\omega} \, e^{jn\zeta} \tag{23}$$

and it results

$$\left| \, i_1 \, e^{-n\gamma} \, \right| \; > \; 1 \quad . \tag{24}$$

This relation can be fulfilled for a large number of junctions at a large rf drive level and a sufficiently small damping constant γ. A small attenuation is also of importance for the step generation. With Eq. (23) the constant-voltage steps (14) take the form

$$H_n = \left| \, J_{m_n} \left(\frac{i_1 e^{-n\gamma}}{\omega(1+\beta_c^2\omega^2)^{1/2}} \right) \right| \quad . \tag{25}$$

Because of the exponential decrease of the junction rf excitation the order of the maximum Shapiro step decreases along the array and, therefore, the contribution of the single junctions to the total quantized voltage also decreases. This leads to an optimum number of junctions which needs not to be exceeded since additional junctions are not excited enough to contribute significantly to the voltage output. Moreover, strongly under-excited junctions ($A_n < 1$) reduce the long-time stability of the reference voltage because of the too small Shapiro steps. The optimum number of junctions can be calculated from the experimental requirement that the attenuation should not exceed 6 dB, that is, the Josephson oscillation amplitudes should not more decrease than by a factor of about two,

$$N_{opt} = \frac{\ln 2}{\gamma} \quad .$$ (26)

In other words, γ limits the number of junctions in succession or the total voltage output of the series connection, respectively. Finally, it should be noted that the single junction stability condition (18) can be generalized for the series array in the following manner:

$$\omega \beta_c^{1/2} > 4.08 \cdot (i_1 e^{-\gamma n})^{-1/4} \quad .$$ (27)

This means that the junctions at the end of the array are driven closer to the chaos than the first ones. With an increasing γ also this kind of instability increases.

5. FREQUENCY DEPENDENCE OF THE ATTENUATION

As the attenuation is one of the fundamental parameters for an optimum operation of a externally driven series array we discuss in more detail the damping constant γ which can be derived from Eq. (21) and the design conditions (20):

$$\gamma = \gamma_s + \gamma_J \quad ,$$ (28)

$$\gamma_s = \frac{1}{2} \left(\frac{C_s}{L_s} \right)^{1/2} \frac{(2\pi f)^2 L_s^2}{R_s} \quad ,$$ (29a)

$$\gamma_J = \frac{1}{2} \left(\frac{C_s}{L_s} \right)^{1/2} \frac{1}{(2\pi f)^2 C^2 R}$$ (29b)

where we have used non-normalized quantities. In the frame of our simplified array model, cf. Fig. 1, the attenuation of the Josephson wave consists of two terms: The first term rising from the stripline is proportional to the square of the drive frequency, the second one is caused by the junctions and inversely proportional to the square of the drive frequency. Furthermore, it is important to note that the latter one depends also inversely on the junction normal-state resistance and not on the subgap resistance [4]. The attenuation due to the junctions becomes of increasing interest for lower drive frequencies.

Because of the different frequency dependence of the two damping components the total attenuation takes a distinct minimum at

$$f_{min} = \frac{1}{2\pi} \left(\frac{R_s}{R C^2 L_s^2} \right)^{1/4}$$ (30)

with

$$\gamma_{min} = \frac{1}{C} \left(\frac{C_s L_s}{R R_s} \right)^{1/2} \quad .$$ (31)

This behavior is illustrated in Fig. 2 where the logarithm of the damping constant is shown as a function of the external drive frequency. The different curves correspond to the PTB array (1) of Ref. 2 and an array of the Friedrich- Schiller-Universität Jena (FSU)[1]; the FSU array

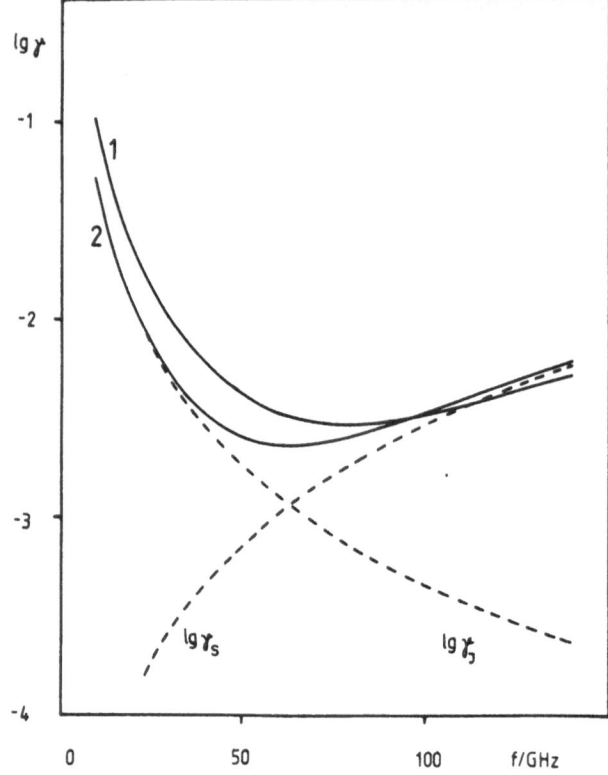

Fig. 2. The logarithm of the damping constant γ as a function of the drive frequency f.

(2) consists of 108 junctions and is driven at 35 GHz, while the PTB one operates with 1474 junctions in succession and a 90 GHz microwave drive. For the FSU arrangement the dashed lines represent separately the attenuation due to the junctions and the stripline, respectively. The intersection of these two graphs determines the minimum attenuation. For frequencies below this minimum the attenuation due to the junction dominates; in the opposite case the main contribution rises from the stripline rf impedance. Calculating, for example, the minimum attenuation of the above two voltage standard arrays due to Eq. (30) we find that the FSU array is driven below its optimum frequency of f_{min} = 63 GHz, while the PTB array operates above its optimum frequency of f_{min} = 78 GHz; the total attenuation for the 1474 junction array estimated from Eqs. (28) and (29) is 4.4 dB, that is, near the practical limit of about 6 dB[4].

6. SUMMARY

If the Josephson junctions are integrated in a superconducting microwave transmission line in a suitable way then the Josephson oscillations may form a traveling wave along the array. This principle acts as an effective method to achieve a nearly uniform co-operative external synchronization of a very large number of Josephson junctions as required in modern cryogenic voltage standards. Our simple stripline-junction model allows to estimate all the essential parameters for the synchronization as well as for the Shapiro step generation.

For a stable and a maximum quantized voltage output the attenuation of the Josephson wave must be as small as possible. The damping constant is composed of two terms rising from the stripline and from the junctions. The first term is proportional to the square of the drive frequency and the second one inversely proportional to just the same quantity. The junction-caused part of attenuation is of increasing interest at lower drive frequencies. As a function of frequency the damping constant takes a minimum at the optimum drive frequency which can be expressed in terms of some simple model parameters.

In practical voltage standard arrangements a sufficiently large rf drive level should be preferred. This ensures not only a good adaptation between the junctions and the microwave transmission line but it allows to operate with lower drive frequencies, for example, in the vicinity of the plasma frequency without any chaotic disturbances. With lower drive frequencies a low-priced voltage standard could be realized.

REFERENCES

1. M. T. Levinsen, R. Y. Chiao, M. J. Feldmann, and B.A. Tucker, An Inverse AC Josephson Effect Voltage Standard, Appl. Phys. Lett. 31:776 (1977).

2. J. Niemeyer, J. H. Hinken, and R. L. Kautz, Microwave-Induced Constant-Voltage Steps at One Volt From a Series Array of Josephson Junctions, Appl. Phys. Lett. 45:478 (1984).

3. C. A. Hamilton, F. L. Lloyd, Kao Chieh, and C. Goeke, A 10-Volt Josephson Voltage Standard, IEEE Trans. Instr. Meas. IM-38:314 (1989).

4. H. G. Meyer, W. Krech, and B. Mikolajczak, Josephson Oscillations in a Series Array Josephson Voltage Standard, J. Appl. Phys. 65:4338 (1989).

5. R. L. Kautz, C. A. Hamilton, and F. L. Lloyd, Series-Array Josephson Voltage Standards, IEEE Trans. Magn. MAG-23:883 (1987).

6. B. Mikolajczak, H. G. Meyer, and P. Seidel, Theoretical Investigations of Microwave-Driven Josephson Junctions in a Large Frequency Range, J. Appl. Phys. 66:735 (1989).

7. F. Müller and H. J. Köhler, Microwave-Induced Voltage Steps at 0.1 V From a Series Array of Josephson Tunnel Junctions, Phys. Status Solidi 106:K165 (1988).

THE SUPERCONDUCTOR INSULATOR SUPERCONDUCTOR MIXER RECEIVER -

A REVIEW

Raymond Blundell

Harvard-Smithsonian
Center for Astrophysics
60 Garden Street Cambridge
Massachusetts 02138 USA

Dag Winkler

Department of Physics
Chalmers University of Technology
S-41296 Göteborg
Sweden

INTRODUCTION

It is now about thirty years since Dayem and Martin[1] observed a step structure in the current-voltage (I-V) curve of a superconductor-insulator-superconductor (SIS) tunnel junction when subjected to microwave radiation at 38 GHz. This step structure was later explained by Tien and Gordon[2] in terms of the photon assisted tunneling of single electrons across the tunnel barrier. Several years later, Tucker developed a full quantum mechanical treatment of heterodyne mixing in the SIS tunnel junction.[3] His theory was used to predict a number of interesting features. In particular, it predicted quantum limited mixer noise and the possibility of mixer conversion gain,[4] not possible through the classical treatment of mixing, and hence the possibility of extremely low-noise receivers. It also predicted that the local oscillator power required for optimum noise performance should be several orders of magnitude lower than that required to correctly operate the Schottky diode mixers usually employed for heterodyne mixing. A complete review of mixing with superconducting tunnel junctions has been given by Tucker and Feldman.[5]

To evaluate the potential performance of a particular tunnel junction type using Tucker's theory, all that is required is a knowledge of its d.c. current-voltage relation. The d.c. bias, local oscillator (L.O.) drive and impedance levels at the mixer's input and output are then used as variable parameters to evaluate or map the performance. In figure 1 we display a d.c. current-voltage curve typical of an SIS tunnel junction used for heterodyne mixing. Also shown in the same figure is the I-V curve in the presence of L.O. power. Photon assisted tunneling steps are clearly visible both below and above the energy gap voltage (V_g). The normal state conductance of the junction is given by G_N. The critical current (I_c) and dropback voltage (V_d) are also shown. In the limit of small intermediate frequency, given by $f_{IF} = |f_s - f_{LO}|$, where f_s is the signal input frequency and f_{LO} the local oscillator frequency, Tucker's theory gives the mixer's output conductance through

$$G_{00} = \sum_{-\infty}^{\infty} J_n^2(\alpha) \frac{d}{dV_{bias}} I_{dc}\left(V_{bias} + \frac{nhf}{e}\right) \tag{1}$$

where α is given by eV_{LO}/hf and V_{LO} is the amplitude of the L.O. voltage. This is simply the slope of the pumped I-V curve. For normal operation this is taken at the center of the first photon step below the gap voltage, with L.O. applied. Similarly, the mixer's signal to signal conductance matrix element may be written

$$G_{11} = \frac{e}{2hf} \sum_{-\infty}^{\infty} J_n^2(\alpha) \left[I_{dc}\left(V_{bias} + \frac{(n+1)hf}{e}\right) - I_{dc}\left(V_{bias} + \frac{(n-1)hf}{e}\right) \right] \tag{2}$$

Nonlinear Superconductive Electronics and Josephson Devices
Edited by G. Costabile *et al.*, Plenum Press, New York, 1991

which, for small α can be considered as the signal input conductance. In this limit it is the slope of the line joining two points on the d.c. I-V curve, one photon step above and one photon step below the operating point. Clearly, from the figure, these conductances are different to the normal state conductance of the junction. The signal input conductance is higher and the IF output conductance is lower than G_N at the optimum d.c. bias point. The electrical circuit in which the junction is placed, or embedded, has to provide the correct admittances at these frequencies. In general both the IF and signal conductances should be nearly real for efficient, low-noise operation. Given that the SIS junction has a large parallel parasitic capacitance associated with the barrier layer, for correct operation it is important to design the millimeter-wave part of the embedding circuit so as to resonate this out at the signal frequency.

BACKGROUND

In 1965 the cosmic background radiation was discovered by Penzias and Wilson.[6] A few years later followed the detection of the spectral lines of formaldehyde at 4.8 GHz,[7] ammonia at 23.7 GHz[8] and water at 22.2 GHz.[9] With attention focussed towards higher frequencies, the search for millimeter-wave rotational transitions of light molecules had begun. The rotational transitions of CO at 115.3 GHz[10] and CN at 113.5 GHz[11] were soon observed. Since this time, the field of millimeter-wave radioastronomy has advanced rapidly. Millimeter-wave radioastronomy has led, in particular, to a better understanding of the colder components of the interstellar medium with the discovery of the giant molecular clouds and the development of the field of astrochemistry. Observations at both millimeter and submillimeter wavelengths are of paramount importance to our understanding in other areas. These include, for example, molecular physics, star formation, the study of planetary atmospheres, and the structure and evolution of galaxies.

One of the main reasons for the rapid evolution of both millimeter and submillimeter radioastronomy has been the continual evolution of the sophisticated techniques required to

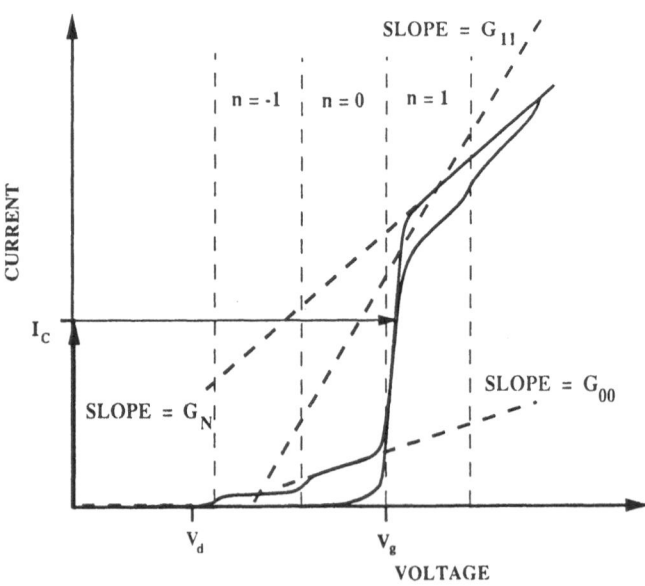

Figure 1. Current-voltage characteristic typical of an SIS tunnel junction used for mixing showing the critical current (I_c), the drop back voltage (V_d), the gap voltage (V_g) and the normal state conductance (G_N). Also shown are local oscillator induced current steps and the junction's input and output conductances.

produce low-noise heterodyne receivers for spectroscopic observation. Spectroscopy is usually performed by dividing the receiver's IF band into channels of width Δf and integrating the noise in each channel separately. The receiver's signal to noise ratio is then given by the Dicke radiometer equation[12]

$$S/N = T_S/T_R \sqrt{\Delta f \, \Delta t} \qquad (3)$$

where T_S is the equivalent black body temperature of the signal, and Δt is the integration time. Clearly, the time required to observe a particular source (i.e. to achieve a given signal to noise ratio) is proportional to the square of the receiver noise temperature, T_R. This should therefore be kept to a minimum for efficient operation. Also, any loss of signal in passing through the Earth's atmosphere effectively increases the receiver's noise temperature resulting in longer integration times.

Figure 2 shows a plot of atmospheric transmission as a function of wavelength typical of a high altitude mountain site selected for submillimeter radioastronomy.[13] The frequencies corresponding to rotational transitions of the CO molecule and the neutral carbon line, both of extreme astrophysical importance, are also indicated. Clearly, relatively good transmission occurs only in certain wavelength ranges or windows. It is somewhat fortuitous that the majority of CO rotational transitions occur in these windows. Receiver development has therefore taken place chiefly around 115 GHz (the 3 mm window), 230 GHz (the 1.3 mm window), and 345 GHz (the 0.87 mm window). A great deal of attention is now being devoted to the CO transition at 460 GHz and the neutral carbon transition at 492 GHz. Work is also proceeding at the higher frequency windows around 690 GHz and 800-900 GHz. Although predominantly driven by a desire to observe these specific transitions, there is a need, partly because of a large interest in extragalactic radioastronomy, to develop good, low-noise receivers to cover the entire extent of the available windows.

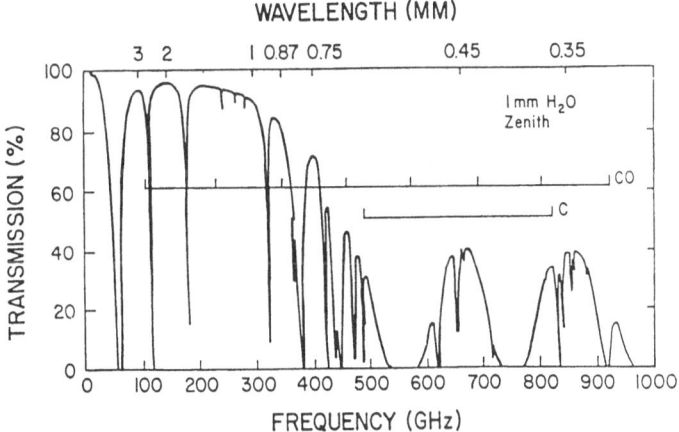

Figure 2. Atmospheric transmission at millimeter and submillimeter wavelengths from the altitude of Mt. Graham (Arizona, USA) for 1 mm precipitable water along the line of sight.

The features predicted by Tucker's theory were extremely attractive to the receiver engineer and many groups quickly developed the technology required to test the theory.[14-18] With the aforementioned design considerations in mind, the first SIS mixer receivers to be used on radio-telescopes were described by Olsson et al.[19] and by Woody et al.[20] The first complete demonstration of quantitative agreement between the theory and experimental measurements was made by Feldman et al.[21] in 1983. Tucker's theory is now well established and both mixer conversion gain and mixer noise close to the quantum limit have been observed in a number of laboratories.[22-28,38] There have also been major advances in the development of SIS junction fabrication technology, the general level of our understanding of the SIS mixer has improved, and receiver noise performance of only several times the quantum limit has now been measured. Because of their lower noise performance, SIS receivers have now replaced the more traditional cooled Schottky diode receivers at many radio-observatories.[29-33,41-43] In the sections that follow, we first present the current state of the art of SIS receiver performance, then we examine the criteria necessary to achieve true submillimeter operation and finally we discuss some of the associated technical difficulties.

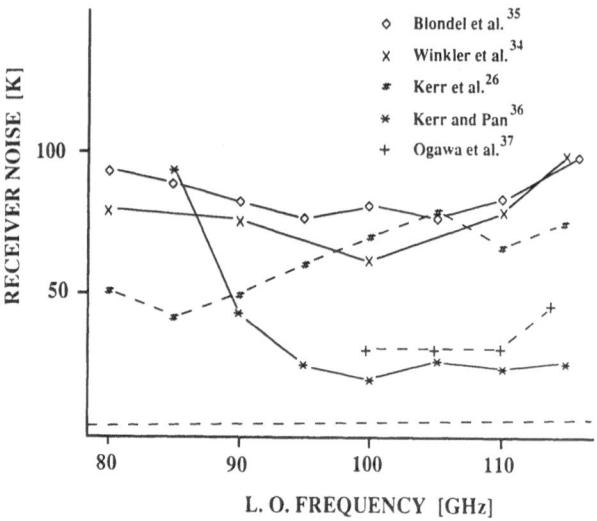

Figure 3. Receiver noise (double sideband) is plotted as a function of local oscillator frequency for a number of SIS receiver systems operating throughout the 3 mm window.

CURRENT STATE OF THE ART

In figure 3 we display some of the best reported receiver noise temperatures as a function of operating frequency for SIS systems built to operate in the atmospheric window centered at about 3 mm wavelength. A line corresponding to a noise equivalent to the quantum limit is also shown in the figure. From this figure, it is clear that there are now many receiver systems offering reasonable, low-noise performance. The fixed tuned mixer receivers of Kerr et al.[26] and Winkler et al.[34] offer low-noise performance throughout most of the atmospheric window with instantaneous bandwidths of about 40%. Low-noise is also offered by the single tuned mixer receivers[35] installed at the interferometer of the Institut de Radio Astronomie Millimétrique (I.R.A.M.) on the Plateau de Bure in the French Alps. We should note the excellent receiver noise performance, about four times the quantum limit, offered by the double tuned mixer receivers described by Kerr and Pan,[36] and more recently by Ogawa et al.[37] These receivers are of similar design and both incorporate a double-tuned waveguide mixer following a design first described by Pan et al.[38] This type of mixer allows almost any desired embedding admittance to be presented to the SIS junction, hence the possibility of low conversion loss and low receiver noise. It is also well suited to both double sideband and single sideband operation.

The 2 mm wavelength range has received attention in only two laboratories, those at the I.R.A.M. and the University of Cologne. The SIS receivers built and tested at these laboratories offer improved performance[39-41] over Schottky diode mixer receivers built for this wavelength range, about ten times the quantum limit.

The atmospheric window centred at about 1.3 mm wavelength contains many molecular transitions of great astrophysical interest and also represents a natural stepping stone towards the design of higher frequency receivers. The first SIS receiver specifically designed for the 1.3 mm waveband was built and tested by Sutton in 1983.[31] Its noise performance, displayed in figure 4, was comparable to or better than Schottky diode receivers operating in the same waveband. More recently, spurred by the success at lower frequencies, a great deal of attention has been given in many laboratories to this wavelength range and numerous SIS receivers have now been developed. The receiver noise offered by the best of these is also reproduced as a function of local oscillator frequency in figure 4. The waveguide receivers

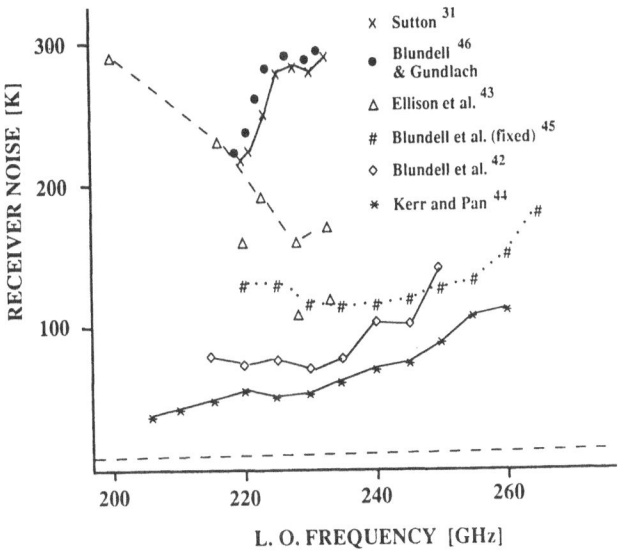

Figure 4. Receiver noise (double sideband) is plotted as a function of local oscillator frequency for a number of SIS receiver systems operating throughout the 1.3 mm window.

built at the I.R.A.M. and Caltech have been in use for several years. Both offer good, low-noise performance,[42,43] about ten times the quantum limit. More recently, excellent low-noise performance, about four times the quantum limit, has been obtained using a receiver constructed at the National Radio Astronomy Observatory.[44] This receiver makes use of a broad-band waveguide mixer which incorporates a series array of inductively shunted SIS junctions and has two short circuit plungers to ensure optimum tuning at the signal frequency. Local oscillator is input to the receiver via a cooled waveguide coupler and low-loss input optics, and an extremely low-noise HEMT IF amplifier completes the design. Also shown in figure 4 is the receiver noise measured using a laboratory test receiver incorporating a waveguide mixer fixed tuned to operate from 220 to 250 GHz.[45] In this case, a large fraction of the receiver noise is known to arise from input loss contributions, so that a receiver noise of only about five times the quantum limit should be possible with a fixed tuned mixer receiver. The results of experiments made with a superconductor-insulator-normal metal (SIN) mixer receiver are also shown in figure 4. In this case, an SIN tunnel junction was simply substituted for an SIS tunnel junction in the same measurement set-up.[46]

Since little work has been done with SIN mixers, this should be considered as an encouraging result. More recently, results of computer simulations have shown that both reasonable gain and low-noise mixing should be possible using SIN junctions, made with current technology, to above 500 GHz.[47,48]

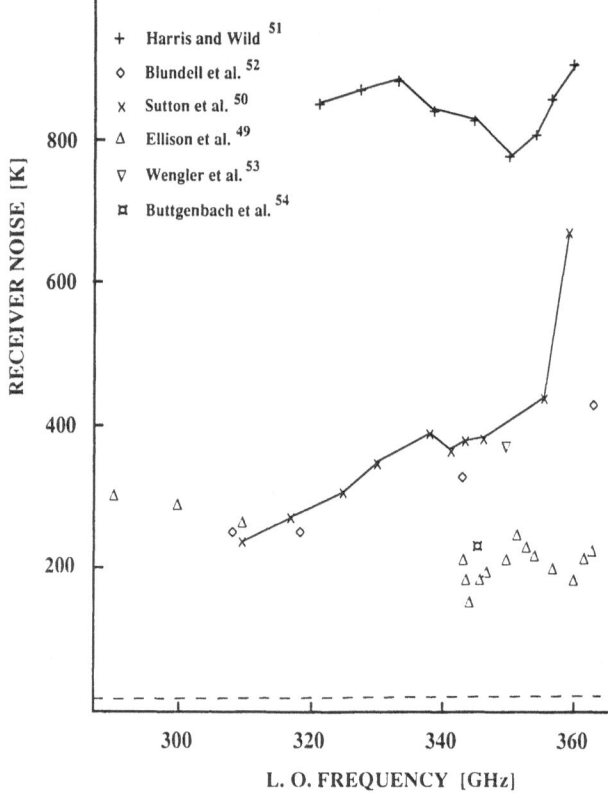

Figure 5. Receiver noise (DSB) is plotted as a function of local oscillator frequency for the few receiver systems operating throughout the 0.87 mm wavelength window.

Most of the development work done above 1.3 mm has been aimed at the production of systems to operate specifically at 345 GHz in the 0.87 mm window. As little work has been published at these wavelengths, all available SIS receiver noise data is included in figure 5. Only three of the receivers listed, all of the waveguide type, have been used on a regular basis for the collection of radio data, those of Ellison,[49] Sutton et al.[50] and Harris and Wild.[51] The others are either laboratory developments,[52] or are of the open structure type[53-54] made for higher frequency operation. The latter apparently offer excellent, low-noise performance over wide bandwidths. However, they should not be thought of as being competitive with waveguide mixer receivers operating at millimeter wavelengths. The lowest noise at 345 GHz, about ten times the quantum limit, is currently offered by the receiver installed at the Caltech Southern Observatory.[49] This receiver incorporates a waveguide mixer with both an E-plane short circuit tuner and a regular backshort tuner to allow correct impedance matching at the signal frequency. To further improve the mixer conversion, an IF impedance matching transformer is also included in the design.

The SIS mixer receiver has been shown to have good, low-noise performance to about 350 GHz. Above this frequency little data is available. There are several possible obstacles that should be considered in the design of SIS mixers for submillimeter wavelength operation. In the sections that follow we will stress some of the desirable features a tunnel junction should possess for operation at submillimeter wavelengths.

Josephson Noise

A finite voltage impressed across an SIS tunnel junction gives rise to the a.c. Josephson effect: an alternating current flows back and forth across the junction at a frequency $2eV/h$, and the drop-back voltage (V_d, shown in Figure 1.) below which the d.c. bias switches back to zero. Josephson noise, or switching noise, can occur when the d.c. bias point is close to V_d or close to an r.f. induced Josephson step.[55,56] Recalling that the optimum d.c. bias voltage for SIS mixing is approximately given by V_g-hf/2e (i.e. the midpoint of the first photon assisted tunneling step below the gap voltage), at high frequencies the lower order Josephson steps approach this value and the width of the unstable region increases. To avoid this region, the d.c. bias voltage must be increased above the optimum value. The Riedel singularity in the Josephson currents at the gap voltage may also be a problem. Josephson switching noise may be reduced through the application of a d.c. magnetic field corresponding to a single flux quantum, ϕ_0=h/2e, in the barrier region. This region is defined as $S=w(\lambda_1+\lambda_2+t)$, where w is the linear dimension of the junction perpendicular to the magnetic field, λ_1 and λ_2 are the London penetration depths in the two electrodes, and t is the thickness of the barrier. Impedance matching considerations indicate that the junction area should be reduced as the operating frequency is increased, and it becomes difficult to couple a full flux quantum into the barrier without impairing the superconducting properties of the electrodes. However, it appears that some control of the Josephson switching noise can be achieved by using a magnetic field with either a series array of larger area junctions or a single junction possessing one long dimension, such as an edge junction.[57] Another possibility to reduce the Josephson switching noise is to include magnetic impurities in the barrier layer.[58] The use of SIN junctions would of course completely eliminate the Josephson noise. However, SIN junctions are less non-linear and have only half of the gap voltage of SIS junctions made with the same superconducting material.

Gap Limitations

Computer simulations using Tucker's quantum mixer theory[3] have shown that the mixer gain is greatly reduced beyond twice the gap frequency, $2f_g=4\Delta/h$.[59] For the standard Nb/Al-oxide/Nb trilayer technology, this implies an upper frequency limit for efficient mixing of about 1.4 THz. For NbN/MgO/NbN junctions, this limit is about 2.4 THz.

To determine the overall receiver performance as a function of junction quality, a theoretical study was made that included both the effects of mixer noise and IF amplifier noise, which appears as effectively multiplied by the mixer's conversion loss. Assuming double sideband operation and ignoring all reactances, calculations were made[60] for three types of junction I-V curves: one of sharp, one of medium and one of dull quality. The results are summarized in figure 6. The quality, or sharpness, of the I-V curve effectively determines the changeover from classical to quantum operation: a mixer incorporating the dull junction remains in the classical regime throughout a large range of the millimeter-wave region, whereas a mixer with the sharp I-V curve is always in the quantum regime and offers the lowest noise. Also, the high frequency limit, $2f_g = 4\Delta/h$, sets in at lower frequencies for the dull and medium quality junctions reflecting an effective reduction in gap voltage.

The effects of pair tunneling, pair breaking in the electrodes and non-equilibrium effects are ignored in Tucker's theory, so that the high frequency limit of $4\Delta/h$ is certainly optimistic. Above the gap frequency, losses in the superconducting electrodes may be significant.[61]

Non-equilibrium Effects

The explanation of the photon assisted tunneling effect[1] in terms of a modulation in the density of states[2,3] by the applied microwave field has been extremely successful in predicting experimental observations. Here, the theories assume an adiabatic modulation of the difference between equilibrium energy levels in the two superconducting electrodes. However, it has been shown that, under certain conditions, changes to the Fermi distribution function could give rise to similar effects.[62] Suppose that a quasiparticle is excited to an energy above the gap edge. It will thermalize with the bath by relaxing to the gap edge and recombine with another quasiparticle to form a Cooper pair. The characteristic time (τ_0) for the relaxation and recombination process is a material dependent quantity. For lower transition temperature materials, τ_0 is usually longer. In Nb, for example, τ_0 is approximately 0.15 ns. Whereas in Al, τ_0 is approximately 440 ns, i.e. three thousand times higher. In a carefully designed experiment, measurements were made at frequencies close to the gap frequency in Al/Al-oxide-In/Al tunnel junctions.[59] Clear non-equilibrium effects were observed in junctions with current densities greater than about 10 A/cm^2. A simple scaling through relaxation-recombination times, the energy gap and the electron density of states at the Fermi level indicates an upper limit, beyond which non-equilibrium effects become significant, of between 200 and 600 kA/cm^2 in the case of Nb tunnel junctions. Fortunately,

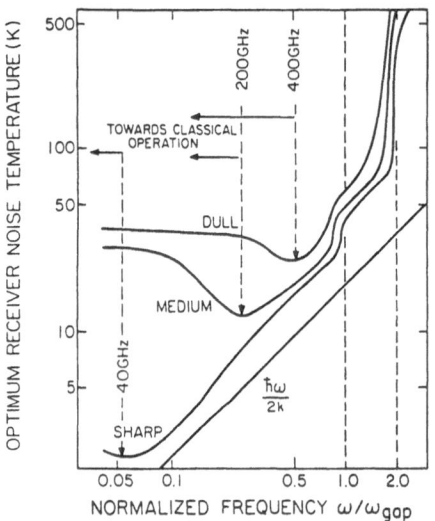

Figure 6. Predicted receiver noise temperature as a function of frequency for three synthetic I-V curves representative of a dull, medium and sharp quality SIS junction, after Feldman.[60]

non-equilibrium effects have not been found to be severe in Nb edge junctions with current densities well above 100 kA/cm^2 and are believed to be of little consequence in NbN, a serious candidate for higher frequency mixer applications.

Critical Current Density

For high frequency applications, the development of small area junctions of good quality has been widely discussed, whereas the related issue of producing high current density devices has received little attention. An $\omega R_N C$ product of about 3-4 has generally yielded good, low-noise mixer results,[63] where ω is the angular frequency and C is the capacitance of the SIS element. Computer simulations using a 5-port model have also indicated that harmonic conversion becomes significant for $\omega R_N C$ products smaller than this.[64] In practice, this number appears to be a compromise between having a high enough junction capacitance to shunt out harmonics of the L.O. and its higher order mixing products and having a sufficiently wide bandwidth for matching purposes at the operating frequency. Rewriting the $\omega R_N C$ product as follows

$$\omega R_N C \;=\; \omega\!\left(\frac{\pi}{4}\frac{2\Delta}{eI_c}\right)(\kappa A) \;=\; \omega\frac{\pi\Delta}{2e}\frac{\kappa}{j_c} \tag{4}$$

where κ is the junction capacitance per unit area, A is the junction area and j_c is the current density. We observe that the $\omega R_N C$ product is independent of the junction size. Furthermore, if $\omega R_N C$ is to be kept constant with frequency, the current density should be proportional to frequency, i.e. $j_c \sim \omega$.

Kerr and Pan[36] pointed out that the optimum $\omega R_N C$ product of 3-4 has chiefly been observed in mixers operating at about 100 GHz, and that there is no reason to assume that this value is frequency independent. Following their arguments, examination of the millimeter-wave circuit suggests that it is the input resistance at the second harmonic ($R_{in,harm}$) which essentially governs the effect of the junction capacitance on harmonic conversion, and not the normal state resistance (R_N) of the junction. Furthermore, they argue that $R_{in,harm}$ is related to the input resistance of the mixer at the fundamental (first harmonic), R_{in}, and that the ratio of $R_{in,harm}$ to R_{in} should be independent of frequency in a well designed mixer operating in the quantum regime. So, if we are to define a frequency independent parameter in the design of SIS mixers, we should use $\omega R_{in}C$.

How then is R_N related to R_{in}? In order to achieve a good match at the signal input port of the mixer, the circuit or source impedance should be made equal the complex conjugate of the mixer's input impedance, i.e. $Z_s = Z_{in}^*$. If we now assume that the reactive component of the mixer's input impedance is resonated out by the circuit at the signal frequency, we may rewrite the condition for a matched input as $R_s = R_{in}$. Using Tucker's theory in its three frequency, low IF approximation, Kerr and Pan[36] investigated the required R_N/R_s ratio to obtain about unity gain (SSB) as a function of frequency. They found that the ratio R_N/R_s required for good operation shows a 1/f dependence. Given that, at 100 GHz, the optimum value for the $\omega R_N C$ product is about 4, they further suggested that the following[36]

$$\omega R_N C \;=\; 4\times100/f[\text{GHz}] \tag{5}$$

should be used as a guide for selecting SIS junctions for efficient mixer operation.

With the aid of equation (2), a simple analytic expresion for the ratio R_N/R_{in} can be derived. In the limit of small L.O. drive, i.e. small α, the only significant terms that remain in the summation (equation (2)) are for n equals 0. We may therefore write

$$G_{11} \approx \frac{(f_g + f/2)/R_N - 0}{2f} \;=\; \frac{2f_g + f}{4fR_N} \;=\; \frac{2+\gamma}{4\gamma R_N} \tag{6}$$

where $\gamma = f/f_g$ (i.e. the signal frequency normalized to the gap frequency). Here we have used $I_{dc}(V_{bias}+ hf/e) = (V_g + hf/2e)/R_N$, i.e. $V_{bias} = V_g - hf/2e$, and $I_{dc}(V_{bias} - hf/e) \approx 0$. Since $R_s = R_{in}$ for the matched case, we have

$$R_N/R_s \;=\; R_N/R_{in} \;=\; G_{11}/G_N \;=\; (2+\gamma)/4\gamma \tag{7}$$

We see that for low frequencies, relative to the gap frequency, the 1/f dependence of R_N/R_s remains. At higher frequencies, the 1/f dependence may be modified to that indicated in equation (7). As the signal frequency is increased beyond the gap frequency, further modification may be required as the optimum bias point will now be located above V_g - hf/2e in order to avoid contributions of negative current flow beyond $-V_g$.[59] Following similar arguments to those of Kerr and Pan,[36] we can now describe the $\omega R_N C$ product through

$$\omega R_N C \;=\; \omega R_s C R_N/R_s \;=\; (2+\gamma)/4\gamma \tag{8}$$

which gives $\omega R_N C = (2+\gamma)/4\gamma = 3.5$, with f = 100 GHz and $f_g \sim 700$ GHz, typical of most lead alloy and niobium junctions used for mixing. Figure 7 is a graphical representation of the '$\omega R_N C$ product rules', equations (5) and (8). Clearly, some modification to the $\omega R_N C$

product rule is required. The expression suggested previously,[36] and the one derived here show a similar functional dependence on frequency, and for practical purposes should be considered as equivalent.

Having made the assumption that $\omega R_N C$ is constant, we indicated above (through equation (4)) that the current density should increase with frequency. Since this basic assumption seems to be incorrect, we should modify this statement to include an increased dependence on frequency, we therefore have

$$j_c \sim f^2 \tag{9}$$

In order to move to higher frequencies therefore, we not only have to make small area junctions, we have the added difficulty of increasing the current density.

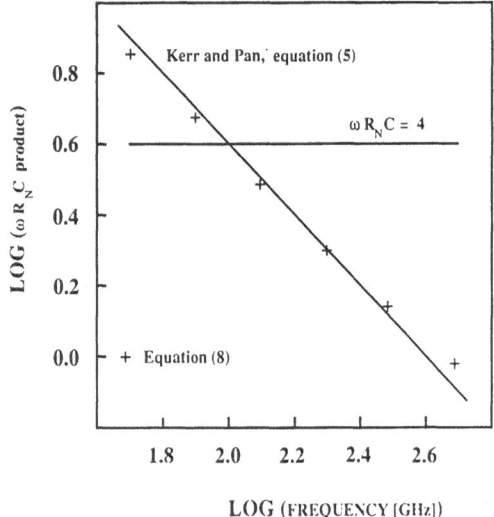

Figure 7. Design 'rule of thumb' for SIS tunnel junctions used in mixing. The $\omega R_N C$ product rule has been modified to include a frequency dependence, $\sim 1/f$.

All Refractory and Semi-refractory Tunnel Junctions

What would be the desirable features of SIS junctions that we would like to employ for submillimeter-wave receivers? First of all they need to have a high current density (≥ 10 kA/cm^2) together with a sharp non-linearity at the gap voltage, and very small leakage currents below the gap voltage. Also, a small area in conjunction with these features is required when using embedding circuitry of about 50 Ω.

A large energy gap is desirable, since the theoretical upper frequency limit for mixing is given by $4\Delta/h$, pair breaking sets in at $2\Delta/h$ and non-equilibrium effects are less serious for larger gap materials. Furthermore, problems with Josephson switching noise will set in at frequencies relative to the gap frequency.

To be able to couple a magnetic flux quantum into the barrier without impairing the superconductivity of the electrodes, small area, long and narrow junctions are desirable. Edge junctions are good candidates for this.

For the practical exploitation of SIS mixers it is also desirable that the junctions are stable in time, can be cycled between 4 K and 300 K without any change in characteristics, and that the fabrication process gives reproducible junction characteristics. The trilayer Nb/Al-ox/Nb process is now a proven technology in these respects.

Table 1. Figures of merit for SIS tunnel junctions. The roll-off frequencies were calculated with a specific capacitance (κ) of 75 fF/μm² for the NbN/MgO/NbN junctions, 45 fF/μm² for the Nb/Al-ox/Nb junctions, and 140 fF/μm² for the edge junctions.

Material	Geometry	j_c[kA/cm²]	V_m[mV]	$f_{\delta V_g}$[GHz]	$f_{4\Delta}$[THz]	Area[μm²]	f_{RNC}[GHz]	Reference
NbN/MgO/NbN	mesa	2.4	65	190	2.4	~1	90	65
NbN/MgO/NbN	cross-line	4	50	170-190		1	120	65
NbN/MgO/NbN	edge	10	35	190	2.4	0.1-0.4	190	65
NbN/MgO/NbN	mesa	0.13-4.3	13	97	2.1	5.6 (?)	120	66
NbN/MgO/NbN	mesa	0.47	18	97	2.4	256	40	67
NbN/MgO/NbN	mesa	<0.1	130	145	2.6	400	18	68,69
NbN/-oxide-/NbN	SNOP	18	17		2.4	6.25-1600	260	70
	edge	9.5	16	97	1.9	100	180	
Nb/Al-ox-Al/Nb	edge	0.01-24	40	24-48	0.97-1.1	0.0022	435	71
Nb/Al-ox/Nb	mesa	3	34	34	1.3	0.5	130	72-74
Nb/Al-ox/Nb	mesa	5	30		1.3	0.5	180	
Nb/Al-ox/Nb	mesa	2.7	20	24-48	1.3	2.2	130	75
Nb/Al-ox/Nb/Al-ox/	SNEP	0.01-0.07	87	24	0.97-1.4		20	76
Nb/Al-ox/Nb/Al-ox	stacked	1	1000	48	1.3	0.45	75	77
	self-al.	0.85	30-35	24	1.4	10	70	78
Nb/Al-ox/Nb	mesa	0.4	70	12	1.3	64	50	79
Nb/Al-ox/Nb	mesa	6.8	11	12	1.4	0.7	210	80
Nb/Al-ox/Nb	SNAP	1.8			1.4	50	100	
Nb/Al-ox/Nb	SNAP	20	20-58	72	1.4	12.6	370	
Nb/Al-ox/Nb	mesa	0.03-0.2	600	7-24	1.4		30	81
Nb/Al-ox/Nb	mesa	0.6-1.3	15-25		1.3-1.4	49	90	82
NbN/Si:H/Nb	SNAP	0.01-0.5	15-35	24-48	1.5-1.8	750 (?)		83
Nb/Si:H/Nb	SNAP	0.025	12	48	1.3			84
NbCN/PbBi	edge	3	150	56	2.0	1	75	85
Nb/Nb₂O₅/PbAuIn	edge	1.3-90	26	24	1.3	0.6-8.1	460	86
Nb/Nb₂O₅/PbBi	edge	2.4-150 (3.5-8.4MA)	15-43 (4.5)	20-40	1.45	0.18	610	57

A summary of junction parameters relevant to high frequency applications for junctions fabricated in recent years is given in table 1. Here, we only include those refractory and semi-refractory junctions which fulfill part or all of the requirements listed above. Given the large volume of available data, the table is necessarily incomplete. However, it should be considered as representative of the different techniques available.

The table is roughly divided into two categories: all refractory Nb and NbN tunnel junctions, and edge type tunnel junctions with soft lead alloy counter electrodes. The V_m-value is the product between the subgap resistance, typically measured at 2 mV for Nb trilayer junctions and at 3 mV for larger gap materials, and the critical current. The voltage width (δV_g) of the current rise at the gap voltage is expressed in terms of a frequency, $f_{\delta V} = e\delta V_g/h$. This frequency gives an indication of where the transition from classical to quantum mixing takes place, see Figure 6. Twice the gap frequency, $2f_g = 4\Delta/h$, should be considered as an absolute maximum theoretical high frequency limit for mixing. In practice, half of this value should be considered as a realistic limit for good, low-noise performance. Finally, in order to understand the role of the gap voltage (V_g) the critical current density (j_c) and the specific capacitance (κ) on mixing we now define the roll-off frequency, as suggested by equation (8):

$$f_{R_NC} = \frac{1}{16\pi R_N C}\{1 + \sqrt{1+64f_g\pi R_N C}\}$$

(10)

Replacing R_N using equation (4) yields

$$f_{R_NC} = \frac{1}{8\pi^2\Phi_0}\left\{\left(\frac{j_c}{\kappa f_g}\right)+\sqrt{\left(\frac{j_c}{\kappa f_g}\right)^2 + 32\pi^2\Phi_0\left(\frac{j_c}{\kappa}\right)}\right\}$$

(11)

where $\Phi_0 = h/2e$ is the flux quantum. Roll-off frequencies obtained using this formula should not be considered as absolute maxima for mixing purposes, but rather as an indication of where the instantaneous bandwidth starts to decrease for tuned mixer mounts.

Reference to the table suggests that although the NbN/MgO/NbN junctions have the highest gap frequencies, current density limitations give rise to low f_{R_NC} values so that wideband operation is reduced to below about 300 GHz.

The all refractory Nb trilayer technology usually gives very high quality devices. The junctions produced using this method generally have high V_m-values and a very strong non-linearity at the gap voltage. Unfortunately, in many cases, the junction quality is reduced with increasing current density and decreasing device area. Martines and Ono[71] have developed an all niobium edge type junction of reasonable quality that has both extremely small area and very high current density. Figure 8 summarizes some of their junction characteristics. In the figure, curve (d) is an example of the highest current density junction made using an all niobium process and curve (c) is an example of an extremely small area junction with high current density and reasonable quality. Both of these curves however show some reduction of the gap voltage when compared to the larger area and lower current density junctions shown as curves (a) and (b). Twice their corresponding gap frequency is a little less than 1 THz. Assuming that the Josephson currents can be effectively quenched in this type of junction, low-noise mixing should be possible to about 500 GHz. Without gap reduction this figure would be 700 GHz. However, reference to table 1 indicates a roll-off frequency below this, so that wideband operation may be impaired.

Given the fact that the maximum roll-off frequency for all refractory trilayer tunnel junctions is currently not higher than about 400 GHz, we should also investigate native barrier tunnel junctions. At the foot of table 1, three examples of edge type tunnel junctions are given. Here, the highest roll-off frequency is above 600 GHz. This is significantly higher than that for the niobium trilayer devices. Since this junction is of the edge type, coupling of a magnetic field to suppress Josephson currents should be possible, so that

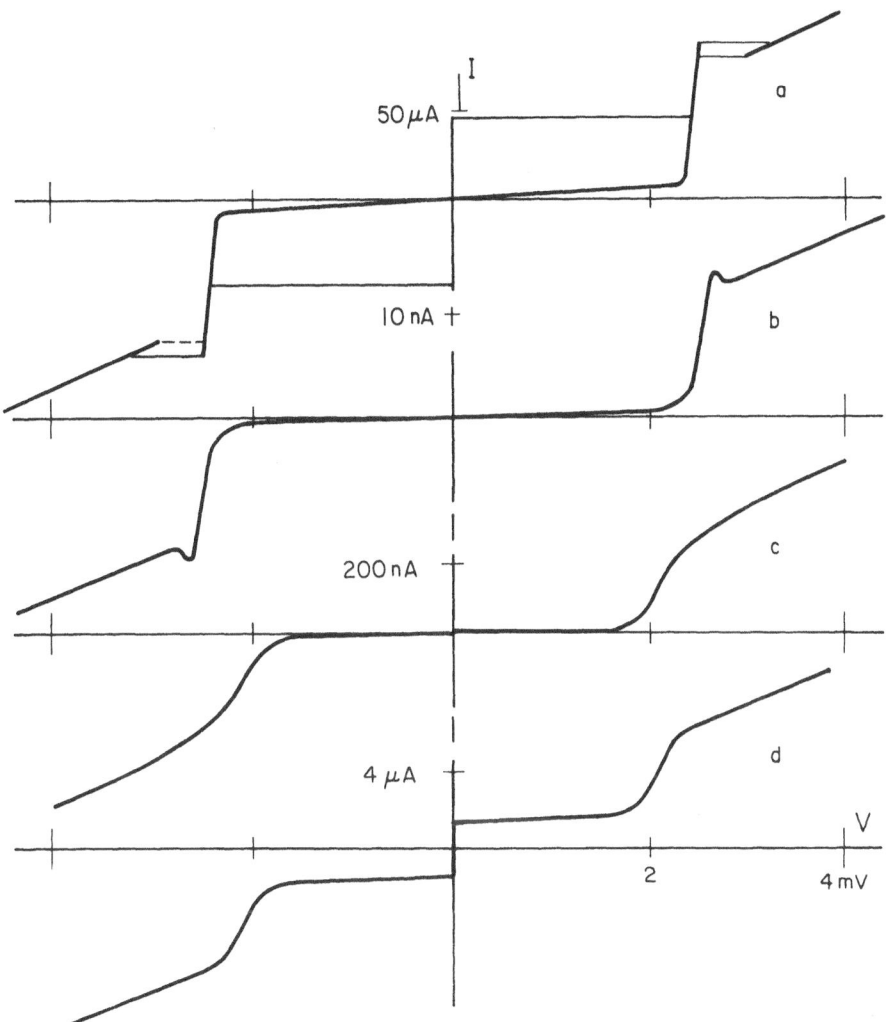

Figure 8. Current-voltage characteristics of (a) overlap and (b) - (d) edge junctions. Junction parameters are: (a) j_c = 1250 A/cm², area = 4 μm², T = 4 K, (b) j_c = 10 A/cm², area = 0.07 μm², T = 4 K, (c) j_c = 4400 A/cm², area = 0.004 μm², T = 1.4 K and (d) j_c = 24,000 A/cm², area = 0.015 μm², T = 4 K. (After Martines and Ono.[71])

wideband operation is expexcted to about 600 GHz. However, the lead alloy counter electrode employed in this junction is not particularly stable at oom temperature so that some extra precautions may be required in the storing and handling of these devices.

For operation at higher frequencies, the larger energy gap superconducting materials should be used for both base and counter electrodes.

CONCLUSIONS

The theoretical predictions of conversion gain and quantum limited noise have both been demonstrated in the laboratory and the general understanding of SIS mixers is good. Receivers based on quasiparticle SIS mixers have shown excellent noise performance up to about 250 GHz and good performance to about 350 GHz. At millimeter wavelengths, SIS receivers have now replaced the cooled Schottky diode receivers used previously at many radio-observatories.

Although measurements were first made close to 400 GHz as long ago as 1983, the development of submillimeter SIS receivers is still in its infancy. This is indicative of the technical problems that remain. These include difficulties associated with the fabrication of high current density, small area junctions and the development of higher, medium T_c super-conducting junctions. Innovation is also required in the development of antenna feeds or coupling structures and in the design and integration of suitable circuitry to allow correct mixer tuning at the signal input. Only then will comprehensive exploitation of the newly completed and proposed submillimeter radio-telescopes take place.

ACKNOWLEDGEMENTS

We would like to thank Marc J. Feldman, Erik L. Kollberg, John M. Martines, C. -Y. Edward Tong and Anthony H. Worsham for many helpful discussions in preparing this manuscript.

REFERENCES

1. A. H. Dayem and R. J. Martin, Quantum interaction of microwave radiation with tunneling between superconductors, Phys. Rev. Lett., 8: 246 (1962).
2. P. K. Tien and J. P. Gordon, Multiphoton process observed in the interaction of microwave fields with the tunneling between superconductor films, Phys. Rev., 129: 647 (1963).
3. J. R. Tucker, Quantum limited detection in tunnel junction mixers, IEEE J. Quantum Electron., QE-15: 1234 (1979).
4. J. R. Tucker, Predicted conversion gain in superconductor-insulator-superconductor quasiparticle mixers, Appl. Phys. Lett., 36: 477 (1980).
5. J. R. Tucker and M. J. Feldman, Quantum detection at millimeter wavelengths, Rev. Mod. Phys., 57: 1055 (1985).
6. A. A. Penzias and R. W. Wilson, A measurement of excess antenna temperature at 4080 Mc/s, Astrophys. J., 142: 419 (1965).
7. L. E. Snyder, D. Bhul, B. Zuckerman and P. Palmer, Microwave detection of interstellar formaldehyde, Phys. Rev. Lett., 22: 679 (1969).
8. A. C. Cheung, D. M. Rank, C. H. Townes, D. D. Thornton and W. J. Welch, Detection of NH_3 molecules in the interstellar medium by their microwave emission, Phys. Rev. Lett., 21: 1701 (1968).
9. A. C. Cheung, D. M. Rank, C. H. Townes, D. D. Thornton and W. J. Welch, Detection of water in interstellar regions by its microwave radiation, Nature, 221: 626 (1969).
10. R. W. Wilson, K. B. Jefferts and A. A. Penzias, Carbon monoxide in the Orion nebula, Astrophys. J., 161: L43 (1970).
11. K. B. Jefferts, A. A. Penzias and R. W. Wilson, Observation of the CN radical in the Orion nebula and W51, Astrophys. J., 161: L87 (1970).

12. R. H. Dicke, The measurement of thermal radiation at microwave frequencies, Rev. Sci. Instr., 17: 268 (1946).

13. R. N. Martin, P. A. Strittmatter, J. H. Black, W. F. Hoffmann, C. J. Hogan, C. J. Lada and W. L. Peters, in: "A Proposal to the National Science Foundation for partial funding of the Submillimeter Telescope," University of Arizona, Steward Observatory, 8 (1989).

14. P. L. Richards T. M. Shen, R. E. Harris, and F. L. Lloyd, Quasiparticle heterodyne mixing in SIS tunnel junctions, Appl. Phys. Lett., 34: 345 (1979).

15. G. J. Dolan, T. G. Phillips, and D. P. Woody, Low-noise 115-GHz mixing in superconducting oxide-barrier tunnel junctions, Appl. Phys. Lett., 34: 347 (1979).

16. S. Rudner and T. Claeson, Arrays of superconducting tunnel junctions as low-noise 10-GHz mixers, Appl. Phys. Lett., 34: 711 (1979).

17. G. J. Dolan, R. A. Linke, T. C. L. G. Sollner, D. P. Woody, and T. G. Phillips, Superconducting tunnel junctions as mixers at 115 GHz, IEEE Trans. Microwave Theory Tech., MTT-29: 87 (1981).

18. S. Rudner, M. J. Feldman, E. Kollberg, and T. Claeson, Superconductor insulator superconductor mixing with arrays at millimeter-wave frequencies, J. Appl. Phys., 52: 6366 (1981).

19. L. Olsson, S. Rudner, E. Kollberg, and C. O. Lindström, A low noise SIS array receiver for radioastronomical applications in the 35 - 50 GHz band, Int. J. Infrared and Millimeter Waves, 4: 847 (1983).

20. D. P. Woody, R. E. Miller and M. J. Wengler, 85-115 GHz receivers for radio-astronomy, IEEE Trans. Microwave Theory Tech., MTT-33: 90 (1985).

21. M. J. Feldman, S. -K. Pan and A. R. Kerr, SIS mixer analysis using a scale model, IEEE Trans. Magn., MAG-19: 494 (1983).

22. A. R. Kerr, S. -K. Pan, and M. J. Feldman, Infinite available gain in a 115 GHz SIS mixer, Physica, 108B: 1369 (1981).

23. W. R. McGrath, P. L. Richards, A. D. Smith, H. van Kempen, R. A. Batchelor, D. E. Prober, and P. Santhanam, Large gain, negative resistance, and oscillations in superconducting quasiparticle heterodyne mixers, Appl. Phys. Lett., 39: 655 (1981).

24. L. R. D'Addario, An SIS mixer for 90 - 120 GHz with gain and wide bandwidth, Int. J. Infrared Millimeter Waves, 5: 1419-1442 (1984).

25. A. V. Räisänen, D. G. Crété, P. L. Richards, and F. L. Lloyd, Low noise SIS mixer with gain for 80-115 GHz, in :"Proceedings of an ESA Workshop on a Space-Borne Sub-Millimetre Astronomy Mission," N. Longdon, ESA Publications Division ESTEC, Noordwijk, Segovia, Spain, 225 (1986).

26. A. R. Kerr, S. -K. Pan, S. Whiteley, M. Radparvar, and S. Faris, A fully integrated SIS mixer for 75-110 GHz, IEEE Microwave Theory Tech., MTT-S Digest: 851 (1990).

27. C. A. Mears, Quing Hu, P. L. Richards, A. H. Worsham, D. E. Prober and A. V. Räisänen, Quantum limited quasiparticle mixers at 100 GHz, submitted to IEEE Trans. Magn., Sept. (1990).

28. An. B. Ermakov, V. P. Koshlets, S. A. Kovtonyuk and S. V. Shitov, Parallel biased SIS arrays for mm wave mixers: main ideas and experimental verification, submitted to IEEE Trans. Magn., Sept. (1990).

29. R. Blundell, K. H. Gundlach, and E. J. Blum, Practical low-noise quasiparticle receiver for 80-100 GHz, Electron. Lett., 19: 498 (1983).

30. S. -K. Pan, M. J. Feldman, A. R. Kerr, and P. Timbie, A low-noise 115 GHz receiver using superconducting tunnel junctions, Appl. Phys. Lett., 43: 786 (1983).

31. E. C. Sutton, A superconducting tunnel junction receiver for 230 GHz, IEEE Trans. Microwave Theory Tech., MTT-31:589 (1983).

32. R. Blundell, K. H. Gundlach, E. J. Blum, J. Ibruegger and H. Hein, IRAM SIS receivers, in: "Int. Symp. on Millimeter and Submillimeter Wave Radio Astronomy", U.R.S.I, Granada (1984).

33. K. H. Gundlach, R. Blundell, J. Ibruegger, and E. J. Blum, SIS quasiparticle mixer receiver for radio-astronomy applications, p. 987, in: "SQUID'85", H. D. Hahlbohm and H. Lübig, Walter de Gruyter & Co., Berlin (1985).

34. D. Winkler, A. H. Worsham, N. G. Ugras, D. E. Prober, N. R. Erikson and P. F. Goldsmith, A 75-110 GHz SIS mixer with integrated tuning and coupled gain, in: "Nonlinear Superconductive Electronics and Josephson Devices," N. F. Pedersen, M.

Russo, A. Davidson, G. Constabile and S. Pagano, eds., Plenum, London (These proceedings).

35. J. Blondel, R. Blundell, M. C. Carter and J. Y. Chenu, unpublished results (1988).

36. A. R. Kerr and S. -K. Pan, Some recent developments in the design of SIS mixers, p. 363, in: "Proc. First Int. Symp. on Space Terahertz Tech.", Michigan (1990).

37. H. Ogawa, A. Mizuno, H. Hoko, H. Ishikawa and Y. Fukui, A 110 GHz SIS receiver for radioastronomy, Int. J. Infrared and Millimeter Waves, 11: 717 (1990).

38. S. -K. Pan, A. R. Kerr, M. J. Feldman, A. W. Kleinsasser, J. W. Stasiak, R. L. Sandstrom, and W. J. Gallagher, An 85-116 GHz receiver using inductively shunted edge junctions, IEEE Trans. Microwave Theory Tech., MTT-37: 580 (1989).

39. R. Blundell, J. Ibrügger, K. H. Gundlach and E. J. Blum, Low-noise 140-170 GHz heterodyne receiver using quasiparticle tunnel junctions, Electron. Lett., 20: 476 (1984).

40. J. Ibrügger, M. C. Carter and R. Blundell, A low-noise broadband 125-175 GHz SIS receiver for radioastronomy observations, Int. J. Infrared and Millimeter Waves, 8: 1573 (1987).

41. B. Vowinkel, K. Eigler, W. Hilberath, K. Jacobs and P. Muller, 125-170 GHz SIS receiver with closed cycle refrigeration system, Int. J. Infrared and Millimeter Waves, 10: 579 (1989).

42. R. Blundell, M. C. Carter and K. H. Gundlach, A low-noise SIS receiver covering the frequency range 215-250 GHz, Int. J. Infrared and Millimeter Waves, 9: 361 (1988).

43. B. N. Ellison and R. E. Miller, A low-noise 230 GHz SIS receiver, Int. J. Infrared and Millimeter Waves, 8: 609 (1987).

44. A. R. Kerr and S. -K. Pan, unpublished results (1990).

45. R. Blundell, E. E. Bloemhof, D. C. Papa and K. H. Gundlach, unpublished results (1990).

46. R. Blundell and K.H. Gundlach, A quasiparticle SIN mixer for the 230 GHz frequency range, Int. J. Infrared and Millimeter Waves, 8: 1577 (1987).

47. C. E. Tong, L. Chernin and R. Blundell, Harmonic mixing in a superconducting tunnel junction, J. Appl. Phys., 68: 4192 (1990).

48. L. Chernin and R. Blundell, Harmonic mixing in a superconductor-insulator-normal tunnel junction receiver, to be published, J. Appl. Phys. Corres. Feb (1991).

49. B. N. Ellison, P. L. Schaffer, W. Schaal, D. Vail and R. E. Miller, A 345 GHz receiver for radioastronomy, Int. J. Infrared and Millimeter Waves, 10: 937 (1989).

50. E. C. Sutton, W. C. Danchi, P. A. Jaminet and R. H. Ono, A superconducting tunnel junction receiver for 345 GHz, Int. J. Infrared and Millimeter Waves, 11: 133 (1990).

51. A. I. Harris and W. Wild, unpublished results (1989).

52 R. Blundell, E. E. Bloemhof, D. C. Papa and K. H. Gundlach, unpublished results (1990).

53. M. J. Wengler, D. P. Woody, R. E. Miller and T. G. Phillips, A Low-noise receiver for millimeter and submillimeter wavelengths, Int. J. Infrared and Millimeter Waves, 6: 697 (1985).

54. T. H. Büttgenbach, R. E. Miller, M. J. Wengler, D. M. Watson and T. G. Phillips, A broad-band, low-noise SIS receiver for submillimeter astronomy, IEEE Trans. Microwave Theory Tech., MTT-36: 1720 (1988).

55. D. Winkler, T. Claeson, and S. Rudner, Josephson interference and its suppression in SIS mixers operating close to the gap frequency, p. 1005, in: "SQUID 85", H. D. Hahlbohm and H. Lübbig, ed., Walter de Gruyter & Co., Berlin (1985).

56. D. G. Jablonski and M. W. Henneberger, Influence of Josephson currents on superconductor-insulator-superconductor mixer performance, J. Appl. Phys., 58: 3814 (1985).

57. D. Winkler, Properties of quasiparticle mixers at frequencies corresponding to the superconducting energy gap, thesis ISBN 91-7032-307-0, Göteborg 1987, Department of Physics, Chalmers University of Technology, S- 41296 Göteborg, Sweden.

58. S. Imai, S. Morita, A. Ishikawa, Y. Takeuti, and N. Mikoshiba, Submillimeter wave response of tunnel junctions with an insulating barrier containing magnetic impurities, IEEE Trans. Mag., MAG-21: 906 (1985).

59. D. Winkler and T. Claeson, High frequency limits of superconducting tunnel junction mixers, J. Appl. Phys., 62: 4482 (1987).

60. M. J. Feldman, Theoretical considerations for THz SIS mixers, Int. J. Infrared and Millimeter Waves, 8: 1287 (1987).

61. R. L. Kautz, Picosecond pulses on superconducting striplines, J. Appl. Phys., 49: 308 (1978)

62. O. Entin-Wohlman, Effect of microwave radiation on tunneling between super-conductors, Phys. Rev. B, 22: 5225 (1980).

63. M. J. Feldman and S. Rudner, Mixing with SIS arrays, in: "Reviews in Infrared and Millimeter Waves," K. J. Button, ed., Plenum, New York 1:47 (1983).

64. S. Withington and E. L. Kollberg, Spectral-domain analysis of harmonic effects in superconducting quasi-particle mixers, IEEE Trans. Microwave Theory Tech., MTT-37: 231 (1989).

65. J. A. Stern, B. D. Hunt, H. G. LeDuc, A. Judas, W. R. McGrath, S. R. Cypher, and S. K. Khanna, NbN/MgO/NbN SIS tunnel Junctions for submm wave mixers, IEEE Trans. Magn., MAG-25: 1054 (1989).

66. G. L. Kerber, J. E. Cooper, R. S. Morris, J. W. Spargo, and A. G. Toth, NbN-MgO-NbN Josephson tunnel junctions fabricated on thin underlayers of MgO, IEEE Trans. Magn., MAG-25: 1294 (1989).

67. M. Radparvar, M. J. Berry, R. E. Drake, S. M. Faris, S. R. Whiteley, and L.S. Yu, Fabrication and performance of all NbN Josephson circuits, IEEE Trans. Magn., MAG-23: 1480 (1987).

68. A. Shoji, M. Aoyagi, S. Kosaka, F. Shinoki, and H. Hayakawa, Niobium nitride Josephson tunnel junctions with magnesium oxide barriers, Appl. Phys. Lett., 46: 1098 (1985).

69. A. Shoji, M. Aoyagi, S. Kosaka, and F. Shinoki, Temperature-dependent properites of niobium nitride Josephson tunnel junctions, IEEE Trans. Magn., MAG-23: 1464 (1987).

70. J. C. Villigier, L. Vieux-Rochaz, M. Goniche, P. Renard, and M. Vabre, NbN tunnel junctions, IEEE Trans. Magn., MAG-21: 498 (1985).

71. J. M. Martines and R. H. Ono, Fabrication of ultrasmall Nb-AlOx-Nb Josephson tunnel junctions, Appl. Phys. Lett., 57: 629 (1990).

72. A. H. Worsham, D. E. Prober, J. H. Kang, J. X. Przybysz, and M. J. Rooks, High quality sub-micron Nb trilayer tunnel junctions for a 100 GHz SIS receiver, submitted to IEEE Trans. Magn., Sept. (1990).

73. H. Dang and M. Radparvar, A process for fabricating submicron all-refractory Josephson tunnel junction circuits, submitted to IEEE Trans. Magn., Sept., (1990).

74. A. W. Lichtenberger, D. M. Lea, C. Li, F. L. Lloyd, M. J. Feldman, R. J. Mattauch, S. -K. Pan and A. R. Kerr, Fabrication of micron size Nb/Al-Al$_2$O$_3$/Nb junctions with a trilevel resist lift-off process, submitted to IEEE Trans. Magn., Sept., (1990).

75. K. H. Gundlach, Fabrication of Nb-Al/Aloxide-Nb tunnel junctions and first mixer experiments in the 100 GHz frequency range, Working Report 203, IRAM, Domaine Universitaire, 38406 St-Martin-D'Hères, France (1990).

76. M .G. Blamire, R. E. Sometkh, G. W. Morris, and J. E. Evetts, Characteristics of vertically-stacked planar tunnel junction structures, IEEE Trans. Magn., MAG-25: 1135 (1989).

77. E. C. G. Kirk, D. G. Hasko, M. G. Blamire, J. E. Evetts, and H. Ahmed, Fabrication routes for sub-micron whole-wafer Nb/Al$_2$O$_3$/Nb tunnel junctions, Presented at Microcircuit Engineering 89, Cambridge (1989).

78. A. W. Lichtenberger, C. P. McClay, R. J. Mattauch, M. J. Feldman, S. -K. Pan, and A. R. Kerr, Fabrication of Nb/Al-Al$_2$O$_3$/Nb junctions with extremeely low leakage currents, IEEE Trans. Magn., MAG-25: 1247 (1989).

79. T. Imamura and S. Hasuo, A submicrometer Nb/Al-ox/Nb Josephson junction, J. Appl. Phys., 64: 1586 (1988).

80. S. Morohashi, S. Hasuo and T. Yamaoka, Self-aligned contact process for Nb/Al-AlOx/Nb Josephson junctions, Appl. Phys. Lett., 48:254 (1986).

81. J. M. Lumley, R. E. Somekh, J. E. Evetts, and J. H. James, High quality all refractory Josephson tunnel junctions for SQUID applications, IEEE Trans. Magn., MAG-21: 539 (1985).

82. M. Gurvitch, M. . Washington, and H. Huggins, High quality refractory Josephson tunnel junctions utilizing thin aluminum layers, Appl. Phys. Lett., 42: 472 (1983).

83. E. J. Cukauskas, M. Nisenoff, D. W. Jillie, H. Kroger, and L. N. Smith, High quality niobium-nitride Josephson tunnel junctions, IEEE Trans. Magn., MAG-19: 831 (1983).

84. H. Kroger, L. N. Smith, and D. W. Jillie, Selective Niobium Anodization Process for Fabricating Josephson Tunnel Junctions, Appl. Phys. Lett., 39: 280 (1981).
85. R. S. Amos, A. W. Lichtenberger, M. J. Feldman, R. J. Mattauch and E. J. Cukauskas, Fabrication of NbCN/PbBi edge junctions with extremely low leakage currents, submitted to IEEE Trans. Magn., Sept. (1990).
86. R. L. Sandstrom, A. W. Kleinsasser, W. J. Gallagher, and S. I. Raider, Fabrication and performance of all NbN Josephson circuits, IEEE Trans. Magn., MAG-23: 1484 (1987).

A 75-110 GHz SIS MIXER WITH INTEGRATED TUNING AND COUPLED GAIN

D. Winkler[†] , A.H. Worsham, N.G. Ugras, and D.E. Prober

Department of Applied Physics, Yale University
P.O. Box 2157 Yale Station, New Haven, Connecticut 06520

N.R. Erickson and P.F. Goldsmith

Five College Radio Astronomy Observatory, University of Massachusetts
Amherst, Massachusetts 01003

ABSTRACT

A broadband waveguide SIS (superconductor-insulator-superconductor) mixer with no adjustable mechanical tuning elements for the 75 - 110 GHz band was constructed and tested. The design is a demonstration of a prototype receiver element in a focal plane array receiver. The large instantaneous bandwidth was accomplished with a broadband waveguide-to-microstrip transition and a scale-modeled microstripline circuitry. The transition to microstripline consisted of a waveguide single ridge 4-step Chebychev transformer. The last step of the ridge connected the waveguide to the microstripline circuit, which contained the SIS element in parallel with a thin film tuning inductor. The operation was double-sideband (DSB). The typical DSB mixer noise temperature, as measured between 79.5 and 110 GHz, was about 35 to 50 K at band center increasing to about 50 - 60 K at the band edges. The lowest mixer noise we obtained was 20 K at 80 GHz for a single untuned junction. For some junction geometries, gain was also observed between 80 to 110 GHz, with a maximum of +3 dB (DSB) at 100 GHz. No impedance transformers were used on the intermediate frequency (*if*) side. For several devices, negative dynamic resistance was observed on the first photon induced step below the gap.

INTRODUCTION

A number of radio telescopes for millimeter wave radio astronomy employ receivers with low noise SIS mixers as front end stages.[1-5] SIS mixers have demonstrated close to quantum limited noise,[6,7] conversion gain, and negative dynamic resistance,[8-11] as predicted by Tucker.[12,13] However, to obtain the ultimate performance, several adjustable mechanical tuning elements are usually needed in these mixer mounts.

For mapping extended regions in space, multiple receivers are used in focal plane arrays to decrease the required observation time.[14,15] Since a focal plane array receiver may require a large number of elements, it is desirable to use fixed tuned elements with large instantaneous

---

[†] Current address: Department of Physics, Chalmers University of Technology, S-412 96 Gothenburg, Sweden.

Nonlinear Superconductive Electronics and Josephson Devices
Edited by G. Costabile *et al.*, Plenum Press, New York, 1991

bandwidth, such that the elements do not have to be retuned for each different observation frequency. Here we report on the design and receiver results of a full-band SIS mixer with integrated tuning. The receiver was designed to cover the band 75 to 110 GHz.

MIXER MOUNT DESIGN

Fig. 1 shows the outline of the mixer mount for one of the microstripline configurations. The signal and the local oscillator (LO) were combined in a low loss broadband cross-guide directional coupler with a -23dB coupling on the LO. A 4-step Chebychev single ridge transformer[16-19] acted as a broadband waveguide-to-microstripline transition and a broadband impedance transformer as well as providing a *dc* -biasing return path(Fig. 1a). The waveguide impedance was transformed down to 50 Ω at the last step, which was connected to a λ/4 long[††] 50 Ω microstripline (pos. A in Fig. 1b). The other end of this line (pos. B) was the base electrode of the SIS element . The counter electrode was connected via a wiring layer to a 90° radial stub (pos. C), which gave a broadband *rf* ground termination at point C. This radial stub functions like an open-circuited λ/4 transmission line.

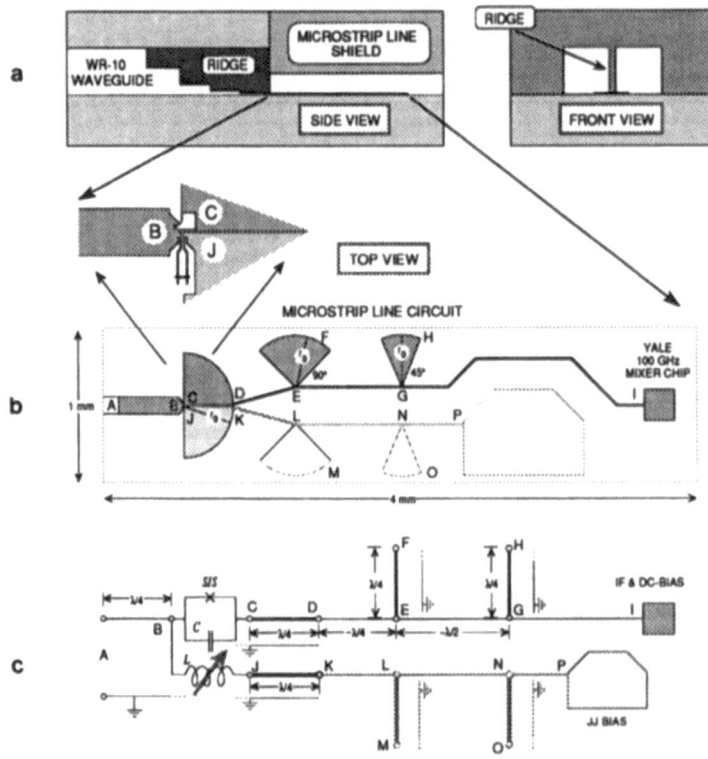

Fig. 1. Mixer mount. a) Waveguide 4-step Chebychev ridge transformer. b) Geometrical outline of the microstripline circuit and c) simplified electrical scheme.

By using a second 90° radial stub next to the one terminating the SIS element, a thin film tuning inductor could be added in parallel to the SIS element (pos. B → J) without shorting the *dc*-bias or the *if*-signal. A low pass filter (section D → I), acts as a *dc*-bias lead and *if* signal transmission line.   As indicated   in Fig. 1b, another low-pass line (K → P) could be

---

[††] This length was chosen to accommodate an additional impedance transformation in case this were needed in the future.

added to the lower radial stub to bias other circuits, such as local oscillators (Josephson,[20] soliton,[21] or flux-flow[22]), or a Josephson tuning inductor.[23] The radial stub gives a broadband and well defined ground near its apex.[24-26] By using radial stubs, it is also easy to provide several *dc*- and *if*-isolated *rf* grounds at nearly the same location.

JUNCTION FABRICATION

Nb/AlO$_x$/Nb trilayer tunnel junctions with areas as small as to 0.5 μm^2, V$_m$~ 40 mV (at 4.4 K), and current densities of between 3000 and 5000 A/cm^2 were used as SIS mixing elements. The devices were fabricated on 50 μm thick quartz substrates using a modified selective niobium insulation process (SNIP).[27,28] The substrates were mounted to thick Si substrates with vacuum grease during the entire fabrication sequence. This mounting procedure allowed the fragile quartz substrates to be processed using standard techniques. In addition, the vacuum grease helped to heat sink the quartz substrate during the metal layer sputter depositions. Details of the device processing are presented elsewhere.[29]

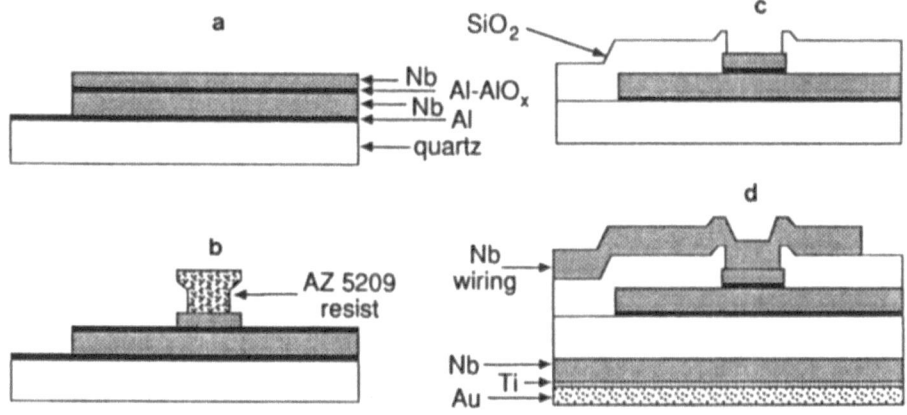

Fig. 2. Schematic sequence of the device fabrication. a.) The trilayer was patterned by liftoff. b.) The top electrode was reactive ion etched. c.) SiO$_2$ isolated the junction area. d.) A wiring layer and backside metalization were deposited.

SCALE MODELING

Scale modeling of SIS mixer mounts by using larger mounts at lower frequencies is a practical way of determining the proper design of the circuit and to verify new concepts. The ridge transformer and the microstripline circuitry were measured and optimized with a 28.4 times larger scale model at 2 - 4 GHz. Table I summarizes the data for the scale measurements.

Table I. Summary of the scale modeling results. The frequencies for the SIS mount are 28.4 times higher, i.e., 2.64 - 3.87 GHz corresponds to 75 - 110 GHz.

Configuration	Microstripline circuit		Total mount
	f [GHz]	VSWR (max)	VSWR(max)
50Ω chip resistor	2.64-3.87	1.33:1	1.38:1
50Ω//1pF chip capacitor	2.64-3.87	3:1	4:1
50Ω//1pF//wire inductor	2.64-3.87	1.8:1	3:1
50Ω//1pF//wire inductor	2.63-3.76		2.1:1

The scale model data was taken with; a) a 50 Ω chip resistor in place for the SIS element, b) a 1 pF chip capacitor in parallel with the 50 Ω chip resistor, c) the same configuration as (b), but with a thin wire inductor added between the input microstripline and the other 90° radial stub (pos. B → J in Fig. 1b).

RESULTS

Fig. 3, curve (a) shows a current-voltage (I-V) characteristic of a 2 junction series array with inductor tuning (a). The normal state resistance (R_n) was 431 Ω for this device. The I-V curve with a 100 GHz local oscillator (LO), is also shown in curve (b). Notice the negative dynamic resistance on the first step below the gap.

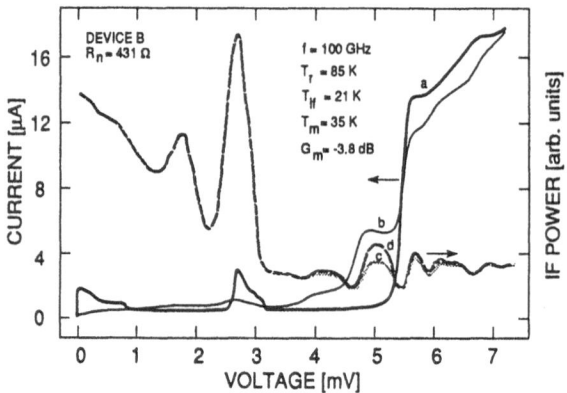

Fig. 3. a) I-V curve of an inductor tuned 2 junction SIS array made in Nb trilayer technology (each junction 0.5 μm² large). b) I-V curve with applied *rf* signal (f_{LO} = 100 GHz). The *if* output power resulting from mixing between the 100 GHz LO-signal and blackbody radiation from a matched internal waveguide load at c) 5.5 K and d) 30 K is also outlined.

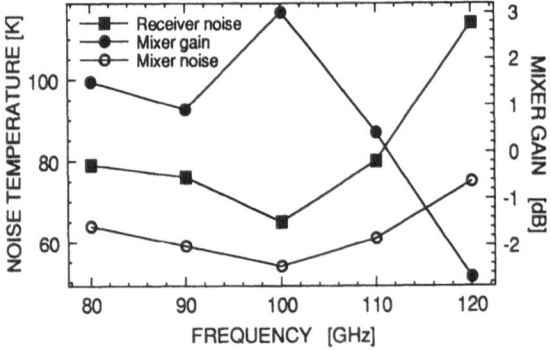

Fig. 4. Performance at different frequencies across the band for device C, an inductor tuned 2 junction series array (each junction 0.5 μm² large). Receiver noise temperature, mixer noise temperature, and mixer gain are shown. The noise of the *if* amplifier was about 21 K in this measurement.

The *if* output power is plotted with an internal waveguide hot/cold load at 5.5 K, curve (c), and 30 K, curve (d) connected to the signal port.

Fig. 4 shows DSB (double sideband) curves for the mixer gain, G_m, the mixer noise temperature, T_m, and the receiver noise temperature, T_r, versus frequency. These data are for device C, another 2 junction series array with $R_n = 125\ \Omega$. Table II summarizes the results for the devices that were measured. Note that for device C, coupled conversion gain (DSB) was obtained between 80 to 110 GHz.

Table II. Summary of the data and results from our measured SIS mixers.

Mixer	Element	R_N [Ω]	Tuning	f_{LO} [GHz]	T_R^{DSB} [K]	T_M^{DSB} [K]	G_M^{DSB} [dB]
Device A	1 jcn 0.5μm²	71	No tuning	80 100	41 (202)	20 39	-3.0 -10
Device B	2 jcns 0.5μm² ea	431	Thin film inductor	80 90 100 110	141 118 85 165	49 40 35 46	-6.4 -5.7 -3.8 -7.5
Device C	2 jcns 0.5μm² ea	125	Thin film Inductor	80 90 100 110 120	79 76 65 80 114	64 59 54 61 75	+1.5 +0.9 +3.0 +0.4 -2.7
Device E	4 jcns 0.5μm² ea	619	No tuning	80 85 90 95 100 105 110	125 119 158 136 136 136 173	60 61 82 72 69 69 83	-7.2 -6.7 -7.9 -7.1 -7.3 -7.3 -8.6
Device I	4 jcns 4μm² ea	124	Thin film Inductor	80 85 90 95 100 105 110	104 65 73 64 66 87 127	62 44 52 47 43 44 49	-5.5 -2.5 -2.5 -1.5 -2.9 -5.6 -8.2

CONCLUSIONS

We have demonstrated a prototype single channel SIS receiver for a focal plane array receiver. This fixed tuned low noise SIS receiver has an instantaneous bandwidth of at least 80 to 120 GHz. Coupled DSB gain was observed for one device over a large frequency band without any *if* impedance transformers. Negative dynamic resistance was observed on the first photon assisted tunneling step below the gap voltage for several of the mixers at around 100 GHz.

We would like to thank J.H. Kang and J.X. Przybysz at the Westinghouse Science and Technology Center and M.J. Rooks at NNF at Cornell University for valuable help and support in device fabrication. Excellent work was done in the machining of the mixer mount and the mm-wave cryogenic hot/cold load by Dick Downing. This work was supported by NSF grant ECS 8604350, AFOSR-88-0270, and by the Swedish National Board for Technical Development.

REFERENCES

1. L. Olsson, S. Rudner, E. Kollberg, and C.O. Lindström, A Low Noise SIS Array Receiver for Radioastronomical Applications in the 35 - 50 GHz Band, Int. J. Infrared and Millimeter Waves 4: 847 (1983).

2. D.P. Woody, R.E. Miller, and M.J. Wengler, 85-115-GHz Receivers for Radio Astronomy, IEEE Trans. Microwave Theory Tech. MTT-33: 90 (1985).

3. S.K. Pan, M.J. Feldman, A.R. Kerr, and P. Timbie, A Low-Noise 115 GHz Receiver Using Superconducting Tunnel Junctions, Appl. Phys. Lett. 43: 786 (1983).

4. K.H. Gundlach, R. Blundell, J. Ibruegger, and E.J. Blum, SIS Quasiparticle Mixer Receiver for Radio-Astronomy Applications, in :"SQUID'85," H.D. Hahlbohm and H. Lübbig, Walter de Gruyter & Co, Berlin, 1985, 987.

5. H. Ogawa, A. Mizuno, H. Hoko, H. Ishikawa, and Y. Fukui, A 110 GHz SIS Receiver for Radio Astronomy, Int. J. Infrared and Millmeter Wave 11: 717 (1990).

6. C.A. Mears, Q. Hu, P.L. Richards, A.H. Worsham, D.E. Prober, and A.V. Räisänen, Quantum Limited Quasiparticle Mixers at 100 GHz, to be published in Appl. Phys. Lett. (1990).

7. S.K. Pan, A.R. Kerr, M.H. Feldman, A.W. Kleinsasser, J.W. Stasiak, R.L. Sandstrom, and W.J. Gallagher, An 85-116 GHz Receiver using Inductively Shunted Edge Junctions, IEEE Trans. Microwave Theory Tech. MTT-37: 580-592 (1989).

8. A.V. Räisänen, D.G. Crété, P.L. Richards, and F.L. Lloyd, Low Noise SIS Mixer with Gain for 80-115 GHz, in :"Proceedings of an ESA Workshop on a Space-Borne Sub-Millimetre Astronomy Mission," N. Longdon, ESA Publications Division ESTEC, Noordwijk, Segovia, Spain , 4-7 June 1986, 255.

9. W.R. McGrath, P.L. Richards, A.D. Smith, H. van Kempen, R.A. Batchelor, D.E. Prober, and P. Santhanam, Large Gain, Negative Resistance, and Oscillations in Superconducting Quasiparticle Heterodyne Mixers, Appl. Phys. Lett. 39: 655 (1981).

10. A.R. Kerr, S.K. Pan, and M.J. Feldman, Infinite Available Gain in a 115 GHz SIS Mixer, Physica 108B: 1369 (1981).

11. L.R. D'Addario, An SIS Mixer for 90 - 120 GHz with Gain and Wide Bandwidth, Int'l J. Infrared Millimeter Waves 5: 1419-1442 (1984).

12. J.R. Tucker, Quantum Limited Detection in Tunnel Junction Mixers, IEEE J. Quantum Electron. QE-15: 1234 (1979).

13. J.R. Tucker, Predicted Conversion Gain in Superconductor-Insulator-Superconductor Quasiparticle Mixers, Appl. Phys. Lett. 36: 477 (1980).

14. N. Erickson, P. Goldsmith, C.R. Predmore, G.A. Novak, and P.J. Visceiso, A 15 Element Imaging Array for 100 GHz, 1990 MTT-S Digest: 973-976.

15. J.M. Payne, Multibeam Receiver for Millimeter-wave Radio Astronomy, Rev. Sci. Instr. 59, 1988: 1911-1919.

16. S. Hopfer, The Design of Ridged Waveguides, IRE Trans. Microwave Theory Tech. MTT-3: 20 (1955).

17. W.J.R. Hoefer and M.N. Burton, Closed-Form Expressions for the Parameters of Finned and Ridged Waveguides, IEEE Trans. Microwave Theory Tech. MTT-30: 2190 (1982).

18. M.V. Schneider, B. Glance, and W.F. Bodtmann, Microwave and Millimeter Wave Hybrid Integrated Circuits for Radio Systems, Bell System Tech. J.: 1703 (1968).

19. R.E. Collin, Theory and Design of Wide Band Multisection Quarter-wave Transformers, Proc. IRE 43: 179 (1955).

20. K.L. Wan, A.K. Jain, and J.E. Lukens, Submillimeter Wave Generation Using Josephson Junction Arrays, IEEE Trans. Magn. MAG-25: 1076-1079 (1989).

21. S. Pagano, R. Monaco, and G. Costabile, Microwave Oscillator Using Arrays of Long Josephson Junctions, IEEE Trans. Magn. MAG-25: 1080-1083 (1989).

22. T. Nagatsuma, K. Enpuku, F. Irie, and K. Yoshida, Flux-Flow type Josephson Oscillator for Millimeter and Submillimeter Wave Region, J. Appl. Phys. 54: 3302-3309 (1983).

23. K.E. Irwin, S.E. Schwarz, and T. Van Duzer, Planar Antenna-Coupled SIS Devices for Detection and Mixing, in :"Digest of the Sixth International Conference on Infrared and Millimeter Waves," K.J. Button, IEEE, New York, 1981, pp. M-4-2.

24. V.P. Vinding, Radial Line Stubs as Elements in Strip Line Circuits, NEREM RECORD: 108 (1967).

25. B.A. Syrett, A Broad-Band Element for Microstrip Bias or Tuning Circuits, IEEE Trans. Microwave Theory Tech. MTT-28: 925 (1980).

26. H.A. Atwater, Microstrip Reactive Circuit Elements, IEEE Trans. Microwave Theory Tech. MTT-31: 488 (1983).

27. A. Shoji, S. Kosaka, F. Shinoki, M. Aoyagi, and H. Hayakawa, All Refractory Josephson Tunnel Junctions Fabricated by Reactive Ion Etching, IEEE Trans. Magn. MAG-19: 827-830 (1983).

28. S. Morohashi, F. Shinoki, A. Shoji, M. Aoyagi, and H. Hayakawa, High Quality Nb/Al-AlOx/Nb Josephson Junctions, Appl. Phys. Lett. 46: 1179-1181 (1985).

29. A.H. Worsham, D.E. Prober, J.H. Kang, J.X. Przybysz, and M.J. Rooks, High Quality Sub-Micron Nb Trilayer Tunnel Junctions for a 100 GHz SIS Receiver. Sept. (1990). Submitted to IEEE Trans. Magn. (1990).

LOW NOISE DC SQUIDS FOR THE 1990S

Mark B. Ketchen

IBM Research Division
Thomas J. Watson Research Center
Yorktown Heights, New York 10598

ABSTRACT

In the 1980s we have seen the successful introduction of planar coupling schemes for dc SQUIDs, the achievement of coupled energy sensitivity near the quantum limit, significant advances in integrated SQUID detectors, and a shift from Pb-alloy to all-refractory technology. The discovery of high-T_c superconductivity in 1986 was soon followed by the demonstration of high-T_cSQUIDs, with bulk and thin-film versions now operating at or above 77K. The 1990s will bring practical low-T_c dc SQUIDs, 0.05mm^2 or less in area, with tightly coupled 2μH input coils, smooth, well-behaved voltage-flux characteristics, very small flux creep related hysteresis, and coupled energy sensitivity at or near the quantum limit at 4.2K. There will be an increased emphasis on the design and development of application specific integrated detectors. All-refractory Nb-based technology will dominate the low-T_c arena and direct coupling to semiconductor capability should quicken the pace and lower the cost of SQUID development and production. High-T_c SQUIDs will follow suit with a significant time delay as our understanding of high-T_c materials and our ability to work with them advances.

INTRODUCTION

The dc SQUID[1] shown in Fig. 1 consists of two Josephson elements[2], each of critical current I_0, in a superconducting loop of inductance L. When the SQUID is biased at non-zero voltage with a constant current, the voltage across the device is periodic in the applied flux, with a period of the flux quantum ϕ_0. Current to be sensed I_i passing through a nearby input coil L_i couples flux to the SQUID via the mutual inductance M_i. A second coil carries a control current I_c used to set the flux bias point and possibly to modulate the SQUID for readout purposes.

A SQUID can be configured to measure minute changes of various physical quantities such as voltage, current, and magnetic field. To make a magnetometer, for example, a pickup loop is connected in series with the SQUID's input coil. A small change in magnetic field induces a persistent current that couples flux to the SQUID. The extreme sensitivity of such a magnetometer (ie. $\sim 10^{-12}G/\sqrt{Hz}$) derives from the fact that the SQUID can resolve a small fraction of ϕ_0, while ϕ_0 is itself a very small quantity.

Nonlinear Superconductive Electronics and Josephson Devices
Edited by G. Costabile *et al.*, Plenum Press, New York, 1991

81

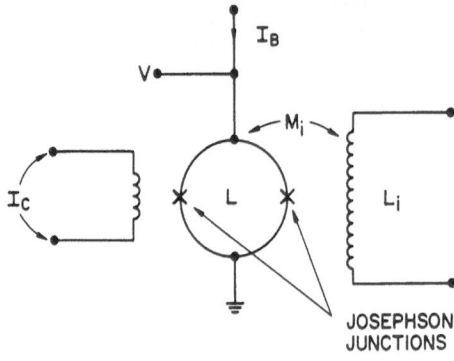

Fig. 1   Schematic of the dc SQUID.

A figure of merit commonly used to compare SQUIDs is the coupled energy sensitivity[3,4],

$$\varepsilon_c = \frac{\varepsilon}{k^2} = \frac{\phi_n{}^2}{2k^2 L} \; , \tag{1}$$

where ε is the intrinsic energy sensitivity, ϕ_n is the flux noise and $k^2 = M_i^2/L_i L$. The intrinsic energy sensitivity in the white noise region is predicted to be[5,6]

$$\varepsilon = \gamma_1 h\!\left(\frac{k_B T}{e I_0 R} \right) + \gamma_2 h \; , \tag{2}$$

where the prefactor $\gamma_1 \approx 3 - 5$, and γ_2, the prefactor of the quantum contribution, is thought to be ~0.2. In the case of tunnel junctions R is the value of the shunt resistor that must be placed across each junction to ensure non-hysteretic behavior[7,8]. The value of ε in the low frequency (typically 1/f) noise regime, while of great importance for many applications, is not easy to predict and tends to depend on the details of the technology in ways that are not always well understood. For today's laboratory produced SQUIDs at the 2μm technology level, values of ε_c in the white noise region at 4.2K in the range of 10-100h are common. Today's best commercially available SQUIDs have $\varepsilon_c \approx 1500$h.

Most early SQUIDs as well as most present and past commercial SQUIDs are inherently three-dimensional, toroids[9] and cylinders[10] being the geometries most commonly used. Such structures, while allowing tight coupling between L and L_i, go through a fabrication process as individuals and do not lend themselves well to miniaturization or large scale production. As an alternative the introduction of planar coupling schemes has allowed the use of integrated circuit fabrication techniques and batch processing to produce high quality SQUIDs with excellent coupling characteristics. An example of such a scheme that will be used as the model in this paper consists of an n-turn planar spiral input coil in close proximity to a washer that serves as the SQUID inductance[11] as indicated in Fig. 2. The washer is interrupted by a slit such that the SQUID loop is completed through two shunted tunnel junctions at its outside edge. SQUIDs of this design fabricated down to the 2μm technology level typically have k^2 in the range of 0.8-0.9.

In most SQUID applications, a remote external circuit is physically connected to the terminals of the input coil of the SQUID. There are, however, some applications in which it is of advantage to integrate the external

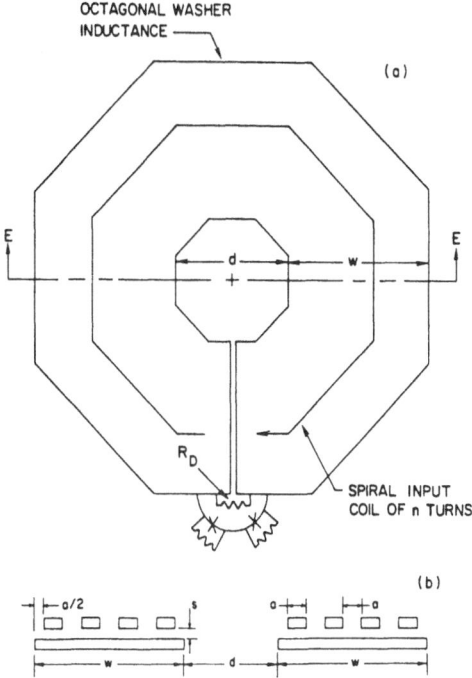

Fig. 2   Octagonal washer dc SQUID design (a) and cross-section EE through
         the washer and input coil (b).

circuit along with the SQUID on a single substrate. There are also situations
in which it is possible to configure the SQUID itself to perform some particular function. Examples of such Application Specific Integrated Detectors,
or ASIDs, include integrated gradiometers[12], miniature susceptometers[13], and
"flexible SQUIDs". Detectors of this kind are becoming increasingly preva-
lent as our fabrication capability matures.

Complex low-T_c fabrication technology has undergone a continual evolution from Pb-alloy, to mixed Nb/Pb-alloy, to all-refractory Nb and NbN.
Nb-AlOx-Nb trilayer-based technology[14] is now the technology of choice for
most digital circuit and analog SQUID applications. This technology is
easier to practice and more robust than previous technologies. It is capable
of producing junctions of superior electrical quality and SQUIDs with excellent noise characteristics. The materials set is sufficiently friendly
to semiconductor technology that there is considerable opportunity to leverage off of existing semiconductor fabrication capability, especially in
such areas as lithography and reactive ion etching.

With the advent of high-T_c superconductivity, the notion of running
SQUIDs in a liquid nitrogen bath has become a reality[15,16]. The white noise
of such devices is in reasonable agreement with theory and comparable to that
of commercially available low-T_c SQUIDs. The 1/f noise, which is not well
understood, is high, typically crossing the white noise at 100s or 1000s of
kHz. Single- and multi-turn high-T_c spiral input coils (20μm linewidth or
greater) have been successfully fabricated and tested,[17,18] however, the
integration of a high-T_c SQUID with such an input coil on a single substrate
is yet to be demonstrated. This will undoubtedly be done in the near future.

NOISE PERFORMANCE OF DC SQUIDS

Design rules for scaling SQUID junctions to obtain very low values of

ε in the white noise region were developed in the 1970s[15] and verified down to the vicinity of the quantum limit on experimental devices as early as 1982[19]. Two key design equations are

$$\beta_L = \frac{2LI_0}{\phi_0} = 1 \tag{3}$$

and

$$\beta_c = 2\pi I_0 R^2 C / \phi_0 \lesssim 1 , \tag{4}$$

where C is the junction capacitance, including parasitics. Equation (3) provides for an optimum critical current modulation depth of ~50% while Eq. (4) ensures non-hysteretic device characteristics[7,8]. Equations (3) and (4) together with Eqs. (1) and (2) provide the basic framework for scaling.

Figure 3 shows the energy sensitivity (either intrinsic or coupled with $k^2 > 0.85$) as a function of frequency for a number of different washer SQUIDs[11,20,21,22,23]. The performance in the white noise region is in good agreement with the theory. Particularly noteworthy are the value of 0.3h for an Nb edge-junction SQUID[22] at 300 mK and the value of 6h recently measured on a Nb-AlOx-Nb SQUID[23] at 4.2K. We still have no comprehensive theory for the low frequency (most often ~1/f) noise of these devices.

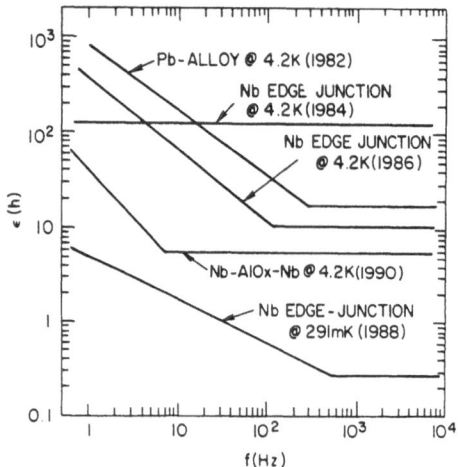

Fig. 3   Energy sensitivity as a function of frequency for a number of low-T_c washer SQUIDs.

Electron traps in the tunnel barrier[24], and critical current fluctuations[21] are often major contributors.

As an example of how performance will evolve through the 1990s we will scale from the performance of the Nb-AlOx-Nb SQUID[23] shown in Fig. 3. This SQUID had L = 50pH, $I_0 = 16\mu A$, R = 14Ω, $I_0 R = 250\mu V$, Josephson current density j_c of 600 A/cm^2, and junction area $\mathscr{A}$ of $3\mu m^2$; with $\beta_c \approx 1$, $C \approx 0.12pF$. For our practical design we will hold L constant at 100 pH which implies $I_0 \approx 10\mu A$. Using our known case and the scaling equations we first calculate a reference case for $1\mu m^2$ junctions from which it will be a straightforward matter to scale further. For our reference case we have $\mathscr{A} = 1\mu m^2$,

$j_c = 10^3 A/cm^2$, $R = 30\Omega$, $I_0 R = 300\mu V$, and $\epsilon(4.2K) = 5h$. The set of equations that then prescribes our scaled design as a function of the junction area $\mathcal{A}$ is:

$$L = 100pH \tag{5}$$

$$I_0 = 10\mu A \tag{6}$$

$$j_c = 10^3 A/cm^2 \left(\frac{1\mu m^2}{\mathcal{A}} \right) \tag{7}$$

$$R = 30\Omega \left(\frac{1\mu m^2}{\mathcal{A}} \right)^{1/2} \tag{8}$$

$$I_0 R = 300\mu V \left(\frac{1\mu m^2}{\mathcal{A}} \right)^{1/2} \tag{9}$$

$$\epsilon(4.2K) = 5h \left(\frac{\mathcal{A}}{1\mu m^2} \right)^{1/2}, \tag{10}$$

where we have assumed that the specific capacitance of the junction is independent of j_c and that parasitic capacitance in parallel with the junctions remains negligible.

To insure smooth, well behaved voltage - flux characteristics, especially with a tightly coupled input coil, it is advisable to place a damping resistor R_D across the SQUID inductance as indicated in Fig 2(a). The value of R_D should be 2-3 times that of R so that the thermal noise associated with R_D will not contribute significantly to ϕ_n[25]. We thus add one final equation to describe our design:

$$R_D = 75\Omega \left(\frac{1\mu m^2}{\mathcal{A}} \right)^{1/2}. \tag{11}$$

The implication of this set of scaling equations is fairly self evident. For 0.01 μm^2 junctions we require $j_c = 10^5$ A/cm^2, R = 300Ω, and $R_D = 750\Omega$. $I_0 R$ will increase to 3mV $\approx$ Vg, and $\epsilon(4.2K)$ will be about 0.5h. Even if we back off a factor of two in resistance values to make $\beta_c \ll 1$ and add considerable process latitude, $\epsilon(4.2K)$ will be of order h. We have no real basis for predicting how the 1/f noise will behave as we scale. Nevertheless the excellent 1/f noise performance of the Nb – AlOx – Nb SQUID at >1μm gives us reason to be optimistic.

COUPLING CONSIDERATIONS

While much work has been devoted to scaling of tunnel junctions to small dimensions little effort has been directed to scaling planar SQUID input coils to sub-μm linewidths. Practical planar SQUIDs typically have input coils with linewidths of 2-5μm, most with input coil inductance well below the industry standard of 2μH. If one has the technology to fabricate deep sub-μm tunnel junctions, one may (or may not) also have the technology to fabricate very fine line input coils. Such coils, assuming they retain good coupling characteristics, will provide some important advantages. For fixed SQUID inductance L and input coil inductance L_i the size of the SQUID reduces rapidly with the linewidth of the input coil. This will lead to a higher density of SQUIDs on the wafer and improved economy of scale. It has been the experience in the semiconductor industry that assuming it can be fabricated at all, a scaled design most often has higher yield than the original. It will also be possible to accommodate a full $2\mu H$ input coil on a physically small washer where parasitics are not a major concern.

The inductance of the washer SQUID can be written as

$$L = L_h + L_t + L_j ,\qquad (12)$$

where L_h is the inductance of the hole, L_t is the inductance of the slit, and L_j is the parasitic inductance associated with the junction region. If L_t is negligible we then have, as originally discussed in 1980[6],[11]

$$M_i = nL_h \qquad (13)$$

$$L_i = n^2 L_n + L_s \qquad (14)$$

$$k^2 = \left(1 - \frac{L_j}{L}\right)\left(1 + \frac{L_s}{n^2 L_h}\right)^{-1} ,\qquad (15)$$

where L_s is the stripline inductance of the coil with respect to the washer. There is an underlying assumption that the current in the input coil is nearly perfectly imaged in the washer and that at the slit the image current associated with each turn of the input coil completes its path by flowing around the inside edge of the washer's hole. L_j is easily kept to a few pH and for designs with linewidths greater than $2\mu m$ the calculated value of L_s is typ-

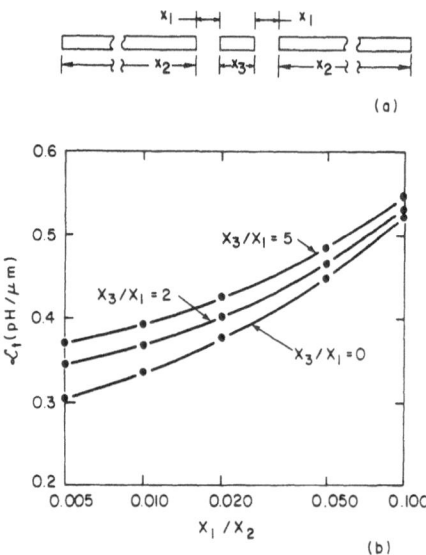

(a)

(b)

Fig. 4 Cross-section through the washer's slit with designation of dimensions (a) and slit inductance per unit length $\mathscr{L}_s$ as a function of various dimensions (b).

ically ~5% of $n^2 L_h$. Equations (13-15) have been verified[11] for a number of different values of n with input coils of linewidth $\gtrsim 2\mu m$.

For a process with only 4-5 levels of lithography it is not possible to fully groundplane the slit, and L_t will not be negligible. It then becomes necessary to modify the expressions for L_i, M_i, and k^2 to include L_t. One way of doing this is to treat the slit as a two dimensional structure with a constant inductance per unit length $\mathscr{L}_t$. Fig. 4(a) shows a cut through the washer orthoganal to the slit and indicates the labeling of the various dimensions. In general there will be a lead from the center of the coil coming out through the slit. For the special case where the washer itself is used to complete the current path from the inside to the outside of the spiral, $x_3 = 0$. Figure 3(b) shows calculated values of $\mathscr{L}_t$ as a function of x_1/x_2 for various values of x_3/x_1. The weak dependencies on dimensions suggest that the assumption of constant $\mathscr{L}_t$ is a reasonable one, although $\mathscr{L}_t$ will actually peak up somewhat at the ends of the slit. There will be some partial groundplaning of the slit by the turns of the nearly spiral input coil. For practical situations we estimate $\mathscr{L}_t$ will be in the range of 0.25-0.35 pH/μm, with little leverage in pushing hard to make the slit dimensions real small. For an 80μm long slit, L_t will then be on the order of 24pH. For a more complex process where the slit can be fully groundplaned, $\mathscr{L}_t$, which will be considerably reduced, can be easily calculated with conventional stripline formulas.

L_t is not a parasitic inductance in the sense of L_j since it partially couples to the turns of the input coil. In the presence of non-zero L_t the expressions for M_i, L_i and k^2 become

$$M_i = n(L_h + L_t/2) \tag{16}$$

$$L_i = n^2(L_h + L_t/3) + L_s \tag{17}$$

$$k^2 = \frac{n^2(L_h + L_t/2)^2}{L\,[n^2(L_hL + L_t/3) + L_s]} \approx \left(1 + \frac{L_j}{L_h} + \frac{L_t}{3L_h} + \frac{L_s}{n^2 L_h} \right)^{-1}, \tag{18}$$

where the last expression in Eq. (18) is valid in the limit of small L_t.

The quantity $L_s/n^2 L_h$ in Eq. (18) is one we must examine carefully. Figure 2(b) shows the cross section EE taken through the washer SQUID in Fig. 2(a), with the vertical dimensions exaggerated and $n = 4$ for clarity. The linewidth and spacing for the coil are both a and the insulator thickness is s. The penetration depths of the coil and washer are λ_1 and λ_2 respectively. The inductance of the octagonal washer of hole size d is approximately $1.05\mu_0 d$ provided the width of the washer w is greater than d[11,27]. In the limit of $s + \lambda_1 + \lambda_2 \ll a$, the inductance per unit length of $\mathscr{L}_s$ of the coil with respect to the washer is given by

$$\mathscr{L}_s = \mu_0 \left(\frac{s + \lambda_1 + \lambda_2}{a} \right) \tag{19}$$

The average length of a spiral turn is $1.05\pi(d + w)$ so the total length of the spiral coil is $1.05\pi n(d + w)$. We can then express $L_s/n^2 L_h$ as

$$L_s/n^2 L_h = 2\pi \left(\frac{d + w}{d} \right) \left(\frac{s + \lambda_1 + \lambda_2}{w} \right), \tag{20}$$

where we have used the fact that $n = w/2a$.

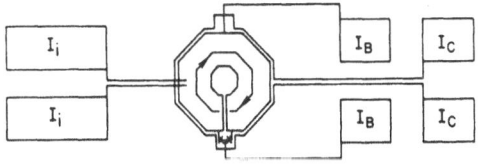

Fig. 5   Layout of a high performance octagonal washer dc SQUID.

As a is made smaller fringing fields become important and Eq. (19) overestimates $\mathscr{L}_S$. On the other hand these fringing fields also start to couple turns directly to each other through leakage flux. The relevant $\mathscr{L}_S$ then becomes the strip line inductance per unit length plus the sum of all the leakage-flux-coupling mutuals. In addition, as the cross-section of the line becomes of order λ_1^2, the kinetic inductance starts to become significant, ultimately dominating at very small dimensions. A careful analysis of this problem[28] shows that Eq.(20) is a valid upper bound on $L_S/n^2 L_h$ over the entire range of a from w down to 0.05 μm! Using Eq. (20) will give modest underestimates of k^2 for intermediate values of a. It also follows that since we have built washer SQUIDs with a = 2.5 μm and n from 10 to 50 or more that work as predicted, the same SQUIDs with a = 0.25 μm should couple just as well or better with L_i greater by a factor of 100.

As a practical example we consider the octagonal washer SQUID depicted in Fig. 5 with d = 60 μm and w = 80 μm. This SQUID will have L_h = 80pH, L_S = 24pH, and L_j = 3pH. The input coil with 0.25 μm lines and spaces will have n=160 turns. Assuming $\lambda_1 = \lambda_2 = 0.086\mu$m and s = 0.15 μm we use Eq. (20) to calculate $L_S/n^2 L_h = 0.06$. We also have $L_t/3L_h = 0.10$ and $L_j/L_h = 0.04$. Using Eqs. (12), (16), (17), and (18) we then calculate L = 107pH, M_i = 14.7nH, L_i = 2.4 μH, and $k^2 \approx 0.85$. The SQUID occupies an area of ~0.05 mm^2, not counting bonding pads. Only marginally less impressive are designs with a = 0.35 μm and a = 0.5 μm corresponding to excimer laser and i-line projection lithography capability.

The small overall size of the washer is advantageous from the point of view of operating in a non-zero magnetic field B. As the aspect ratio of the washer decreases, magnetic flux penetration related problems such as hysteresis and certain temperature dependencies will moderate. Near the edges of the washer we can expect magnetic fields of order B(d + w)/t, where t is the washer thickness[29]. Our compact design minimizes (d+w). In addition it is an advantage to use a technology in which the washer is fabricated last in the process sequence, giving one the option of increasing t. This will decrease the magnetic field at the washer edges even more, further decreasing deleterious flux penetration effects.

APPLICATION SPECIFIC INTEGRATED DETECTORS

As examples of ASIDs we will briefly review integrated gradiometers, miniature susceptometers, and a new type of integrated detector, the flexible SQUID. The layout of a model first derivative gradiometer[30] with series wound pickup loops is shown in Fig. 6. The SQUID for such a detector must itself not respond to a uniform applied magnetic field. Two-hole or figure-eight

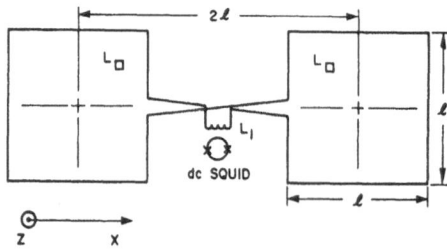

Fig. 6   First derivative dc SQUID gradiometer.

SQUIDs with integrated dual spiral input coils are viable options. Provided the interconnect inductance is much less than the inductance of each pickup loop $L_\square$, it can be shown that with $L_i \approx 2L_\square$, the gradient noise is

$$\left.\frac{\partial B_z}{\partial x}\right|_n = \frac{\phi_n}{k\ell^3} \sqrt{\frac{2L_\square}{L}} \; \alpha \; \sqrt{\varepsilon_c} \, /\ell^{5/2} \; , \tag{21}$$

where ℓ is as indicated in Fig. 6. For a washer SQUID with $k^2L = 85\text{pH}$ and $\varepsilon_c = 5\text{h}$ we calculate $\left.\frac{\partial B_z}{\partial x}\right|_n = 3 \times 10^{-12}\,\text{G/cm}\sqrt{\text{Hz}}$ for $\ell = 1\text{cm}$ and $5 \times 10^{-13}\,\text{G/cm}$ $\sqrt{\text{Hz}}$ for $\ell = 2\text{cm}$. Such performance makes this type of detector extremely attractive for biomagnetic and magnetic anomaly detection applications.

Figure 7 shows the layout of an integrated miniature SQUID susceptometer with two pickup loops configured in opposite sense over a hole in the groundplane[31]. The center-tapped field coil with contact pads well away from

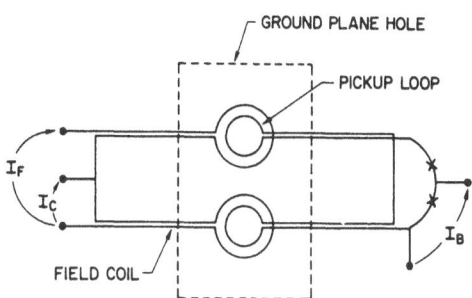

GROUND PLANE HOLE

PICKUP LOOP

I_F

I_C

I_B

FIELD COIL

Fig. 7   Miniature dc SQUID susceptometer.

the hole makes a single turn around each of the pickup loops. In the absence of a sample the field current I_F should couple no flux to the SQUID. In the presence of a sample the SQUID will measure changes in the sample's magnetic moment which, for example, may be induced by application of I_F or polarized light[32].

The figure of merit for such a device is given by[31]

$$S_n = \phi_n \left(\frac{r_0}{r_e} \right) \; , \tag{22}$$

where S_n has the units of spins (of moment μ_B) per $\sqrt{\text{Hz}}$, r_0 is the radius of the pickup loop, $r_e = e^2/mc^2$ is the classical electron radius, and ϕ_n has the units of $\phi_0/\sqrt{\text{Hz}}$. The first version of this device[13] had square pickup loops $\sim 17.5\mu\text{m}$ across, each with inductance of about 30pH. This is equivalent to a circular pickup loop with $r_0 = 10\mu\text{m}$. The SQUID had a flux noise ϕ_n of $8.4 \times 10^{-7}\,\phi_0\,/\sqrt{\text{Hz}}$ ($\varepsilon = 27\text{h}$) at 4.2K and $5.6 \times 10^{-7}\,\phi_0/\sqrt{\text{Hz}}$ ($\varepsilon = 12\text{h}$) at 1.8K. The corresponding values of S_n are 3000 spins / $\sqrt{\text{Hz}}$ and 2000 spins / $\sqrt{\text{Hz}}$ respectively! Deep sub-μm technology and operation at temperatures of a few hundred mK should reduce S_n to a few spins/ $\sqrt{\text{Hz}}$. Detectors of this type will be used increasingly in the study of the static and dynamic properties of very small samples.

As a final example of an ASID we consider the proposed flexible SQUID detector shown in Fig. 8. Part of this SQUID's inductive loop is configured as a cantilever structure that extends outward over a hole in an insulating layer above the groundplane. The drive current I_D is applied through L_D, the cantilever portion of L. If the loop bends slightly such that its cantilever portion deflects by an amount Δt, then there is change in the applied flux $\phi_a = I_D L_D$ given by

$$\Delta\phi_a = \phi_a \frac{\Delta t}{(\lambda_1 + \lambda_2 + t)} \quad , \qquad (23)$$

where λ_1 and λ_2 are now the superconducting penetration depths of the groundplane and SQUID loop respectively, and t is the distance from the underside of the SQUID loop to the top surface of the groundplane. The amplitude noise of the deflection Δt_n can be expressed as

$$\Delta t_n = \frac{\phi_n}{\phi_a} (\lambda_1 + \lambda_2 + t) \quad . \qquad (24)$$

For an optimized 20 pH SQUID with $\phi_n \approx 10^{-7}\phi_0/\sqrt{Hz}$ ($\varepsilon \approx h$), and $\lambda_1 = \lambda_2 = t = 0.1\mu m$, $\Delta t_n \approx 3 \times 10^{-14}$ cm/$\sqrt{Hz}$. For a sufficiently high mechanical Q this will be the order of the thermal noise of the cantilever structure. Integrated

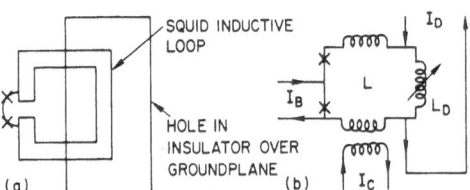

Fig. 8   Flexible dc SQUID sensor.

micromechanical sensors of this type would be of potential use for the measurement of acceleration in general and for the measurement of force, pressure, flow, gas absorption and adsorption, etc. in low and ultra-low temperature environments.

LOW-T_c PROCESS TECHNOLOGY

We will make no attempt to do a comprehensive review of low-T_c process technology but will instead focus on some of the key aspects of all-refractory trilayer based fabrication approaches. Over the past several years trilayer technologies have become dominant for low-T_c fabrication. Such technologies allow one to move away from lift-off patterning and into high resolution lithography and semiconductor state-of-the-art reactive ion etching (RIE) and planarization.

In a trilayer technology the first step is to fabricate a large Josephson junction that is the size of the entire wafer. For most applications the technology of choice has become Nb-AlOx-Nb where the sandwich structure is fabricated in-situ without breaking vacuum[14]. Typically a low stress Nb layer is sputter deposited on an oxidized silicon wafer, followed by the sputter deposition of a thin Al layer. The Al is thermally oxidized to form the tunnel barrier, and then the final low stress Nb layer is sputtered. Such trilayer wafers become the starting material for a subsequent multi-level-of-lithography fabrication process.

The minimum number of levels of lithography required to fabricate a SQUID from a trilayer is four. By way of example we will briefly review the Selective Niobium Anodization Process, or SNAP[33]. After application of the first resist stencil RIE or plasma etching through the entire thickness of the trilayer sandwich is done to define a pattern of interconnected trilayer areas. The resist is then stripped and a second resist stencil is applied to define the location of the actual junctions. The interconnected trilayer areas are next anodized down to the AlOx tunnel barrier to form a Nb_2O_5 dielectric layer that is self-aligned to the resultant junction pedestals. This resist is stripped and a third stencil with a liftoff profile is applied for resistor definition. Resistors, Ti or PdAu for example, are then evaporated and lifted off. After another Nb deposition the last resist stencil is applied and RIE or plasma etching is used to pattern the final Nb level to wire up components. The SQUID's input coil can be of the base electrode material and the washer of the wiring-up level or visa versa. Junctions fabricated by this and similar processes have been demonstrated to have exceptionally low subgap leakage for j_c up to several thousand A/cm^2. It is presently an active area of research to push this to the $5-10\times10^4$ A/cm^2 level.

An extremely critical part of the fabrication process is the junction definition step. In practice it has been difficult to pattern junctions below $1\mu m^2$ in area with wet anodization. A more sophisticated approach is to define the counter-electrode and then the base-electrode by a two-step RIE process. Various techniques can then be used to apply a low dielectric constant insulator (such as SiO_2) that is planarized and self aligned to the junction pedestals[34]. Schemes such as this should be scalable down to $0.1\mu m$ or below and will be key to the fabrication of future ultra-high performance SQUIDs.

Advanced lithography, as driven by semiconductor technology, will provide the capability to define structures down to $0.5\mu m$ (i-line, 5X), $0.35\mu m$ (excimer laser, 5X), and $0.1\mu m$ (electron -beam). The fact that the Nb-AlOx-Nb trilayer does not pose a contamination threat to semiconductor technology will permit a much closer coupling to that technology than has been the case in the past. If one is willing to pay the price of additional complexity a groundplane, small area contacts, and additional wiring levels can be integrated into the process in a relatively straightforward fashion.

HIGH-T_c SQUIDs

With the discovery of high-T_c superconductivity has come the potential for practical high-performance SQUIDs operating at 77K. Such devices would enormously reduce the refrigeration overhead associated with SQUID operation and would greatly facilitate, for example, the introduction of multichannel systems requiring close spacing between superconducting pickup loops and the room temperature environment. As with low-T_c SQUIDs, a useful 77K SQUID must have acceptable noise performance and must, in general, be equipped with a well coupled input coil of reasonable inductance. Although three-dimensional bulk devices may offer an interim solution at some level, in the long term an integrated planar technology needs to be developed. We will briefly review the present state-of-the-art of the noise performance of thin-film 77K SQUIDs and discuss the status of input coil integration.

Shown in Fig. 9 are the energy sensitivities as a function of frequency of several high-T_c SQUIDs [16,35] representative of today's best 77K performance. Also shown for comparison are data from several 4.2K SQUIDs. None of the 77K SQUIDs had input coils and the data represent intrinsic energy sensitivities. All of the 4.2K SQUIDs had input coils and that data represents coupled energy sensitivities. With an operation temperature higher by a factor of 18 one expects from Eq.(2) that the energy sensitivity in the white noise region will be higher by a factor of 18 for 77K SQUIDs, all other things being equal. The best results for polycrystaline TlBaCaCuO SQUIDs and bicrystal YBCO SQUIDs are in line with this prediction. The 1/f noise, however, is high and not well understood. In the case of the bicrystal YBCO SQUIDs this noise clearly originates in the Josephson elements and not in the epitaxial YBCO films. It is important that we understand this noise and

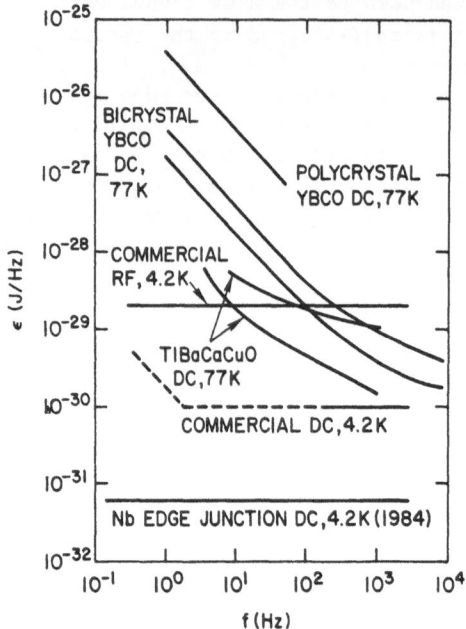

Fig. 9   Energy sensitivity as a function of frequency for various low-T_c and
high-T_c SQUIDs. For comparison with Fig. 3, 10^{-31}J/Hz $\approx$ 150h.

hopefully figure out how to reduce it. In spite of the high low-frequency
noise, the noise of the best TlBaCaCuO SQUID is lower than that of a representative commercial rf SQUID above 10Hz, and there are applications for
which 77K SQUID noise performance is already adequate.

    To actually use a 77K SQUID most applications require an input coil,
and the integration of such a coil has proved to be a non-trivial exercise.
Processing technology required to fabricate a multi-level coil structure,
including cross-overs, on top of or in conjunction with a high-T_c SQUID has
not yet been demonstrated, although significant progress in this direction
has been made. The approach has been to fabricate an input coil structure,
in some cases along with magnetometer or gradiometer pick-up loops, on one
substrate and then to sandwich this together with a high-T_c or low-T_c SQUID
fabricated on a separate substrate. Using this approach k^2 of 0.81 was
demonstrated for a single level YBCO input coil coupled to a TlBaCaCuO SQUID
at 77K. This configuration was used to demonstrate a magnetometer with 5
times the magnetic field sensitivity of the bare SQUID. In another
experiment[18] a multilevel 10-turn input coil with cross-overs was demonstrated using an innovative fabrication approach. Verification of the operation of this structure at 77K was shown with the coil in close proximity
to a 4.2K SQUID. 77K SQUIDs with integrated input coils will undoubtedly
follow in the not-too-distant future.

CONCLUSIONS

    We have reviewed the current status of dc SQUID technology and discussed
how the technology will evolve through the 1990s. There is easily room for
an order of magnitude or more improvement in energy sensitivity of low-T_c
SQUIDs, and basic planar coupling concepts now well verified at the 2.5μm
level will continue to apply down to dimensions of 0.1μm or less. In general
as lithographic dimensions are reduced to the 0.1-0.25 μm regime, we will

find SQUIDs that are smaller, quieter, friendlier, and better matched to input and output than their $>1\mu m$ counterparts. The use of ASIDs such as integrated dc SQUID gradiometers will become widespread and a host of new integrated detectors will appear. Fabrication technology based on the trilayer concept, primarily Nb-AlOx-Nb, will continue to evolve with significant benefit from closer coupling to semiconductor technology than in the past. Our understanding of high-T_c materials will continue to improve. If the 1/f noise can be lowered, high-T_c SQUIDs will replace their low-T_c predecessors in many applications for which commercial low-T_c SQUIDs are now used. For ultra-low noise applications, however, SQUID operation at or below 4.2K will always provide an advantage.

ACKNOWLEDGMENTS

This work was partially conducted under the auspices of the Consortium for Superconducting Electronics.

REFERENCES

1. R.C. Jaklevic, J. Lambe, A.H. Silver, and J.E. Mercereau, Phys. Rev. Lett. 12, 159 (1964).

2. B.D. Josephson, Phys. Lett. 1, 251 (1962); Adv. Physics 14, 419 (1965).

3. V. Radhakrishnan and V.L. Newhouse, J. Appl. Phys. 42, 129 (1971).

4. J.H. Claassen, J. Appl. Phys. 46, 2268 (1975).

5. C.D. Tesche and J.Clarke, J. Low Temp. Phys. 29, 301 (1977).

6. M.B. Ketchen, IEEE Trans. Magn. MAG-17, 387 (1981).

7. W.C. Stewart, Appl. Phys. Lett 12, 277 (1968).

8. D.E. McCumber, J. Appl. Phys. 39, 3113 (1968).

9. R.P. Giffard, R.A. Webb, and J.C. Wheatley, J. Low Temp. Phys. 6, 533 (1972).

10. J. Clarke, W.J. Goubau, and M.B. Ketchen, J. Low Temp. Phys. 25, 99 (1976).

11. J.M. Jaycox and M.B. Ketchen, IEEE Trans. Magn. MAG-17, 400 (1981); M.B. Ketchen and J.M. Jaycox, Appl. Phys Lett. 40, 736 (1982).

12. M.B. Ketchen, W.M. Goubau, J. Clarke, and G.B. Donaldson, J. Appl. Phys. 44, 4111 (1978).

13. M.B. Ketchen, T. Kopley, and H. Ling. Appl. Phys. Lett. 44, 1008 (1984).

14. M. Gurvitch, M.A. Washington and H.A. Huggins, Appl. Phys. Lett. 42, 472 (1983).

15. J.E. Zimmermann, J.A. Beall, M.W. Cromar, and R.H. Ono, Appl. Phys. lett. 51, 617 (1987).

16. R.H. Koch, W.J. Gallagher, B. Bumble, and W.Y. Lee, Appl. Phys. Lett. 54, 951 (1989).

17. B. Oh, R.H. Koch, W.J. Gallagher, V. Foglietti, G. Koren, and A. Gupta, Appl. Phys. Lett. 56, 2575 (1990).

18. F.C. Wellstood, J.J. Kingston, and J. Clarke, Appl. Phys. Lett., 56, 2336 (1990).

19. D.J. Van Harlingen, R.H. Koch, and J. Clarke, Appl. Phys. Lett. 41, 197 (1982).

20. C.D. Tesche, K.H. Brown, A.C. Callegari, M.-M. Chen, J.H. Greiner H.C. Jones, M.B. Ketchen, K.K. Kim, A.W. Kleinsasser, H.A. Notarys, G. Proto, R.H. Wang, and T. Yogi, IEEE Trans. Magn. <u>MAG-21</u>, 1032 (1985).

21. V. Foglietti, W.J. Gallagher, M.B. Ketchen, A.W. Kleinsasser R.H. Koch, S.I. Raider and R.L. Sandstrom, Appl. Phys. Lett. <u>49</u>, 1393, (1986).

22. D.D. Awschalom, J.R. Rozen, M.B. Ketchen, W.J. Gallagher, A.W. Kleinsasser, R.L. Sandstrom, and B. Bumble, Appl. Phys. Lett. <u>53</u>, 2108 (1988).

23. M.B. Ketchen, M.Bhushan, S.B. Kaplan, and W.J. Gallagher, "Low Noise dc SQUIDs Fabricated in $Nb - Al_2O_3 - Nb$ Trilayer Technology", to be published in the proceedings of the 1990 Applied Superconductivity Conference.

24. C.T. Rogers and R.A. Buhrman, W.J. Gallagher, S.I. Raider, A.W. Kleinsasser, and R.L. Sandstrom, IEEE Trans Magn. <u>MAG-23</u>, 1658 (1987)

25. V. Foglietti, W.J. Gallagher, M.B. Ketchen, A.W. Kleinsasser, R.H. Koch, and R.L. Sandstrom, Appl. Phys. Lett. <u>55</u>, 1451 (1989).

26. W.H. Chang, J. Appl. Phys. <u>50</u>, 8129 (1979).

27. M.B. Ketchen, W.J. Gallagher, A.W. Kleinsasser, S. Murphy and J.R. Clem in <u>SQUID '85</u>, edited by H.D. Hahlbohm and H. Lubbig (Walter de Gruyter, Berlin, New York, 1985), 865.

28. M.B. Ketchen, "Design Considerations for dc SQUIDs Fabricated in Deep Sub-μm Technology", to be published in the proceedings of the 1990 Applied Superconductivity Conference.

29. R.P. Huebener, R.T. Kampwirth, and J.R. Clem, J. Low Temp. Phys. <u>6</u>, 275 (1972).

30. M.B. Ketchen, J. Appl. Physics <u>58</u>, 4322 (1985).

31. M.B. Ketchen, D.D. Awschalom, W.J. Gallagher, A.W. Kleinsasser, R.L. Sandstrom, J.R. Rozen, and B. Bumble, IEEE Trans. Magn. <u>MAG-25</u>, 1212 (1989).

32. D.D. Awschalom, J. Warnock, and S. vonMolnar, Phys. Rev. Lett. <u>58</u>, 812 (1987).

33. H. Kroger, L.N. Smith and D.W. Jillie, Appl. Phys. Lett. <u>39</u>, 280 (1981).

34. S. Nagasawa, H. Tsuge and Y. Wada, IEEE El. Dev. Lett. <u>9</u>, 414 (1988).

35. R. Gross, P. Chaudhari, M. Kawasaki, M.B. Ketchen, and A. Gupta, Appl. Phys. Lett. <u>57</u>, 727 (1990).

36. Commercial SQUID data shows representative performance for S.H.E./B.T.I. rf and dc SQUIDs, Biomagnetic Technologies, Inc., 4174 Sorrento Valley Blvd., P.O. Box 210079, San Diego, CA 92121.

OPERATION OF THIN FILM DC-SQUIDs MADE OF HIGH-Tc SUPERCONDUCTORS

V. Foglietti

IESS-CNR, Via Cineto Romano 42 Roma Italy

R.H.Koch, W.J. Gallagher, R.B. Laibowitz and B. Oh

IBM T.J.Watson Research Center, Yorktown Heights, NY, USA

Introduction

Superconducting Quantum Interference Devices (SQUIDs) seem to be one of the most probable early applications of high temperature superconductivity. Many reported devices operate at 77 K [1,2,3,4] and noise measurements made on high-Tc dc-SQUIDs demonstrate that the sensitivity is not so far from that of commercially available helium temperature SQUIDs. However further technological developments are required to make this sensitivity useful for real applications. In particular, practical devices will need well controlled Josephson elements and well matched input transformers for coupling to useful signals.

A classic three layer Josephson junction is extremely difficult to fabricate with the high-Tc materials due to the very low coherence length $\xi_c < \xi_{ab} < 2nm$. Also alternative weak link structures [5] relying on linear dimensions of order a few ξ_{ab} are very hard to fabricate, even with the most sophisticated lithographic tools available today. Fortunately, however, the small coherence length also results in naturally occurring Josephson coupling at large angle grain boundaries in polycrystalline high-Tc films. Both ac and dc Josephson effects are observed, with a small value of the shunt resistance. The intrinsic low value of the McCumber parameter make these grain boundary weak links particularly attractive used as active elements in dc-SQUID devices. At the present stage of development 77 K dc-SQUIDs fabricated with random grain boundaries in single layer polycrystalline $TlBaCaCuO$ material (the phase 2223) have demonstrated the best energy sensitivity, [1] although there has been considerable progress in deliberately fabricating Josephson elements as well. One direction of approach has been to deliberately isolate single large angle grain boundary weak links in SQUID structures. The most controlled way of doing this with $YBaCuO$ films [2] has been by using an epitaxial film grown on a $SrTiO_3$ bicrystal. It has also been possible to isolate single large angle grain boundaries in $YBaCuO$ films grown with special microstructures [6]. Recently promising techniques of deliberately making Josephson junctions have also been demonstrated in the form of "edge junction" and "step edge" weak links [7,8]. These artificial weak links entail a multilevel film growth technology, and these device structures have the advantage of allowing supercurrent transport in

Nonlinear Superconductive Electronics and Josephson Devices
Edited by G. Costabile *et al.*, Plenum Press, New York, 1991

95

both the base and counter electrodes to flow along the a-b planes of the superconducting crystal structures, where the coherence length is much longer than it is in a direction parallel to the c axis. With each of these recent deliberate junction approaches with $YBaCuO$ films very nice SQUID transfer characteristics have been demonstrated, but no devices have demonstrated noise as low as that in the polycrystalline $TlBaCaCuO$ devices.

As far as the coupling structures needed for practical high-Tc SQUIDs, there is considerable difficulty in achieving a reliable multilevel technology due again to the inherent anisotropy of the high temperature superconductors. This leads to the requirement of a high degree of compatibility between the design of the coil, the materials, the deposition technique, and the patterning method. Progress in fabrication of multiturn thin film coils made of multilevel $YBaCuO$ films makes this material a promising candidate for coils [9,10]. Combined with the noise performance of the $TlBaCaCuO$ dc-SQUIDs, this result makes particularly attractive a hybrid solution where these SQUIDs are efficiently coupled to a multi-level $YBaCuO$ input coil grown on a separate substrate. Already efficient coupling between a $TlBaCaCuO$ SQUID sandwiched with a separately grown single level $YBaCuO$ has been demonstrated [11].

The strong efforts in research activity to get reliable high-Tc SQUID devices are well justified by the impact that these devices could have in biomedical and military applications. This paper will be entirely devoted to the present status of dc-SQUID devices fabricated with high-Tc materials. The second section of this paper will be devoted to a brief discussion on design considerations for high-Tc SQUID devices. The dc-SQUIDs made from polycrystalline films will be discussed in the third section of this paper, describing the devices developed at the IBM laboratories. The third section will discuss the progress in device fabrication using artificial weak links. Finally, coupling considerations and future trends of these devices, focusing on possible biomedical applications, will be discussed in the last section.

Design considerations

The design considerations for 77 K SQUIDs rely on the well established theory developed for 4.2 K devices. That theory is based on the resistively shunted junction (RSJ) model which we can reasonably assume to be valid for the 77 K weak links. To maintain phase coherence across the weak links the Josephson coupling energy must be greater than the thermal fluctuation energy:

$$\Phi_o i_o / 2\pi >> KT \tag{1}$$

where Φ_o is the flux quantum and i_o is the critical current of the weak link. The magnetic energy associated with the flux quantum Φ_o must also be greater than the thermal energy, thus avoiding hopping mechanism between adjacent fluxon states that would destroy the periodic response of the SQUID:

$$\Phi_o^2 / 2L >> KT \tag{2}$$

where L is the SQUID inductance. Equation 1 puts a lower limit on the weak link critical current for a SQUID working at 77 K:

$$i_o >> 4\mu A. \tag{3}$$

Equation 2 results in an upper limit for the SQUID inductance :

$$L \ll 1nH. \tag{4}$$

also for operation at 77 K. These limitations are compatible with the optimization condition on the screening parameter β_L :

$$\beta_L = 2Li_o/\Phi_o \simeq 1 \tag{5}$$

as derived from the SQUID theory. The second condition from the theory is on the McCumber parameter β_c :

$$\beta_c = 2\pi i_o R^2 C/\Phi_o < 1 \tag{6}$$

where R is the shunt resistance of the weak links and C is the capacitance. This condition is naturally achieved because of the low value of the shunt resistance R, which manifests itself in the universally observed non-hysteretic current-voltage characteristics of high-Tc weak links.

The conclusion from these considerations as well as from simulations carried out for achievable device parameters at 77 K is that the expression of the white energy sensitivity, determined from the 4 K SQUID theory [12]:

$$S_\epsilon \simeq 9KTL/R \tag{7}$$

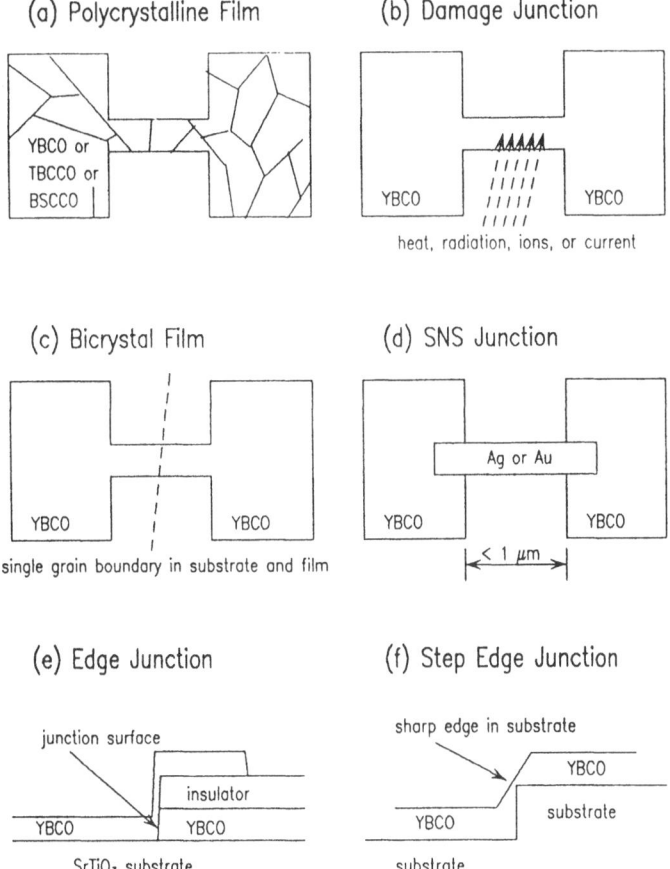

Fig. 1 Possible types of weak links that may be used as the active elements in dc-SQUID devices.

and all the other quality factors are still valid at 77 K, provided that the SQUID parameters are close to the optimum values.

Natural grain boundaries high-Tc dc-SQUIDs

The first high-Tc dc-SQUID were made at IBM using polycrystalline $YBaCuO$ films with small grain sizes made either by e-beam deposition or by sputtering, both with high temperature post anneals [13]. They consisted of large superconducting pads connected by parallel lines with dimensions ranging from 5 to 40 μm long and wide. These lines, shown in Figure 1a, form an array of weak connections, one for each grain boundary crossed by the patterned line. The action of the dominating weak connections, having the lowest values of critical current, could possibly determine a voltage-flux characteristic with a dominant periodicity, thus satisfying the major requirement of a SQUID device. Figure 2 shows a current versus voltage characteristic of one patterned line. The gradual onset of voltage and the low critical current, with respect to the unpatterned film, are typical weak link related effects.

More recent studies at IBM have used polycrystalline $TlBaCaCuO$ in single level "washer" type design with an inductance L ranging from 80 up to 150 pH [14]. The "washer" design is slightly modified introducing two constrictions with dimensions comparable with the grain size of the film ($\simeq 2x2\mu m^2$). Figure 3 is a SEM micrograph of one of these devices. The upper side shows the hole of the washer, $40x40\mu m^2$, the lower side shows the constriction region. The plate-like morphology of the surface make it possible to hypothesize the locations of the underlying grain boundaries which may be acting as dominant weak links.

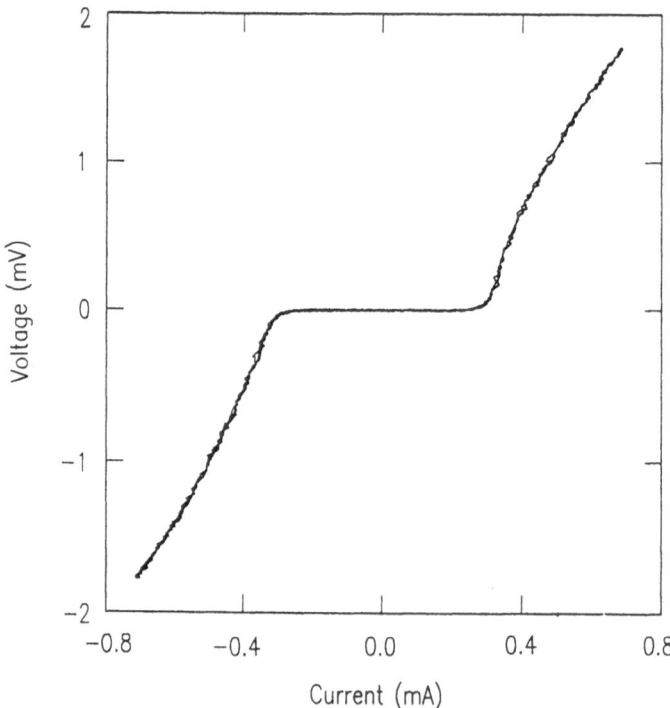

Fig. 2 Typical current-voltage characteristic of a stripe patterned from a polycrystalline film.

The *TlBaCaCuO* films used in the SQUIDs were deposited on yttrium-stabilized zirconia using a symmetric rf-diode sputtering system with two identical targets facing each other [15]. The substrate was held outside the discharge region facing the center-line between the targets. Subsequent to the deposition the films were wrapped in gold foil with a $Tl_2Ba_2Ca_2Cu_3O_y$ pellet and annealed at high temperatures ($\approx 900°C$) in sealed ampules. The films had terraced textures (evident in Figure 3) indicating grains ranging in size up to 10μm. Resistive transitions showed a substantial drop near 120 K and were usually complete above 105 K. Device patterning was done by argon ion milling through photoresist mask. Contacts were made with sputtered or evaporated gold following an argon preclean.

Figure 4 shows a typical flux-voltage characteristic at 77 K for a *TlBaCaCuO* dc-SQUID biased at the maximum critical current. Two characteristics slightly shifted are obtained by sweeping the magnetic flux back and forth. Significant hysteresis of the voltage across the SQUID as the applied magnetic flux is varied was observed in all the polycrystalline devices. It cannot simply be attributed to direct effects of flux trapping and untrapping through the grain boundaries of the superconducting material forming the device by the fact that the hysteresis actually "leads" on the back and forth sweep rather than "lags". The type of hysteresis was seen in single lines of high Tc films, and the same type of leading hysteresis has been reported in the critical current of Nb, NbN and In microbridges. In a very elegant way Walton et al. [16] showed that this type of hysteresis is a consequence of the strong modulation of the order parameter in the mixed state of a metallic superconductor. The theory cannot be directly applied to a non mixed state. However the physical mechanism of nascent and exiting vortices from a superconducting material, that is responsible for the hysteresis in metallic superconductors, can reasonably be expected to be present also in the grain boundaries

Fig. 3 SEM micrograph showing the center portion of a *TlBaCaCuO* SQUID and an enlargement of the link region. The center hole in the SQUID is 40 x 42 μm^2 and the constriction has minimum width of 4μm.

of a polycrystalline high-Tc material. The hysteresis can be quantified as :

$$h = \epsilon / \Delta B \qquad (8)$$

where ΔB is the peak-to-peak amplitude of the applied field sweep and ϵ is the shift in the $V - \Phi$ curves. The h value is strongly dependent on the geometry of the device and on the material composing it. It decreases rapidly as the superconducting area forming the device decreases (i.e., the washer area decreases). Typical levels of h obtained with small washer geometries are $\simeq 5 \cdot 10^{-3}$, not nearly as small as the 10^{-6} values that are typical of low-Tc SQUIDs.

In flux-locked loops in which SQUIDs are usually operated in practical situations, hysteresis is less of a concern. The feedback coil in the flux-locked loop applies a uniform field to the SQUID, cancelling the applied field [17]. Any hysteresis in the output will only be related to the size of the magnetic error signal of the feedback loop and will result in a loss of linearity of the system response. The flux-locked loop lead to an effective decrease of h of two orders of magnitude, when the magnetic field is applied perpendicular to the substrate. Hysteresis remains a problem when a magnetic field is applied in the plane of the substrate, since the feedback loop cannot cancel the field in this direction.

The noise performance of SQUIDs can be measured in different ways. The more complete and accurate noise measurement procedure is to measure the voltage noise spectral density S_V and the responsivity $\partial V / \partial \Phi$, for every bias point determined by the bias current, the dc-flux and the temperature [1]. The flux noise spectral density is then derived from:

$$S_\Phi = S_V / (\partial V / \partial \Phi)^2 \qquad (9)$$

The more straightforward noise measurement, requiring less time in optimizing bias conditions, is the conventional flux-locked loop. This technique permits an easy approach to the measurement of the flux noise spectral density S_Φ versus temperature, giving results that agree within a factor two with the more accurate type of measurement [17]. The flux-locked loop is very important and practically unavoidable in usual applications (e.g. Biomagnetism) requiring dynamic ranges

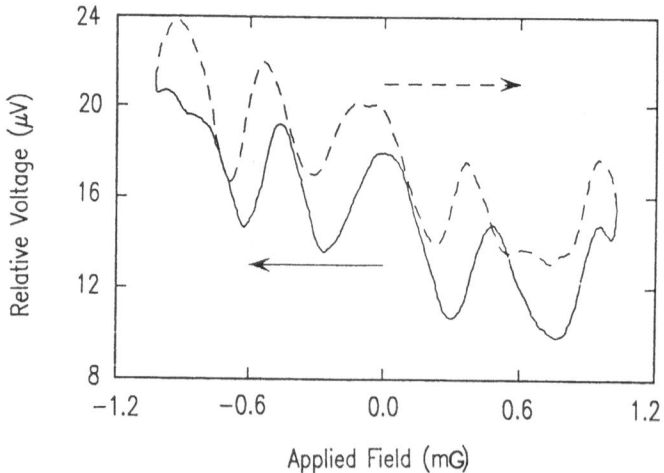

Fig. 4 Voltage versus flux characteristic of a 77 K $TlBaCaCuO$ dc-SQUID. The two characteristics slightly shifted are obtained by sweeping the magnetic flux back and forth.

orders of magnitude higher than the intrinsic one of the dc-SQUID. Figure 5 plots the SQUID output at 77 K versus time in the flux-locked loop over approximately 11 hours of measurement. The upper curve was taken immediately after cooling the SQUID from room temperature in a field of 1 mOe and the lower curve after 1 day of stabilization at 77 K. No drift or exponential decay was observed. Small steps in the output, which are not shown, corresponding to fractionally coupled flux jumps were observed with a rate of about 2 jumps per day. The noise seen in the figure reflects the 1/f noise of the SQUID. The energy sensitivity in locked and unlocked mode is shown in fig. 6. For comparison the typical noise performance of low temperature commercial SQUIDs are also shown.

The decrease in large fluctuations observed in the lower curve reflects about a factor three reduction of the low-frequency noise power below 100 Hz, as shown in Figure 6.

Figure 7 shows the noise performance versus temperature for several polycrystalline high-Tc SQUIDs. The flux noise spectral density is measured at 10 Hz. The noise performance of the $TlBaCaCuO$ devices is almost two orders of magnitude lower than the other SQUID fabricated with different polycrystalline materials. The difference is in the higher responsivity $\partial V/\partial \Phi$ of the thallium-based SQUIDs, more than $60\mu V/\Phi_o$ in some devices measured at 77 K. This high values of responsivity is probably related to better quality of the grain boundary weak links in thallium based devices. The reason for this is not understood at present. The voltage noise spectral densities are quite similar in all the devices fabricated with different materials. They have a 1/f like spectrum which not arise simply from critical current fluctuations, but from some unknown source. A recent theory in-

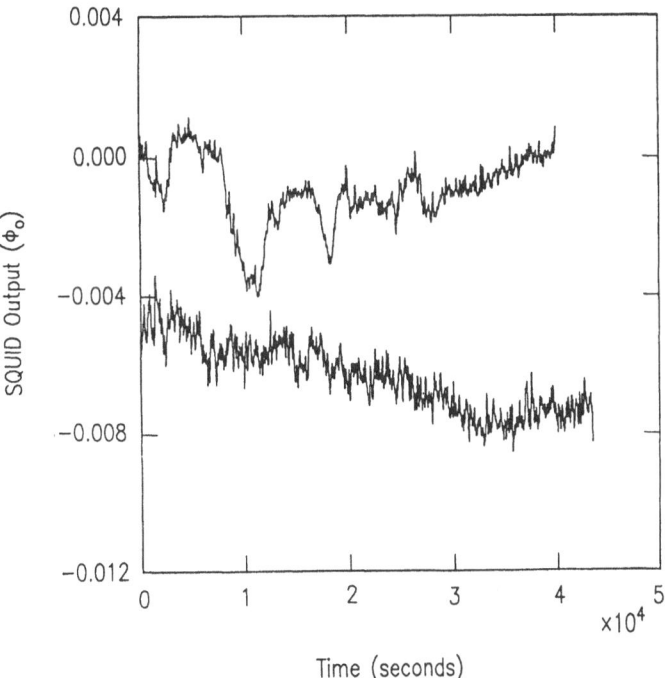

Time (seconds)

Fig. 5 SQUID output versus time in flux-locked-loop over approximately 11 hours of measurements at 77 K. The upper curve was taken immediately after cooling the SQUID. The lower curve was taken after 1 day of stabilization in liquid nitrogen.

troduced a new source of noise, due to multiple elastic scattering due to disordered mobile vortices [18]. This effect is related to the smallness of the coherence length in the high-Tc superconductors. The theoretical model predicts a 1/f magnetic flux noise that the experimental evidence seems to rule out, however if we assume the scattering localized in the grain boundary weak links, where there is a higher density of disordered vortices, the scattering of quasiparticles could mainly result in voltage fluctuations across the weak link, consistent with the experimental results.

Artificial weak links

While devices made of polycrystalline films are very interesting for demonstrations and basic studies, it will be extremely difficult to fabricate such devices reproducibly enough for incorporation in significantly more complex structures. In contrast, most modern low-Tc devices are made using highly reproducible and well characterized tunnel junctions of materials like Nb or NbN. These devices have uniform properties within a run, have predictable and reproducible properties from run to run, and their characteristics and circuit parameters are well understood [19]. On the contrary polycrystalline high-Tc devices suffer from a low yield within good runs (approximately 5 % work well at 77 K) and poor reproducibility from run to run. It is thus extremely important for real applications, to develop a truly reliable technology with good controls over the electrical parameters of the weak links. There is a world-wide effort to to fabricate such weak links both for basic

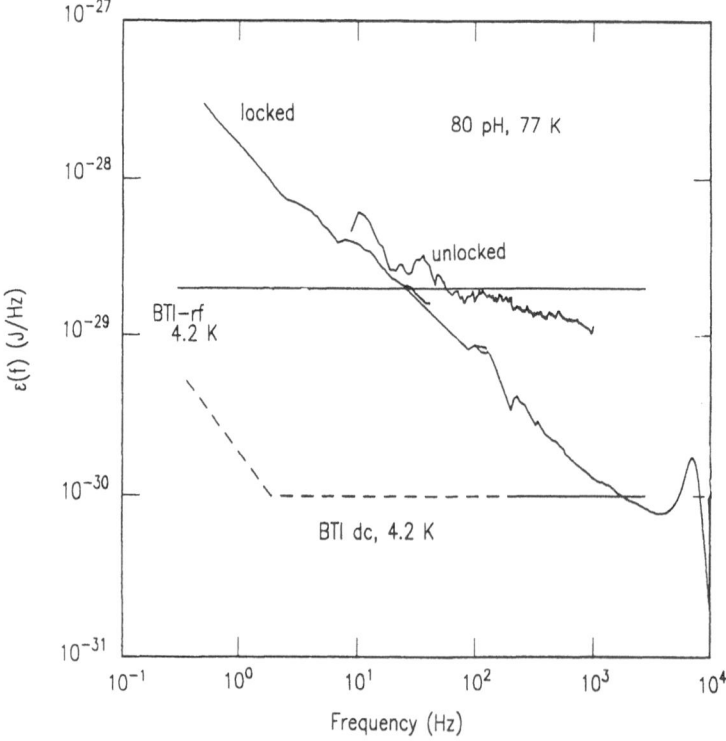

Fig. 6 Energy sensitivity in locked and unlocked modes. For comparison the typical noise performance of low temperature commenrcial SQUIDs are also shown.

understanding as well as for instrumentation and other possible applications. If the quality of the weak link is the major concern then a high value of the $R_N I_o$ product is required, this is a very difficult challenge for high-Tc superconductors. However for SQUID applications, high quality Josephson tunnel junctions are not a fundamental requirement. In low Tc SQUIDs, the McCumber parameter is usually deliberately lowered by putting normal metal shunts in parallel with the Josephson junctions.

Figure 1 shows a number of possible weak links types that might be used as the active elements in dc-SQUID devices. Among these, the grain boundary junctions (GBJ) made with bicrystal films are engineered structures which are closest to the random polycrystalline dc-SQUIDs which have worked well at 77 K. $YBaCuO$ SQUIDs made with this approach on $SrTiO_3$ bicrystal substrates have shown almost ideally periodic voltage-flux characteristics, low high frequency noise, good resistance to thermal cycles, and good control over the critical current [2]. While the high frequency (white) noise performance in these is close to the that of the Thallium-based polycrystalline devices, the bicrystal devices have about an order of magnitude more noise power at low frequencies. It is not clear how compatible this technology can be made with with multilevel coil structures.

The "edge junction" structure is particularly attractive for SQUID fabrication because it entails a multilevel technology that appears to be quite compatible with the requirements of insulating crossovers between superconducting layers. The name "edge junction" is based on the geometry of the device and does not necessarily imply that these devices operate via tunneling of paired electrons through a barrier. Figure 1e shows a schematic of the first experimentally demonstrated high-Tc edge junction structure [7]. Both the base and counter electrodes are grown epitaxially on the same substrate with their c-axis normal to the substrate. In this

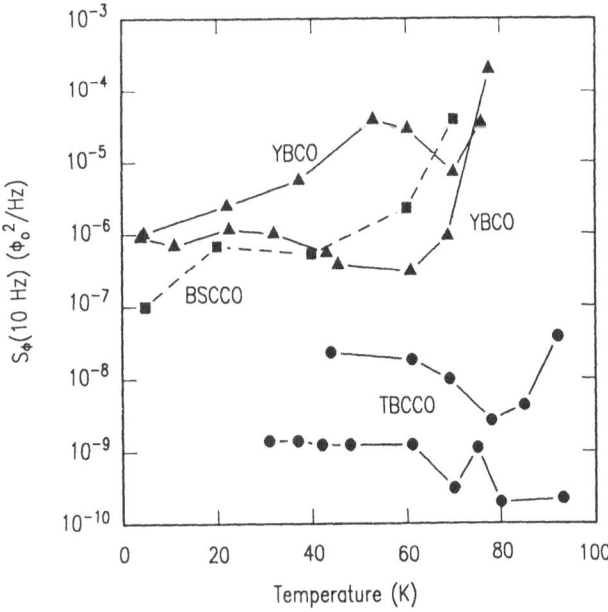

Fig. 7 Flux noise spectral density of several polycrystalline high-Tc SQUIDs.

way, transport from film to film is along the a-b plane direction which is both the longer coherence length direction and the higher current density direction. This direction for current flow is clearly more suitable than that in a sandwich junction, where the transport direction would be along the c-axis direction for typical electrode films which have c-axis normal texture. A scanning electron micrograph of a fabricated high-Tc edge junction SQUID is shown in Figure 8. The $YBaCuO$ base electrodes in these edge junctions was deposited by laser ablation and covered with a thick insulating film of evaporated BaF_2 or laser ablated $PrBa_2Cu_3O_{7-\delta}$. Afterwards, the $YBaCuO$-insulator bilayer was lithographically patterned and then ion milled to expose the base electrode edge. After a plasma oxifluoridation step, which may both regenerate the edge surface and leave a thin "barrier," the counter electrode was deposited, also by laser ablation. Device processing was completed by ion milling the counter electrode and then lifting off a contact level after the appropriate photolithography. The shape of the I-V curves in these edge junctions showed the sharp voltage onset that similar to that in the natural grain boundary devices where current flow was believed have been dominated by a single weak grain boundary. The observation that excessive current through the device can irreversibly change the shape of the curve supports the view that Josephson coupling is occurring through many metallic shorts, each several coherence lengths in size, connecting the base and counter electrode. Figure 9 shows a voltage versus flux characteristic for an edge junction SQUID working at 52 K. The hysteresis in these curves is not visible in this plot and was about 50 parts per million. The temperature working range for these device has been recently extended up to 70 K using a complete in situ process, with also great benefits in terms of yield and reproducibility of the electrical parameters [20]. The other weak link structures

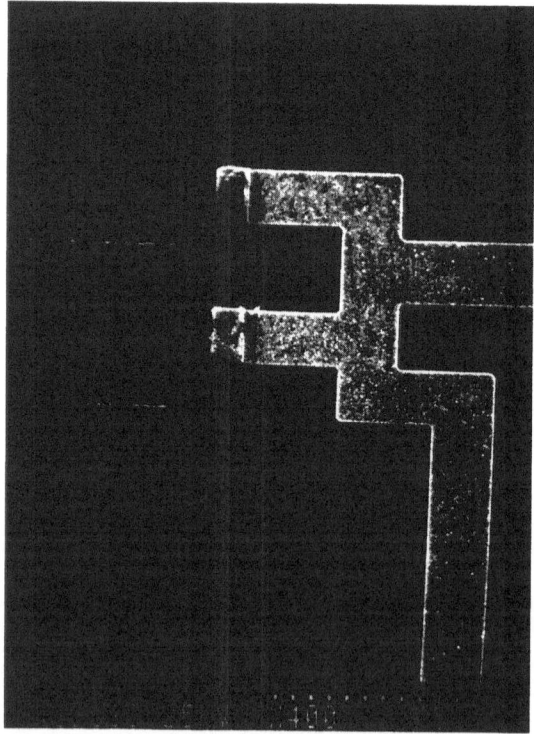

Fig. 8 SEM micrograph showing a dc-SQUID made with 'edge junction' weak links.

shown in Figure 1 have been proposed at many different laboratories and are being pursued. At this point all that can be said is that all are promising candidates to be used in SQUID devices.

Future trends

The bare dc-SQUID may be a particular good sensor of magnetic field if a special design is used. In this design the magnetic field is directly coupled to the SQUID inductance made of many loops in parallel, thus reducing the SQUID inductance without decreasing the pick-up sensing area. Often the experimental situation is not compatible with the direct coupling, in these cases, in order to improve the field sensitivity, a transformer coupled dc-SQUID is unavoidable. A dc-SQUID tightly coupled with an input coil is the major challenge in the near future. The simplest approach is to make a high Tc SQUID on one substrate and a high-Tc single level input coil on a second substrate and then sandwich the two substrates together. Such an experiment, performed at IBM, found a coupling coefficient $\alpha \simeq 0.7$ without any increase in flux noise associated with the input coil[11]. A single level input coil results in only a modest improvement in magnetic field sensitivity of about 5. However the highly effective coupling demonstrated by sandwiching implies that input coils and high-Tc SQUIDs can be developed and fabricated separately, waiting for a more reliable technology able to integrate everything on a single substrate. Several groups are now developing multiturn input coils working at 77 K [10,21]. The coils are constructed with a three layer crossover technology usually using $SrTiO_3$ as an insulator between the two superconducting layers. The goals of this active area of research are twofold: to construct a sensitive magnetometers from high-Tc SQUIDs and to achieve a substantial technological advancement by developing a sophisticated multilayer high-Tc circuit.

Many fields will certainly benefit from sensitive magnetometers working at 77K. Biomagnetic applications for high-Tc SQUIDs are of particular interest. Coupling an existing tallium based SQUID to a multiturn input coil matched with a pick-up coil of $1cm^2$ gives a field sensitivity of $100fT/\sqrt{Hz}$ at 10 Hz. This sensitivity is enough for cardiac magnetic studies, while neuromagnetic signals needs

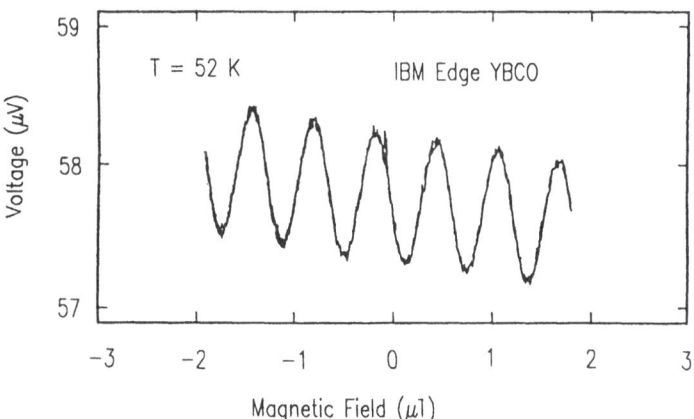

fig. 9 Voltage versus flux characteristic of a dc-SQUID with 'edge junction' weak links, the characteristic is measured at 52 K.

an order of magnitude lower sensitivity [22]. Furthermore the dewar system would be greatly simplified, enabling the closer proximity of the detector to the magnetic source and improving the signal to noise ratio. SQUIDs working at 77 K could also be used together with CMOS readout circuits working at the same temperature and in close proximity to the SQUIDs. This will result in a terrific simplification of the systems, considering that the dominant direction of Biomagnetism is the emergence of 100 channel instrumentation. A high-Tc biomagnetic system will certainly be more flexible, economic, and easy to use than low-Tc systems, and this will happen in a reasonably short time of 5 or 10 years, the time required for technology to provide high yield and good controllability of the devices.

References

[1] R.H. Koch, W.J. Gallagher, B. Bumble and W. Lee, Appl. Phys. Lett. 54, 2259 (1989).

[2] R.Gross, P. Chaudhari, M. Kawasaki, M.B. Ketchen and A. Gupta, Appl. Phys. Lett. 57, 727 (1990).

[3] D.W. Face, J.M. Graybeal, T.P. Orlando and D.A. Rudman, Appl. Phys. Lett, 56, 1493 (1990).

[4] R.H. Hausser, M.Diegel and R. Rogalla, Appl. Phys. Lett 52, 844 (1988).

[5] see for example: K.K. Likharev, *Dynamics of Josephson junctions and circuits*, Gordon and Breach Publisher, New York.

[6] S.E. Russek, D.K. Lathrop, B.H. Moeckly, R.A. Buhrman, D.H. Shin and J. Silcox, Appl. Phys. Lett. 57, 1155 (1990).

[7] R.B. Laibowitz, R.H. Koch, A. Gupta, G. Koren, W.J. Gallagher, V. Foglietti, B. Oh and J.M. Viggiano, Appl. Phys. Lett. 56, 686 (1990).

[8] R.W. Simon, J. B. Bulman, J.F. Burch, S.B. Coons, K.P. Daly, W.J. Dozier, R. Hu, A.E. Lee, J.A. Luine, C.E. Platt, S.M. Schwarzbek, M.S. Wire, and M.J. Zani, Applied Superconductivity Conference Paper ES-2, Snowmass Colorado, September, 1990.

[9] J.J. Kingston, F.C. Wellstood, P. Lerch, A.H. Miklin and J. Clarke, Appl.Phys.Lett. 56, 189 (1990).

[10] F.C. Wellstood, J.J. Kingston and J.Clarke, Appl. Phys. Lett. 56, 2336 (1990).

[11] B. Oh, R.H. Koch, W.J. Gallagher, V.Foglietti, A. Gupta, G. Koren, W.Y. Lee, Appl. Phys. Lett. 56, 2575 (1990).

[12] J. Clarke in *Superconducting Electronics*, edited by H. Weinstock and M. Nisenoff, Springer New York 1989, p.87.

[13] R.H. Koch, C.P. Umbach, G.J. Clark, P. Chaudhari and R.B. Laibowitz, Appl. Phys. Lett. 51, 200 (1987).

[14] W.J. Gallagher, R.H. Koch, V. Foglietti, B. Oh and W.Y. Lee, in 1989 *International Superconductivity Electronic Conference* , *Extended Abstracts* (Japan Society od Applied Physics, Tokyo, Japan, 1989), p.477.

[15] W.Y. Lee, V.Y. Lee, J. Salem, T.C. Huang, R. Savoy, D.C. Bullock, and S.S.P. Parkin, Appl. Phys. Lett. 52, 329 (1988).

[16] B.L. Walton, B. Roseblum and F. Bridges, Phys. Rev. Lett. 32, 1047 (1974).

[17] V. Foglietti, R.H. Koch, W.J. Gallagher, B. Bumble, B. Oh and W.Lee, Appl. Phys. Lett. 54, 2259 (1989).

[18] L. Wang, Y. Zhu, H.L. Zhao and S. Feng, Phys. Rev. Lett. 64, 3094 (1990).

[19] R.L. Sandstrom, A.W. Kleinsasser, W.J. Gallagher and S.I. Raider, IEEE Trans.Magn. MAG-23, 1984 (1987).

[20] G. Koren, E. Aharoni, E. Polturak, and D. Cohen, (preprint).

[21] B. Oh, R.H. Koch, W.J. Gallagher, V. Foglietti, R.F. Robertazzi, A. Gupta, and W.Y. Lee (unpublished).

[22] G.L. Romani, S.J. Williamson, L. Kaufman, "Biomagnetic Instrumentation", Rev.Sci.Instr. 53, 1815 (1982).

CHARACTERIZATION OF RF-SQUIDS

FABRICATED FROM EPITAXIAL YBa$_2$Cu$_3$O$_7$ FILMS

G.J. Cui, K. Herrmann, Y. Zhang, C.L. Jia,
Ch. Buchal, J. Schubert, W. Zander,
A.I. Braginski and Ch. Heiden

Institute for Thin Film- and Ion Technology
Forschungszentrum Jülich (KFA), D-5170 Jülich, FRG

INTRODUCTION

The purpose of our work was to evaluate rf-SQUIDs with weak-link microbridges engineered in epitaxial high-temperature-superconductor (HTS) YBa$_2$Cu$_3$O$_7$ (YBCO) films. The rf-SQUID was also a convenient tool to evaluate the feasibility of alternative approaches to weak link fabrication. In the absence of HTS Josephson tunnel junctions, weak links having reproducible critical current (I_c) and resistance (R_n) values set by design could be used in commercial HTS SQUIDs and other integrated circuits.

We chose the rf- over dc-SQUIDs due to the simplicity of rf-SQUID fabrication and contactless testing. Also, in spite of the lower energy resolution of HTS dc-SQUIDs (see e.g. the review by Clarke[1]), the field sensitivity of rf-SQUID magnetometers could be sufficient for many applications. The necessary requirement, however, is to reduce the low frequency 1/f flux noise significantly below the level observed in polycrystalline YBCO containing many grain boundaries which all contribute to the noise. Nearly single-crystalline, epitaxial films exhibit relatively low 1/f noise.[1] Further noise reduction can be expected to follow improvements in the crystalline perfection of the film.

Until recently, thin film YBCO one-hole rf-SQUIDs were fabricated from post-annealed, polycrystalline or textured films.[2,3] These SQUIDs utilized the natural grain-boundary junctions present in wide microbridges and exhibited temperatures of operation, T_{op}, up to 55 or 65K, significantly lower than the film critical temperature, $T_c > 85K$. The low-frequency 1/f noise was high, $S_v = 10^{-3}$ to 10^{-2} $\Phi_o Hz^{-1/2}$ at 10 Hz. The authors attributed it to random fluxoid motion at grain boundaries. Similar results were obtained in this laboratory with textured YBCO film rf-SQUID microbridges which contained grain boundaries.[4]

Very recently, operating rf-SQUIDs with weak links fabricated by several different means in epitaxial, *in-situ*

Nonlinear Superconductive Electronics and Josephson Devices
Edited by G. Costabile *et al.*, Plenum Press, New York, 1991

109

grown YBCO were demonstrated by the TRW group.[5] A narrow disordered stripe across a patterned microbridge was created by local in-diffusion of Al or by a focussed ion-beam irradiation. Functional weak links were also obtained in microbridges patterned over steps milled in single-crystal substrates. The exact nature of all these microbridges was not elucidated. In each case, however, the film's critical current density (J_c) was locally reduced by two or three orders of magnitude and, consequently, rf-SQUIDs operating between 4.2 and 60K could be obtained.

In our work, we have been investigating two types of weak-links: (1) step-edge and (2) locally ion-implanted microbridges. In the first case, we hoped to find an alternative for inducing (seeding) grain-boundary Josephson junctions which would be more viable technologically than the use of bi-crystal substrates by Gross et al..[6] They fabricated dc-SQUIDs from epitaxial YBCO, with grain-boundary junctions present only at the bi-crystal boundary, and demonstrated highest T_{op} = 87K and lowest 1/f noise attained to date in any YBCO SQUID. In the second case, our approach was to use local ion implantation of Ar^+ as a patterning tool for flux-flow microbridges.[7] The characteristic length scale in such microbridges is the perpendicular penetration depth, $\lambda_\perp$, rather than the coherence length. In both cases, we intended to produce engineered, low-noise weak links operating at 77K.

SQUIDS WITH STEP-EDGE JUNCTIONS

Fabrication

Steps were fabricated in (100) $SrTiO_3$ substrates 10x10 mm^2 by Ar^+ ion milling through a metal mask.[8] Parallel steps extended over the whole substrate width. A layer of Nb, 300 to 400 nm thick, was deposited on the substrate and patterned photolithographically by either lift-off or reactive ion etching (RIE) with SF_6. The latter approach tended to produce better defined and cleaner mask edges, as determined by scanning electron microscopy (SEM). Several step heights, between h = 20 and 200 nm, were fabricated. After milling, the Nb mask was removed by RIE and the substrate annealed in flowing O_2 at 1000°C for 1 hour, to reduce the surface damage caused by ion milling.

Epitaxial, c-axis oriented YBCO films were *in-situ* deposited on (100) $SrTiO_3$ by pulsed laser deposition (PLD). Film thicknesses were d = 60 to 240 nm, with most results obtained using 60 - 70 nm thick films. Typical films had a T_c = 87 to 89K and critical current density at 77K in the 10^6 A/cm^2 range.

Sixteen one-hole SQUID structures were patterned by optical lithography and Ar^+ ion-beam etching through an AZ 1512 resist mask on each 10x10 mm^2 substrate. The mask was aligned to position all microbridges over the substrate steps as shown in Fig. 1 for a single SQUID. Each rf-SQUID structure consisted of a 1.4 x 1.4 mm^2 flux concentrator with a central hole of 100x100 μm^2 corresponding to an inductance of approximately L = 150 pH. The step-edge microbridge was 10 μm long and nominally 2 μm wide. A 0.65 mm wide step was also present in each flux concentrator.

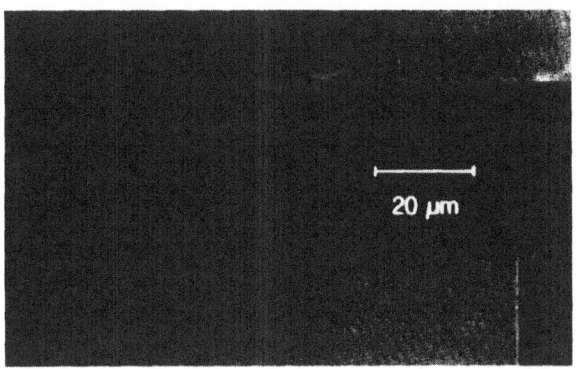

Fig. 1. Optical photograph of a 2 μm
wide YBCO microbridge positioned over
a step in the (100) SrTiO₃ substrate.

To obtain dc current-voltage (I-V) characteristics of the
SQUID microbridge, after determining its rf-characteristics,
the flux concentrator was photolithographically cut and gold
contacts deposited by lift-off.

Measurements

The measured SQUID was loosely coupled to a 40-turn,
movable 1 μH tank-circuit coil wound of 0.06 mm copper wire
and positioned over the SQUID hole. The tank circuit frequency was 13 MHz. Superimposed low frequency current provided
the external flux modulation. The V-Φ characteristics (at
several V-I_{rf} steps) were determined, in the unlocked mode,
over a range of temperatures. The temperature in a He-flow
cryostate was measured before and after the rf-measurements
using platinum (T > 30K) and germanium (T < 30K) thermometers
positioned adjacent to the sample. During the actual SQUID
measurement, the thermometers were disconnected from the
external voltmeter to avoid spurious signals. The SQUID noise
was measured between 1 and 10^5 Hz using a HP 3562A dynamic
signal analyzer. The dewars used for measurements had a
Mumetal or Mumetal and Pb shielding. However, noise spectra
exhibited strong peaks at 50 Hz and its harmonics, indicating
that shielding was still inadequate.

Results

In rf-SQUIDs fabricated from 60 - 70 nm thick films, the
triangular and periodic V-Φ SQUID pattern could be measured at
all substrate step heights, from 200 to 20 nm, corresponding
to a range of d/h between 1/3 and 3, approximately. In the
absence of the step-edge, no such SQUID performance was
observed. At d/h = 2 - 3, the V-Φ pattern appeared only at
high temperatures, above 60 - 70K and persisted to near 80K,
while at d/h = 1/3 the T_{op} range was typically 4.2 to 30K,
with a considerable scatter. In addition to this low temperature range, a much stronger but nonperiodic SQUID response
was also observed at d/h < 1 when the temperature increased
above 65 - 70K and approached T_c. The effect of step height,
at a nearly constant film thickness, is consistent with an

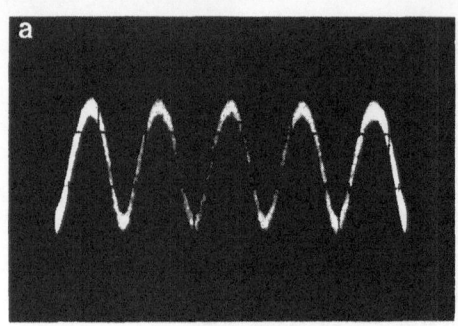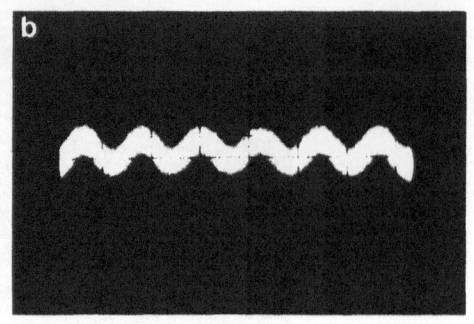

Fig. 2.   Traces of V vs Φ at 50K (left) and 77K (right) in a SQUID with d/h = 1/3, h = 200 nm.

$I_c(T)$ much higher at d/h >> 1 than at d/h << 1 at a reference temperature of, say, 4.2K and a typical temperature dependence of microbridge $I_c = I_o(1-T/T_c)^n$, $n \geqslant 1$.   The nonperiodic response could be tentatively associated with the random array of junctions in the 0.65 mm wide step in the SQUID loop located opposite to the microbridge.

Figure 2 shows, as an example, V-Φ traces obtained on several steps of the V-I_{rf} characteristic at 50 and 77K.   The value of the transfer function $\delta V/\delta\Phi$ was 5 $\mu V/\Phi_o$ at 50K and 1.6 $\mu V/\Phi_o$ at 77K.   The broadest temperature range of SQUID response, 4.2 to 82K, was obtained when d/h = 1 and h = 120 nm.   The best fabrication yield and reproducibility of T_{op} values in the 16 rf-SQUIDs on the chip was also obtained at d/h = 1, h = 120 nm.   In the best case, both the yield and reproducibility approached 100% while values of $\geqslant$50% were typical.   In contrast, at d/h = 1/3 to 2/3 the reproducibility was poor.   The "optimum" values of d/h and h, however, cannot be considered as firm guidelines.   Although a large number of SQUIDs (>100) was characterized, we do not yet have enough systematic data to claim optimization.

Direct measurements of step-edge junction I-V characteristics in a SQUID has been performed after cutting of the loop and deposition of metal contacts which involved several photolithographic process steps.   Until present, this resulted invariably in the destruction of the microbridge superconducting properties.   The measured low-temperature resistances were of the order of 10 ohm and may have been altered (increased) after the additional patterning steps.   Consequently, we do not yet have any direct data on the values of $I_c(T)$ and $R_n(T)$ and their dependence upon d/h and h.   The I_c could only be estimated indirectly from $\beta_L = 2\pi I_c L/\Phi_o = 1$ at a temperature where the SQUID operation changed from the dissipative to dispersive mode.   In SQUIDs with broad T_{op} ranges, this gave I_c of 2 μA at temperatures between 40 and 75K.   With a conservative assumption of linear $I_c(T)$, i.e. n = 1, one obtains a low-temperature I_{co} of the order of 4-10 μA, $V_c = I_{co}R_n <$ 0.1 mV and a reduction of the material J_c by a factor of 10^3.

Low frequency noise measurements were performed on a few step-edge SQUIDs over a range of temperatures between 45 and 77K.   In all measured SQUIDs, the white noise voltage level at 77K was $S_{V}^{1/2} = 4-5 \times 10^{-4}$ $\Phi_o Hz^{-1/2}$, with the instrumentation limit at $\leqslant 2 \times 10^{-4}$ $\Phi_o Hz^{-1/2}$.   The crossover to frequency-dependent noise was sample-dependent, typically between 100 and 500 Hz, with 30 Hz being the lowest recorded crossover frequency near 77K. Over the temperature range (45 to 77K), the $1/f$ noise data points at 1 Hz were scattered between 0.3 and 1×10^{-2} $\Phi_0 Hz^{-1/2}$, due, at least in part, to insufficient signal averaging, and appeared temperature-independent. Above 10 Hz, the frequency dependence was weak and could be approximated by $1/f^\alpha$ where $\alpha \leqslant 1/2$.

Microstructure

High-resolution transmission electron microscopy (HREM) of YBCO/SrTiO$_3$ step sections with $d/h = 1/3$ permitted us to obtain some insight into the nature of the step-edge junction. Photographs were obtained of steps with angles of $\theta = 65°-70°$, $44°-45°$ and $30°-35°$ where θ is the angle with the film/substrate plane. As the available pictures are not yet of print quality, we defer their publication and present, instead, a schematic drawing (Fig. 3) illustrating the observed positioning of Cu-O planes at the step. Remarkably, at all observed angles, the film thickness across the step was not reduced significantly.

On a steep step, $\theta = 70°$ (Fig. 3a), a c-axis film grew on the SrTiO$_3$ step surface epitaxially, with atomic planes parallel to those in the cubic substrate and perpendicular to Cu-O planes in the planar film. Two 90° grain boundaries were induced across the common (103) plane near the edges, each at

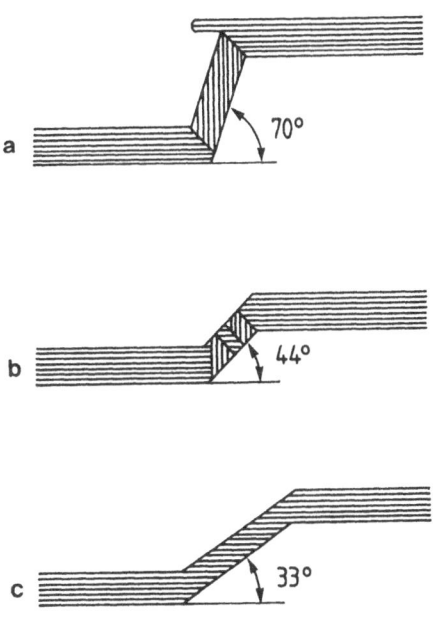

Fig. 3. Schematic representation of Cu-O planes and grain boundaries observed by HREM in step sections with different step angles θ: (a) $\theta = 70°$, (b) $\theta = 44°$, (c) $\theta = 33°$.

45° to Cu-O planes. A planar "overhang" over the steep "cliff" was also seen.   When $\theta \cong 45°$ (Fig. 3b), several such grain boundaries were induced, as both growth directions were equally favorable. A 90° (010) grain boundary parallel to one set of Cu-O planes and orthogonal to the other has also been observed. At the low θ angle value (Fig. 3c), the film grew epitaxially with Cu-O planes parallel to the substrate surface and no grain boundaries.

We associate the SQUID operation with the presence of induced grain boundaries and believe that our functioning microbridges contained two or more grain boundary Josephson junctions connected in series. At sufficiently steep angles, $\theta > 45°$, the grain boundary formation and the resulting junction properties could have been relatively reproducible since the growth mode on a steep step surface should not depend upon the exact angle value. However, when $\theta = 45°$, a variability of grain boundary configurations and resulting junction properties would be unavoidable. Obviously, at too shallow step angles no junctions would be formed. We noted that ion milling through a resist mask did not produce sharp steps and, indeed, no working step-edge junctions resulted from this technique. The HREM data showed clearly that sharp, steep edges will be necessary for junction reproducibility. Until we obtain additional microstructural information for a range of d/h and h values, we can only speculate that, due to somewhat different growth rates at different angles of atomic flux incidence, the arrangement of grain boundaries at the step may be altering with the increasing film thickness (d/h) and that in sufficiently thick films (large d/h) the planar growth direction will eventually dominate, thus shunting the grain-boundary junctions present near the film/substrate interface.

SQUIDS WITH ION-IMPLANTED MICROBRIDGES

Ion Implantation of YBCO

The effects of ion implantation into YBCO films are not yet well understood. It was amply shown, however, that the generated lattice damage and the introduction of dopant ions result in a reduction of T_c and J_c.[9] To form weak links, one can either use MeV ion irradiation[9] to strongly reduce the microbridge J_c with only a mild reduction in T_c or to use implantation as a patterning tool. In this case, one strives to suppress T_c in the implanted volume of the film[10] without affecting any superconducting properties of the unimplanted part.

We attempted to form a weak link constriction by thinning the microbridge's superconducting crossection, i.e. suppressing superconductivity in its upper part, presumably reducing the T_c and J_c at intermediary depths but leaving the deepest part of the film undamaged through the selection of ion energy and the resulting implantation depth.[11] We did not know the depth dependence of T_c in our undamaged films but assumed that it was reduced within some distance from the interface with the substrate.

Fabrication and Measurements

The one-hole rf-SQUID patterns described above were

transferred into epitaxial, c-axis YBCO films dc magnetron sputtered on (100) $SrTiO_3$. Chemical etching in diluted H_3PO_4 was used for patterning. The film thickness was 200 nm.

Narrow stripes were patterned and locally ion-implanted by Ar^+ ions. Open slits were patterned across each micro-bridge by direct writing with 20 kV e-beam (Philips SEM 525M) in a PMMA 950K resist mask, 300 nm thick. This thickness was sufficient for protecting the underlying YBCO.[11] The slits were written to be 100-200 nm wide but upon the resist development the measured stripe widths varied between 200 and 500 nm, due to difficulties in focussing and electron scattering. However, the stripe width was still comparable with $\lambda_\perp$. The 200 nm thick films were implanted with a constant dose of 100 keV of Ar ions: 5×10^{12} cm^{-2}.[11] The total implanted depth range was of the order of 100 nm.[11]

SQUID measurements were performed as described above for step-edge devices.

Results and Discussion

Of the 24 ion implanted SQUIDs, only four exhibited a periodic rf-SQUID response over a range of temperatures. Of these, the best sample was characterized in detail. The V-Φ trace was well defined, periodic and nearly triangular on up to 10 steps of the I_{rf}-V curve and over a broad temperature range between 4.2 and 60K. With increasing temperature, the mode of operation was changing from dissipative to dispersive, as defined by $I_c(T)$. As in step-edge SQUIDs, the low-temperature I_c, estimated from β_L, was of the order of 10 μA. Un-implanted structures did not exhibit any SQUID response.

Figure 4 shows an example of the I_{rf}-V and corresponding V-Φ traces at 4.2K. A value of $\beta_L = 5$ was extracted from these data.[12] The output signal decreased with increasing

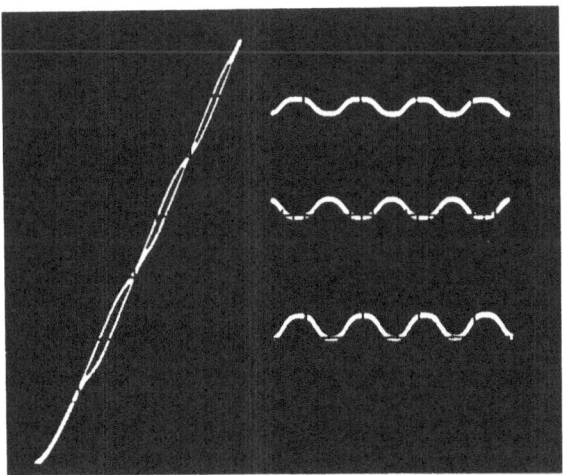

Fig. 4. The I_{rf}-V curve and V-Φ pattern (first three steps) measured in an rf-SQUID with ion-implanted weak link at T = 4.2K.

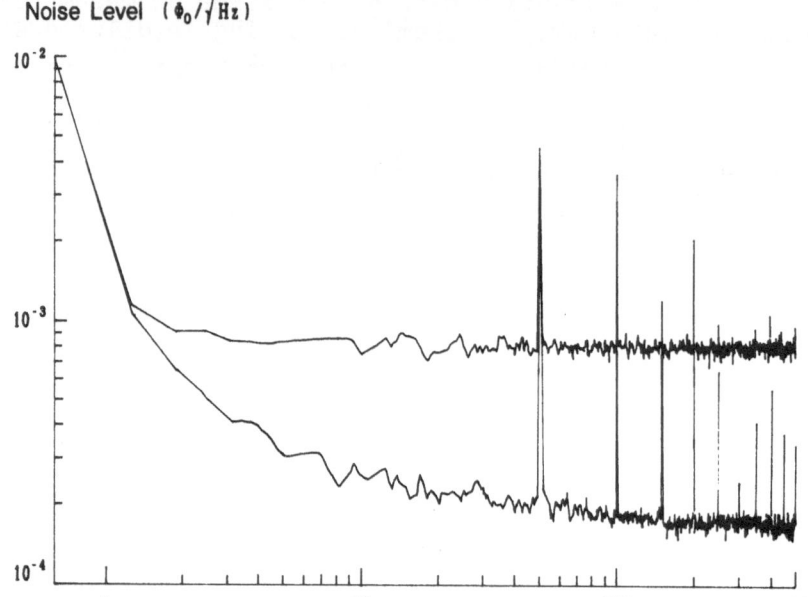

Noise Level ($\Phi_0/\sqrt{Hz}$)

Frequency (Hz)

Fig. 5. Low frequency noise spectra in an an
rf-SQUID with ion-implanted weak link measured
at two temperatures: (a) 4.2K and (b) 34K.

temperature so that noise measurements could only be performed
between 4.2 and 45K. The intrinsic SQUID white noise at 4.2K,
estimated from the shape of the I_{rf}-V curve[13] was S_v = 9x10-5
$\Phi_O Hz^{-1/2}$, a factor of 5 lower than the measured value and
below the instrumentation limit. Figure 5 shows two frequency
spectra of noise measured at 4.2 and 34K. It is seen that the
1/f noise was nearly temperature independent, as in the case
of SQUIDs with step-edge junctions. A comparison between the
1/f noise results for the two SQUID types was only possible in
the vicinity of 40K since the measuring temperature ranges
were different. The comparison showed clearly that the 1/f-
type noise level was several times higher in SQUIDs with step-
edge microbridges.[14] This higher level could be tentatively
attributed to grain boundaries in the 0.65 mm wide step in the
flux concentrator. A planned trivial change in the geometrical design of step-edge patterns will eliminate the broad
edge and clarify this point.

The ion implantation results presented above are consistent with those obtained at TRW by focussed ion beam irradiation.[5] The weak links obtained were of the SS'S flux-flow
type. This follows from the inferred microbridge I_c value
which was up to 10^3 lower than that in the unimplanted film
while the implantation range could reduce the effective
superconducting cross-section by only a factor of 2 or so.
Clearly, the material's J_c was significantly reduced in the
whole volume beneath the implanted stripe with, probably, an
attending reduction in T_c.

116

CONCLUSIONS

The step-edge approach permitted us to fabricate rf-SQUIDs with 90° grain-boundary Josephson junctions induced in the YBCO epitaxial microbridge near the two step edges. Such double junctions, capable to operate at and slightly above 77K, can emulate the single grain-boundary junctions on bicrystal substrates and represent a technologically promising alternative. The fabrication yield and reproducibility of operating temperatures can be surprisingly high. Local ion implantation in epitaxial YBCO produced rf-SQUIDs with SS'S flux-flow weak links. The highest operating temperature was only 60K and the fabrication yield discouragingly low. Both types of epitaxial rf-SQUIDs had 1/f noise lower by an order of magnitude than that reported earlier for rf-SQUIDs fabricated from polycrystalline YBCO.

ACKNOWLEDGEMENT

We thank Dr. H.-M. Mück (KFA) for help with the e-beam lithography and Dr. S.S. Tinchev (FUBA) for performing the preliminary thin film rf-SQUID experiments at KFA.

REFERENCES

1. J. Clarke, Principles and Applications of SQUIDs, Cryogenics, 77:1208 (1989).
2. K. Betts et al, High T_c Thin Film and Device Development, Trans. IEEE on Magnetics 25: 965 (1989).
3. K.P. Daly et al., Characterization of a High Temperature Superconducting Oxide Thin-Film rf-SQUID, Trans IEEE on Magnetics 25: 1305 (1989).
4. S.S. Tinchev, unpublished results, 1990.
5. R.W. Simon, Progress Toward a YBCO Circuit Process, Proc. 2nd Conf. Sci & Tech. of Thin Film Superconductors, Denver, April 30 - May 4, 1990 (in print).
6. R. Gross et al., Low Noise $YBa_2Cu_3O_{7-\delta}$ Grain Boundary dc SQUIDs, Appl. Phys. Lett. 57: 727 (1990).
7. K.K. Likharev, Superconducting Weak Links, Rev. Mod. Phys. 51: 101 (1979).
8. The use of metal masks was suggested to us by R.W. Simon, TRW.
9. A.E. White et al., Controlled Reduction of Critical Currents in YBa2Cu3O7 Thin Films, Appl. Phys. Lett. 53: 1010 (1988).
10. G.J. Clark et al., Effects of Radiation Damage in Ion-Implanted Thin Films of Metal-Oxide Superconductors, Appl. Phys. Lett. 51: 139 (1987).
11. J.P. Biersack, in "Ion Beam Modification of Insulators", P. Mazzoldi and G. Arnold, eds., Elsevier, Amsterdam, pp. 1-56 and p. 11.
12. Y. Zhang, Ph.D. Dissertation, Univ. of Giessen, 1990.
13. L.D. Jackel and R.A. Buhrman, Noise in the rf SQUID, J. Low Temp. Phys. 19: 201 (1975).
14. G.-J. Cui et al., Properties of rf-SQUIDs Fabricated from Epitaxial $YBa_2Cu_3O_7$ Films, Proc. LT-19 HTS Satellite Conf., Cambridge 1990 (in print).

SUPERCONDUCTIVE ANALOG ELECTRONICS FOR SIGNAL PROCESSING

APPLICATIONS

Jonathan B. Green

Lincoln Laboratory
Massachusetts Institute of Technology
Lexington, Massachusetts 02173

INTRODUCTION

Superconductivity has many attributes that make it useful in the development of devices and components for signal processing applications. This chapter discusses these attributes with several examples presented to illustrate the use of superconductive components for wideband signal processing. The chapter is organized into four sections. The first section provides a brief overview of generic signal processing as accomplished by linear filters. The following three sections illustrate the use of superconductivity to construct three different types of such filters. All structures described have been developed with a view toward analog signal processing of signals with multigigahertz bandwidths.

ANALOG SIGNAL PROCESSING

Even in this era of digital computing, modern-day radar and communications systems still rely on analog signals to transmit information. In fact, electromagnetic wave propagation is inherently an analog process. For this reason, a need still exists for processing such waveforms. Of course, the initial stage of signal processing might involve analog-to-digital conversion, but this would then require all subsequent information processing to occur in the digital domain.

Substantial enhancements to system performance can often be achieved by performing initial waveform processing in the analog domain, prior to eventual analog-to-digital conversion for input to a digital signal processor. Indeed, in certain applications a completely analog set of waveform processing algorithms can suffice, without resorting to digital techniques.

The heart of waveform processing in linear systems is a filter with a defined impulse response. Such a filter can be constructed based on the transversal filter architecture shown in Figure 1. The essence of the structure is a tapped delay line, with the option of weighting the n individual taps prior to subsequent summation of the time-delayed replicas of the original input signals. The summation of the weighted, time-delayed replicas of the original input signal can occur over the tapped outputs themselves. The mathematics of such filters

Nonlinear Superconductive Electronics and Josephson Devices
Edited by G. Costabile *et al.*, Plenum Press, New York, 1991

119

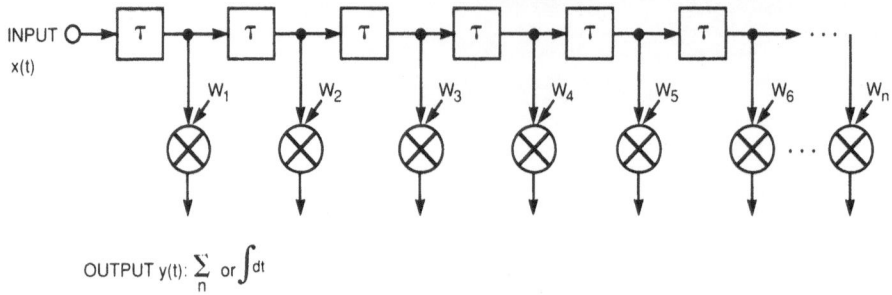

Fig. 1. Architecture of a tapped delay line transversal filter.

can easily be seen to describe the convolution of the filter's tap weights W_n with the original input signal $x(t)$, which is given by

$$y(t) = \sum_n W_n x(t - n\tau) \, , \qquad (1)$$

where $y(t)$ is the filter output and τ is the delay increment between successive taps. Alternatively, the summation of weighted tap outputs can occur as an integration of these outputs individually, in time. This case is discussed later in the section on the time-integrating correlator.

In either of the above cases, the requirements of a transversal filter include the need for a delay line that can adequately pass the input signal (waveform to be processed) without undue amplitude or phase distortion. Additionally, a means of sampling the signal without distortion of the delay line response must be available. To perform the weighting of the tap outputs, it is desirable to have elements that can vary at least the amplitude of the tapped output, if not also the phase. Finally, a means for summation of the resulting weighted, time-delayed replicas of the input signal, either in time or across the taps, must exist.

The technology of superconductivity is able to provide a means for accomplishing all of the above functions. For delay of the input waveforms without signal distortion, superconductive delay lines can be made that are nearly lossless and dispersionless.[1] Similarly, by using techniques that are well known to designers of conventional microwave integrated circuits, wideband proximity couplers can be constructed that serve to transfer a reduced-amplitude portion of the input waveform to the weighting elements.

In order that Eq. (1) remain valid in the structure of choice, it is necessary that significant energy not be lost at any of the taps. In fact, the relationship between the insertion loss IL of the delay line (defined as the output power relative to the input power given in dB), and the strength C (also in dB) of the couplers that will sample the input signal at the tapping points, is given by

$$IL = 10 \times m \times \log\left(1 - 10^{-C/10}\right) \, , \qquad (2)$$

where m is the number of couplers (taps) along the length of the delay line. It is important to note that the relationship in Eq. (2) holds only if each tap feeds a matched load so that no energy can couple back into the delay line. The relationship between C and m described by

Eq. (2) is plotted in Figure 2 for several values of tolerable delay line insertion loss. Clearly, the device designer must sacrifice signal strength at the input to the mixing elements when designing a device with a large number of taps. As an example, for a device with 128 taps the couplers must be no greater than -28 dB in strength for the delay line insertion loss to be kept below -1 dB.

DISPERSIVE DELAY LINE

By utilizing the low-loss properties of superconducting stripline, it is possible to fabricate a device capable of performing the convolution of a wideband input signal with a fixed (nonprogrammable) filter response. As shown in Figure 3, the device architecture in this case consists of a cascaded array of backward-wave couplers. These couplers act as a means for tapping the input signal. The coupler's output signal is a propagating wave in the opposite direction on the neighboring line. [It should be noted that in this structure energy can be returned to the input line, so Eq. (2) does not hold.] The frequency response of the couplers is given by Oliver[2] as

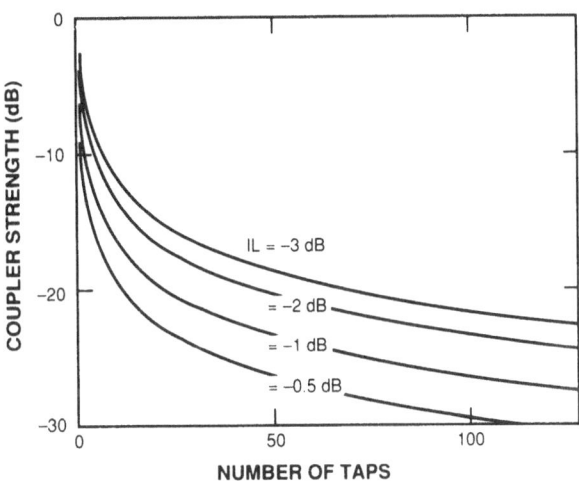

Fig. 2. Allowed coupler strength (in dB) for a tapped delay line transversal filter as a function of the number of taps.

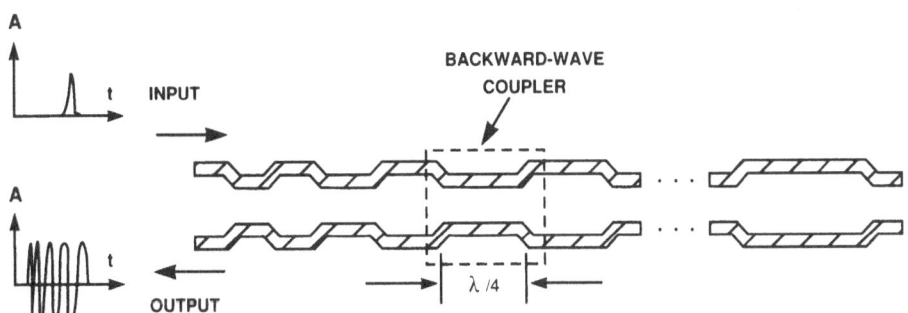

Fig. 3. Arrangement of cascaded backward-wave couplers used in a dispersive delay line structure.

$$|H(\omega)| = \frac{\kappa}{\sqrt{1-\kappa^2}} \sin\left(\omega \frac{\sqrt{\varepsilon_r}}{c} l_c\right), \tag{3}$$

where ε_r is the dielectric constant of the propagation medium, c is the velocity of light in free space, and l_c is the coupler length. The strength of the couplers is related to the constant κ and is determined by the proximity of the neighboring transmission lines. This can be tailored along the device length, as will be discussed later. Additionally, we see from Eq. (3) that the coupler response is maximized when its length is an odd number of quarter wavelengths, which is expressed as

$$l_c = (2n - 1)\lambda/4 \tag{4}$$

where λ is the electromagnetic wavelength in the propagation medium.

Therefore, by appropriately defining the pattern of the neighboring transmission lines, one can design a device that has the impulse response of a weighted chirp: a linearly frequency modulated waveform of the form

$$h(t) = W(t)e^{-j2\pi\left[f_0 t + (B/2T)t^2\right]} \quad \text{for } -T/2 \le t \le +T/2 . \tag{5}$$

Here, f_0 is the center frequency of the waveform, B is the waveform bandwidth, and T is the chirp duration. The weighting function W(t) is determined by the variation of coupler strength along the device length, while the argument of the exponential in Eq. (5) is determined by the length and spacing of the backward-wave couplers.

The dispersive delay lines are referred to as "passive," since they have no "active" elements such as transistors or Josephson junctions. The filter response of these devices is fixed and determined at the time of device fabrication by the photolithographic patterning of the coupled transmission lines. The devices can be made with a high degree of accuracy, as can be seen from the typical frequency response characteristics shown in Figure 4. The measured response of a so-called flat-weighted device[3] is shown in Figure 4(a), and the response of a device whose individual coupler strengths have been appropriately tailored to give a Hamming-weighted function[4] is shown in Figure 4(b). Such weighting functions are well known to the radar signal processing community as they are useful in instances where sidelobe reduction is important in order to separate nearby targets of disparate size.

Dispersive delay lines can be used for generation of wideband waveforms as well as for recompression of these waveforms into a collapsed pulse. Figure 5 shows the operation of such waveform generation and pulse compression. An impulse is applied to the flat-weighted device whose frequency response characteristics are shown in Figure 4(a). The output of this device is a 38-ns chirped waveform, which is then input to the Hamming-weighted device illustrated in Figure 4(b). The output of the Hamming-weighted device is the recompressed pulse shown in Figure 5. The narrow pulse width shown in Figure 5 illustrates the wideband capability of these devices.

One application of these dispersive delay lines is in spectrum analyzer subsystems where the spectrum analysis is accomplished using the chirp transform algorithm. In this algorithm, the input signal to be analyzed is first multiplied by a chirp, then convolved with a chirp, and finally postmultiplied by a chirp. (This last multiplication is only necessary if

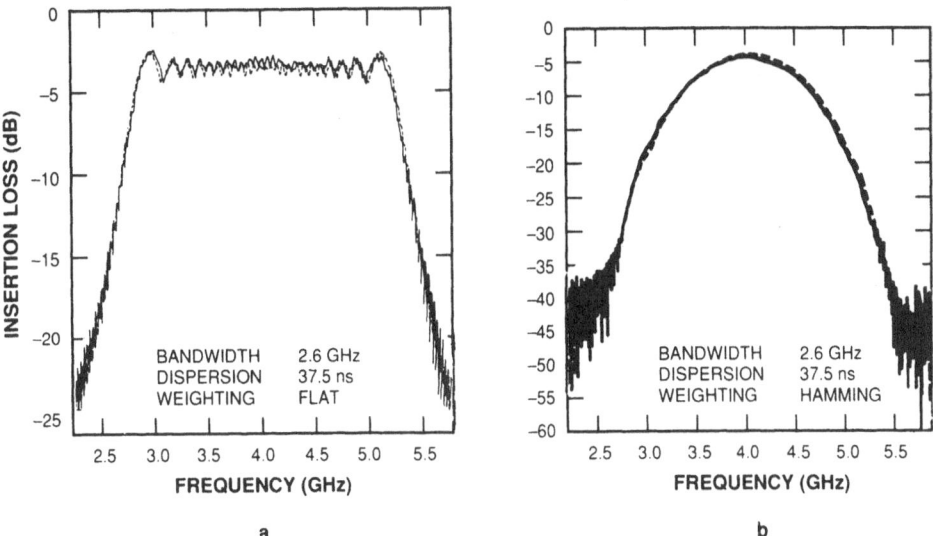

Fig. 4. Measured frequency response of a pair of superconductive dispersive delay lines. (a) Flat-weighted dispersive delay line. (b) Hamming-weighted dispersive delay line.

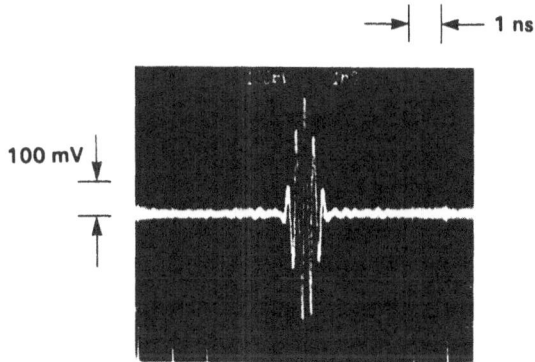

Fig. 5. Compressed pulse response observed from the output of a Hamming-weighted superconductive dispersive delay line. The input signal was derived from a matched flat-weighted dispersive delay line.

one desires phase as well as amplitude information.) Since the dispersive delay lines can generate broadband chirps as well as convolve their inputs with their chirped impulse response, as described above, they are ideal components for systems that perform nearly instantaneous spectrum analysis over multigigahertz bandwidths.[5]

The dispersive delay lines described above are the most mature of the analog signal processing devices so far developed. Additional efforts have included the development of such devices with even larger time-bandwidth (TB) products. One of these efforts has stressed the extension of the transmission line lengths using a combination of stacked substrates.[6] In this case, careful concatenation of delay lines fabricated on 127-μm substrates led to a device with 112-ns dispersion.

Further efforts have focused on the use of thinned substrates. The coupling of adjacent transmission lines is dependent on the ratio of the separation between the lines to the substrate thickness. Therefore, one can maintain the desired coupling and isolation of the lines, yet pack more delay line length onto a given substrate area, by simultaneously scaling both their separation and distance from the ground plane. Preliminary work toward this end has achieved the goal by a combination of electrostatic bonding of a niobium-coated wafer to a pyrex support and etching back a thick wafer down to a 15-μm silicon thickness. The etch stop and resulting final thickness were determined by a resistivity-selective etchant that did not react with a deposited epilayer.[5]

TIME-INTEGRATING CORRELATOR

The superconductive time-integrating correlator falls into the class of active devices in that it utilizes Josephson junctions for the function of analog multiplication and for the digital logic circuitry. Consequently, the fabrication of the device is quite a demanding process. The application of such a time-integrating correlator is often found in spread-spectrum transmission systems, e.g., radar or communications,[7] where the recognition and synchronization of waveforms are needed.

A schematic diagram of the time-integrating correlator is shown in Figure 6. The basic device structure consists of a tapped delay line, with each tap feeding an individual circuit comprising a mixer, an integrator, and a comparator (MIC). The outputs from each comparator serve as the input signals to a digital address encoder. Note the similarity between Figure 6 and the generic transversal filter architecture in Figure 1. The time-integrating correlator serves the function of an electronically programmable filter where the tap weights W_n are dynamically changing, set by the reference waveform streaming into the device simultaneously with the signal being processed.

In the time-integrating correlator,[8] the output of each of the delay line taps consists of time-delayed replicas of the two analog input signals. These signals then enter a circuit that performs both the multiplication of these two waveforms and the integration of the resulting product signal over time. It is this integration of the multiplication of two time-shifted waveforms that accomplishes the desired correlation function. At each tap (channel) along

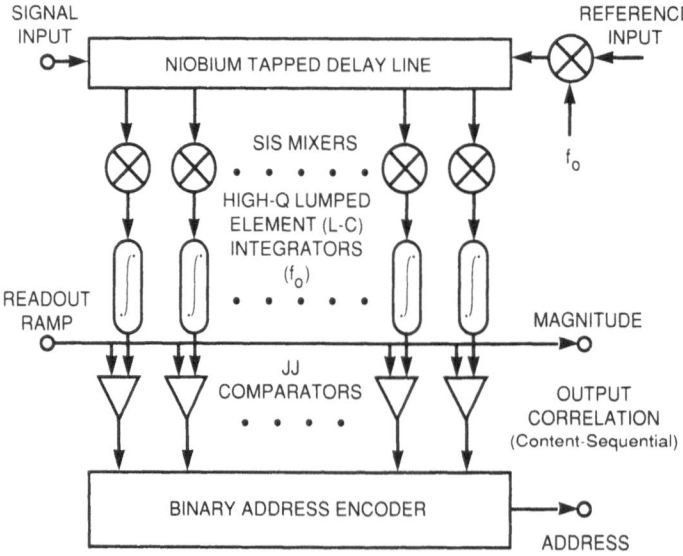

Fig. 6. Block diagram of the superconductive time-integrating correlator.

the device, the correlation function is sampled for different relative time shifts of the two original input waveforms.

In order to perform the multiplication process, an array of superconductor-insulator-superconductor (SIS) tunnel junctions is biased at the point of maximum nonlinearity on the quasiparticle I-V curve. The input signals that feed this mixer array are deliberately frequency offset from each other prior to entering the correlator device. The offset frequency f_0 is a few tens of megahertz. If the input signals are correlated, then the output of the mixer (analog multiplier) will have a strong component at f_0. Following the mixer array is a high-Q discrete element L-C resonator designed with a resonant frequency equal to f_0. The high-Q resonator allows linear integration (to within 1 dB) of currents at frequency f_0 from the mixer stage to occur over time durations up to $Q/4\pi f_0$.

An electrical schematic of the complete MIC circuit is shown in Figure 7. In order to read out the samples of the correlation function thus computed (by integrating the multiplied, time-shifted replicas of the original two input signals), the Josephson junction (JJ) comparator circuit in Figure 7 is utilized. The operation of the circuit is as follows. After the time-integration period is complete, the input signals to the device are turned off and a linearly increasing ramp current is immediately applied to the comparator circuit embedded in the resonator loop. This ramped current initially flows through the junction labeled JJ_c to ground while this junction is in the zero-voltage state.

Since the resonator is of high Q, the integrated current (representing the correlation sample to be measured) persists for some time and therefore adds to the ramp current flowing through JJ_c. When the sum of these currents exceeds the critical current of JJ_c, this junction switches to the voltage state, as seen in Figure 8. Therefore, *a priori* knowledge of the injected ramp current enables us to infer the value of the correlation sample (integrated resonator current). Qualitatively, the situation is as follows. If the correlation sample is a low value, the integrated resonator current will be small, thus requiring a larger

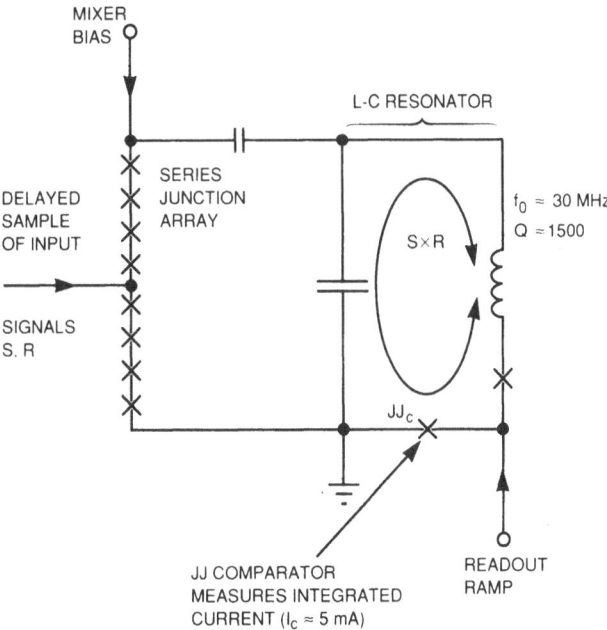

Fig. 7. Electrical schematic of an MIC cell comprising one channel of the time-integrating correlator. S and R indicate the signal and reference inputs, respectively.

injected ramp current to cause the comparator to switch to the voltage state. Conversely, for larger correlation values, less ramp current and thus less time are required to induce the switching of the comparator junction.

The method described above also allows determination of the polarity of the integrated signal, as well as its magnitude. The dotted curve in Figure 8 represents the case where the integrated signal is negative. Comparing this to the solid curve, which shows the situation for a positive correlation sample, it can be seen that the absolute timing of the point at which the comparator junction switches will be offset by an amount equal to $1/2f_0$. Therefore, by knowing the injected ramp slope and the critical current of the comparator junction, one is able to determine both the magnitude and sign of the correlation sample.

In a multichannel correlator device, the comparator circuits in the individual channels are controlled by a common injected ramp current. From the above discussion, then, it can be seen that the comparator in the channel with the largest correlation sample switches to the voltage state prior to all of the others. This is useful, as the peak of the correlation function is the most important data in many applications.

In order to determine exactly which channel is reporting its output at any instant, the comparators' switching voltages feed the inputs to an address encoder. The address encoder then provides a binary representation of the channel position that is generating an output. As described above, the channel position of the correlation peak is directly related to the relative time-offset between the two input signals. As illustrated in Figure 9, the address encoder is constructed out of cross-coupled OR-gates, each gate designed using the 4JL logic family.[9] The address encoder in Figure 9 has seven inputs. Two such encoders are integrated in a complete 14-channel time-integrating correlator.

In summary, operation of the correlator is divided into two phases. First, samples of the correlation function are obtained during the integration period while the wideband input

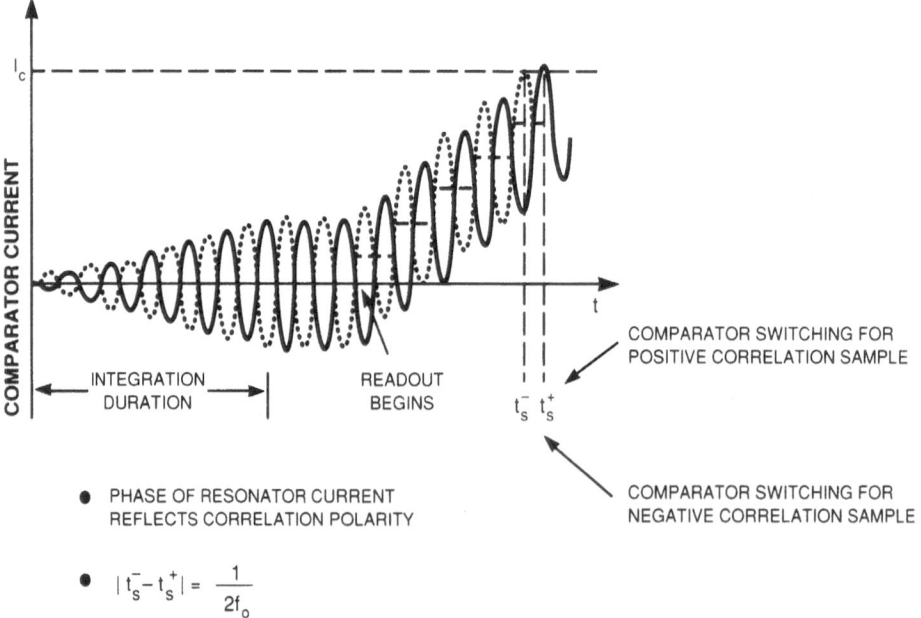

Fig. 8. Current flowing through a Josephson junction comparator during correlation and readout. The dotted curve represents a negative polarity correlation sample, and I_C represents the critical current of JJ_C.

signals are input to each end of the tapped delay line. Second, after the integration period, the values of the computed correlation samples are read out via the application of a linearly increasing current ramp to the comparator circuits embedded in each of the channel resonators. Thus, the output is obtained as a set of ordered pairs: one element of each pair represents a digital word indicating which correlation sample is being reported (channel position), while the other element is a time representative of the magnitude and sign of that particular correlation sample.

The integration of wideband analog and digital circuitry in superconductive technology is successfully demonstrated by the correlation output shown in Figure 10. This

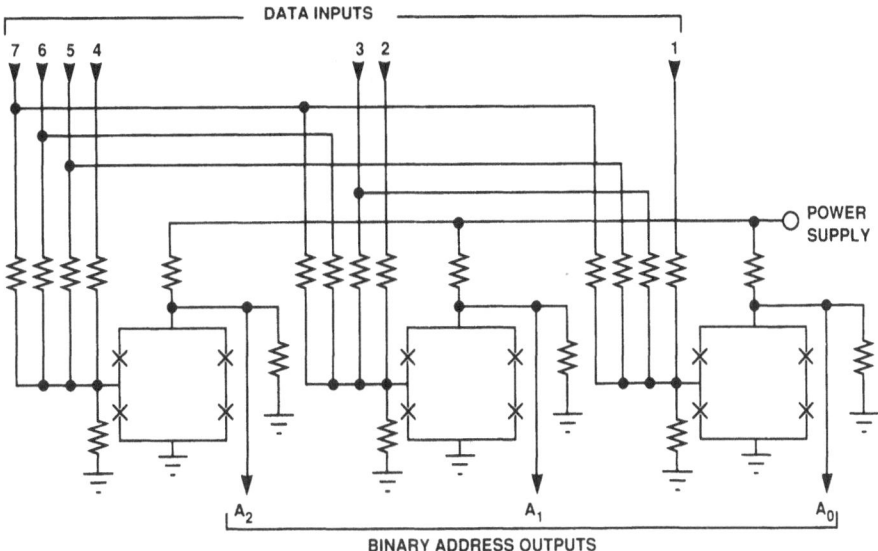

Fig. 9. Electrical schematic of a seven-input binary address encoder using cross-coupled 4JL OR-gates.

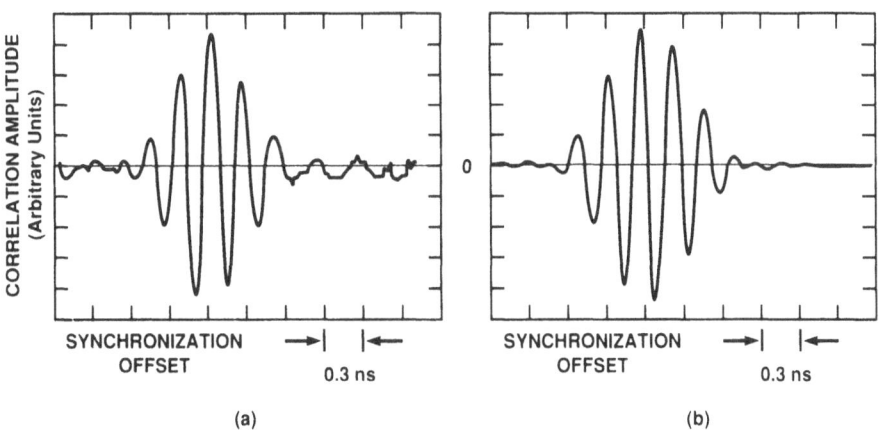

Fig. 10. Autocorrelation function of a wideband (2.6 GHz) Hamming-weighted chirp waveform.
(a) Measurement obtained with the superconductive time-integrating correlator. (b) Analytical calculation.

test of the correlator was accomplished using the superconductive dispersive delay lines described above to generate the necessary wideband waveforms for measuring the cross-correlation function. The duration of the chirped waveform provided by the dispersive delay line was 38 ns and spanned a bandwidth of 2.6 GHz centered at 4 GHz. By design, the output spectrum of the device was weighted with the familiar Hamming function in order to suppress the sidelobe level of its autocorrelation function. In order to synthesize a longer waveform of a duration on the order of microseconds, the dispersive delay line was given an input that consisted of a burst of impulses at a rate of approximately 20 MHz.

The resulting burst of wideband chirps was split into two paths, with one path going through a continuously adjustable delay element. The two waveforms thus formed were then input to the multichannel correlator and the correlation function was measured in channel 4, by mapping out the timing of the appropriate address encoder output as a function of the adjustable delay between the two input waveforms. The results of this experiment are shown in Figure 10, compared with the output one would get based on the ideal mathematical autocorrelation function of a wideband Hamming-weighted chirp waveform. Figure 10 shows excellent agreement between the measured function and that expected by theory. Thus, this experiment validated the operation of the correlator over bandwidths beyond 2.5 GHz.

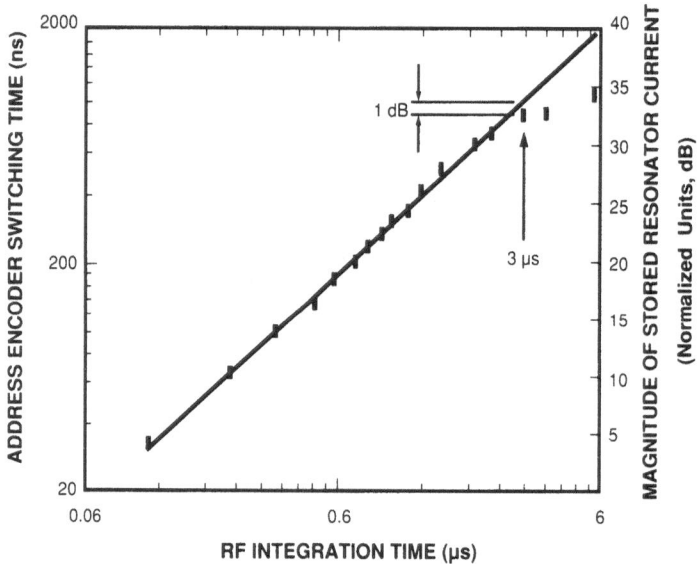

Fig. 11. Resonator contents as a function of input signal duration as inferred by measuring address encoder output timing. Integrator (resonator) compression is defined at the 1-dB saturation point.

An additional advantage of a *time*-integrating correlator is its ability to process waveforms whose duration is independent of the physical length of the tapped delay line. Indeed, as hinted previously, the maximum waveform duration is limited by the quality factor of the integrating resonator. This fact is demonstrated in Figure 11, showing measurements of the linearity of signal integration and readout. In this experiment, the input signals to the correlator were formed from a gated CW source centered at a frequency of approximately 3 GHz. As before, the source signal was split into two paths and presented to the correlator inputs. The timing of the output of the address encoder was then monitored as a function of the length of the gated CW input signals. Since the address encoder timing is directly related to the integrated resonator current, this experiment enabled a measurement of the saturation in the resonator due to its finite Q.

This data is plotted in Figure 11 and demonstrates the ability to integrate over signals up to 3 μs in duration, as determined from the 1-dB compression point shown in the figure. This long integration time coupled with the broadband capability demonstrated in Figure 10 leads to the conclusion that such a device can support waveforms with TB products greater than 6000. This implies a signal processing gain in excess of 37 dB.

The correlator device is capable of performing the cross-correlation function between two analog waveforms with bandwidths in excess of 2.5 GHz. It is constructed as a substantial integration of superconductive components including a wideband long delay line, tunnel junction mixers, high-Q L-C resonators and superconductive digital logic, all fabricated on the same substrate. Such integration of superconductive components can significantly advance the capabilities of high performance radar and communications systems.

MOSFET-TAPPED DELAY LINE

The time-integrating correlator described above is an active device in that it incorporates Josephson junctions. Yet another active device can be conceived of, namely, one that integrates superconductivity with semiconductor electronics. Such an integration might be especially attractive when using the high-T_c family of superconductors.

The diagram in Figure 12 shows how a marriage of superconductive and semiconductor technologies might achieve an implementation of the transversal filter architecture described in Figure 1. In this case, each tapped output from the low-loss superconductive delay line is input to a semiconductor field-effect transistor, with the transistor acting to control the weighting strength of the individual taps. The means for implementing this can be controlled by the appropriate voltage on the gates of the transistors. Different gate voltages correspond to different transistor-channel conductances and therefore to different transistor attenuation factors.

A distinction can be made among the means of control one has over the tap weights in the transversal filters discussed so far. In the dispersive delay lines described earlier the effective tap weights W_n were determined at the time of device fabrication, set by the photolithographic patterning of the strengths and positions of the backward-wave couplers. By contrast, the time-integrating correlator allowed for a programmable set of tap weights. However, this was accomplished by a second electrical waveform passing through the device at the same time as the waveform being processed, so the waveform controlling the tap weights (labeled reference in Figure 6) needed to be of the same wide bandwidth as the input signal waveform.

The device structure shown in Figure 12 also allows for an electronically variable filter function. The difference is that the tap weights are set *prior to* the application of the input signal to be processed. Therefore, the bandwidth of the signals used for tap-weight control can be greatly reduced compared with the wide bandwidth of the input signal. In other words, wide-bandwidth, flexible signal processing can be controlled by lower-bandwidth programming waveforms.

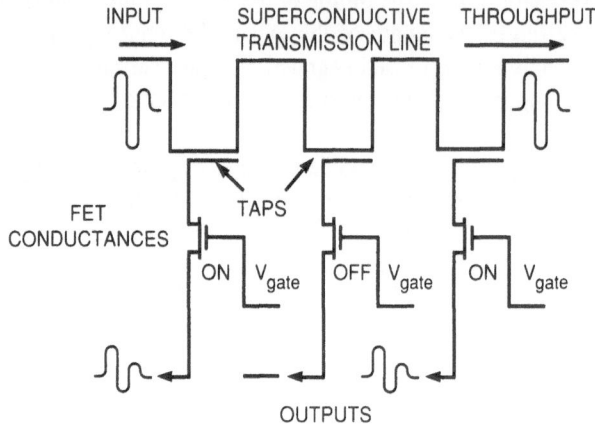

Fig. 12. Block diagram of the MOSFET-tapped transversal filter architecture. In order to achieve the convolution between the input signal and the transistor weighting factors, one would sum the transistor outputs together.

Experimental confirmation of this device concept has been reported[10] using silicon MOSFETs fabricated with all-niobium metallizations. Several issues related to device operation and fabrication arise when merging superconductive and semiconductor technologies. One concern is the use of low-T_c superconductors, such as niobium, which necessitate operation of the semiconductor circuits at 4.2 K. Bipolar transistors, being minority carrier devices, fail to operate at such low temperatures because of carrier freeze-out. In other words, there is not enough thermal energy to ionize the donors and acceptors.

Carrier freeze-out can also affect the operation of MOSFET transistors[11] if the sources and drains are too lightly doped. Sufficient doping of these regions to levels where the Fermi level enters either the conduction or valence bands can ensure that carrier freeze-out will not adversely impact device performance. Since a MOSFET transistor is a majority carrier device, then, operation can proceed in the normal fashion. For an n-channel MOSFET, phosphorus doping levels greater than 10^{18} cm^{-3} are sufficient to allow device operation at 4.2 K.

The concept of using such an n-channel silicon MOSFET as a voltage-programmable tap weight for wideband signals is demonstrated in Figure 13. Here, a short, broadband impulse was applied to the source of the transistor. The transistor's output was monitored using a sampling oscilloscope and is displayed in Figure 13. For application of a gate voltage suitable to turn the transistor off, we see that very little energy appears at the transistor's output (drain). As the gate voltage is increased, the channel conductance of the transistor also increases, and the amplitude of the impulse seen at the output rises. It is important to note in Figure 13 that the time scale of the input impulse is around 500 psec, indicating the very wide bandwidth of the signal being controlled by the dc potential on the transistor's gate electrode.

In order to quantify the allowable range of tap weight values achievable by modulation of the control transistor's gate voltage, a high frequency CW signal was input to the source of the transistor, and the transistor's insertion loss was measured as a function of applied gate potential. The results of this measurement are shown in Figure 14, where the gate-controlled attenuation of the transistor is clearly noted. Comparing the maximum output of the transistor to the minimum output obtained with the transistor in the off state, it is observed that the range of possible tap weights spans 18 dB. This confirms

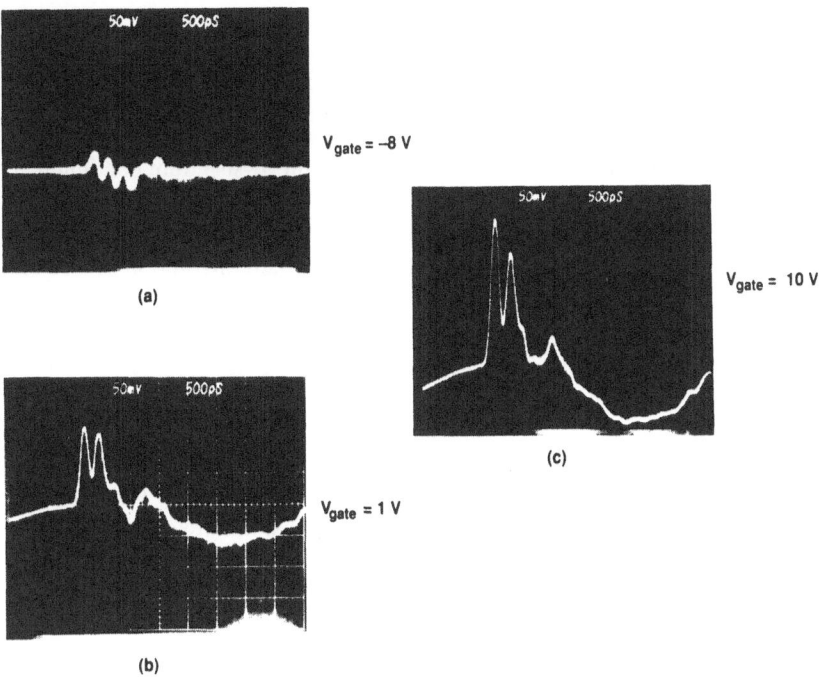

Fig. 13. Time domain transmission test of the modulation of channel conductance by the control of the gate potential for (a) transistor in the off state, (b) transistor conductance partially on, and (c) transistor in the full on state.

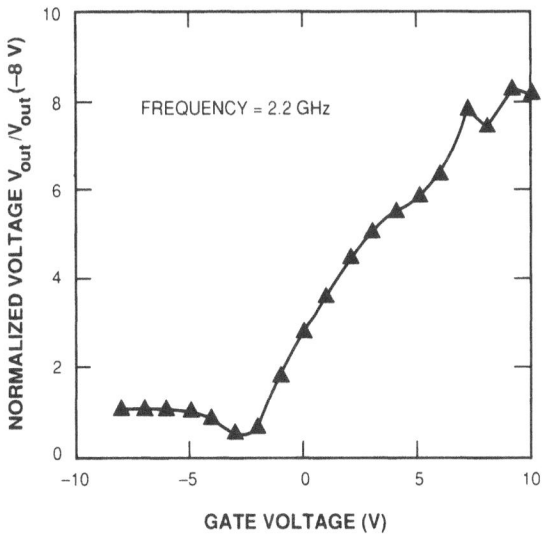

Fig. 14. Measured high frequency tap-weight control range for an n-channel silicon MOSFET with niobium metallizations, operated at 4.2 K.

the value of using such transistors fabricated with niobium metallizations and operating at 4.2 K as tap-weight control elements in a programmable transversal filter.

CONCLUSIONS

The ability to recognize weak signals in the presence of interference is valued in many applications. In particular, radar signal processing and spread-spectrum communications require such waveform analysis on signals with bandwidths in the range of several gigahertz and beyond. Superconductive components offer the circuit designer various tools with which to construct devices capable of performing the convolution or correlation between waveforms. These components include low-loss delay lines, tunnel junctions used as efficient mixers, and tunnel junctions used as elements in logic circuits. Additionally, it is possible to further increase device functionality by the integration of semiconductor components with the superconductive ones.

Three possible implementations of this technology have been reviewed in order to illustrate the advances in systems performance that might be obtained by the application of superconductivity. This chapter is not meant to be exhaustive in its suggestions for device designs. Rather, it is hoped that from examples of what has been already done, systems/circuit scientists will be spurred to develop still other applications of superconductivity to high performance architectures.

ACKNOWLEDGMENTS

The author gratefully acknowledges the work of present and past members of the Analog Device Technology Group at M.I.T. Lincoln Laboratory, with a special mention of appreciation to Alfredo Anderson, Manjul Bhushan, Maureen Delaney, Mark DiIorio, Richard W. Ralston, Stanley Reible, Lawrence Smith, and Richard S. Withers. This work was supported by the Department of the Air Force, the Department of the Navy, and DARPA.

REFERENCES

1. R. L. Kautz, "Miniaturization of normal-state and superconducting striplines," J. Res. Natl. Bur. Stand. 84:247 (1979).
2. B. M. Oliver, "Directional electromagnetic couplers," Proc. IRE 42:1686 (1954).
3. M. S. DiIorio, R. S. Withers, and A. C. Anderson, "Wideband superconductive chirp filters," IEEE Trans. Microwave Theory Tech. 37:706 (1989).
4. F. E. Nathanson, Radar Design Principles: Signal Processing and the Environment, McGraw-Hill, New York (1969).
5. R. S. Withers, "Wideband analog signal processing," in Superconducting Devices, S. Ruggiero, ed., Academic, San Diego (1990).
6. R. S. Withers, A. C. Anderson, J. B. Green, and S. A. Reible, "Superconductive delay line technology and applications," IEEE Trans. Magn. MAG-21:186 (1985).
7. R. C. Dixon, Spread Spectrum Systems, Wiley, New York (1976).
8. J. B. Green, A. C. Anderson, and R. S. Withers, "Superconductive wideband analog signal correlator with buffered digital output," SPIE 879:71 (1988).
9. H. Nakagawa, E. Sogawa, S. Takada, and H. Hayakawa, "Operating characteristics of Josephson direct-coupled four-junction logic (4JL) gate," Bull. Electrotech. Lab., 27 Sept. 1984, p. 257.
10. M. A. Delaney, R. S. Withers, A. C. Anderson, J. B. Green, and R. W. Mountain, "Superconductive delay line with integral MOSFET taps," IEEE Trans. Magn. MAG-23:791 (1987).
11. A. K. Jonscher, "Semiconductors at cryogenic temperatures," Proc. IEEE 52:1104 (1964).

SUPERCONDUCTING TUNNEL JUNCTIONS AS NUCLEAR PARTICLE

DETECTORS

A. Barone, C. Camerlingo*, R. Cristiano*, B. Ivlev°, S. Pagano*, G. Peluso, G. Pepe+, and U. Scotti di Uccio+

Dipartimento di Scienze Fisiche, Università di Napoli "Federico II", Naples, Italy
+Sez. di Napoli dell' Istituto Nazionale di Fisica Nucleare,Naples, Italy
*Istituto di Cibernetica del Consiglio Nazionale delle Ricerche, Arco Felice, Italy
°Landau Institute for Theoretical Physics, USSR Academy of Sciences, Moscow, USSR

INTRODUCTION

In reviewing the large variety of possibilities offered by superconductivity in the small scale applications, it is often neglected the important role which can be played by superconducting tunnel junctions as radiation detectors in physics and astrophysics. The growing interest of such devices in this context lies in the perspective of outstanding performances, partially already demonstrated, in the high energy spectrometry, in the fast discrimination, in the spatial resolution. In this article we shall confine our attention to the first topic. As far as the drawbacks of cryogenic requirements are concerned, it should be point out that conventional semiconductor junction detectors require liquid nitrogen not only during operation but also for storage since those devices are not thermally cyclable. In the other hand helium cryostat technology is dramatically improving and, if confined to a temperature of the order of a few degrees (1-4 K), cryocoolers extremely compact within today can provide the necessary cryogenic environment for space based equipment.

In the next Section general considerations are given concerning the energy resolution obtainable by superconducting junctions and early experimental results. Recent theoretical investigations are discussed in Section 3; in Section 4 niobium-based junction detectors will be analyzed

ENERGY SPECTROSCOPY

The proposal of using superconductors as radiation detectors dates back almost half century[1]. After a variety of experiments employing superconducting strips, the exploitation of the peculiar performances of Superconducting Tunnel Junction (STJ)[2] opened new possibilities in the field. While the voltage developed across a current carrying superconductor (thick or thin film) as a consequence of the impinging radiation could serve as an energy threshold discriminator, the junction structure offered the possibility for energy spectrometry. That is, in the former case one can get an on-off response due to the transition of a section of the film from the superconducting to the

Nonlinear Superconductive Electronics and Josephson Devices
Edited by G. Costabile *et al.*, Plenum Press, New York, 1991

133

normal state whereas the tunnel junction can operate in a proportional regime, namely it can produce signals proportional to the energy lost by the radiation.

The interest of the superconductor for high energy resolution lies in its small energy gap ($\Delta \approx 1$ meV) compared to that of a semiconductor ($\Delta \approx 1$ eV). Indeed let us consider the two quantities: E the energy lost by the radiation in the sensitive volume of the detector and ε the minimum ionizing energy. The latter represents the energy necessary to create the pair of "free carriers" , namely positive and negative ions in a gas chamber, electron and hole in a semiconductor detector. In the superconductor what is involved is the creation of excited quasi-particles as consequences of pair breaking. Thus, the smaller value of the gap and consequently of ε for a superconductor with respect to a semiconductor (the ratio is about 10^{-3}) allows the creation of a larger number of pairs carriers $N = E/\varepsilon$ reducing thereby the statistical fluctuations and improving the energy resolution. More precisely, assuming a Poisson distribution of fluctuations we have for the energy resolution R the expression :

$$R = \frac{\Delta E_{FWHM}}{E} = 2.35 \sqrt{\frac{F\varepsilon}{E}} \tag{1}$$

where ΔE_{FWHM} is the Full Width Half Maximum of the energy peak and F the Fano factor which can take values between 0 and 1.

In the case of a conventional Silicon detector the energy resolution is :

$$R = \frac{\Delta E_{FWHM}}{E}(Si) = 2.35 \sqrt{\frac{0.31 \cdot 3.61 eV}{5.89 \ KeV}} = 3.2\% \tag{2}$$

where the 5.89 KeV X-ray (MnKα) of the Fe^{55} source is considered. The corresponding ideal resolution for a Sn-Sn$_x$O$_y$-Sn superconducting tunnel junction would be:

$$R = \frac{\Delta E_{FWHM}}{E} (Sn) = 2.35 \sqrt{\frac{0.58 meV}{5.89 \ KeV}} = 0.07\% \tag{3}$$

where the conservative assumption of F=1 has been made and the $\varepsilon = \Delta_{Sn}$ has been assumed.

As far as the mechanism of the radiation detection in superconducting system is concerned, while addressing the reader to a quite comprehensive reference[3], we shall here summarize the essential aspects. The ionizing radiation interacts with the superconductor following the usual mechanisms of the energy loss which depend on the nature of the radiation (charged particles, X-ray, etc.), and of the material.

Whatever will be the primary interaction of the radiation with the matter soon after a breaking of Cooper pairs into quasi-particles and phonon excitation will take place. The resulting picture can be described in terms of a three-fluid system (Cooper pairs, quasi-particles and phonons) in non-equilibrium conditions. It is worth nothing that while in semiconductor detectors phonons are essentially lost, in the case of a superconductor phonons of energy up to 50 meV $\gg \Delta$ remain into play creating in turn other quasi-particles through pair breaking. The whole picture can be quite rigorously described in terms of sophisticated complex equations of the non-equilibrium superconductivity. However, a rather simplified treatment can be given by resorting to the Rothwarf-Taylor equations[4]:

$$\frac{\partial N}{\partial t} = R_i + 2\Gamma_{ph} N_\omega - R_r N^2$$

$$\frac{\partial N_\omega}{\partial t} = \frac{1}{2} R_r N^2 R - \Gamma_{ph} N_\omega - \Gamma_{esc} \Delta N_\omega \tag{4}$$

Table 1. summary of experimental results on STJs as X-ray detectors (the data refer to 5.89 KeV X-ray).

Junction Type	FWHM (eV)	R (%)	T (K)
Sn-based STJ[11]	41	0.7	0.4
Sn-based STJ[9]	88	1.5	0.35
Sn-based STJ[10]	65	1.1	0.32
Nb-based STJ[12]	250	4.2	1.4
Nb-based STJ[13] Laser light pulse	125 84	2.1	0.4
Nb-based STJ[20]	30	0.5	0.4

The left members of the two equations give the variations in the number of quasi-particles and phonons respectively. The former is given by the two contributions of the injection rate of quasi-particles produced directly by the interaction (R_i) and that resulting by the pair breaking due to the phonons ($2\Gamma N_\omega$) diminished by the quasi-particles recombination (at a rate $-R_r N^2$). The latter is given by the phonons produced by the recombination minus those lost for pair breaking and those lost by phonon escape.

Leaving this argument to Section 3, let us now outline the main results obtained so far by using superconducting junctions as energy spectrometers. After the very preliminary experiments discussed in the Ref.2, more advanced results were reported still using Sn-Sn_xO_y-Sn junctions but of a higher quality and using lower operation temperature (0.32 K)[5].

Experimental results using Nb based junctions were then reported and discussed in connection with the perspective of the possible use in the determination of the upper bound of the neutrino mass[6]. A value of ε down to about 20 meV was also obtained[7]. Further experiments were performed on Sn-Sn_xO_y-Sn junctions with excellent results[8-11] (for a comprehensive review the reader is referred to Ref. 3). A FWHM down to 41 eV for the MnKα was obtained[11].

Although extremely good these results, as stated above were realized by using Sn-based junctions which have serious problems of stability against thermal cycling and therefore great attention is presently dedicated again to the refractory Nb-based structures. In this context interesting results have been obtained[12-13,20]. Before proceeding in both instrumentation problems and physical aspects of non-equilibrium superconductivity underlying the radiation interaction with the junction it is useful to summarize the main results outlined so far. This is given in Table 1.

RELAXATION PROCESSES IN A STJ

The shape, the rise-time and the decay-time of the current pulse originated in the STJ by the incident radiation depend on the external electronic circuit and on the internal relaxation processes which bring the STJ far from equilibrium. Considering these latter, as already pointed out in Sec. 2, the most important are:

inelastic scattering of quasi particles and phonons;

- escape of phonons into the environment of the junction (substrate, helium, etc);
- tunneling of quasi-particles across the junction;
- recombination of quasi-particles to form Cooper pairs under the emission of a phonon with an energy $E \geq 2\Delta$;
- pair breaking by 2Δ-phonons;
- diffusion of quasi-particles into the junction electrodes.

The inelastic scattering of quasi-particles and phonons is the main channel of energy relaxation and it is described through the rate τ_{ph}^{-1}. This process yields a population of excess quasi-particles mainly with an energy of the order of Δ before the recombination process becomes effective. In the limit $T \ll \Delta$ and for quasi-particle energy $\varepsilon \gg \Delta$, the expression for τ_{ph}^{-1} can be approximated as:

$$\tau_{ph}^{-1} = \frac{c \cdot \omega}{v_F}$$

(6)

where c is the sound velocity and v_F is the Fermi velocity.

The phonon escape rate has been only evaluated from oversimplified models. Assuming an energy independent transmittivity at the boundaries α, τ_{esc}^{-1} is given by[14]:

$$\tau_{esc}^{-1} = \frac{\alpha c}{2d}$$

(7)

where d is the junction thickness.

Taking into account the expressions for τ_{esc}^{-1} and τ_{ph}^{-1}, we notice that $\tau_{esc}^{-1} < \tau_{ph}^{-1}$ for the high energy phonons, while $\tau_{ph}^{-1} < \tau_{esc}^{-1}$ for phonons with energy of the order of the superconducting gap.

In the low temperature region ($T \leq 0.2 \div 0.3$ T_c), the recombination process of the quasi-particles is not effective. In fact, the recombination rate τ_R^{-1} in the limit of low temperatures and low phonon frequencies is given by the following expression[15]:

$$\tau_R^{-1} = \lambda \frac{\Delta^3}{\omega_D^3} \sqrt{\frac{2T}{\pi\Delta}} \, e^{-\frac{\Delta}{T}}$$

(8)

where λ is the electron-phonon coupling constant and ω_D is the Debye energy. In these conditions the non-equilibrium electrons will collect in a region of the order of T near the gap edge, and the only relaxation process allowed for the system is the tunneling across the barrier at a rate[15]:

$$\tau_T^{-1} = (e^2 R_{NN} N_0 A d)^{-1} \left[1 - \left(\frac{\Delta_1}{\Delta_2 + V} \right)^2 \right]^{\frac{1}{2}}$$

(9)

where R_{NN} is the normal state resistance, $A d = V(0)$ is the effective volume of the pick-up electrode and N_0 is the single spin normal density of states at Fermi energies.

The energy collected by the excited quasi-particles will be approximatively equal to the energy E_{rel} released by the incident radiation. In fact, it is possible to neglect the small contribution of the high energy phonons which escape from the detector, since $\tau_{ph}^{-1} > \tau_{esc}^{1}$ in this case. Moreover, due to the slow recombination rate, the population of the 2Δ-phonons will not be far from the thermal equilibrium, so that also escape of the low energy phonons is negligible. Keeping in mind this physical picture, the Eliashberg kinetical equations[16] for the non-equilibrium quasi-particle distribution $f(\varepsilon)$ can be solved, obtaining:

$$f(\varepsilon) \approx f_0(\varepsilon) + c(r,t) \cdot e^{-\frac{\varepsilon - \Delta}{T}}$$
(10)

where $c(r,t)$ is the solution of the following equation:

$$\frac{\partial c}{\partial t} - \check{D}\nabla^2 c + \tau_R^{-1}\left(c + \frac{1}{2} \cdot e^{\frac{\Delta}{T}} \cdot c^2\right) + \tau_T^{-1} \cdot c = 0$$
(11)

in which $\check{D} = \sqrt{2T/\pi\Delta} \, D_e$ is an effective diffusion coefficient for the quasi-particles. Notice that in Eq.11 a source term due to the pair-breaking effect has been neglected, as it is proportional to the small number of 2Δ-excess phonons.

The boundary conditions of Eq. 11 are given by the conservation of the energy:

$$E_{rel} = E_{el} = \int d^3p \, \frac{1}{(2\pi)^3} \cdot 2\Delta \cdot \int d^3r \cdot c \, (\vec{r},t) \cdot e^{-\frac{\varepsilon - \Delta}{T}}$$
(12)

and by the relation

$$\nabla c \cdot \vec{n} = 0$$
(13)

on the boundary of the superconductive electrode.

Integrating Eq. 11 over the volume of the pick-up electrode, taking into account the condition (Eq. 13) and omitting the quadratic term in Eq. 11, we obtain the following equation for the spatial average of $c(t)$:

$$\frac{\partial \langle c \rangle}{\partial t} + \frac{\langle c \rangle}{t} = 0$$
(14)

with

$$\tau^{-1} = \tau_R^{-1} + \tau_T^{-1}$$
(15)

The non-equilibrium part of the tunnel current is given by:

$$\delta I = I_0(V) \cdot \langle c \rangle \, e^{\frac{\Delta}{T}}$$
(16)

where $I_0(V)$ is the equilibrium current of the STJ evaluated at the bias V. The total electronic charge Q passing through the junction is therefore

$$Q = I_0(V) \int \langle c \rangle(t) \, e^{\frac{\Delta}{T}} dt$$
(17)

In the linear regime, the signal produced by the detector is proportional to the absorbed energy

$$Q = \frac{e\, E_{rel}}{\varepsilon} \qquad\qquad (18)$$

where Q is the produced charge, e is the electronic charge and ε is the mean energy required to produce an excitation. The value of ε depends on details of the interaction mechanisms of the system quasi-particles-phonons-Cooper pairs in the detector. Using Eq. 12, we obtain the following expression for ε:

$$\varepsilon = \frac{1}{I_0(V)} \frac{\sqrt{\pi}\,(2\Delta)^{\frac{3}{2}}\, T^{\frac{1}{2}}\, N_0 V(0)}{\tau}\, e^{-\frac{\Delta}{T}} \qquad\qquad (19)$$

Assuming typical values of the experimental parameters for a $Sn\text{-}Sn_x O_y\text{-}Sn$ junction of Ref.10, we obtain $\varepsilon \approx \Delta$ for $T < 0.2\ T_c$. This estimate is fully in agreement with the determination of Kurakado[17] of the upper limit of the mean energy loss per excess quasi-particle in a superconductor ($\varepsilon_{max} \approx 4\,\Delta$).

The measured values of ε are generally higher[18] also if the contributions due to spurious effects (diffusion of quasi-particles outside the junction area, electronic noise, etc.) are negligible or otherwise evaluated. A possible explanation is that the effective recombination time is shorter than that given by Eq.3 in the limit of low temperatures, as suggested also by the experimental evidence[11].

NIOBIUM BASED STJ

In this section niobium-based STJs will be investigated, with special attention to their limit features.

The choice of niobium as junction electrode comes mainly from the need of a reliable detector operation, also after repeated thermal cycles between room temperature and the cryogenic environment.

The modern development of the fabrication technology of full niobium junctions, using a trilayer technique makes possible the fabrication of very high quality niobium-aluminum oxide-niobium junctions with a nearly ideal current-voltage (I-V) characteristic[19]. The high quality of the I-V characteristic is an essential feature for an X-ray STJ detector, since in its normal operation the junction is biased in the subgap voltage region. Any deterioration of the properties of the junction electrode and of the insulating tunnel barrier will result in an increased leakage current, which increases the background noise and reduces the sensitivity of the detector.

A further advantage of using niobium electrodes is that, due to its relatively high critical temperature ($T_c \approx 9.2$ K) , the detector can be operated at low reduced temperatures (T =1 K implies $T/T_c = 0.11$) without the need of the complicated cooling techniques necessary below 1 K. This fact is particularly important for airborne applications (e.g. satellites) where size and weight of the payload are important factors.

According to Eq. 1 the intrinsic energy resolution of a niobium-based STJ is expected to be :

$$\Delta E_{FWHM} \approx 7\, eV \qquad\qquad (20)$$

However the experimental results reported so far have not yet reached this limit. The best result reported so far is from Kurakado et al.[20] They have reported a noise level, measured supplying a test pulse to the STJ, as low as 30 eV at a temperature of 0.4 K (see Table 1).

At higher temperatures the energy resolution is expected to decrease because of the fast increase of the quasi-particles recombination rate, and also for the increase of the noise due to the fluctuations of the number of thermally excited quasi-particles.

The quasi-particle recombination time t_R (Eq. 8) can be rewritten as[21]:

$$\tau_R = \tau_0 \left[\pi \frac{T}{T_c} \right]^{-\frac{1}{2}} \left[\frac{2\Delta(0)}{kT_c} \right]^{-\frac{5}{2}} e^{\frac{\Delta(0)}{kT}}$$

(21)

where τ_0 is a characteristic time for the material considered (for Nb : $\tau_0 = 1.49$ 10^{-10} s, $T_c = 9.2$ K, $\Delta(0) = 1.53$ meV). τ_R has to be compared with the tunneling time τ_T given by Eq. 9, which can be rewritten as[22] :

$$\tau_T = e^2 N(0) A d R_{nn}$$

(22)

For a Niobium-based STJ is obtained :

$$\tau_T = 5 * 10^{-9} A[\mu m^2] d[\mu m] R_{nn}[\Omega]$$

(23)

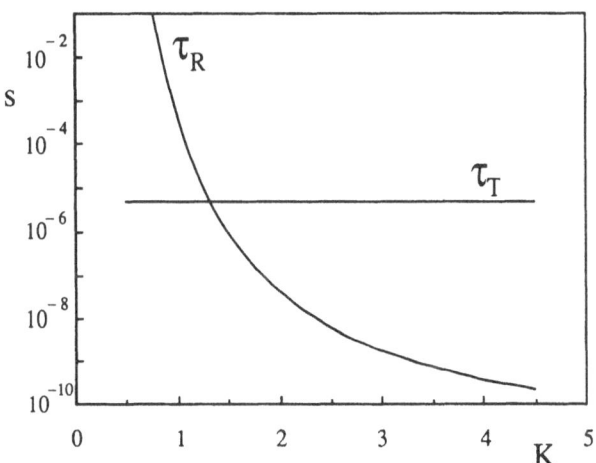

Fig. 1   Comparison between the relaxation time τ_R and the tunnel time τ_T, for a Nb-based junction. The junction parameters are : A=100μm^2, d=1μm, and R_{NN}=10Ω.

In Fig. 1 is shown a comparison between τ_R and τ_T at temperatures below 4.2 K for a typical Nb-Al$_2$O$_3$-Nb junction where it is clearly shown that at temperatures below 1.5 K the quasiparticles tunneling time is much shorter then the recombination time, thus allowing the collection of most of the quasiparticles generated. In Ref.22 it has been pointed out the possibility that a phonon created by the reconbination of two quasiparticles may break another

Cooper pair before escaping from the superconductor. As a consequence the recombination time has to be multiplied by a factor F_T called phonon trapping factor, given by :

$$F_T = 1 + \frac{\tau_{\gamma l,t}}{\tau_B}$$

(24)

where $\tau_{\gamma l,t}$ is the average phonon escape time for the longitudinal and transverse phonons, and τ_B the phonon pair-breaking time. The presence of this factor can considerably increase the effective recombination time, thus increasing the operating temperature range for the detector.

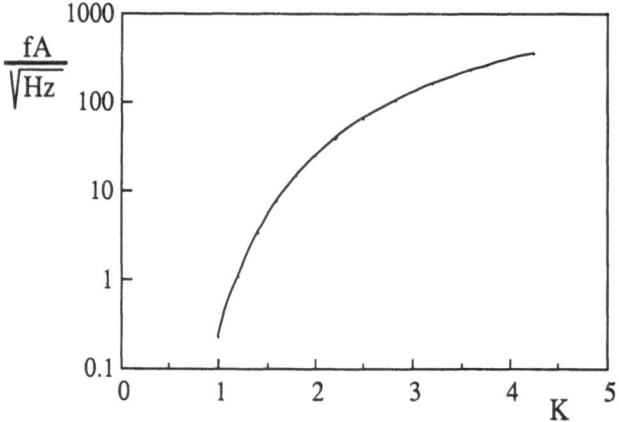

Fig. 2   Temperature dependence of the spectral current noise density generated by the junction considered in Fig. 1, biased at $V_o = 1\,mV$.

The other main effect of the temperature on the energy resolution is the noise due to the fluctuations of the thermally excited quasiparticles when the detector is biased in the subgap voltage region. This noise has a white spectrum (at frequencies $f \ll V_o\, e/h$, V_o being the bias voltage)   with amplitude :

$$\langle i_{qp}^2 \rangle = 2\,e\,I_{qp}(V_o)\,coth\left(\frac{e V_o}{2 k T}\right)$$

(25)

where I_{qp} is the   junction quasi-particle current at the bias point. The corresponding bias current is given by :

$$I_{qp} = \frac{1}{R_n n}\int_0^\infty N(V)\,N(V - V_o)\,\left[f(V - V_o) - f(V)\right]d V$$

(26)

where the density of states N(V) is :

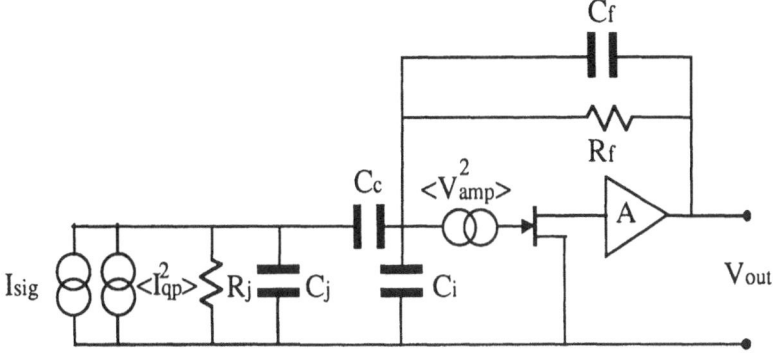

Fig. 3   Typical charge amplifier circuit for an STJ detector

$$N(V) = \frac{|V|}{\sqrt{\left(V^2 - \frac{\Delta^2}{e^2}\right)}} \; \Theta(e|V| - \Delta)$$

(27)

and $f(V) = (1 + e^{\,eV/KT})^{-1}$ is the Fermi distribution. Considering an actual Nb-Al_2O_3-Nb junction as an ideal BCS Nb-Insulator-Nb junction,

I_{qp} can be evaluated from Eq. (26) and ,choosing as working voltage Vo = 1 mV, the temperature dependence of the spectral density of the noise current can be obtained by means of Eq. (25), as it is shown in Fig. 2.

In order to consider the effect of noise on the detector performances, a typical charge amplifier circuit connected to the STJ will be considered, as shown in Fig. 3. R_j and C_j represent the junction dynamical resistance and capacitance respectively, C_c is a dc voltage decoupling capacitor, C_i represents the total input capacitance (FET + cables), C_f and R_f are the feedback capacitor and resistor respectively. Some typical values for these components are : C_c = 100 nF, R_f = 300 MΩ, C_f = 1 pF. The amplifier gain is typically 1000. C_j ranges around 40-80 fF/μm^2 while R_j depends on the bias point, the junction area and the temperature. A value of 10^6 Ω is reasonable for an high quality junction with an area of 100 μm^2.

By a simple network analysis, the equation for the output voltage V_{out} is:

$$\frac{dI_{sig}}{dt} = \alpha V_{out} + \beta \frac{dV_{out}}{dt} + \gamma \frac{d^2 V_{out}}{dt^2}$$

(28)

where

$$\alpha = -\frac{A+1}{AR_j C_c R_f}$$

(29a)

$$\beta = -\left[\frac{C_c + C_i + C_f(A+1)}{AR_j C_c} + \frac{(C_j + C_c)(A+1)}{AR_f C_c} \right]$$

(29b)

141

$$\gamma = -\left[\frac{(C_j + C_c)(C_c + C_i + C_f(A+1))}{AC_c} - \frac{C_c}{A}\right] \qquad (29c)$$

$$I_{sig} = \frac{Q_0}{\tau} e^{-\frac{t}{\tau}} \qquad (29d)$$

I_{sig} represents the current pulse generated by the excess quasiparticles.

The study of Eq.(28) when the noise sources are included presents some difficulties. In the present work we will follow the approach introduced by Twerenbold[24]. It consists in considering negligible the FET input capacitance C_i, the feedback characterized only by the capacitance C_f and excluding from the circuit analysis the junction dynamical resistance R_j. With these assumptions Eq. (28) reduces to one having only the γ term. This approximation, according to Twerenbold, allows to write an expression for the voltage signal output produced by the noise sources, which are the amplifier and the junction respectively. The former is:

$$\langle V_s^2 \rangle = \langle V_{amp}^2 \rangle + \frac{A_f}{\nu} \qquad (31)$$

where

$$\langle V_{amp}^2 \rangle = 4\,KTR_{eq}^{FET} \qquad (32a)$$

$$R_{eq}^{FET} \approx \frac{0.7}{g_m} \approx 700\,\Omega \qquad (32b)$$

g_m is the input FET mutual transconductance, and A_f is the magnitude of the FET 1/f noise. Typical values for a FET at T = 150 K are :

$$\langle V_{amp}^2 \rangle \approx 5.8 * 10^{-18} \frac{V^2}{Hz} \quad and \quad A_f \approx 5.8 * 10^{-15} V^2 \qquad (33)$$

It is worth noting that for frequencies above 1 KHz the 1/f contribution to the amplifier noise is negligible.

Introducing the equivalent noise charge Q_N, it is possible to show that

$$Q_N^2 = Q_s^2 + Q_f^2 + Q_p^2 \qquad (34)$$

where

$$Q_s^2 = 0.39\,\tau^{-1}(C_j + C_f)^2 \langle V_{amp}^2 \rangle \approx (38\,e)^2 \qquad (35a)$$

$$Q_f^2 = 1.23\,(C_j + C_f)^2 A_f = (5\,e)^2 \qquad (35b)$$

$$Q_p^2 = 0.39\,\tau \langle i_{qp}^2 \rangle \qquad (35c)$$

where C_j = 8 pF and C_f = 1pF have been used. Moreover the output voltage of the amplifier is normally filtered by a shaping circuit to achieve a pulse with a suitable shape for the analysis as well as to improve the signal to noise performances. The time constant of the shaping circuit τ has been fixed at 5 μs.

Table 2. Equivalent noise charge Q_N, accumulated charge Q, and energy resolution relative to a 5.89 KeV X-ray, at two different temperatures for an ideal Nb-based junction

T (K)	Q_N (e)	Q (e)	ΔE (eV)
4.2	3000	$2.31 \ 10^4$	1800
1.0	45	$3.85 \ 10^6$	7

It is worth noting that, since Q_s is proportional to C_j+C_f, a small area junction will improve the noise performances of the amplifier. Concerning Q_p, the equivalent noise charge is:

$$Q_p = 3000 \ e \qquad \text{at } T= 4.2 \ K \qquad\qquad (36a)$$

$$Q_p = 2 \ e \qquad \text{at } T= 1.0 \ K \qquad\qquad (36b)$$

It can be seen that at T=4.2K this noise source is dominant, whereas at T=1.0K the amplifier is the principal noise source. Fig.2 also shows that below 1 K the intrinsic junction noise become very small so that is the front end electronic which finally limits the energy resolution.

Rewriting Eq.(1) in terms of the accumulated charge Q, and of the charge fluctuation Q_N , it is obtained for the energy resolution ΔE :

$$\frac{\Delta E}{E} = 2.35 \frac{\sqrt{\left(Q_N^2 + F Q\right)}}{Q} \qquad\qquad (37)$$

where the accumulated charge can be expressed as:

$$Q = \frac{E}{\varepsilon} \frac{\tau_T^{-1}}{\tau_T^{-1} + \tau_R^{-1} F_T^{-1}} \qquad\qquad (38)$$

In the worst case of F = 1, and assuming $\varepsilon = \Delta(0)=$ 1.53 meV and $F_T = 100$,[22] the results reported in Table 2 are obtained.

It appears from the Table 2 that a Nb STJ-based detector achieves its limit energy resolution of 7 eV with 5.89 keV X-Ray radiation only at T = 1.0 K.

It should be noticed however that $\Delta(0)$ = 1.53 meV for Nb as well as F =1 lead to an underestimate of the detector performances.

Moreover, in spite of the fact that the intrinsic junction noise value was computed considering an ideal junction having a BCS like I_{qp}, a more realistic evaluation that accounts also for the leakage currents should lead to a Q_N value 5 times larger,[23] a value that still should not significatively affect the resolution at low temperatures.

In conclusions, the analysis performed indicates that Nb based junctions operating at temperatures of the order of 1 K, at the actual available state of the art in the fabrication technology, should be adequate candidates as detectors for energy spectroscopy.

Measurements reported in literature (see Table 1) exhibit an energy resolution greater than that theoretically expected. This can be due either to the inadequate nonequilibrium picture, which could result in an overestimate of the accumulated charge, or to a simplified circuital analysis which could also underestimate the noise contributions. However it must be noticed that the junctions so far used in experiments were not of the same good quality as the one whose parameters have been used in the calculations.

ACKNOWLEDGEMENTS

The authors wish to thank L. Frunzio and R. Monaco for providing the experimental data of an high quality Nb-based junction, and C. Nappi for helping in calculating the junction noise spectrum.

Financial support from the National Research Council of Italy through the Progetto Finalizzato "Superconductive and Cryogenic Technologies" is gratefully acknowledged.

REFERENCES

[1]   D.H. Andrews, R.D. Fowler and M.C. Williams, Phys. Rev. 76, 154 (1944)
[2]   G.H. Wood and B.L. White, Appl. Phys. Lett. 15, 237 (1969); Can J. Phys. 51, 2032 (1973)
[3]   A. Barone "Superconductive Tunneling Detectors" World Scientific Publ. (1988) see also A. Barone, S. De Stefano, K.E. Gray Nucl. Inst. Meth. A235, 254 (1985)
[4]   A. Rothwarf and B.N. Taylor, Phys Rev Lett. 19, 27 (1967)
[5]   M. Kurakado, J. Appl. Phys. 55, 3185 (1984)
[6]   A. Barone, G. Darbo, S. De Stefano, G. Gallinaro, A. Siri, R. Vaglio, S. Vitale, Proc. LT17, pag 933, Elsevier Science Publ. (1984)
[7]   A. Barone, G. Darbo, S. De Stefano, G. Gallinaro, A. Siri, R. Vaglio, S. Vitale, Nucl. Instr. Meth. A234, 61 (1985)
[8]   D. Twerembold, Europhysics Lett. 1, 209 (1986)
[9]   H. Kraus, Th. Pet...eins, F. Probst, F. W. von Feilitzsch, R.L. Mossbauer, V. Zaoek and E. Umlauf, Europhysics Lett. 1, 161 (1986)
[10] D. Twerembold and A. Zhender, J. Appl. Phys. 61, 1 (1987)
[11] Rothmund and A. Zehnder in "Superconductive Tunneling Detectors" , ed. by A. Barone, World Scientific Publ. (1988), p. 52
[12] P. Garé,et al. IEEE Trans. on Magn., 25, 1351 (1989)
[13] M. Kurakado, A. Matsumura and T. Kasminaga, Tech. Dig. 8th Sensor Symposium, p.247 (1989)
      M. Kurakado, A. Matsumura, Sensors and Actuators, A21-A23,p.33 (1990)
[14] C.C. Chi, M.M.T Loy, and D.C. Cronemayer, Phys. Rev. B23, 124 (1981)
[15] B. Ivlev, G. Pepe, and U. Scotti di Uccio, Nucl. Instr. Meth. A (in press)
[16] G.M. Eliasberg, Sov. Phys. JEPT 29, 698 (1969)
[17] M. Kurakado and H. Hazaki, Nucl. Instr. Meth. 185, 141 (1981)
[18] D. Twerembold, in "Superconductive Tunneling Detectors" , ed. by A. Barone, World Scientific Publ. (1988), p. 38
[19] A.W. Lichtemberger et al., IEEE Trans. Magn. 25, 1247 (1989)
      S. Morohashi, and S. Hasuo, J. Appl. Phys, 61, 4835 (1987)
      K. Ishibashi, et al. IEEETrans. Magn. 25, 1354 (1989)
[20] M. Kurakado, T. Takahashi and A. Matsumura, Appl. Phys. Lett., 57, 1933, (1990)
[21] S.B. Kaplan et al., Phys. Rev. B14, 4854 (1976)
[22] E. Hebrank et al., Nucl. Instrum. and Meth., A288, 541 (1990)
[23] This estimate has been obtained from a very high quality junction (Vm≈90mV at T=4.2K) recently obtained in our laboratory.

EFFECT OF LOSSES ON THE OUTPUT VOLTAGE OF A FLUX-FLOW TYPE

JOSEPHSON OSCILLATOR

Yongming Zhang

Department of Physics
Chalmers University of Technology
S412 96 Göteborg, Sweden

INTRODUCTION

A flux-flow type Josephson oscillator (FFO), which utilizes the uni-directional flux flow (a moving vortex array) in a long Josephson junction has been consided for use as a mm and submm wave oscillator in integrated superconducting receiver systems mainly because it gives high output power and wide bandwidth.[1-3]

In order to fully utilize the high output power of an FFO, it is usually operated near the velocity-matching (VM) condition, i.e. the velocity of the flux flow u approaches that of the electromagnetic wave $\bar{c}$ propagating along the long Josephson junction. The corresponding dc I-V curve shows a resonant like current step which is called a *velocity-matching step* (VM step). The maximum height of the current step γ_m occurs when $u=\bar{c}$ (Fig.1), which is defined as the *height of the VM step*. It is the maximum normalized bias current under which an FFO can be operated in the flux-flow state. The height of the VM step γ_m is an important parameter for the practical application and the theoretical study of such an oscillator.[4]

In order to explore the dependence of FFO properties on junction dampings, we have simulated the effects of quasiparticle tunneling loss α and surface loss β on the height of the VM step γ_m, on the spatial distribution of the ac voltage amplitude $V_{ac}(x)$ along the junction, and on the output voltage V_{osc} of the junction at the VM condition. Here, V_{osc} is the ac voltage amplitude at the output end of an FFO with an open-circuit boundary condition.[2] It is proportional to the square root of the output power of the oscillator. The simulations were performed using the perturbed sine-Gordon equation(PSGE) with a uniformly distributed bias current γ which corresponds to a long overlap Josephson junction. The importance of including a surface damping term in the mathematical model is discussed. The dependence of the damping α on the critical current density j_c of a junction is illustrated for some typical junction materials. It gives the possible j_c ranges for designing an FFO.

Nonlinear Superconductive Electronics and Josephson Devices
Edited by G. Costabile *et al.*, Plenum Press, New York, 1991

145

MATHEMATICAL MODEL AND COMPUTATION TECHNIQUES

The mathematical model describing the flux-flow type Josephson oscillator, which is formed by a long, narrow junction, is the perturbed sine-Gordon equation (PSGE)

$$\phi_{xx} - \phi_{tt} - \sin\phi = \alpha\phi_t - \beta\phi_{xxt} - \gamma \tag{1}$$

with appropriate boundary conditions for an overlap junction[3,4]

$$\phi_x(0,t) + \beta\phi_{xt}(0,t) = -\beta_e \tag{2a}$$

$$\phi_x(L,t) + \beta\phi_{xt}(L,t) = -\beta_e - \phi_t(L,t)/R_{load} \tag{2b}$$

Here, ϕ is the usual Josephson phase variable, x is the distance along the junction normalized to the Josephson penetration depth λ_J, and t is time normalized to the inverse of the Josephson plasma frequency ω_J. The subscript x and t denote spatial and time derivatives. The γ term represents a uniformly distributed bias current density j_B normalized to the Josephson critical current density j_c, i.e. $\gamma = j_B/j_c$. The terms $\alpha = G/\omega_J C$ and $\beta = \omega_J \mathcal{L}/R_S$ represent quasiparticle loss and surface loss, respectively. α is typically between 0.01 and 0.3, while β is typically between 0.001 and 0.03 in experiments depending on the junction material and the temperature.[5] Here C is the capacitance per unit area, G is the effective shunt conductance per unit area, $\mathcal{L}$ is the inductance per unit length and R_S is the electrode surface resistance per unit length. The constant β_e, which determines the boundary conditions at both ends of the junction of normalized length L, is the normalized external dc magnetic field in the $-y$ direction.[4] R_{load} is the terminal resistance normalized to the characteristic impedance of the junction.[3] In this paper, we limit our discussion to the open-circuit boundary condition, which means R_{load} is equal to infinity in Eq.(2b).

In order to calculate Eq.(1) numerically, we divide the long junction into lots of small sections of length $\Delta x = L/N$, where N is the number of sections. We reduce Eq.(1) into a set of difference equations by using[4] an implicit finite-difference scheme*. These difference equations were solved with a tridiagonal algorithm. In our simulations, we have used a fixed time step $\Delta t = 0.05$ with the length of each space section $\Delta x = 0.1$ for given sets of α, β, γ, and β_e. The criterion for a steady-state solution is that the dc voltage V_{dc}, which is equal to $<\phi_t>_t$ in the normalized units, is uniform for all sections within an error of 1%. Here $<...>_t$ means the average in time.

RESULTS AND DISCUSSION

Velocity-Matching Step

Fig.1 shows the calculated normalized dc I-V curves, and the relationships between the bias current γ and the output voltage V_{osc} of an FFO for

---

*At boundary points, x=0 and L, we use a backward explicit finite-different scheme in time, and the implicit finite-different scheme in space to reduce Eq.(1) and Eqs.(2) into difference equations.

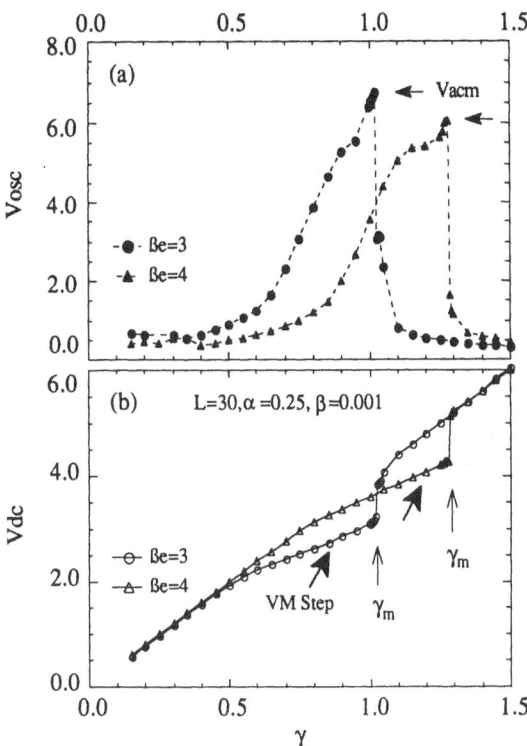

Fig.1 (a) Relationship between the ac output voltage V_{osc} and the
bias current γ of a flux-flow type Josephson oscillator.
(b) The normalized dc I-V curves. Curves are drawn for two
different external magnetic field β_e=3, and 4 with L=30,
α=0.25 and β=0.001 kept constant.

different external magnetic fields, β_e=3 and 4, with L=30, α=0.25 and
β=0.005 kept constant. V_{osc} is the ac voltage amplitude at the output end,
$V_{osc}=V_{ac}(L)=Maximum(\phi_t(L)-<\phi_t(L)>_t)$. The bold black arrows in Fig.1b indicate
the regions of VM steps for β_e=3 and 4. Note that the step is in the horizontal (γ) direction.

For a given β_e, a VM step arises in the dc I-V curve and the maximum
height of this current step, γ_m, occurs when $u=\bar{c}$. For $\gamma<\gamma_m$, i.e. $u<\bar{c}$, V_{osc}
increases with γ; it reaches its maximum V_{acm} (Fig.1a) when $\gamma=\gamma_m$, i.e. at the
velocity-matching condition. For $\gamma>\gamma_m$, V_{osc} becomes very small because the
motion of flux flow switches to a zero-fluxon oscillation. In the I-V curve
this switch corresponds to a jump from the VM step to the McCumber-Stewart
background curve which approaches asymptotically the ohmic line $\gamma=\alpha<\phi_t>_{xt}$
of an FFO. Here, $<...>_{xt}$ means the average both in time and space. Fig.1a
shows that the output ac voltage amplitude is strongly dependent on the bias
current. V_{osc} reaches its maximum value V_{acm} when the oscillator is biased at
the velocity matching condition, $\gamma=\gamma_m$. This emphasizes the importance of
biasing such an oscillator at the velocity-matching condition in order to
obtain a maximum output power from the oscillator.

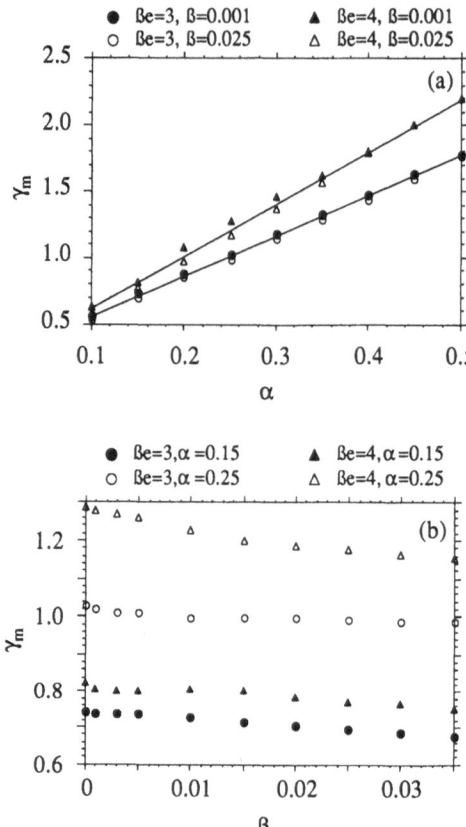

Fig.2 (a)The VM step height γ_m as a function of quasiparticle tunneling loss α. The solid lines represent Eq.(4) with a=0.20,
b=0.05 for β_e=3, and a=0.20, b=0.01 for β_e=4. (b)γ_m as a
function of surface loss β. The junction length L is equal to
30 (λ_J) in our simulation.

Height of the Velocity-Matching Step

Detailed numerical simulations gave relationships between damping parameters and the VM step height γ_m using L=30 and β_e=3 or 4. The results presented here on γ_m are important data in order to save computer time in the simulation of electromagnetic properties of an FFO, such as the the output voltage, at the velocity-matching condition. Our calculations were carried out for the following cases:

(a) α changing from 0.10 to 0.50 with β=0.001 and 0.025 (Fig.2a);

(b) β changing from 0 to 0.050 with α=0.15 and 0.35 (Fig.2b).

It is evident from Fig.2 that the surface damping β affects γ_m only slightly for a given β_e (i.e., for a given oscillation frequency). The quasi-particle damping α and the external magnetic field β_e have a large effect on γ_m. We find that with β kept constant, γ_m increases nearly in proportion to α, i.e.,

$$\frac{d\gamma_m}{d\alpha} = k' \qquad , \qquad (3)$$

with $k'=\beta_e$ for β=0.001 and 0.025 (Fig.2a); and that with α kept constant, γ_m decreases slightly with increasing β (Fig.2b).

An approximate expression for γ_m can be derived[4] as

$$\gamma_m = a + \alpha (1+b) \beta_e \qquad (4)$$

where a is the value of the dc supercurrent averaged in space $<j_{s,dc}/j_c>_x$ which depends on α, β and β_e. Here $<...>_x$ means the average in space. a is usually around 0.1 to 0.3. b is the influence of the self-field on the dc voltage and is usually less then 10% of the external magnetic field. The expression of Eq.(4) fits our numerical results well.

Spatial Distribution of the AC Voltage Amplitude

At the velocity matching condition, i.e. $u=\bar{c}$ ($\gamma=\gamma_m$), the flux flow motion is unidirectional. The ac voltage amplitude V_{ac} is near zero at the end (x=0) where the vortices are nucleated, and it increaess gradually in the flux flow direction and reaches a maximum V_{acm} at the output end (x=L).

Fig.3 shows the spatial distribution of the ac voltage amplitude along the junction at different damping parameters. We observe that both dampings α and β strongly affect (decrease) the maximum ac output voltage V_{acm} which is $V_{osc}(L)$ at the velocity matching condition. All V_{acm} values are indicated by arrows in Fig.3. The same applied magnetic field β_e is kept for all the simulations in this figure which means all these curves actually correspond to the same output frequency.

The Maximum Output Voltage

For L=30, the detailed numerical simulations give the relationship between dampings and the maximum output voltage V_{acm}. It is the output ac voltage amplitude at the velocity matching condition, i.e., $V_{acm}=V_{ac}(L)$ at

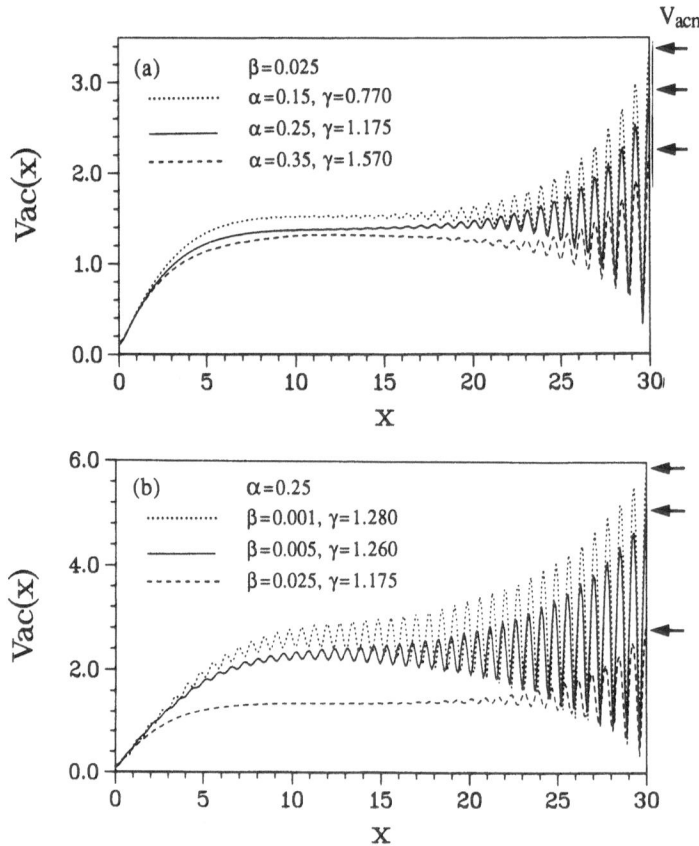

Fig.3 Effects of losses on the spatial distribution of the ac
voltage amplitude along the junction. (a)Curves are drawn for
different quasiparticle losses α with $\beta=0.025$ kept constant.

(b)Curves are drawn for different surface losses β with $\alpha=0.25$
kept constant. All are biased at their velocity matching
condition, $\gamma=\gamma_m$. Other parameters are L=30 and β_e=4. It should
be noted that γ_m depends on the losses.

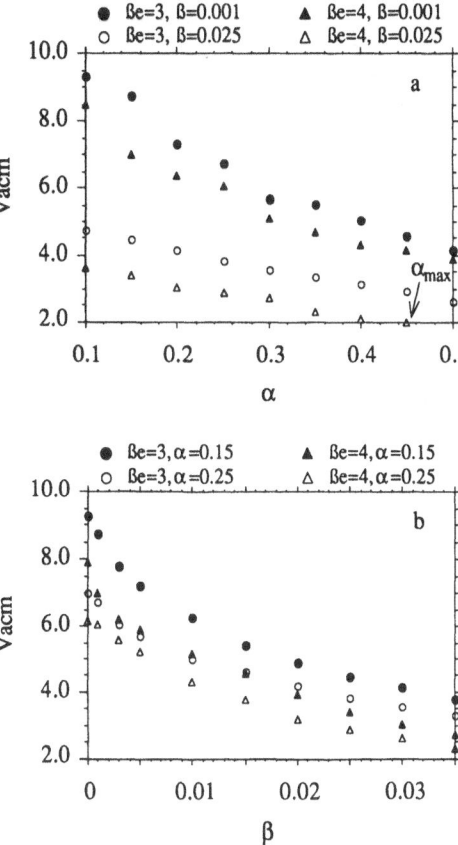

Fig.4 (a) The maximum ac output voltage V_{acm} as a function of quasi-particle tunneling loss α. (b) V_{acm} as a function of surface loss β. The external magnetic fields correspond to $\beta_e=3$ and 4 with L=30 kept constant. α_{max} in Fig.4(a) represents the maximum possible tunneling loss of an FFO for $\beta_e=4$ and $\beta=0.025$.

$\gamma=\gamma_m$. It is evident from Fig.4 that V_{acm} decreases with increasing dampings and increasing β_e. The surface damping β affects the value of V_{acm} greatly, although it affects the value of γ_m slightly; the flux-flow oscillation almost disappears if $\beta>0.035$.

This result on V_{acm} points out that the effect of surface damping on flux flow motion is quite high and it should not be excluded in the mathematical models[3] of an FFO. Actually, at high external magnetic field β_e, which corresponds to a high working frequency, this term has a larger effect than at low frequency (small β_e). It has a large damping effect on the flux-flow propagation along the long Josephson junction.

For an FFO, a suitable damping α should be chosen to keep both the travelling wave mode dominating in the junction[2] and the output voltage large enough at the velocity matching condition. The minimum α value can be estimated by using[2] $\alpha L \geq 3$. It equals to 0.1 for L=30. The maximum α value depends on β and on the magnetic field β_e; it is usually around 0.4. α_{max} in Fig.4(a) represents the maximum tunneling loss of an FFO for $\beta=0.025$ and $\beta_e=4$.

Possible Current Density Ranges for Designing an FFO

The quasiparticle loss α depends on the junction current density j_c. It can be rewritten as

$$\alpha = \frac{1}{I_c R_n} \left(\frac{j_c \phi_o}{2\pi C}\right)^{1/2} . \qquad (5)$$

It depends on junction parameters $I_c R_n$ (which in turn is proportional to the superconducting energy gap in a tunnal junction) and C.

Fig.5 shows α vs j_c for some typical junction materials (Table 1). It gives the possible j_c ranges for designing an FFO, knowing that α should be in the range of about 0.1 to 0.4. A high $I_c R_n$ product is quite important for an FFO working at high frequency.

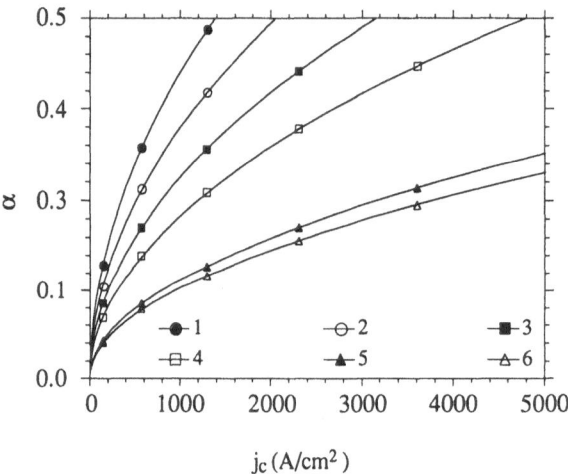

Fig.5 The dependence of quasiparticle loss α on junction current density j_c of some typical junction materials (see Table 1 for parameters).

Table 1. Physical qualities of typical junctions

No.	Junction Type	IcRn(mV)	C(fF/μm^2)	Ref.
1	Pb-Alloy	1.15	14	(1)
2	Pb-Alloy	1.40	14	(1)
3	Nb/α-Si/Nb	1.30	25	(6)
4	Nb/α-Si/Nb	1.60	25	(6)
5	Nb/Al-Al$_2$O$_3$/Nb	1.68	60	(6)
6	Nb/Nb oxide/Nb	1.20	140	(6)

CONCLUSION

The effect of losses (α, β) on the output voltage V_{osc} of a flux-flow type Josephson oscillator formed by a long overlap Josephson junction has been studied. The output ac voltage V_{osc}, which is proportional to the square root of the output power of the oscillator, reaches its maximum V_{acm} at the velocity matching condition $\gamma = \gamma_m$. V_{acm} is strongly dependent on the quasiparticle tunneling loss α and the surface loss β. It decreases with increasing dampings (α, β) and increasing external magnetic field β_e. The surface damping β affects the value of V_{osc} greatly, although it affects the height of the VM step γ_m slightly; the flux-flow oscillation almost disappears for $\beta > 0.035$. It is important to include a surface damping term ($-\beta\phi_{xxt}$) in the mathematical model (Eq.1) for describing an FFO. The relation between the quasiparticle loss α and the critical current density j_c depends upon junction material parameters, especially upon the I_cR_n product. A high I_cR_n product is important in designing such an oscillator. The possible j_c ranges for an FFO is determined by knowing that α should be in the range of about 0.1 to 0.4.

ACKNOWLEDGEMENTS

Useful discussions with Prof.Tord Claeson and Dr.Dag Winkler are greatly appreciated. This work was supported by the Swedish Board of Technical Development.

REFERENCES

1. T.Nagatsuma, K.Enpuku, F.Irie and K.Yoshida, Flux-Flow Type Josephson Oscillator for Millimeter and Submillimeter Wave Region, J.Appl.Phys.54: 3302(1983).
2. T.Nagatsuma, K.Enpuku, K.Yoshida and F.Irie, Flux-Flow-Type Josephson Oscillator for Millimeter and Submillimeter Wave Region.II.Modeling, J.Appl.Phys.56:3284(1984).
3. T.Nagatsuma, K.Enpuku, K.Sueoka, K.Yoshida and F.Irie, Flux-Flow-Type Josephson Oscillator for Millimeter and Submillimeter Wave Region. III.Oscillation Stability, J.Appl.Phys.58:441(1985).
4. Y.M.Zhang and P.H.Wu, Numerical Calculation of the Height of Velocity-Matching Step of Flux-Flow Type Josephson Oscillator, J.Appl.Phys., to be published in the November 1, 1990 issue.
5. A.Davidson and N.F.Pederson, Effect of Surface Losses on Soliton Propagation in Josephson Junction, Appl.Phys.Lett.48:1306(1986).
6. H.Hayakawa, Josephson LSI Technology and Circuits, in: "Superconducting Electronics", H.Weinstock and M.Nisenoff eds., NATO ASI Series, Vol.F59, Springer Press, Berlin (1989).

THIN FILM SURFACE RESISTANCE MEASUREMENTS

FOR SUPERCONDUCTING CAVITY APPLICATIONS

A. Andreone[1], C. Attanasio[2], A. Di Chiara[1],
L. Maritato[2], A. Nigro[2], V. Palmieri[3],
G. Peluso[1], R. Preciso[3] and R. Vaglio[2]

[1]Dipartimento di Scienze Fisiche, Università di Napoli, and INFN
Sezione di Napoli, Napoli, Italy
[2]Dipartimento di Fisica, Università di Salerno, Salerno, Italy
[3]Laboratori Nazionali di Legnaro, INFN, Legnaro (Pd), Italy

INTRODUCTION

First attempts to use rf superconductivity for particle accelerators are dated more than thirty years ago [1 − 2]. Since then many efforts have been made in trying to develop high field rf superconducting cavities in order to achieve higher energies in both electron and ion accelerators. The main reason to utilize superconducting resonators is related to their capability of sustaining rf electrical fields of several MV/m with high quality factor. Sheet metal cavities of Niobium are planned to be installed in different storage rings such as LEP, DESY, KEK, CEBAF, etc. OFHC copper cavities internally coated with superconducting thin films (few micron thick) have also given very good results [3 − 4]. This last configuration takes advantage of the high copper thermal conductivity together with the possibility of reduced costs. Electroplated Lead has been used in the past but more recently sputtering techniques have given the possibility of obtaining good quality Nb [5] coated cavities which have revealed higher performances (better quality factor at high accelerating fields) in respect to the Nb bulk cavities. Moreover the sputtering application gives the chance of using non conventional superconducting materials for accelerating cavities [6 − 8]. On the other hand some cavities, coated with high T_c material as NbN, Nb_3Sn [9] and $(NbTi)N$ [10] thin films, have shown poorer performances, in respect to what expected by the BCS theory, with a strong dependance of the quality factor on the rf field amplitude. This behavior is essentially caused by the so called residual losses [11 − 12] the origin of which can be manyfold and, at this time, is not yet completely explained. The understanding of the causes of such losses is essential for further progresses in this field. A model in which surface losses are mainly caused by grain boundaries, in polycrystalline Nb and Nb−alloys coated cavities has been recently proposed [13 − 14]. The availability of a suitable experimental set-up without the need of using large resonating structures, can give the possibility of easily testing new coating materials also in the light of this model. Measurements of the surface resistance of superconducting samples by using a microstrip resonator are currently in progress.

Nonlinear Superconductive Electronics and Josephson Devices
Edited by G. Costabile *et al.*, Plenum Press, New York, 1991

155

SUPERCONDUCTING RESONANT CAVITIES

An important parameter used in evaluating cavity performances is the quality factor Q, defined as

$$Q = \omega \frac{U}{P} \qquad (1)$$

where U is the energy stored in the cavity and P is the dissipated power on the cavity walls. The energy U can be written as

$$U = \frac{\mu_0}{2} \int_V H^2 dv$$

while

$$P = \frac{R_s}{2} \int_\Sigma H^2 ds$$

where H is the magnetic field inside the cavity, V is its volume, Σ is the cavity walls surface, R_s is the surface resistance and ω is the operating frequency. From the previous relations the quality factor can be written as $Q = \Gamma/R_s$, where Γ is a factor depending only on the geometry of the resonator and on the operating frequency. This clearly shows that in order to obtain high Q, low R_s are needed. For a normal metal the surface resistance is given by

$$R_s = \left(\frac{\omega \mu_0}{2\sigma_n}\right)^{1/2} = \frac{1}{\delta} \frac{1}{\sigma_n} \qquad (2)$$

where σ_n is the dc conductivity of the metal and δ is the skin penetration depth. This formula is not valid anymore in the "anomalous skin effect" limit [15], where the electronic mean free path becomes larger than δ. In this region R_s becomes indipendent on the dc conductivity and this leads to an intrinsic limit in obtaining high Q values even using very pure materials at low temperatures. For a copper cavity, operating at 500 MHz, Q nearly equal to 10^5 can be obtained. For superconducting materials equation (2) is still valid if we introduce an effective conductivity σ_{eff} which is a complex quantity related to dissipation due to both normal and superconducting electrons. Because the surface conductivity of a superconductor is very high, Q values up to $10^9 - 10^{10}$ can be obtained for bulk Niobium cavity at 500 MHz at the temperature of 4.2 K. However, for high accelerating fields, the performances of bulk cavities are limited by the thermal quench, due to local surface defects. Infact, normal zones on the surface of the cavity walls can start to dissipate and, because of the poor thermal conductivity of Niobium, the area of these zones can rapidly increase giving rise to a sudden decrease of the quality factor. In order to reach high accelerating fields with low power losses, the thermal properties of Niobium used for the realization of bulk cavities have been improved by in vacuum annealing procedures. The cleanliness during the process was also been pointed out to be very important to avoid dust inclusion inside the cavity. Generally Nb [5] sputter coated cavities have better surface properties along with the higher copper thermal conductivity. At the present the best Nb sputter coated cavities have shown quality factors of $\sim 3 \cdot 10^9$ at accelerating fields of $\sim 7\ MV/m$ (Q at zero field of $6 \cdot 10^9$) with their performances essentially limited by the electron loading. Morover this technique in principle allows to coat a copper cavity using other superconducting materials with better rf properties. It is then important to establish useful criteria to choose a superconductor for these applications. The surface resistance of a superconductor is well desribed by the expression

$$R_s = R_s(BCS) + R_{res} \qquad (3)$$

where R_{res}, related with the residual losses, is temperature indipendent and can have different frequency dependences. $R_s(BCS)$ can be obtained from the microscopic

BCS theory. In the local dirty limit (short mean free path), if $T \leq T_c/2$ and $\omega \ll 2\Delta/\hbar$, we obtain [8]

$$R_s(BCS) = \frac{1}{2\lambda}\frac{\sigma_1}{\sigma_2^2} \sim \omega^{2-\alpha}\rho_n^{1/2}e^{-1.76(T_c/T)} \tag{4}$$

Here λ is the London penetration depth, σ_1 and σ_2 are respectively the real and imaginary part of σ_{eff}, ρ_n is the normal state resistivity, T_c is the critical temperature and T is the bath temperature; α is a numerical parameter which can assume values between 0 and 0.5. Equation (4) shows that, to obtain low surface resistance, materials with good metallic behavior (low ρ_n) and high T_c values are requested. Another important requisite for a superconducting material to be used in accelerating cavities is the high critical magnetic field, related to T_c by

$$H_c \sim \gamma^{1/2}T_c$$

where γ is the "Sommerfeld constant" related to the density of states at the Fermi energy. Low sensitivity to radiation damage and low secondary emission coefficient should also be considered as an important property for materials to be used for such application. Based on these considerations materials like NbN [6], $MoRe$ [7], and $(NbTi)N$ [8] have been proposed as coating materials for accelerating cavities. Recently $(NbTi)N$ [10] sputter coated cavities have been realized at the CERN laboratories in Geneva. They have shown higher values of the quality factor Q in small accelerating fields E_a with respect to Niobium cavities even tough the Q versus E_a curves present a stronger slope. However the improvement of the deposition techniques can limit this effect. This behavior, probably related with the granular nature of these films, has been qualitatively explained by modelling a polycrystalline film as a network of superconducting grains coupled via Josephson junctions [13 − 14]. In this case rf losses at the grain boundaries are the main source of residual resistance appearing in equation (3). Morover for the residual losses is foreseen a square frequency dependance in this model. The use of a microstrip resonator is then important to easily check these ideas without coating the large operating cavities.

SURFACE RESISTANCE MEASUREMENT

The quality factor Q of a microstrip resonator is given by

$$\frac{1}{Q} = \frac{1}{Q_0} + \frac{1}{Q_D} + \frac{1}{Q_R} \tag{5}$$

where Q_0, Q_D, Q_R, are respectively the quality factors related to the surface resistance of the resonator, to the polarization in the dielectric and to emission of radiation. More precisely we have for a linear resonator

$$Q_0 = \frac{\mu_0\omega\sqrt{\epsilon}}{2\eta\,\alpha_c(w/h)} \tag{6a}$$

$$Q_R = \frac{3\epsilon\pi^2\eta\,Z(w/h)}{8\mu_0^2\omega^2h^2} \tag{6b}$$

$$Q_D \sim (tan\delta)^{-1} \tag{6c}$$

where w and h are the width of microstrip and height of the substrate, ϵ is the substrate effective dielectric constant and $\eta=120\,\pi$ ohm. Functions α_c and Z are explicitly given by [16]

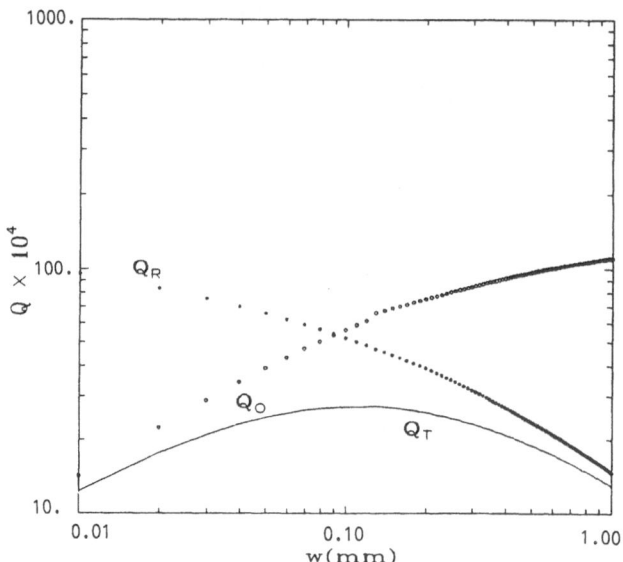

Fig.1 Quality factor for a linear resonator against the microstrip width. Q_0 and Q_R are the conductive and the radiative quality factors defined in the text and $1/Q_T = 1/Q_0 + 1/Q_R$.

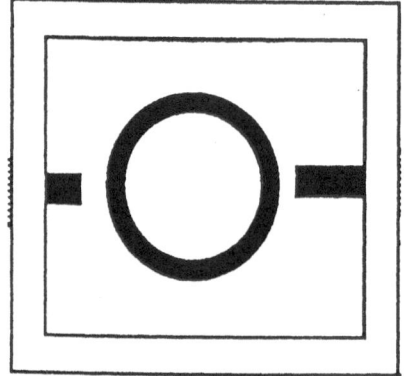

Fig.2 Schematic view of the ring resonator together with the input and the output antennas.

$$Z\left(\frac{w}{h}\right) = \begin{cases} \frac{1}{2\pi} ln\left[8(h/w) + 0.25(w/h)\right] & (w/h \le 1) \\ \\ \left\{(w/h) + 1.393 + 0.667ln\left[w/h + 1.444\right]\right\}^{-1} & (w/h \ge 1) \end{cases} \tag{7a}$$

and

$$\alpha_c\left(\frac{w}{h}\right) = \begin{cases} 1.38 A \frac{R_s}{hZ(w/h)}(32 - (w/h)^2)/(32 + (w/h)^2) & (w/h \le 1) \\ \\ \frac{6.1\times10^{-5}A}{h} R_s Z(w/h)\epsilon\left[w/h + (0.667w/h)/(w/h + 1.444)\right] & (w/h \ge 1) \end{cases} \tag{7b}$$

where α_c is given in dB/unit length and A is a quantity related to the strip thickness [16]. Because Q_0, is inversely proportional to R_s, if the other mechanisms give a small contribution (substrate having $tan\delta$ less than 10^{-6} should be used), a measure of the quality factor determines the surface resistance of the superconductor. The expressions for Q_0 and Q_R have been calculated for a Niobium sample (R_s=0.4 $\mu\Omega$) on a sapphire substrate ($tan\delta$ less than 10^{-7} [17]) having ϵ=11 and h=0.127 mm at the frequency of 1 GHz. We have seen (figure 1) that in order to have $Q_0 \sim 10^5$, lower than Q_R, at less a microstrip with a width of 100 μm should be used. However we should point out that expressions given above for the functions $Z(w/h)$ and $\alpha_c(w/h)$ refer to a linear resonator. A ring resonator, schematically shown in figure 2, could further reduce the radiative losses [18]. In our laboratory preliminary measurements of Nb surface resistance at the liquid helium temperature are in progress. The cavity is made of a Nb ring resonator deposited by magnetron sputtering on a sapphire substrate 0.127 mm thick. The ground plane is made of bulk Niobium and the requested definition of the ring geometry has been obtained by means of a photolithografic process. The resonant peaks in the curve of the ratio between the input and output power versus frequency gives a direct measurement of the microstrip quality factor Q. With this method quality factor up to $10^5 - 10^6$, which are typical for good quality Niobium thin film resonators using a microstrip geometry, can be detected. The accuracy of the method can be further improved by performing measurements in which surface resistance of the investigated material is compared with that of a reference ring resonator.

ACKNOWLEDGMENTS

This work has been supported by the Istituto Nazionale di Fisica Nucleare through the program "SPARTA" .

REFERENCES

[1] H. Piel in Proceedings of the 12th International conference on High-Energy Accelerators, Chicago, 1983.

[2] M. Tigner and H. Padamsee in CLNS-82/553, Cornell 1982, AIC Conference of SLAC Summer Accelerator School.

[3] C. Benvenuti, N. Circelli and M. Hauer, Appl. Phys. Letts., 45, 583 (1984).

[4] G. Arnolds-Mayer and W. Weingarten, IEEE Trans. Magn., MAG − 23(2), 1620 (1987).

[5] C. Benvenuti, D. Bloess, E. Chiaveri, N. Hilleret, M. Minestrini and W. Weingarten, in Proceedings of the Third Workshop on RF Superconductivity, K.W. Shepard Ed., (ANL, Argonne, Illinois, 1988), pag. 445.

[6] E.J. Cukauskas, W.L. Carter and S.D. Quadri, J. Appl. Phys., 57, 2538 (1985).

[7] A. Andreone, A. Barone, A. Di Chiara, F. Fontana, F. Mascolo, V. Palmieri, G. Peluso, G. Pepe and U. Scotti Di Uccio, J. of Superconduc., 2, 4 (1989).
A. Andreone, A. Barone, A. Di Chiara, F. Mascolo, V. Palmieri, G. Peluso, G. Pepe and U. Scotti Di Uccio, IEEE Trans. Magn., MAG − 25(2), 1972 (1989).

[8] R. Di Leo, A. Nigro, G. Nobile and R. Vaglio, J. Low. Temp. Phys., 78, 41 (1990).

[9] M. Peineger, M. Hein, N. Klein, G. Müller, H. Piel and P. Thüns, in Proceedings of the Third Workshop on RF Superconductivity, K.W. Shepard Ed., (ANL, Argonne, Illinois, 1988), pag. 503.

[10] C. Benvenuti, in Proceedings of the Fourth Workshop on RF Superconductivity, Y. Kojima Ed., (Tsukuba, Japan, 1989), pag. 869.

[11] H. Piel, in Proceedings of the CERN Accelerator School, "Superconductivity in Particle Accelerators" S. Turner Ed., (CERN, Geneva,1989), pag. 149.

[12] J. Halbritter, in Proceedings of the Second Workshop on RF Superconductivity, H. Lengeler Ed. (CERN, Geneva, 1984), pag. 427.

[13] C. Attanasio, L. Maritato and R. Vaglio, Phys. Rev. B, (1990), to be published.

[14] C. Attanasio, L. Maritato and R. Vaglio, IEEE Trans. Magn., to be published.

[15] G.E.H. Reuter and E.H. Sondheimer, Proc. Roy. Soc., A195, 336 (1948).

[16] K.C. Gupta, R. Garg and I.J. Bahl, "Microstrip Lines and Slotlines", Artech House, pag. 88.

[17] A.C. Anderson, R.S. Withers, S.A. Reible and R.W. Ralston, IEEE Trans. Magn., MAG − 19(1), 485 (1983).

[18] J.J. Jimenez and J.J. Guijarro, Rev. Phy. Appliquee, 8, 279 (1973).

NOISE IN SUPERCONDUCTING JOSEPHSON JUNCTIONS

J. Bindslev Hansen

Physics Laboratory I/MIDIT
The Technical University of Denmark
DK-2800 Lyngby, Denmark

1. INTRODUCTION

This is a review of the noise properties of superconducting Josephson junctions, i.e. superconducting tunnel junctions and superconducting metallic weak links (point contacts, microbridges). For **tunnel junctions** theoretical models have been worked out and they agree fairly well with experimental results obtained on well-characterized tunnel barriers. For **metallic weak links** no general theory exists yet, and calculations of the noise properties based on the phenomenological Resistively-Shunted Junction (RSJ) model predicts white noise power levels which are 2-100 times smaller than the measured values.

Noise sets the ultimate limits of performance for all electrical devices. We shall here review the noise properties of superconducting Josephson junctions which are of importance in applications of superconducting electronics in limiting quantities like the detector sensitivity and the oscillator linewidth.

This review covers the noise properties of junctions only, while other contributions to this book address the noise in applications of superconducting junctions.

1.1. Noise from dissipation

The fluctuation-dissipation analyses of Nyquist, and Callen and Welton[1,2] show how noise is generated in a given physical system, e.g. a Josephson junction, through the interaction between the junction and a heat bath, i.e., through the dissipation in the system. **Noise is the stochastic component of the junction response** and since **dissipation** is the ability of the junction to convert ordered energy into disordered energy, the two phenomena, noise and dissipation, are intimately connected. **Ordered energy** in the **junction** is through **dissipative processes** ("friction") transferred to the **heat bath** where it is **randomized** ("disorder") and **fed back to the junction as fluctuations** (the heat bath may be modelled as a very large collection

Nonlinear Superconductive Electronics and Josephson Devices
Edited by G. Costabile *et al.*, Plenum Press, New York, 1991

161

of harmonic oscillators)[3]. This connection between **friction (dissipation)** and **noise (fluctuations)** is formulated in the fluctuation-dissipation theorem, see below.

Theoretically, a linear system in thermal equilibrium with voltage bias is considered. The impedance $Z(\omega) = R(\omega) + iX(\omega)$ which is given by $V(\omega) = Z(\omega)I(\omega)$ determines the dissipation in the system:

$$\overline{P(\omega)} = \overline{I(\omega)V(\omega)} = \overline{V^2(\omega)}/Z(\omega) = \frac{1}{2}\overline{V^2(\omega)}R(\omega)/|Z(\omega)|^2 .$$

The current fluctuations in the system are given by the mean square deviation:

$$\overline{|\Delta I|^2} = \overline{|I(t) - \bar{I}|^2} .$$

In practice, the spectral density of the noise power, $S_I(\omega)$, is measured:

$$S_I(\omega) \quad = \quad \lim_{t \to \infty} \left(\frac{\overline{|I(\omega)|^2}}{2t} \right) \tag{1}$$

where $I(\omega)$ is the Fourier component at ω:

$$I(\omega) = \frac{1}{\sqrt{2\pi}} \int_{-\infty}^{\infty} (\Delta I) e^{i\omega t} dt .$$

Integrating $S_I(\omega)$ over all frequencies we get the mean square deviation:

$$\overline{(\Delta I)^2} \quad = \quad \int_{-\infty}^{\infty} S_I(\omega) d\omega .$$

The fluctuation-dissipation theorem as derived by Callen and Welton[2] can be expressed as a relation between $S_I(\omega)$, the spectral density of the current noise power, and the dissipation $R(\omega)$ at temperature T:

$$S_I(\omega) \quad = \quad \frac{1}{\pi R(\omega)} \frac{\hbar\omega}{2} \coth\left(\frac{\hbar\omega}{2k_B T} \right) \tag{2}$$

1.2 How to measure noise

Equation (1) gives the recipe for measuring the spectral noise power density, see the schematic in Fig. 1:

1) use **a narrowband filter** on the current response to
   find the Fourier component $I(\omega)$;
2) then use **a square law detector** (rectifier) followed by
3) an **averaging circuit** to get $\overline{|I(\omega)|^2}$ (which must be normalized).

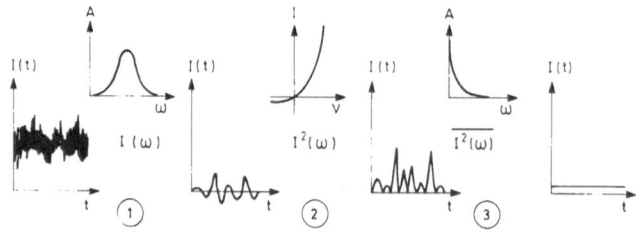

Fig. 1.   Pictorial presentation of the measurement procedure for the current noise power spectral density, see Eq. (1).

In the time regime, the auto-timecorrelation function, $F_I(\tau)$, may be introduced. $\overline{F_I(\tau)} = \overline{I(t+\tau)I(t)}$ is related to the noise power density through the Fourier transformation (Wiener-Khintchine's theorem):

$$S_I(\omega) = \frac{2}{\pi} \int_0^\infty \overline{F_I(\tau)} \cos(\omega\tau) d\tau.$$

1.3. Three noise regimes: $1/f^\alpha$, "white", and quantum

In general, the measured current noise power, $S_I(\omega)$, depends on frequency. Figure 2 serves to illustrate the different noise regimes which are typically found in a conducting channel:

1) at **low** frequencies: **excess noise** $1/f^\alpha$ with $\alpha \approx 0.8 - 1.2$
   (flicker noise or 1/f noise);
2) at **intermediate** frequencies: **"white" noise** (Nyquist-Johnson
   thermal noise or shot noise) and,
3) in principle, at **very high** frequencies: **quantum noise**
   from zero-point fluctuations (hf/2 fluctuations). In practice, however, the system response is never instantaneous but limited by an RC or L/R characteristic time, τ defining a characteristic frequency, $f_{cut-off} = 1/(2\pi\tau)$. The system response is therefore **suppressed at high frequencies** and the quantum noise is usually not seen. $f_{cut-off}$ is given by the smaller of $f_{RC} = 1/(2\pi RC)$ and $f_{RL} = R/(2\pi L)$.

2. NOISE REGIMES IN JUNCTIONS

2.1. $1/f^\alpha$ noise

The origin of the **excess low frequency noise** is not fully understood neither in low-T_C tunnel junctions and weak links nor in the present low-quality high-T_C Josephson SNS junctions and microbridges based on ceramic cuprates. This form of noise has generally a complicated functional dependence on parameters like

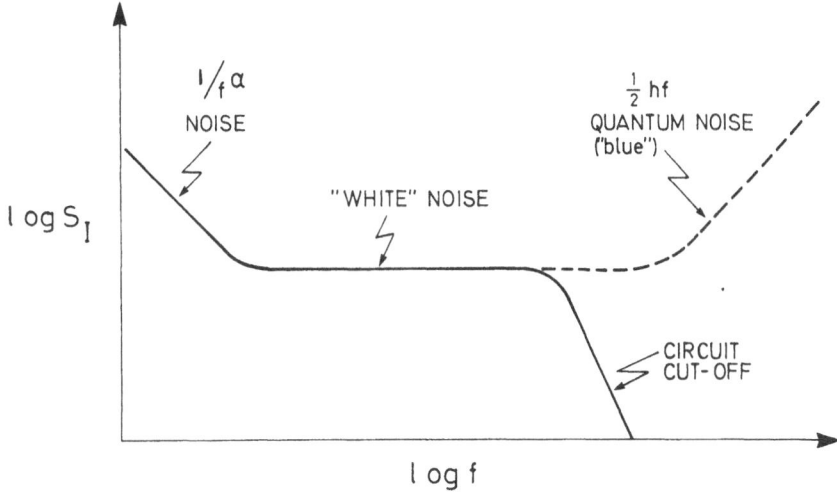

Fig. 2.   Double-logarithmic schematic of the noise power vs. frequency for a conducting channel.

material, size, history, voltage, and temperature, see refs. 4 and 5 for recent reviews. The most successful model so far is valid for SIS tunnel junctions[6]. It is based on the existence of electron states ("traps") in the imperfect (oxide) barrier, implying that the tunneling conductance depends on the electron occupancy of traps with different activation energies, each trap contributing a Lorentzian noise peak. A smooth distribution of trap energies sums up to a $1/f$ noise spectrum. Strong evidence for this model is provided by the observation of distinct Lorentzian noise peaks in nearly perfect barriers whose low frequency conductance is dominated by a few traps. Random telegraph switching noise generated by a single trap has been observed in the tunneling conductance[7].

2.2. Energy scales for thermal, shot and quantum noise

In the following we will neglect the $1/f^{\alpha}$ noise and focus the attention on **thermal noise, shot noise and quantum noise** in Josephson elements. There are at least four important energy quantities involved:

1) the Josephson coupling energy, $E_J = \hbar I_c / 2e$

2) the thermal noise energy, $\qquad E_T = k_B T$

3) the potential energy, $\qquad\quad E_P = eV$

4) the photon energy, $\qquad\qquad E_f = \hbar \omega = hf$

To convert between units we note that the four energies E_J, E_T, E_P and E_f are equal for the following values of I_c, T, V, and f:

$$0.05 \ \mu A \ \approx \ 1 \ K \ \approx \ 0.1 \ mV \ \approx \ 25 \ GHz.$$

Summary of the expressions for the current noise power spectral density, $S_I(f)$, of a **linear system**:

a) in thermodynamic equilibrium,
the **thermal noise** limit: $k_B T \ \gg \ eV, \ hf$

b) out of thermodynamic equilibrium with voltage bias V,
the **shot noise** regime: $eV \ \gg \ k_B T, \ hf$

c) in the photon noise limit:
the **quantum** regime: $\quad hf \ \gg \ k_B T, \ eV.$

2.3. Case a): Linear system in thermal equilibrium (k_BT >> eV)

Nyquist and Johnson[1] found for the classical case ($k_B T \gg hf$):

$$S_I(f) = \frac{4 k_B T}{R} \quad for \quad k_B T \ \gg \ eV, hf.　\qquad (3)$$

Callen and Welton[2] considered the quantum-mechanical case and derived Eq.

(2) for the **generalized Nyquist noise**:

$$S_I(f) = \frac{2hf}{R}\coth(hf/2k_BT) = \frac{4}{R}\left\{\frac{hf}{\exp(hf/k_BT)-1} + \frac{1}{2}hf\right\} \qquad (2^*)$$

$$\text{"Planck"} \qquad\qquad \text{"blue"}$$

where the second way of writing the expression separates the low frequency thermal (Planck) noise from the high frequency noise hf/2. As expected, Eq. (2*) reduces to the classical limit Eq. (3) for $k_BT >> hf$, but in the **quantum limit** of dominant photon energy (**case c** above) it gives:

$$S_I(f) \quad = \quad \frac{2hf}{R} \quad for \quad hf \gg k_BT, \quad k_BT \gg eV \qquad (4)$$

the **zero-point fluctuations** (also sometimes called "blue" noise). We note again that the divergent $S_I(f)$ is limited at high frequencies by the finite circuit response time.

2.4. Case b): Linear system out of thermodynamic equilibrium: Shot noise in tunnel junction with voltage bias

For a tunnel junction with **voltage control** (biased at constant $\overline{V}$) Rogovin and Scalapino[8] derived the following expression for the **quasiparticle shot noise** (**case b**):

$$S_{I_{qp}}(f) = e\left\{ I_{qp}\left(\overline{V} + \frac{hf}{e}\right)\coth\left(\frac{e\overline{V} + hf}{2k_BT}\right) + I_{qp}\left(\overline{V} - \frac{hf}{e}\right)\coth\left(\frac{e\overline{V} - hf}{2k_BT}\right)\right\} \qquad (5)$$

Zorin et al. have obtained the corresponding expression[9,10] for the general case of arbitrary source impedance.

In the **high frequency** limit, $hf \gg e\overline{V}$, Eq. (5) agrees with the generalized Nyquist expression, Eq. (2):

$$S_{I_{qp}}(f) \quad = \quad \frac{2hf}{R}\coth\left(\frac{hf}{2k_BT}\right) \qquad for \quad hf \gg e\overline{V} \qquad (2^*)$$

and in the **low frequency** limit, $e\overline{V} \gg hf$, Eq. (5) reduces to the following "white" noise:

$$S_{I_{qp}} \quad = \quad 2eI_{qp}\coth\left(\frac{e\overline{V}}{2k_BT}\right) \qquad for \quad e\overline{V} \gg hf. \qquad (6)$$

In the low temperature, low frequency limit, $e\overline{V} \gg hf, k_BT$, Eq. (6) yields the expression for Schottky's shot noise:

$$S_{I_{qp}} \quad = \quad 2eI_{qp}(\overline{V}) \qquad for \quad e\overline{V} \gg hf, \quad k_BT. \qquad (7)$$

Note that Eq. (6) reduces to the Nyquist thermal noise expression in the classical limit, $k_BT \gg e\overline{V}$:

$$S_{I_{qp}} = \frac{4 k_B T I_{qp}}{\overline{V}} = \frac{4 k_B T}{R_{qp}}, \quad for \quad k_B T \gg e\overline{V} \gg hf. \quad (3^*)$$

where by definition $R_{qp} = \overline{V} / I_{qp}$.

2.5 Shot noise in the supercurrent ?

Since the supercurrent, I_p, across a Josephson tunnel junction is a result of coherence between the two Cooper pair wavefunctions of the two superconducting electrodes there is no dissipation involved and hence no fluctuations. Only in case of interaction between the Cooper pair wavefunctions and a **dissipative environment** will fluctuations be introduced into the supercurrent. When the Cooper pair condensate couples to Josephson plasmons in a lossy junction or to the photon field in a lossy cavity, the dissipation introduces shot noise in the supercurrent and also in the pair-quasiparticle interference current. These noise currents have been discussed by Rogovin and Scalapino[8], the expressions are analogous to Eq. (5).

2.6 Summary of the noise regimes

Low frequency excess noise, "$1/f noise$",
$$S_I(f) = F(V, T, size, material, history)/f^{\alpha} \quad with \quad \alpha \approx 0.8 - 1.2$$

Thermal noise ("white"),
$$k_B T \gg hf, eV \quad : \quad S_I = 4 k_B T / R.$$

Shot noise ("white"),
$$eV \gg k_B T \quad : \quad S_{I_{qp}} = 2 e I_{qp}.$$

Quantum noise ("blue),
$$hf \gg k_B T, \quad eV \quad : \quad S_I(f) = \frac{4}{R}\left(\frac{hf}{\exp(hf/k_B T) - 1} + \frac{1}{2}hf\right) \approx 2hf/R$$

In addition, for **driven Josephson junctions**, **deterministic chaos** produces extremely large noise power, often corresponding to an effective noise temperature of more than 10^6 K. Chaotic noise may also be seen in combination with thermal fluctuations, for example as thermally induced probing of unstable bias points[11].

3. NOISE IN JOSEPHSON JUNCTIONS

3.1 Noise effects in Josephson elements

Experimentally, the effects of noise on Josephson junctions are **observed in the following ways:**

1) **Direct noise measurements using a low-noise detection system, e.g. a SQUID.**

2) **Measurements of noise induced switching or rounding of the I-V characteristics at I_c.**

3) **Measurements of the linewidth of the radiation at the Josephson frequency, $f_J = 2eV/h$.**

4) **For driven junctions, measurements of the noise rounding of high frequency phase-locked current singularities, Shapiro current "steps", and of photon-assisted tunneling, PAT, "steps" in the I-V characteristics.**

3.2. Theoretical models of Josephson junctions

The full quantum-mechanical microscopic model of a Josephson tunnel junction as derived by Werthamer[12] is a complex non-linear integro-differential equation which may be solved numerically[13]. The inclusion of stochastic terms to describe the effects of noise would require very large computer resources and to our knowledge this has not yet been done. As far as superconducting weak links (microbridges and point contacts) are concerned, no general microscopic model has been worked out yet, although progress has been made[14]. In this situation, most researchers in the field have resorted to the simplest, and also for other purposes most widely used phenomenological model of Josephson junctions: the Resistively-Capacitively-Shunted Junction (RCSJ) model. In this model the critical current, I_c, the damping $\approx 1/R$ and the capacitance, C, are phenomenological parameters which do not depend on frequency. The model is a good approximation to the classical limit, $k_B T \gg e\overline{V}$, it does not include shot noise. Josephson junctions are, in general, low impedance devices and therefore current controlled. For a small junction (without spatial variation) the RCSJ model with a dc bias I and a noise current $\tilde{I}_N = \sqrt{\overline{(\Delta I)^2}}$ leads to the following Langevin equation for the Josephson phase, ϕ :

$$\phi_{tt} \quad + \quad \alpha\phi_t \quad + \quad \sin\phi \quad = \quad i \quad + \quad \tilde{i}_N(t) \tag{8}$$

where the subscript, t, denotes differentiation with respect to time, which is normalized to the inverse maximum plasma frequency, $\omega_{po}^{-1} = (\sqrt{2eI_c/(\hbar C)})^{-1}$. $\alpha = 1/\sqrt{\beta_c} = \omega_{RC}/\omega_{po} = (RC\omega_{po})^{-1}$ is the damping parameter (β_c is the Stewart-McCumber parameter). Currents are normalized to the unperturbed critical current, $i = I/I_c$ and $\tilde{i}_N(t) = \tilde{I}_N/I_c$. In the RCSJ model the current noise is simply the Nyquist "white" noise as given by T and R: $S_I = 4k_B T/R$ and the auto-timecorrelation function is

$$\overline{F_I(\tau)} = \overline{I(t+\tau)I(t)} = \frac{2k_B T}{\pi R}\int_0^\infty \cos\omega\tau d\omega = 2k_B T\delta(\tau)/R.$$

Equation (8) is a non-linear second order differential equation with a stochastic driving term. It is in general not analytically tractable, not even for $\tilde{I}_N = 0$. The equation also describes a variety of mechanical systems, of which the most intuitively appealing is the motion in a gravity field of a small ball on a tilted washboard immersed in a viscous fluid with Brownian noise. The bias current, i, controls the tilt of the potential $U = -E_J(\cos\phi + i\phi)$, see Fig. 3. For $\alpha < 1$ the motion is underdamped resulting in **hysteretic** behavior (unless the noise dominates), whereas for $\alpha > 1$ the dynamics are overdamped and the force-velocity curve (the I-V characteristics) is **non-hysteretic.**

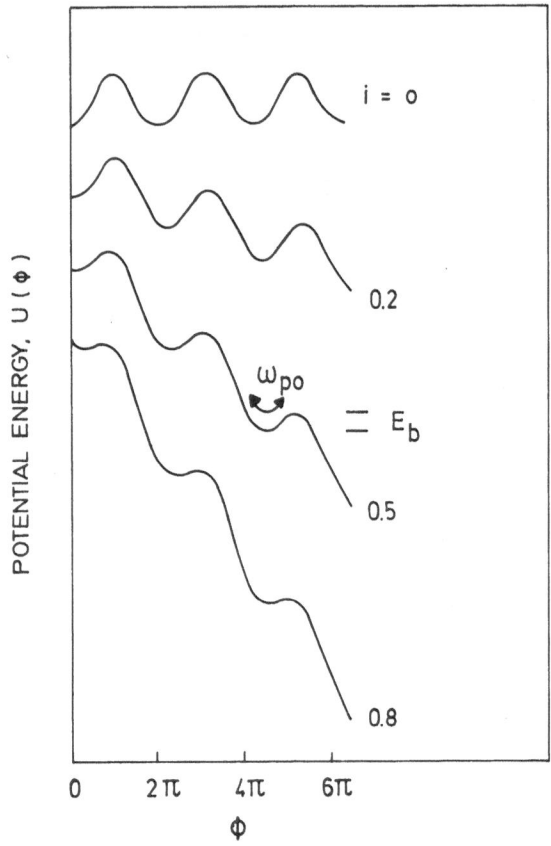

Fig. 3.   The tilted wash-board potential of the RCSJ-model, Eq.(8): The plot shows the potential energy $U(i) = -E_J(\cos\phi + i\phi) = -(\hbar I_c/2e)(\cos\phi + i\phi)$ vs. the Josephson phase ϕ for a current biased junction with i = I/I_C. The potential barrier, $E_b(i)$ is indicated for i = 0.5. The natural resonance frequency in the zero-voltage state is the plasma frequency, $\omega_p = \omega_{po}\sqrt{\cos\phi_o} = \sqrt{2eI_c\cos\phi_o/\hbar C}$.

3.3. Thermally activated escape from the zero-voltage state, $\overline{V} = 0$

Noise has the effect of perturbing the dc I-V characteristics around the critical current. In the limit where thermal fluctuations dominate, $E_J > k_BT \gg eV, hf$ the perturbation of the $\overline{V} = 0$ state may be treated within Kramers general reaction rate theory[15].

168

Within this theory Ambegaokar and Halperin (AH)[16] treated the **overdamped** RSCJ-model in the **C = 0 limit** (valid for point-like junctions like microbridges and point contacts) and found analytical expressions for the thermal rounding of the I-V characteristics, see Fig. 4a, as a function of the noise parameter $\Gamma = E_J / E_T = \hbar I_c / 2 e k_B T$, the ratio between the Josephson coupling energy, E_J, and the energy of the thermal fluctuations. Fig. 4b reproduces recent data on microbridges of the ceramic superconducting cuprate Y-Ba-Cu-O (T_c = 92 K). As seen, the AH model provides excellent fits to the experimental I-V curves[17,18].

In the **underdamped** regime ($\alpha < 1$), noise induces premature switching from the metastable state at $\overline{V} = 0$ to the stable state at $\overline{V} \approx 2\Delta / e$ for $i \leq 1$. This stochastic behavior may be captured by recording the bias current for a large number of switching events thereby getting a good estimate of the switching probability density, $P(i)$[19]. Assuming adiabatic conditions, the lifetime $\tau(i)$ in the metastable state may be found from $P(i)$ and the noise energy may be deduced, $E_{noise} = k_B T_N$, where T_N is the noise temperature. In Kramers transition state theory (intermediate damping regime), the thermal activation rate (also called the escape rate) is given by

$$\tau^{-1}(i) = \omega_p(i) \exp(E_{noise} / E_b(i)) = \omega_p(i) \exp(k_B T_N / E_b(i))$$

where $E_b(i)$ is the **height of the energy barrier**, see Fig. 3, and $\omega_p(i) = \omega_{po} \sqrt{\cos\phi} = \omega_{po}(1 - i^2)^{1/4}$ is the plasma frequency, the natural **attempt frequency** in the potential well of the $\overline{V} = 0$ state. This type of measurement on underdamped small Josephson tunnel junctions was first reported by Fulton and Dunkleberger[19], see Fig. 5. It has been developed further by many groups and analyses of the inverse "retrapping" process in the potential well have also been

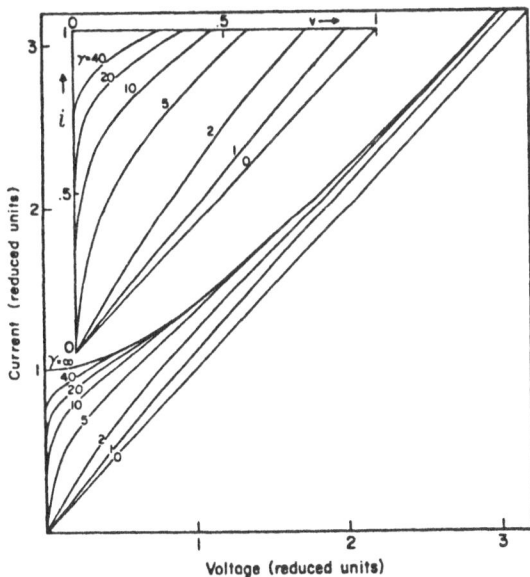

Fig. 4a.

The effect of thermal noise on an **overdamped** Josephson junction. Calculated thermally noise-rounded I-V characteristics of an overdamped (C=0) RSJ-model junction for varying noise parameter, here reproduced from Ambegaokar and Halperin (AH), Ref. 16. $\gamma = 2\Gamma = 2 E_J / k_B T = I_c \hbar / e k_B T$ is the noise parameter. Current axis: $i = I / I_c$, voltage axis: $V / I_c R$.

included[20]. Experimentally, the analysis of the escape rate in underdamped junctions is a convenient way of measuring the unperturbed critical current, I_c and the effective noise energy, E_{noise}, coupled to the junction.

In the **extreme underdamped** limit $\alpha \ll 1$ the dynamics within the potential well becomes important because the equilibration rate in the well may be smaller than the escape rate.

In microwave-driven junctions, thermal fluctuations perturb the induced **photon assisted tunneling, PAT, steps** and the induced **phase-locked (Shapiro) steps** in the same way as the $\overline{V} = 0$ state around I_c. For $eV > k_B T$ shot noise becomes important, see below.

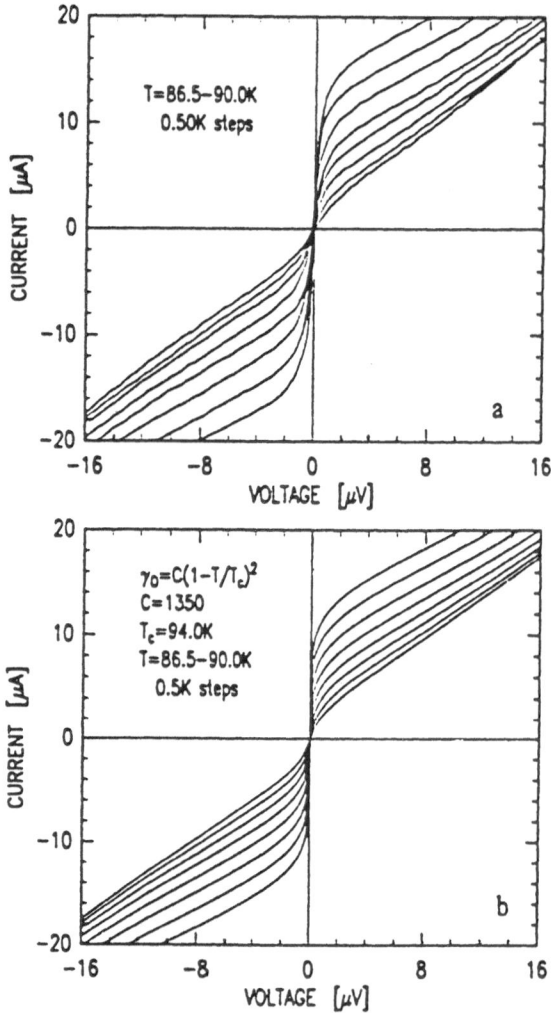

Fig. 4b.   Sets of experimental and theoretical I-V curves reproduced from Gross et al., Ref. 18. **Upper part:** measurements performed on an artificial grain-boundary Y-Ba-Cu-O junction close to T_c. **Lower part:** I-V curves calculated on the basis of the AH theory, cf. Fig. 4a.

So far we have only been concerned with junctions in the **classical** regime. At low temperature, Josephson elements with high plasma frequency are dominated by **quantum** fluctuations. For $\hbar \omega_{po} \gg k_B T$ quantum tunneling behavior, i.e., Macroscopic Quantum Tunneling, **MQT**, and Macroscopic Quantum Coherence,

MQC, should be observable. MQT has been reported by a number of groups[21,22] for Josephson junctions and SQUIDs at millikelvin temperature where the rate of thermal activation out of the potential well of the $\overline{V} = 0$ state is negligible.

3.4. Quasiparticle shot noise

The quasiparticle shot noise, $S_I = 2eI_{qp}$, is an out-of-equilibrium transport effect in superconducting tunneling structures, see Eq. (5). It stems from the discrete, "drop-like" nature of the dissipative quasiparticle tunneling process. It is observable for high voltage bias, $E_P = eV > k_B T$. At 4.2 K this implies that $V > 0.4$ mV. The effect of the quasiparticle shot noise may be observed in the broadening of the Josephson linewidth and the rounding of high order PAT steps and Shapiro steps in the dc I-V curve. An example of the latter is shown in Fig. 6a for the 4th Shapiro step at 5 mV in a high resistance (R = 176 Ohm), low capacitance (C = 10 fF) tunnel junction pumped at 604 GHz by a far-infrared laser[23]. The fit shown is based on ref. 24 with an equivalent noise temperature, $T_N(eq.) = 2eIR/4k_B = 25 \pm 5$ K, and an unperturbed half-step size of $2.5 \pm 0.3 \mu A$ It is interesting to compare the magnitude of this **shot noise** (as given by the equivalent noise temperature) with the effect of **ohmic heating** as measured by the effective noise temperature, $T_N = 15 \pm 5$ K, found from a fit to the rounding of a similar 5 mV phase-locked step in a **metallic** point contact of niobium, Fig. 6b[25]. There is **no** shot noise in **clean** metallic weak links (granular systems with some combination of direct conduction and tunneling structures exhibit both heating and shot noise).

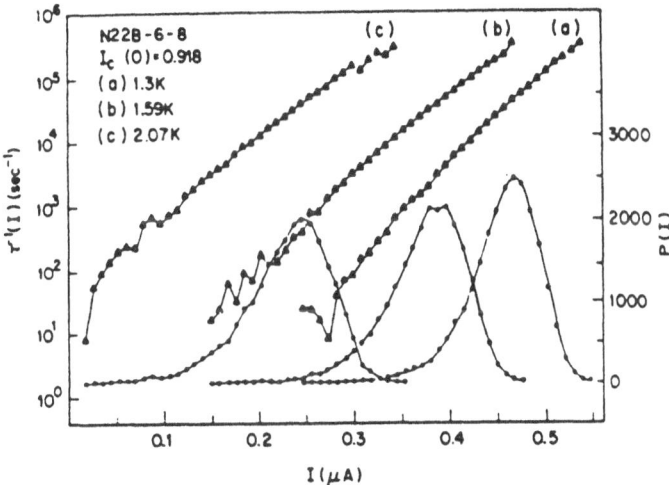

Fig. 5.   The effect of thermal noise on an **underdamped** Sn/SnOx/Sn Josephson junction, from Fulton and Dunkleberger, Ref. 19. Experimental switching point distributions, P(I) vs. I, at three different temperatures and the inferred switching rate curves, $\tau^{-1}(I)$. With the use of Kramers' reaction rate theory[15] the data gives information about E_{noise} and the potential barrier $E_b(i)$, i.e., the effective noise temperature $E_{noise} = k_B T_N$ and the unperturbed critical current.

For a clean three-dimensional metallic constriction the **heating model** of Tinkham et al.[26] gives

$$T_N = \frac{1}{2}(T_{bath} + T_{max}) = \frac{1}{2}\left(T_{bath} + \sqrt{T_{bath}^2 + \left(\frac{\sqrt{3}\,eV}{2\pi k_B}\right)^2}\right) \approx \frac{\sqrt{3}\,eV}{4\pi\,k_B}$$

$$\text{for } T_{max} \gg T_{bath}$$

from which the inequality

$$S_I = \frac{4k_B T_N(\text{eq})}{R} \approx \sqrt{\frac{3}{\pi}}\,eI < 2eI \qquad \text{for } eV > \frac{2\pi}{\sqrt{3}}k_B T \approx 3.7\,k_B T$$

is derived[23]. Note that in the shot noise limit, $eV > 3.7\,k_B T$ the **heating** in a **clean metallic weak link** produces **less** noise than the **shot noise** in a comparable **tunnel junction** (biased at the same current) by a factor of about $\sqrt{\frac{3}{2\pi}} \approx 0.28$.

3.5. Quantum noise

The high frequency quantum noise, the second term of Eq. (2*), may be measured by a sensitive high frequency detector. A small, low-capacitance Josephson tunnel junction is ideal for this purpose. Using a low inductance loop to connect the mixer-detector Josephson junction to a normal metal resistor of 0.05 to 0.7 Ω, Koch

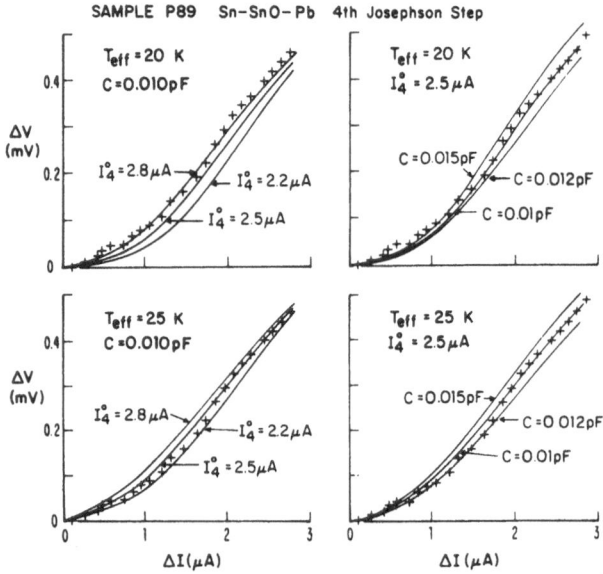

Fig. 6a. Experimental data and theoretical fits for the 4th phase-locked Shapiro "step" in a 176 Ω Sn/SnOx/Sn **tunnel junction** at 604 GHz, Danchi et al., Ref. 23. The measured I-V curve, plotted from the center of the step, is indicated by the crosses. Theoretical fits for various choices of the parameters, the effective noise temperature, $T_N = T_{eff}$, the maximum unperturbed half-step size, I_4^o, and the junction capacitance, C, are shown as solid lines. Best values: $T_N = 25$ K, $I_4^o = 2.5$ μA, C = 10 fF.

et al.[27], measured the quantum high frequency noise in the resistor. The detection scheme relied on resonant non-linearities in the junction I-V curve in the range $\overline{V}$ = 20 μV to 1 mV (corresponding to a Josephson frequency, fJ, of 10 to 500 GHz) which were used to mix down the noise around fJ to low frequency. Neglecting the junction capacitance, the measured low frequency noise at f << fJ is (to first order in the mixing terms)[28,29]:

$$S_I(f) \;=\; \frac{4k_B T}{R} \;+\; \frac{2e\overline{V}}{R}\left(\frac{I_c}{I}\right)^2 \coth\left(\frac{e\overline{V}}{k_B T}\right) \tag{9}$$

where the first noise term is the "direct" thermal current noise power in the resistor and the second term comes from the mixed-down high frequency noise in R. In the experiments which confirmed the validity of Eq. (9), the second term exceeded the first "direct" noise term by an order of magnitude at T = 1.6 K and fJ = 500 GHz[27].

3.6 Linewidth measurements

For a Josephson **tunnel** junction biased at a voltage $\overline{V}$, Dahm et al.[30] derived Eq. (5) and used it to find the linewidth of the Josephson radiation arising from

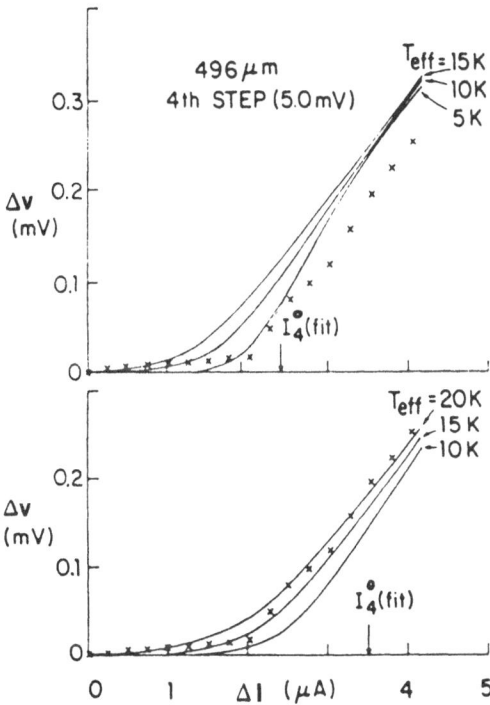

496 μm
4th STEP (5.0 mV)

$T_{eff} \doteq 15K$
10K
5K

Δv (mV)

I_4^o(fit)

$T_{eff} = 20K$
15 K
10 K

Δv (mV)

I_4^o(fit)

0   1 ΔI (μA)   4   5

Fig. 6b.

As Fig. 6a but for a **metallic weak link**, a Nb point contact, pumped at 604 GHz (λ = 496 μm) with fits based on the RSJ-model (C = 0), from Weitz et al., Ref. 25. Best values: $T_N = T_{eff}$ = 15 K and I_4^o = 3.5 μA .

fluctuations in the dissipative quasiparticle current, I_{qp}. The mean square of the voltage deviation is simply given by:

$$\overline{(\Delta V)^2} \;=\; R_D^2 \int_0^{\infty} S_{I_{qp}}(\omega)\,d\omega$$

where R_D is the dynamic resistance at the bias point, $R_D = dV/dI$. Using frequency modulation theory and the low frequency limit $hf \ll e\overline{V}$ (cf. Eq. (6)), the linewidth becomes

$$\Delta \omega_{qp} = \left(\frac{2e}{\hbar}\right)^2 R_D^2 e I_{qp}(\bar{V}) \coth\left(\frac{e\bar{V}}{2k_B T}\right). \tag{10}$$

In the low voltage, **thermal limit** $k_B T \gg e\bar{V}$ this expression reduces to:

$$\Delta f_{qp} = \Delta \omega / 2\pi = \left(\frac{2e}{h}\right)^2 4\pi k_B T \frac{R_D^2}{\bar{V}} I_{qp} = \frac{4\pi k_B T}{\Phi_o^2} \frac{R_D^2}{R_S} \tag{11}$$

where $\Phi_o = h/2e$ is the flux quantum and $R_S = \bar{V}/\overline{I_{qp}}$ is the static (dc) quasiparticle resistance at the bias point.

As mentioned above, the coupling of the junction to a lossy resonator introduces an additional pair current contribution to the shot noise (and therefore also to the linewidth)[8,31]:

$$\Delta \omega_p = \left(\frac{2e}{\hbar}\right)^2 R_D^2 2e I_p(\bar{V}) \coth\left(\frac{e\bar{V}}{k_B T}\right) \tag{12}$$

which, as before, in the thermal noise limit (low voltage and frequency) becomes:

$$\Delta f_p = \left(\frac{2e}{h}\right)^2 4\pi k_B T \frac{R_D^2}{\bar{V}} I_p. \tag{13}$$

Combining Eqs. (11) and (13), Dahm et al. found for the total linewidth,

$$\Delta f = \frac{4\pi k_B T}{\Phi_o^2} R_D^2 \frac{I_{qp} + I_p}{\bar{V}}. \tag{14}$$

Dahm et al. measured the radiation linewidth at 10 GHz using tin and lead tunnel junctions biased at resonant self-induced steps (first Fiske step) in the I-V curve at T = 1.4 to 4.2 K[30]. No conclusive evidence was found for the existence of the $\Delta \omega_p$ term, Eq. (10) (in these experiments $\Delta \omega_p$ should have been almost constant on the step). The observed linewidth varied linearly with the quasiparticle current, $I \approx I_{qp}$, as expected, but the measured linewidth (in the range 0.1 to 10 MHz) was typically 2-3 times larger than predicted by Eq. (11).

We note that the experiments reported in ref. 30 were carried out on rather large junctions (300×800 $\mu m^2 \approx 0.6 \times 1.6$ λ_J^2), where λ_J is the Josephson penetration depth (the screening length in the junction). This was the reason for the resonant (Fiske) spatial modes. Full statistical analyses of thermal fluctuations in **spatially extended tunnel junctions** have been carried out[32,33]. Long and narrow Josephson tunnel junctions, also called Josephson transmission lines, JTLs, which in the underdamped regime work as fluxon (soliton) oscillators with extremely narrow linewidth, were investigated theoretically and experimentally. A one-dimensional Josephson transmission line is governed by the perturbed, thermal sine-Gordon equation[33]:

$$-\phi_{xx} + \phi_{tt} + \sin\phi = i - \alpha\phi_t + \beta\phi_{xxt} + \tilde{i}_N(x,t) \tag{15}$$

where the two loss parameters α and β stem from the shunt conductance and the surface loss, respectively. The expression for the linewidth of the mth harmonic of the fundamental frequency $f_1 = (\bar{V}/\Phi_o)/2$ radiated from **one end** of the soliton oscillator biased on the lowest order self-induced resonant current singularity (the first zero-field step) is similar to Eq. (11)[32]:

$$\Delta f_{1,m} = m^2 \frac{\pi k_B T}{\Phi_o^2} \frac{R_D^2}{\bar{V}} I. \tag{16}$$

The observed linewidth at 4.2 K from 10 GHz Josephson soliton oscillators is typically from 5 kHz to 1 MHz, a factor of 2-3 above the value predicted from Eq. (16)[32,34]. Direct measurements using low-noise dc SQUIDs have revealed high levels of low frequency noise (1 Hz to 25 kHz) in Josephson soliton oscillators with $S_V \approx 4k_B T R_D$ (no factor of R_D^2 / R_S as expected)[35]. Analog or numerical simulations of this complex spatio-temporal non-linear system in the presence of noise are needed. For two-junction dc SQUIDs, simulations with two uncorrelated voltage noise sources of $S_V = 4k_B RT$ have been carried out[36]. In that case, mixing and down-conversion of the noise around the Josephson frequency and its higher harmonics lead to a total noise of $S_V \approx 16 k_B RT$. Similar contributions to the low frequency noise spectrum is expected for spatially extended Josephson junctions like the soliton oscillators.

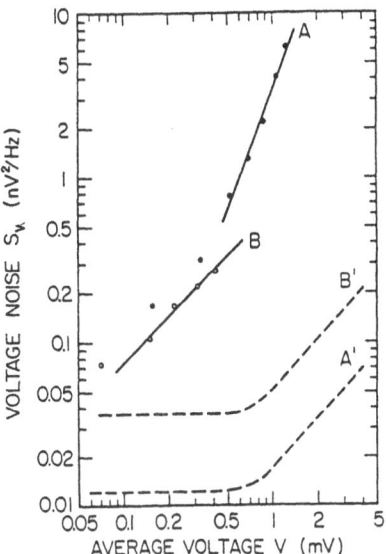

Fig. 7.   Figure reproduced from Carroll et al., Ref. 39, showing measured and calculated low frequency voltage noise power versus bias voltage for two Nb point contacts. The voltage noise was measured by an rf SQUID. The solid lines A and B represent the best fits to the data points (circles). The dashed lines A' and B' are the corresponding theoretical predictions. With increasing bias voltage, the predicted cross-over from the thermal noise to the shot noise regime is seen.

Within the RSJ-model for current-controlled **overdamped** (C = 0) **metallic weak links** of normal resistance R_N, Likharev et al.[28,37] found an expression for the linewidth, similar to Eq. (11), except for a second order term coming from noise mixed-down from the Josephson frequency, cf. Eq. (9) ($k_B T \gg hf$) :

$$\Delta f = \frac{4\pi k_B T}{\Phi_o^2 R_N} R_D^2 \left(1 + \frac{1}{2}\left(\frac{I_c}{I}\right)^2 \right) \qquad \text{for } I \geq I_c. \qquad (17)$$

175

The **measured linewidths** of the Josephson radiation and the **measured low frequency noise** from point contacts and microbridges **exceed the predicted value by factors of 2 to 100**[38,39]. An example of S_V vs. V curves for Nb point contacts is shown in Fig. 7. This discrepancy is not surprising: **firstly**, many weak links are ill-defined mixtures of metallic and granular tunneling structures; **secondly**, as mentioned in the introduction, a general microscopic theory of metallic superconducting Josephson weak links is still lacking.

4. CONCLUSIONS

Noise sets the ultimate limits of resolution and sensitivity for superconducting electronics. This review covers only noise in Josephson junctions. Other contributions to this book address the effects of this noise in applications (e.g. in SQUIDs).

The various noise contributions, $1/f^\alpha$, **thermal (Nyquist), shot and quantum noise** as well as the corresponding regimes of parameters have been discussed.

The **excess** $1/f^\alpha$ **noise** is only partly understood, best for tunnel junctions, least for metallic weak links.

The high frequency **quantum** noise is only observable in extreme cases. A few studies have been reported[27,40], more are needed.

As far as **thermal** noise and **shot** noise are concerned, we note that in many cases, Josephson junctions are operated in non-linear regimes, where the governing Langevin-type equations are intractable. The description of noise phenomena in Josephson junctions has therefore by and large been restricted to simple linearized circuits.

For well-characterized superconducting **tunnel junctions**, the agreement between theory and experiments is generally acceptable (within factors of 2 to 3). An outstanding problem is the experimental verification of the existence of shot noise in the pair current (for junctions coupled to lossy resonators) and in the pair-quasiparticle interference current.

For superconducting **metallic weak links** (microbridges and point contacts), however, the discrepancy between observed values and theoretical predictions is large (factors of 2 to 100). Improved theoretical models and measurements on well-characterized weak link structures are called for.

Finally, we note that applications of Josephson tunnel junctions in dc SQUIDs and in parametric amplifiers have proved that is possible, in practice, to operate such devices in the extreme low-noise regime close to the quantum limit [41,40].

Acknowledgment

I thank Mogens R. Samuelsen, Arthur Davidson, Christopher J. Lobb, Alexander Zorin and Thorsten Holst for discussions.

References

1. H. Nyquist, "Thermal Agitation of Electric Charge in Conductors", Phys. Rev. 29:614 (1927), 32:110 (1928); and J. B. Johnson, "Thermal Agitation of Electricity in Conductors", Phys. Rev. 32:97 (1928).

2. H. B. Callen and T. A. Welton, "Irreversibility and Generalized Noise", Phys. Rev. 83:34 (1951).

3. A. O. Caldeira and A. Leggett, "Quantum Tunneling in a Dissipative System", Ann. of Phys. (New York), 149:374 (1983).

4. P. Dutta and P. M. Horn, "Low-Frequency Fluctuations in Solids: 1/f Noise", Rev. Mod. Phys. 53:497 (1981).

5. M. B. Weissman, "1/f Noise and other Slow, Nonexponential Kinetics in Condensed Matter", Rev. Mod. Phys. 60:537 (1988).

6. C. T. Rogers and R. A. Buhrman, "Conductance Fluctuations and Low Frequency Noise in Josephson Junctions", IEEE Trans. Magn. MAG-19:453 (1983).

7. C. T. Rogers, R. A. Buhrman, W. J. Gallagher, S. I. Raider, A. W. Kleinsasser, and R. L. Sandstrom, "Electron Trap States and Low Frequency Noise in Tunnel Junctions", IEEE Trans. Magn. MAG-23:1658 (1987).

8. D. Rogovin and D. J. Scalapino, "Fluctuation Phenomena in Tunnel Junctions", Ann. of Phys. (New York) 86:1 (1974).

9. A. B. Zorin, "Fluctuations in Finite-Capacitance Josephson Tunnel Junctions", Fiz. Nizk. Temp. (Sov. J. Low Temp. Phys.) 7:709 (1981); and A. B. Zorin, "Voltage Fluctuations in Josephson Tunnel Junctions", Physica B 108:1293 (1981).

10. A. B. Zorin, K. K. Likharev, and S. I. Turovets, "Dynamics of Tunnel Josephson Junctions with Finite Riedel Peak", IEEE Trans. Magn. MAG-19:629 (1983).

11. N. F. Pedersen and A. Davidson, "Chaos and Noise Rise in Josephson Junctions", Appl. Phys. Lett. 39:830 (1981); and D. C. Cronemeyer, C. C. Chi, A. Davidson, and N. F. Pedersen, "Chaos, Noise, and Tails on the IV-Curve Steps of rf-Driven Josephson Junctions", Phys. Rev. B 31:2667 (1985); and A. Davidson, B. Dueholm, and M. R. Beasley, "Experiments on Intrinsic and Thermally Induced Chaos in an rf-Driven Josephson Junction", Phys. Rev. B 33:5127 (1986).

12. N. R. Werthamer, "Nonlinear Self-Coupling of Josephson Radiation in Superconducting Tunnel Junctions", Phys. Rev. 147:255 (1966).

13. A. A. Odintsov, V. K. Semenov, and A. B. Zorin, "Specific Problems of Numerical Analysis of the Josephson Junction Circuits", IEEE Trans. Magn. MAG-23:763 (1987).

14. T. M. Klapwijk, G. E. Blonder, and M. Tinkham, "Explanation of the Sub-harmonic Energy Gap Structure in Superconducting Contacts", Physica 109/110B: 1657 (1982); and G. E. Blonder, M. Tinkham, and T. M. Klapwijk,

"Transition from Metallic to Tunneling Regimes in Superconducting Microstructures: Excess Current, Charge Imbalance, and Supercurrent Conversion", Phys. Rev. B 25:4515 (1982); and G. B. Arnold, "Superconducting Tunneling without the Tunneling Hamiltonian", J. Low Temp. Phys. 68:1 (1987).

15.   H. A. Kramers, "Brownian Motion in a Field of Force and the Diffusion Model of Chemical Reactions", Physica, 7:284 (1940); see also: P. Hänggi, P. Talkner and M. Borkovec, "Reaction-Rate Theory: Fifty Years after Kramers", Rev. Mod. Phys. 62:251 (1990).

16.   V. Ambegaokar and B. I. Halperin, "Voltage due to Thermal Noise in the dc Josephson Effect", Phys. Rev. Lett. 22:1364 (1969).

17.   M. Tinkham, "Resistive Transition of High-Temperature Superconductors", Phys. Rev. Lett. 61:1658 (1988).

18.   R. Gross, P. Chaudhari, D. Dimos, A. Gupta, and G. Koren, "Thermally Activated Phase Slippage in High-T_c Grain-Boundary Josephson Junctions", Phys. Rev. Lett. 64:228 (1990).

19.   T. Fulton and L.N. Dunkleberger, "Lifetime of the Zero-Voltage State in Josephson Tunnel Junctions", Phys. Rev. B 9:4760 (1973).

20.   E. Ben-Jacob, D. J. Bergman, B. J. Matkowsky, and Z. Schuss, "The Lifetime of Non-Equilibrium Dissipative Steady States", in "SQUID'85", H. D. Hahlbohm and H. Lübbig, eds., Walter de Gruyter, Berlin (1985); and R. Christiano and P. Silvestrini, "Decay of the Running State in Underdamped Josephson Junctions", J. Appl. Phys. 59:1401 (1986).

21.   M. H. Devoret, J. M. Martinis, and J. Clarke, "Measurement of Macroscopic Quantum Tunneling out of the Zero-Voltage State of a Current Biased Josephson Junction", Phys. Rev. Lett. 55:1908 (1985); and J. M. Martinis, M. Devoret, and J. Clarke, "Experimental Tests for the Quantum Behavior of a Macroscopic Degree of Freedom: The Phase Difference across a Josephson Junction", Phys. Rev. B 35:4682 (1987).

22.   R. F. Voss, and R. A. Webb, "Macroscopic Quantum Tunneling in $1\,\mu m$ Nb Josephson Junctions", Phys. Rev. Lett. 47:265 (1981); and L. D. Jackel, J. P. Gordon, E. L. Hu, R. E. Howard, L. A. Fetter, D. M. Tennant, R. W. Epworth, and J. Kurkijärvi, "Decay of the Zero-Voltage State in Small Area High Current Density Josephson Junctions", Phys. Rev. Lett. 47:697 (1981).

23.   W. C. Danchi, J. Bindslev Hansen, M. Octavio, F. Habbal, and M. Tinkham, "Effects of Noise on the dc and Far-Infrared Josephson Effect in Small-Area Superconducting Tunnel Junctions", Phys. Rev. B 30, 2503 (1984).

24.   P.A. Lee, "Effects of Noise on the Current-Voltage Characteristics of a Josephson Junction", J. Appl. Phys. 42:325 (1971).

25.   D. A. Weitz, W. J. Skocpol, and M. Tinkham, "Far-Infrared Frequency Dependence of the ac Josephson Effect in Niobium Point Contacts", Phys. Rev. B 18:3282 (1978).

26.   M. Tinkham, M. Octavio, and W. J. Skocpol, "Heating Effects in High-Frequency Metallic Josephson Devices: Voltage Limit, Bolometric Mixing and Noise", J. Appl. Phys. 48:1311 (1977).

27.   R. H. Koch, D. J. van Harlingen, and J. Clarke, "Measurements of Quantum Noise in Resistively Shunted Josephson Junctions", Phys. Rev. B 26:74 (1982).

28.   K. K. Likharev and V. K. Semenov, "Fluctuation Spectrum in Superconducting Point Junctions", Zh. Eksp. Teor. Fiz. Pis´ma Red. 15:625 (1972) [JETP Lett. 15:442 (1972)].

29.   R. H. Koch, D. J. van Harlingen, and J. Clarke, "Quantum-Noise Theory for the Resistively Shunted Josephson Junction", Phys. Rev. B 26:2132 (1980).

30.   A. J. Dahm, A. Denenstein, D. N. Langenberg, W. H. Parker, D. Rogovin, and D. J. Scalapino, "Linewidth of the Radiation Emitted by a Josephson Junction", Phys. Rev. Lett. 22:1416 (1969).

31.   M. J. Stephen, "Noise in the ac Josephson Effect", Phys. Rev. 182:531 (1969); and M. J. Stephen, "Theory of a Josephson Oscillator", Phys. Rev. Lett. 21:1629 (1968).

32.   E. Joergensen, V. P. Koshelets, R. Monaco, J. Mygind, M. R. Samuelsen, and M. Salerno, "Thermal Fluctuations in Resonant Motion of Fluxons on a Josephson Transmission Line: Theory and Experiment", Phys. Rev. Lett. 49:1093 (1982).

33.   M. Salerno, E. Joergensen, and M. R. Samuelsen, "Phonons and Solitons in the "Thermal" sine-Gordon System", Phys. Rev. B 30:2635 (1988); and M. Salerno, M. R. Samuelsen and H. Svensmark, "Thermal sine-Gordon system in the Presence of Different Types of Dissipation", Phys. Rev. B 38: 593 (1988).

34.   J. Bindslev Hansen, Yu. Ya. Divin, J. Mygind, and M. R. Samuelsen, "Measurements of Frequency and Linewidth of the Radiation Emitted from Overlap Josephson Junctions", in "SQUID '85", H. D. Hahlbohm and H. Lübbig, eds., Walter de Gruyter, Berlin (1985).

35.   J. Bindslev Hansen, T. Holst, F. Wellstood, and J. Clarke, "Low Frequency Noise in Resonant Soliton Oscillators", Appl. Sup. Conf. 1990 (to be published in IEEE Trans. Magn. MAG-27, 1991).

36.   C. D. Tesche and J. Clarke, "DC SQUID: Noise and Optimization", J. Low. Temp. Phys. 29:301 (1977); and C. D. Tesche and J. Clarke, "DC SQUID: Current Noise", J. Low Temp. Phys. 37:397 (1979).

37.   A. N. Vystavkin, V. N. Gubankov, L. S. Kuzmin, K. K. Likharev, V. V. Migulin, and V. K. Semenov, "S-c-S Junctions as Non-Linear Elements of Microwave Receiving Devices", Rev. Phys. Appl. (Paris) 9:79 (1974)

38.   G. Vernet and R. Adde, "Linewidth of the Radiation by a Josephson Point Contact", Appl. Phys. Lett. 19:195 (1971); and J. Mygind, N. F. Pedersen,

O. H. Soerensen, B. Dueholm, T. F. Finnegan, J. Bindslev Hansen, M. T. Levinsen, and P. E. Lindelof, "Josephson Radiation Emitted from a Microbridge at 9, 35 and 69 GHz", in "SQUID'80", H. D. Hahlbohm and H. Lübbig, eds., Walter de Gruyter, Berlin (1980).

39. K. R. Carroll and H. J. Paik, "Noise in a Point Contact dc SQUID", J. Low Temp. Phys. 75:187 (1989).

40. R. Movshovich, B. Yurke, P. G. Kaminsky, A. D. Smith, A. H. Silver, and R. W. Simon and M. V. Schneider, "Observation of Zero- Point Noise Squeezing via a Josephson-Parametric Amplifier", Phys. Rev. Lett. 65:1419 (1990); and R. Movshovich, B. Yurke, P. Kaminsky, A. D. Smith, A. H. Silver, and R. W. Simon, "Vacuum Noise Squeezing at Microwave Frequencies using a Josephson Parametric Amplifier", Appl. Sup. Conf., 1990 (to be published in IEEE Trans. Magn. MAG-27, 1991).

41. M. B. Ketchen and R. F. Voss, "An Ultra-Low Noise Tunnel Junction dc SQUID", Appl. Phys. Lett. 35:812 (1979); and M. B. Ketchen, D. D. Awschalom, W. J. Gallagher, A. W. Kleinsasser, R. L. Sandstrom, J. R. Rozen and B. Bumble, "Design, Fabrication and Performance of Integrated Miniature SQUID Susceptometers", IEEE Trans. Magn. MAG-25:1212 (1989).

SWITCHING DYNAMICS OF JOSEPHSON JUNCTIONS IN THE PRESENCE OF NOISE

Paolo Silvestrini

Istituto di Cibernetica del C.N.R,
80072 Arco Felice (Naples) Italy

I. INTRODUCTION

In the present paper we collect some our recent results concerning some aspects of the metastability of the zero-voltage state in hysteretical Josephson junctions. We shall consider the dynamics of the junction both in the presence of the classical thermal fluctuations and in a quantum picture as well.

Thermally activated processes are characterized by the decay of the zero-voltage with an escape rate provided, at a first glance, by the usual Kramers result[1]

$$\Gamma = A_t \, \omega_0 \, \exp(- U_0/KT)$$

where U_0 is the barrier height of the free energy of the junction represented by the well known washboard one-dimentional potential. The oscillation frequency of the metastable state is indicated by ω_0, and A_t is the Kramers prefactor which depends on the damping of the junction. This simple result can be obtained within a model describing the junction dynamics in terms of an equivalent, current biased, circuit consisting of the junction capacitance C and a resistor R (assumed to be constant) in parallel with a "superconductor" element (the well known RSJ model).

We actually have considered the effect of the voltage dependent junction resistance R_j on the decay probability[2], removing hence the strong assumption of the RSJ model of an ohmic shunt resistance (arbitrary chosen) to describe the junction dissipation. In fact some our experimental data clearly show the relevance of the intrinsic voltage-dependent dissipation due to the presence of quasiparticles[3]. This has been obtained analyzing our experimental data within a slightly modified RSJ model, consisting of the same equivalent circuit in which, however, the dissipative element is allowed to be voltage dependent. This latter aspect is relevant, because allows to obtain a damping level which is decreasing with the temperature, and hence, at low temperatures, we are able to study, for the first time, extremely underdamped systems.

Nonlinear Superconductive Electronics and Josephson Devices
Edited by G. Costabile *et al.*, Plenum Press, New York, 1991

181

The study of this aspect of the thermal regime is also particularly relevant in the quantum limit, because the macroscopic quantum behaviour of the junction can depend critically upon the dissipation level. In fact, in a quantum picture we must consider the possibility of metastability by barrier penetration, characterized by a tunnel probability. As first pointed out by Caldeira and Legget[4], the macroscopic nature of the tunneling implies an important role of the coupling to the environment, namely the effect of dissipation. A further important consequence of the quantum effects is the existence of energy levels in the well. Indeed a resonance in absorption of microwave have given evidence of different energy eingenstates[5].

We give a further insight upon the effects of the presence of quantized energy levels (ELQ) on the supercurrent decay . Starting from the master equation developed by Larkin and Ovchinnikov[6], we investigate the non-stationary conditions[7]. That is conditions where the escape process induces a fast reduction of particles trapped in the metastable state and, as a consequence, the occupancy probability of the energy levels is far from the equilibrium. The effects of ELQ is in our case very relevant without the need of external microwave irradiation, as consequence of a suitable choice of the working parameters, such as the sweeping rate and the effective dissipation. In this case the escape rate is an oscillating function of bias current and the oscillation amplitude increases as the viscosity decreases, the effects being greater when the temperature T is higher than the crossover temperature T_0. The curves obtained predict the results of new possible experiments.

II. THERMAL REGIME: ASPECTS OF DISSIPATION EFFECTS

In this paragraph, we discuss some implications of dissipation in the supercurrent decay, restricting ourselves to the classical thermal activation mechanism. As well known the relevant variable in describing the behaviour of a Josephson junction is the quantum phase difference θ between the two superconductors forming the junction. This variable cannot be decoupled from microscopic degrees of freedom, the quasiparticles excitations, and this circumstance produces dissipation. This latter is quite complicate to describe in real junction taking into account the various aspects of the problem in its full complexity. At the first glance, we can distinguish between intrinsic mechanisms of dissipation, related to the presence of quasiparticles excitations, and influences of the external circuit biasing the junction. Obviously this latter aspect depends strongly on the experimental configuration, and, in some experimental situation, the junction dissipation may be dominated by any external resistive shunt or load line[8].

The situation in which an intrinsic resistance produces the effective damping seems to be more desirable than when an external resistance dominates. In fact, in the former case we are dealing with a well defined system, the junction, at a well defined temperature. In the latter, we are dealing with a composite system with not well spatially defined elements, the external circuit, with some possible spurious interference. Moreover, the intrinsic dissipation sets the lower limit for the junction dissipation.

A general expression for the total current of a superconducting tunnel junction can be obtained within the microscopic theory. Its general expression is rather complicate, but in the case of time independent voltage V across the junction, the tunneling current can be cast in the form[9]

$$I(V,T) = I_j(V,T)\sin(\theta) + [\sigma_1(V,T)\cos(\theta) + \sigma_0(V,T)] \, V \qquad 1)$$

The first term $I_j(V,T)\sin(\theta)$ describes processes in which phase coherent tunneling of Cooper pairs occurs, and for $V=0$ represents the d.c. Josephson current. The dissipative term $I_{qp} = \sigma_0(V,T) \, V$ represents the quasiparticle tunneling. The phase dependent dissipative term $\sigma_1(V,T)\cos(\theta)V$, could be interpreted as describing a quasiparticle tunneling process which involves a concomitant destruction and creation of pairs on the two superconductors forming the junction, therefore involving phase coherence effects. It is interesting to observe that this term has been ignored in the RSJ model used to describe the junction dynamics in the presence of noise. In fact many features of the problem can be accounted for in terms of a simple lumped circuit model in which the distributed capacitance and the intrinsic tunneling conductance are considered as lumped elements in parallel with a nonlinear Josephson element. As a further simplification, it is typically considered the simple, but unrealistic, case of both the quasiparticle resistance R and Josephson current I_j independent of the voltage. As well known, within this model, the current-carrying states of a Josephson junction are described by a langevin-like equation for the phase θ, which in adimentional units assumes the form[10]:

$$\ddot{\theta} + \varepsilon\,\dot{\theta} = -\frac{dU}{d\theta} + \sqrt{\varepsilon}\,\xi(t) \qquad 2)$$

where $\varepsilon = [\omega_j RC]^{-1}$, $U(\theta) = -(\alpha\theta + \cos\theta) + \text{const}$, $\omega_j = \sqrt{\dfrac{2\pi I_c}{\Phi_0 C}}$. Here α is the bias current normalized to the critical one I_c, $\alpha = I/I_c$, and Φ_0 is the magnetic quantum flux. The dots indicate the derivation with respect to the normalized time $\underline{t} = \omega_j t$. The properties of $\xi(\underline{t})$ are

$$<\xi(\underline{t})> = 0; \quad <\xi(\underline{t})\,\xi(\underline{t}')> = \frac{4}{\gamma}\,\delta\,(\underline{t} - \underline{t}')$$

where $\gamma = \dfrac{\Phi_0 I_c}{\pi k T}$.

The behaviour predicted by eq. 2) is easily understood in terms of the mechanical analog, i.e., that of a Brownian particle of unitary mass performing its motion in the washboard potential $U(\theta)$, where θ is the position of the particle. The voltage across the junction is related to the speed of the particle, $V = \dfrac{\phi_0\omega_j}{2\pi}\,\dot{\theta}$.

For $\alpha < 1$, $U(\theta)$ shows a series of minima (potential wells) separated by an energy barrier U_0 ($U_0 = -\alpha\pi + 2[\alpha\sin^{-1}\alpha + \sqrt{1-\alpha^2}\,]$), and the $V=0$ state can be visualized as a particle which is trapped in a well and is performing oscillations in the potential minimum.

The success of the RSJ model in explaining experimental data can be considered "surprising", if one consider the approximations contained in it. We can first note that the model is essentially based on the subdivision of the total current in three simple terms as given by eq.1, which can be obtained within the microscopic theory only for time independent voltage[9]. This primary assumption is not consistent with the model because the voltage, although zero in average in the V=0 state, is oscillating at the plasma frequency. These oscillations correspond, in the mechanical analog, to the oscillations of the brownian particle at the bottom of the well. This last aspect leads, in the microscopic theory, to rather complicate expressions for the total current, with superconductive and dissipative parts no longer expressible in terms of simple functions, and to an inconsistency of the RSJ model. It is however clear that the amplitude of the oscillations corresponds to a very low voltage (which can be extimated in a few tenth of nV), and hence it is unlikely to be responsible of large effects in the junction dynamics. It is then reasonable to neglect this aspect if you do not require a quantitative agreement, in all the details, between theory and experiments. The problem remains anyway open, and would require some more probing.

A more relevant limit of the RSJ model is actually the assumption of an ohmic resistor to describe the junction dissipation. It is in fact clear that the dissipative element is, in the real junction, due to the combination of different, non linear mechanisms. As a consequence the dissipation level is strongly voltage and temperature dependent. A voltage dependent resistance is, however, in contradiction with the assumption of an ohmic dissipative element, and leads to an ambiguity in the definition of the resistance R to put in the theory (namely, if one ohmic relevant resistance can be assumed, which voltage must one consider?). Because of that in each experiment a somewhat arbitrary fitting resistance has to be assumed. However the few experiments[3,8,11-15] on the subject, regarded as a whole, do not give an unique answer, rather different authors reach different conclusion about the effective resistance which fits the data by theory. This is a natural consequence of the fact that all the approximations and the limits of the RSJ model are, by force, included in this arbitrary fitting resistance. Moreover, one can also adjust the theory to the data with an other free parameter which necessarily must be extracted from the data, the critical current. The use of free fitting parameters make much easier a quantitative agreement of data with the theory, but the limits of the model may come out in the form of subtle inconsistency of the fitting parameters[16].

In order to avoid one of approximations of the RSJ model, we have studied the effect of the voltage dependent junction resistance R_j on the critical current decay of Josephson junctions. We have obtained a simple expression for the lifetime which takes into account the non linear dependence of R_j on the voltage, whatever it is[2]. Obviously, this does not remove all the approximations of the model, but allows to study the ideal case of dissipation completely dominated by the quasiparticle tunneling resistance. In this case is in fact possible to obtain an analytical expression to compare with the experiments. We have then analyzed our experimental data in the light of the new theory.

The details of the theory are reported in ref.2. We discuss here only the main results.

We start considering a slightly modified RSJ model, consisting of the same equivalent circuit in which the dissipative element is allowed to be voltage dependent.

The motion equation of the junction in adimentional units assume still the form of the Langevin equation Eq.1) for the phase θ, with a velocity dependent friction parameter $\varepsilon = \varepsilon(\dot{\theta})$. The Fokker-Planck equation associated is then more complicate than in the case of constant ε[17,18]. However, in underdamped systems, it can be solved by a first-passage time technique[17,18] to obtain the life-time τ of the metastable state in the form of a double integral along the energy variable[2]. It is obvious that τ will depend on the actual form of the R vs V dependence, but we can obtain an analytical expression in the relevant case of dissipation due to the quasiparticle tunneling[2]. In symmetrical junctions and in the limit case of $V \ll \Delta$, the quasi particle tunneling resistance can be expressed as[19,9]

$$R_{qp} = \frac{R^*}{\ln(\frac{\mu}{V}) - 1} \qquad (3)$$

Here Δ is the superconductor gap, $\mu \equiv \min(kT/e, \Delta)$. For our convenience we also defined the temperature dependent ohmic resistance $R^* = \frac{4R_N kT}{e\Delta} \cosh^2[\frac{e\Delta}{2kT}]$ where R_N is the junction normal resistance measured on the I-V curves at $V > \Delta$.

For dissipation dominated by the quasiparticles tunneling, the expression for the lifetime can be cast in the simple form[2]

$$\tau = \frac{F(U_0)}{\eta_1} \qquad (4)$$

where $F(E) = \sum_{n=1}^{\infty} \frac{[\gamma/2E]^n}{n!n}$; $\eta_1 = \varepsilon^* n(E_1)$; $\varepsilon^* = [\omega_j R^* C]^{-1}$ and $n(E)$ is an adimentional parameter with a logarithmic dependence on E:

$n(E) = \ln[\frac{2e\mu}{\hbar\omega_j} \frac{1}{\sqrt{2E}}] - 0.81$. The value E_1 is implicitly defined by the

relationship $E_1 = \frac{1}{\gamma n(E_1)}$. Note that, for $\gamma/2E \gg 1$, $F(E) = \frac{\exp(\frac{\gamma}{2}E)}{\frac{\gamma}{2}E}$.

It is interesting to notice that the expression for the lifetime of the zero voltage state Eq.4) is identical to the one obtained within the RSJ model, in the case of a constant resistance[10], provided one assumes an effective dissipation η_1. We can then conclude that an intrinsic dissipation due to the quasiparticles tunneling produces the RSJ results, as one ohmic relevant resistance could be assumed. This resistance corresponds to the quasiparticle tunneling resistance at the voltage $V_1 = \dfrac{\Phi_0 \omega_j}{2\pi} \sqrt{2E_1}$. This voltage is typically of the order of a few nV in experimental situations. Note that the value of the effective resistance is then smaller than R* by a factor $n(E_1)$ ($n(E_1) \cong 10$ in typical experimental situations).

A comparison of theory with experiments, can be done measuring the statistical distributions of switching current values from the V=0 to the V≠0 state(related to the lifetime of the V=0 state) of an underdamped junction[3]. The fitting of the data leaving η_1 as a free parameter allows to get information on the effective resistance R_{eff}[3]. This must be performed at various temperature in order to get, with the aid of the temperature dependence of the fitting resistance, a check on the consistence of the model. In fact, although the correlation between the fit parameters does not allow a very precise determination of R_{eff} at each temperature, R_{eff} shows a well defined behaviour as a function of the temperature[3]. An independent measurement of the quasiparticle resistance at the very low voltage V_1 is very complicate. The useful physical quantities for the comparison are provided by the subgap resistance R_{qp} measured on the I-V curves at a somewhat higher voltage $V_{qp} >> V_1$, as well as by a theoretical extrapolation by Eq.3) at lower voltage;

$$R_{V1} = R_{qp} \, \frac{\ln(\dfrac{\mu}{V_{qp}}) - 1}{\ln(\dfrac{\mu}{V_1}) - 1} \tag{5}$$

We have taken data in extremely underdamped Nb-NbOx-Pb Josephson junctions. In Fig. 1) we show the temperature dependence of R_{eff} as obtained from the switching current distribution in extremely underdamped structures, as well as the quasiparticles resistance R_{40} measured on the I-V characteristics at V=40 μV. We have also plotted the extrapolation from V_{qp}=40 μV to the "effective" voltage V_1 (as given by Eq.5). The quantities are plotted in a semilogarithmic scale versus T^{-1}.

It is evident for all these resistances the same typical exponential temperature dependence due to the presence of quasiparticles thermally activated above the superconducting gap Δ. This circumstance clearly shows that R_{eff} is dominated by an intrinsic quasiparticle mechanism as the other subgap resistances. In fact the temperature dependence of R_{eff} very well fits Eq. 3) with Δ=1.3 mV (the niobium and lead gaps are almost equal, and this value is very reasonable). The shift between R_{40} and R_{eff} indicates that the effective resistance in the activation process corresponds to a voltage much smaller than 40 μV, as expected from our calculations. The small remaining shift between R_{eff} and R_{V1} cannot, however, be ascribed completely to the experimental uncertainty. It would be in fact surprising to obtain within this simple model an agreement between theory and experiment quantitative in all the details (for instance, the interference term is not included at all). The good qualitative agreement obtained here indicates the importance of the quasiparticle resistance in the junction dissipation, as well as that the relevant resistance must be taken at a very low voltage.

As a further consideration, we must point out that the high values of resistance obtained at low temperatures allowed us to study extremely underdamped systems characterized by the condition $\eta_1/\omega_0 < (\pi\gamma U_0)^{-1}$. Beside the intrinsic interest of this aspect in the thermal limit, this event opens new important possibilities to study experimentally macroscopic quantum effects. We will give an example in the next paragraph.

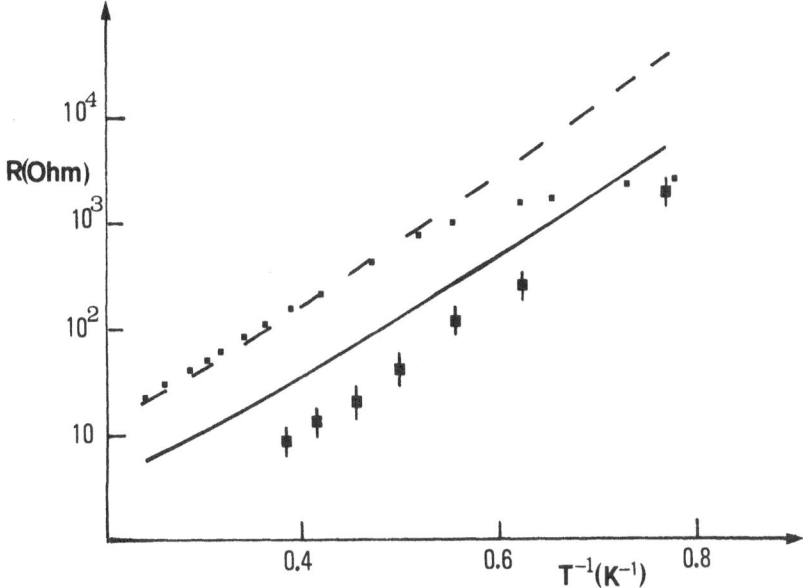

FIG.1 *Some relevant junctions resistances versus the reverse temperature. The squares are the effective resistance values as obtained from data on the switching current distributions. We also report the dynamic resistance values (dots) measured on the I-V characteristics at V=40 μV and a fit of them by eq.3) with Δ= 1.3 mV (dashed line), as well as the extrapolation at the effective voltage V_1 (R_{V1}, solid line).*

III. QUANTUM PICTURE

In a quantum picture we must consider the presence of energy levels in the potential well of the washboard potential associated to a Josephson tunnel junction. In the weak friction limit, there are in fact sharp and well separated energy levels inside the potential well, and the dynamic of the system must be described by the kinetic equation for the probabilities ρ_j of finding the particle at the j-th level[6,7,20,21,22]

$$\frac{\partial \rho_j}{\partial t} = \sum_k (w_{jk}\rho_k - w_{kj}\rho_j) - \gamma_j \rho_j \tag{6}$$

where w_{jk} is the probability for transition from the k-th into the j-th level due to the interaction with the heat bath, and γ_j is the probability of

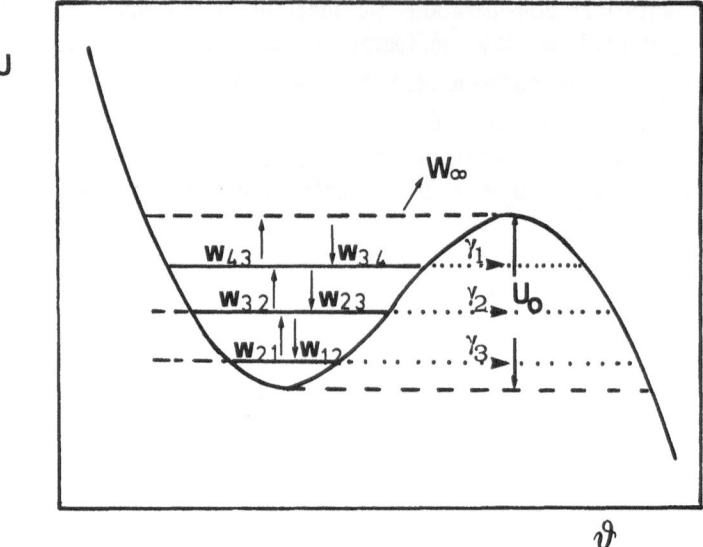

Fig. 2 *Sketch of the junction dynamics in a quantum picture.*

tunneling through the barrier which strongly depends on the energy level position. A sketch of the dynamics is shown in the fig. 2.

The positions of the levels, the transitions between levels and the tunneling probabilities can be calculated with the aid of the expressions reported in Ref.6. For levels not very close to the top of the potential barrier it is possible to calculate their positions and the tunneling rate. The levels positions E_n are given by the condition

$$S(E_n) = \hbar\pi (n + 1/2)$$

where $S(E)$ is the action in the classically accessible region

$$S(E) = \frac{\pi}{4} \beta_0 \sqrt{mU} (x_2 - x_3)^2 \sqrt{x_1 - x_3} \ F\left(-\frac{1}{2};\frac{3}{2};3;\frac{x_2-x_3}{x_1-x_3}\right) \tag{7}$$

Here F is the hypergeometric function and $x_3 < x_2 < x_1$ are the roots of the cubic equation

$$x^3 - \frac{3}{2}x^2 + \frac{E}{2U} = 0 \tag{8}$$

where $\beta_0 = \sqrt{1 - (I/I_c)^2}$, $m = \frac{\hbar^2 C}{e^2}$, $U = \frac{\hbar I_c \beta_0^3}{3e}$ and C is the junction capacitance.

The tunneling rate from the *n-th* level is given by the following expression:

$$\gamma(E_n) = \frac{\delta E \ exp \ [-2So(E_n)/\hbar]}{\hbar\sqrt{2\pi \ \Gamma(n+1)}} \ exp \ \{(n+1/2) \ [ln(n+1/2) - 1]\} \tag{9}$$

188

where δE is the level spacing, Γ is the gamma function and $S_O(E)$ is the action in the classically inaccessible region:

$$S_O(E) = \frac{\pi}{4}\beta_0 \sqrt{mU}\ (x_1 - x_2)^2\ \sqrt{x_1 - x_3}\ \ F\left(-\frac{1}{2};\frac{3}{2};3;\frac{x_1 - x_2}{x_1 - x_3}\right) \tag{10}$$

To solve Eq. 6 we need now only the expression for the transition probability through levels. We have that only the transitions to the neighbour levels are significant. For deep levels we have[6]:

$$w_{j-1,j} = \frac{\delta E}{e^2 R}\ [1 + coth(\frac{\delta E}{2KT})]\ |<j|\phi|j-1>| \tag{11}$$

with $|<j|\phi|j-1>| = -\dfrac{\pi^2 \beta_0 (x_2 - x_3)}{2k^2 K^2(k)sinh[\pi K'(k)/K(k)]}$, $k=\sqrt{\dfrac{x_2 - x_3}{x_1 - x_3}}$

where $K(k)$ and $K'(k)$ are the complete elliptic integrals. x_i are in this case the solution of Eq.(8) where $E = \dfrac{E_j + E_{j-1}}{2}$

Close to the top of the potential barrier we must care about the level width. The transition probability density in an interval dE is still given by Eq.(11) but where the matrix element is:

$$|<j|\phi|E>| = \frac{9\pi\beta_0^2 (E_j - E_{j-1})\ (E - E_j)^2}{(\hbar\omega_j)^4\ 2m\ sinh^2[\dfrac{\pi(E - E_j)}{\hbar\omega_j}]}\ |B|^2 \tag{12}$$

with $|B| = \dfrac{2m}{\pi}\ exp(2\pi y)\ |1 + \dfrac{\sqrt{2\pi}}{\Gamma(1/2 + iy)}\ exp[\dfrac{\pi}{2}\ y + i\dfrac{6\sqrt{6}}{5\hbar}\beta_0\ \sqrt{mU} +$

$+2iy\ ln(6\beta_0\ (2m\omega_j\ /\hbar)^{1/2}\ )]\ |$

where $y = \dfrac{(E - U)}{\hbar\omega_j}$, ω_j is the plasma frequency and $2\phi=\theta$. The maximum of this expression determines the level position E_{j+1}, while its width determines the tunneling probability $\gamma(E_{j+1})$. This equation is also used for transitions to energy values above the potential barrier[6].

The system is initially trapped in a relative minimum of the potential, but it can escape from the well via either macroscopic quantum tunneling or thermal activation. In this case a switching from the $V= O$ to the $V\neq O$ state is observed in the junction.

Experimentally the bias current is a periodic function of time, and the transitions from the $V= O$ to the $V\neq O$ state are repeatedly observed at random values of I smaller than I_c. The physical quantity directly measured is the distribution $P(I)$ of the current switching values. In statistical terms we can say that the "total" probability of finding the particle inside one of the potential wells at the initial time is the certainty, $\rho(t=0) =1$. However there exists a certain probability of escape from the well leading to a $\rho(t)$ which is a decreasing function of time (obviously, since I is increasing with time at a certain sweeping rate, ρ can be seen as a function of the current as well). The barrier height is vanishing at $I=I_c$. It is easy to correlate the measured quantity $P(I)$ with $\rho(I)$:

$$P(I) = - \frac{d\rho}{dI} \qquad (13)$$

Although in non-stationary conditions the lifetime of the $V=0$ state looses its classical meaning, since it now depends not only on the current, but also on the initial condition as well as on the rate of change of the potential, it is however useful to define a "generalized" total escape rate $\Gamma(I)$, in analogy with the quasi-stationary case, as

$$\Gamma(I) = - \frac{dlg[\rho(t)]}{dt} = - \frac{dlg[\rho(I)]}{dI} \frac{dI}{dt} \qquad (14)$$

At the initial time, the statistical populations of energy levels are distributed according to the Boltzmann distribution

$$\rho_j (0) = A \; exp \left(- \frac{Ej}{KT}\right) \qquad (15)$$

where ρ_j is the population of the j-th level or the probability to find there a particle. It is clear that $\rho(t) = \Sigma_j \; \rho_j \; and \; \Sigma_j \; \rho_j (0) = 1$.

$A = 1 / \Sigma_j \, exp \left(-\frac{Ej}{k_B T}\right)$ is a normalization factor.

All the curves reported here are obtained by numerical solution of the master equation (6). It has been solved with the Boltzmann distribution Eq.15 as initial condition, namely the system is initially in thermal equilibrium in a very deep potential well corresponding to the initial current value $I_0 = I(t=0) < I_c$[23]. Starting from this time we numerically follow the time evolution of the levels population, taking into account that also the current is changing with the time[7].

We present here simulated P(I) and $\Gamma(I)$ curves at various temperatures, for some experimentally significative junction parameters, and for different values of the sweep rate dI/dt.

A Effects of the ramp rate

The presence of quantized energy levels can be revealed by a fast sweep of the junction bias[7], in order to enhance non-stationary effects on the statistical population of the energy levels. The effect can be here summarized by the following argument: during the escape process the population of the upper levels undergoes to a fast escape rate and this tends

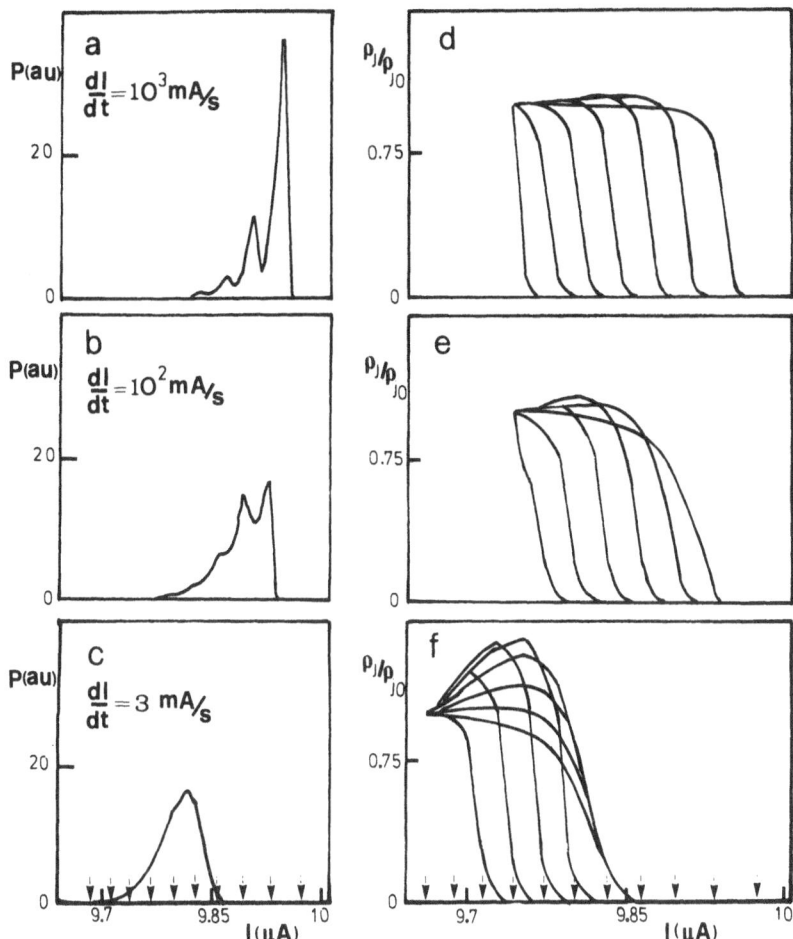

Fig. 3 *Switching current distributions P(I) (left column) and the corresponding levels population evolution (right column) reported as a function of the current I at various values of the bias frequency dI/dt : a) and d), dI/dt= 10^3 mA/sec; b) and e), dI/dt= 10^2 mA/sec; c) and f) dI/dt= 3 mA/sec. The levels populations ρ_j are normalized to the initial Bolzmann population $\rho_{jo}=\rho_j(t=0)$, in order to have all of them in the same scale. The simulation is performed for the following junction parameters: I_c= 10 µA; C= 5 pF; R= 100 kΩ; T= 200 mK. The arrows indicate the current values for which the upper energy level approaches the top of the energy barrier.*

to push the system out of the thermal equilibrium. If the rate of change of the bias current is slow respect to typical relaxation time of the system (related to transition probability between levels) the process is adiabatic and we are in quasi-stationary conditions. That is at each value of current, the diffusion process from the lowest levels will be able to refill the upper levels keeping a quasi-stationary distribution inside the well. The non stationary conditions come out as a consequence of a fast sweep of the junction bias[7], which in turn produces a rapid change in the potential shape and a fast reduction of the potential barrier U_0. If the sweep rate is so fast that the refilling from the lowest levels by thermal diffusion has not enough time to take place, the upper levels will be depopulated and once almost emptied can no longer contribute to the escape. So that until the next level approaches energy values suitable for activation, there is a rapid decrease of the decay rate. This occurs periodically as a level is emptied and yields, in the current switching distributions of the junction, to an oscillatory behaviour which is a clear manifestation of the presence of quantized energy levels.

This considerations can be visualized in fig. 3.

In the left column (figures a,b, and c) the switching current distributions are reported for various, decreasing values of the sweep rate. In the right column the corresponding populations of the levels in the well are reported as a function of the bias current, in order to visualize the evolution of the levels populations in the various cases. The arrows indicate the current values at which the upper level reach the barrier top. For the higher value of the sweep rate the oscillatory behaviour is clearly shown, with many minima of P(I) related to the position of the levels. Looking at the corresponding evolution of system, it is clear that the diffusion process from the lower levels does not occur, and the populations remain essentially constant, except for the level close to the barrier top which undergoes a fast reduction due to the escape process. Reducing the rate of change of the potential barrier, the diffusion from the lower levels has more time to refill the upper level, and this tends to destroy the oscillatory behaviour in the escape rate. At the lowest sweeping rate, the oscillatory behaviour disappeared completely, corresponding to a large thermal diffusion from the bottom of the well which keeps a quasistationary situation in the well.

Let us now give a look at some other features of this peculiar effect.

B Temperature dependence of the quantum effects

The temperature dependence of the effect is exemplified in fig.4. In discussing the figure, the bath temperature T must be compared with the cross-over temperature $T_0 = \hbar\omega_0/K$, that in our simulation is ranging between 100-140 mK, depending on the bias current. It is first shown the situation at a temperature $T < T_0$ (T=100 mK) . In this case almost all the switchings occur at currents values corresponding to the presence of only one level in the well. In fact, at low temperatures almost all the particles are initially in the ground state, the transition element w_{ij} between levels are small and there is no time to populate other levels when you run at fast speed. All the particles are "frozen" in the ground state, so that one cannot see the typical oscillations due to the levels discreteness. These fluctuations are evident increasing the temperature, namely for $T > T_0$. They are more evident in the lower dissipation junction, because the transitions probabilities between levels are smaller and practically uninfluential. In this conditions the area under each peak is close to be proportional to the initial population of the corresponding level. Increasing the temperature we can observe a larger number of oscillations since the initial Bolzmann population is dealt out over many levels. It is obvious that this is true only up to values of T not so large to permit many transitions between levels and, as a consequence, a repopulation of the upper levels. The upper limit for T to observe the oscillations due to the level discreteness depends on the damping level, which also influences the w_{ij} elements. The curves shown refer to a set of junction parameters which leads to a cross-over temperature $T_0 \cong 100$ mK, but the analysis of the results allows to extend the conclusions to different junction parameters with different T_0 values.

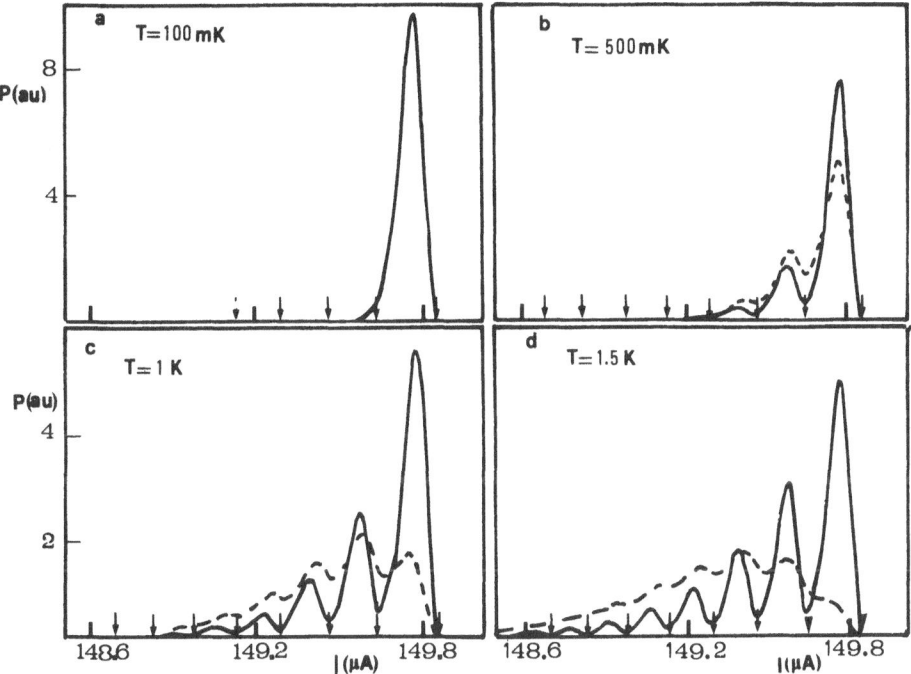

FIG. 4 . *Current switching distributions P(I) vs I as obtained by numerical integration of the system 2) for a junction with $I_C = 150 \ \mu A$ and $C = 5$ pF. The sweep rate is here $dI/dt = 10^5$ mA/sec. Two damping resistance values, $R = 1 k\Omega$ (dashed lines) and $R = 100 \ k\Omega$ (solid lines), have been considered at different temperatures: a) $T = 100$ mK; b) $T = 500$ mK; c) $T = 1$ K; d) $T = 1.5$ K. For this junction parameters the cross-over temperature is $T_0 = 100-140$ mK in the considered bias range. The arrows indicate the current values for which an energy level approaches the top of the energy barrier. Note that in a) (i.e. $T = 100$ mK), the dashed and solid lines are coinciding.*

C Dependence on the effective resistance

The effect of the damping level is exemplified in fig.5. In the higher resistance junction, it is evident the oscillatory behaviour. The oscillation amplitude reduces as the resistance decreases; in fact an increasing of dissipation leads to higher transition probabilities between levels and to a faster thermalizzation. The diffusion from the lower levels is able to restore a quasiequilibrium situations, which tends to destroy the oscillatory behaviour. The curve for R= 200 Ω does not present, in fact, any evident effects due to the energy levels discreteness.

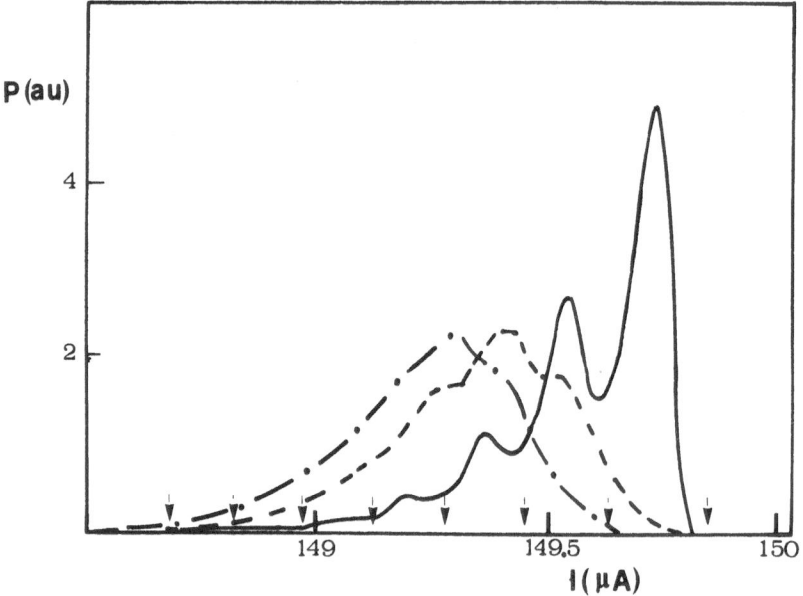

FIG. 5 .*Switching current distributions P(I) at T = 700 mK, for various effective resistance values: R=5 KΩ (solid line); R= 500 Ω (dashed line) and R= 200 Ω (dash-dotted line). Here dI/dt=3·10⁴ mA/sec, I_C=150 μA and C = 5pF.*

Conclusions

We have discussed some our data about the thermal induced escape out of the zero-voltage state of a Josephson junction removing the main assumption of the resestively shunted junction model. We have in fact included in the analysis the voltage dependent dissipation due to the quasiparticles tunneling. The analytical expression for the lifetime obtained in this way has been compared with our experimental results, providing an answer to the open question of what resistance dominates the junction dissipation. We have clearly shown the relevance of the quasiparticles current in the junction damping, and this intrinsic mechanism dominates, in our experimental configuration, up to rather high resistance values.

The extremely low dissipation level which can be obtained in this way, allows an experimental study of a new effect related to the quantum behaviour in the presence of dissipation (macroscopic quantum effects). We have shown that a fast sweep of the current biasing the junction, which in turn induces nonstationary conditions in the escape process, can modify the shape of the current switching distributions P(I), which become an oscillating function of the current. The effect, very large without the need for microwave irradiation, is due to the presence of discrete energy levels in the potential well. We have then investigated the dependence, in nonstationary conditions, of the current switching distributions both on the temperature and on the dissipation level. As a peculiar aspect of this behaviour, we stress that the effect of the energy levels discreteness becomes evident at temperatures larger than the cross-over temperature, and it is greater at decreasing damping values. The curves shown predict the results of new possible experiments. Experimental effort in this direction is in progress.

ACKNOLEDGMENTS

The author wishes to thank Antonio Barone, Yuri Ovchinnikov, Kenneth Gray, Andrei Varlamov, Sergio Pagano, Roberto Cristiano and Berardo Ruggiero for useful discussions and suggestions. This work has been partially supported by the National Research Council of Italy under the Progetto Finalizzato " Superconductive and Cryogenic Technologies".

REFERENCES

1) H.A. Kramers, Physica 7, 284 (1940). It is impossible to mention all the theoretical works after Kramers. See for instance M. Buttiker, E.P. Harris and R. Landauer, Phys. Rev. **B28**, 1268, (1983), and references therein.

2) P. Silvestrini, J. Appl. Phys. **68**, 663 (1990).

3) P. Silvestrini, O. Liengme and K.E. Gray, Phys.Rev. **B37**, 1525 (1988);   P. Silvestrini, R. Cristiano, S. Pagano, O. Liengme and K.E. Gray, Phys. Rev. Lett. **60**, 844 (1988).

4) A.D. Caldeira and A.J. Leggett, Phys. Rev. Lett. **46**, 211 (1981). For a general account on the problem see , for instance, A.D. Caldeira and A.J. Leggett, Ann. Phys. (N.Y.) **149**, 374 (1983); Antony J. Leggett, *Direction in Condensed Matter Physics*- Vol.1, p.187, ed. by G.Grinstein and G. Mazenko, (World Scientific, Singapore,1986) and references therein.

5) J.M. Martinis, M.H. Devoret, and J. Clarke, Phys. Rev. **B35**, 4682, (1987); M.H. Devoret, D. Esteve, J.M. Martinis, A.N. Cleland, and J. Clarke, Phys. Rev. **B36**, 36 (1987).

6). A.I. Larkin, Yu.N. Ovchinnikov, Zh. Eksp. Teor. Fiz. **91,** 318 (1986) [Sov. Phys. JETP **64**, 185 (1986)].

7) P. Silvestrini,Yu. N. Ovchinnikov and R. Cristiano, Phys. Rev. **B41**, 7341 (1990).

8) M. H. Devoret, J.M. Martinis, D. Esteve and J. Clarke, Phys. Rev. Lett. **53**, 1260, (1984);   M.H. Devoret, J.M. Martinis and J. Clarke, Phys. Rev. Lett. **55**, 1908, (1985);   J.M. Martinis, M.H. Devoret and J. Clarke, Phys.

Rev. **B35**, 4682, (1987);   A.N. Cleland, J.M. Martinis and J. Clarke, Phys. Rev. **B37**, 5950, (1988).

9)   A. Barone and G. Paternò, *-Physics and Applications of the Josephson Effect-* (Wiley, New York, 1982).

10) See, for instance, A. Barone, R. Cristiano and P. Silvestrini, J. Appl. Phys. **58**, 3822, (1985).

11) R.F. Voss and R.A. Webb, Phys. Rev. Lett. **47**, 265, (1981).

12) L. D. Jackel, J.P. Gordon, E.L. Hu, R.E. Howard, L.A. Fetter, D.M. Tennant, R.W. Epworth and J. Kurkijarvi, Phys. Rev. Lett. **47**, 697, (1981).

13) S. Washburn, R.A. Webb, R.F. Voss and S.M. Faris, Phys. Rev. Lett. **54**, 2712, (1985).

14) J.R.Kirtley, C.D. Tesche, W.J. Gallagher, A.W. Kleinsasser, R.L. Sandstrom, S.I. Raider, and M.P.A. Fisher, Phys. Rev. Lett. **61**, 2372 (1988).

15) A.T. Jonson, C.J. Lobb, and M. Tinkam, Phys. Rev. Lett. **65**, 1263 (1990)

16) R. Cristiano and P. Silvestrini, Il Nuovo Cimento **D10**, 869 (1988).

17) C.W. Gardiner - *Handbook of Stochastic Methods* - ed. by Hermann Haker , (Springer, Berlin, 1983), chap. 9.

18) R.L. Stratonovich - *Topics in the Theory of Random Noise* - (Gordon and Breach, New York, 1967), vol. I, chap. 4.

19) A.I. Larkin and Yu. N. Ovchinnikov, Sov. Phys. JETP **24**, 1035, (1967).

20) K.S. Chow, D.A. Browne and V. Ambegaokar, Phys. Rev. **B37**, 1624 (1988).

21) P. Kopietz and S. Chakravarty, Phys Rev. **B38**, 97 (1988).

22) Paolo Silvestrini, Phys. Lett. **A** (to be published).

23) It is clear that the result in principle depends on the choice for I_0. however, whatever is I_0 so that we have many levels in the well at $t=0$, the results does not depend on I_0 at all.

CORRELATED SINGLE ELECTRON TUNNELING IN ULTRASMALL JUNCTIONS

T.Claeson[1], P. Delsing[1], D. Haviland[1], L. Kuzmin[1,2], and K.K. Likharev[2]

[1]Physics Department, Chalmers University of Technology, S-412 96 Göteborg, Sweden
[2]Physics Department and Laboratory of Cryoelectronics, Moscow State Univ., Moscow 119899 GSP, USSR

INTRODUCTION

Tunneling of electrons in a macroscopic tunnel junction is a stochastic process governed by several parameters. Of main importance is the tunneling barrier - its width and height, intermediate states in the barrier, dielectric constant, etc - and the applied voltage. Also the density of electron states of the electrodes (including the possible effect of superconductivity) and environmental parameters like the temperature and the bias circuit come into play. However, the number of tunneling electrons per unit time is very large in a usual tunnel junction and we need not care about single tunneling events. Modern micro-electronic fabrication technology makes it possible to realize small tunnel junctions such that the tunneling of a single electron requires a significantly large energy.

This has several consequences, e.g. a Coulomb blockade of tunneling at low bias before a threshold voltage is exceeded, correlation in time between successive tunneling events, and correlation in space when tunneling through two, or more, junctions in series.

The single electron tunneling (SET) effects may find a number of technical applications like a new current standard and other standards, ultra-sensitive electrometers, and fast response transistors that may be packed more closely together than conventional components. However, many technical problems have to be solved such as realizing sufficiently separated and sharp steps in the I-V curve at microwave irradiation, electrostatic coupling to the device without excessive capacitive loading, and fabrication of even smaller, more uniform and rugged junctions in large numbers. Another hampering factor is the need of using low temperature, preferably well below 1 K, to study or utilize SET effects. Smaller junctions may raise the temperature range of applicability but, on the other hand, the higher voltage and power levels accompanying smaller devices may be a disadvantage in densely packed ULSI circuits.

The environment is of great importance for the SET properties of a tunnel junction. A low impedance that is shunting the junction may completely mask any SET effects. However, the dependence may even be utilized to probe the high frequency excitations of the circuitry surrounding a tunneling junction (or point contact). We will show that it is even possible to obtain information regarding such a basic, and much debated, issue as the time of tunneling.

This lecture will discuss the different aspects of single electron tunneling and the conditions that have to be met for them to occur. To study the time and space correlations, we found it helpful to use arrays of series coupled junctions. The occurence of charge solitons on these arrays is discussed. A few possible applications will be presented. In particular, analogs to the well known Josephson effects and applications will be stressed, in order to connect to concepts discussed in other contributions to this Workshop.

Nonlinear Superconductive Electronics and Josephson Devices
Edited by G. Costabile *et al.*, Plenum Press, New York, 1991

Continuous vs. discrete charging. The charge transfer through the leads to a tunnel junction is continuous. The surface charge at the junction changes continuously. The electric field and the capacitor charge of the junction, Q ($=V/C$) are also continuous variables. The change of stored charge due to tunneling of an electron e, however, is a discrete event.

Voltage variations due to tunneling. The change in junction voltage $\Delta V = e/C$ (where C is the tunnel junction capacitance, see fig. 1 for notations) at a tunneling event is very small for typical tunnel junctions. For a junction of size, e.g., 10 μm x10 μm it is of the order of 0.1 μV, and is usually completely masked by thermal fluctuations, even at low temperatures (1 K corresponds to a fluctuation voltage of 86 μV). As a result, the only noticeable consequence of the discreteness of charge transfer will be the well known shot noise. We will now discuss considerably smaller junctions, of the order of 0.1 μm x 0.1 μm or less.

Coulomb blockade. In ultrasmall tunnel junctions, the capacitive charging energy, $Q^2/2C$, may become the dominant energy. If $|Q| < e/2$, a tunneling event ($\pm e$) would lead to a higher capacitive energy, see fig. 2. Only for $|Q| > e/2$, i.e. $|V| > V_{threshold} = e/2C$, tunneling will be favorable. No current will pass the tunneling channel for a bias voltage less than the threshold, or the Coulomb blockade, value[1-10].

Relaxation oscillations. With a small bias current forced through the junction, the charge will build up to e/2, an electron will tunnel, changing the junction charge to -e/2, the charge builds up again, etc. Relaxation oscillations of both the charge and the voltage ($V=Q/C$) occur as illustrated in fig. 3. The oscillation frequency[1-6] is determined by the current $<I> = nef$, n being an integer. The behavior is reminiscent of the one of a leaking faucet building up water drops that fall with regular intervals. The I-V curve, and the oscillations, will be affected by parameters like temperature, tunnel resistance, and environment. The oscillator linewidth is broadened by a shunt admittance. The oscillation will be well defined at small bias but the amplitude decays as the curent is increased[1]. Fig. 4 shows how the oscillation is broadened with increased bias. In all cases, the extrapolations of the I-V curves at large positive and negative biases will be offset by a voltage $2V_{off} = 2 \cdot e/2C$.

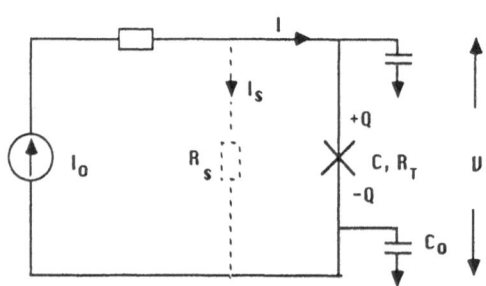

Fig.1.Equivalent circuit of a current biased tunnel junction.

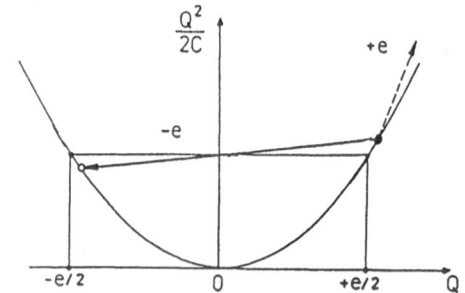

Fig.2. Change of the charging energy, Q2/2C, resulting from the tunneling of a single electron. First when |Q|>e/2, it becomes advantageous for an electron to tunnel, $\Delta Q = \pm e$. This is shown by the arrow for Q>e/2.

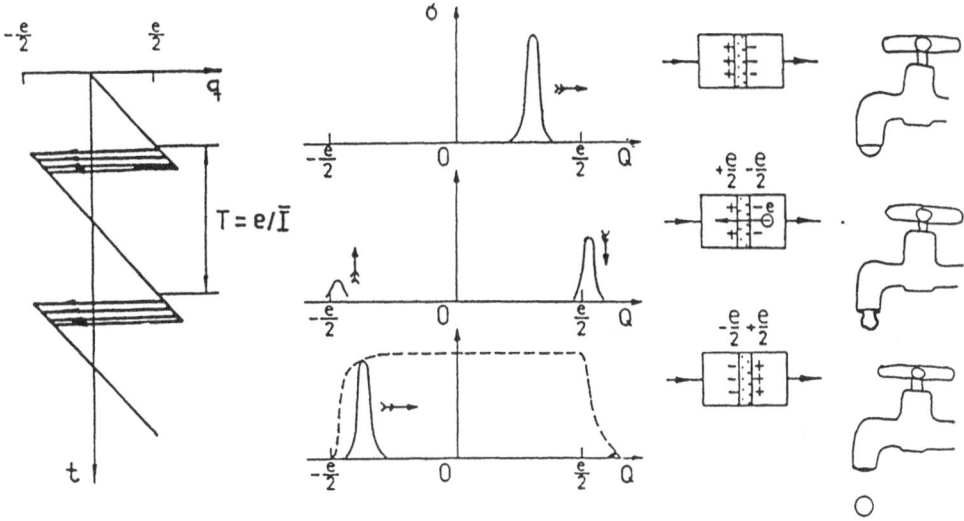

Fig. 3.   The relaxation oscillation in the building up of charge (and voltage)
across an ultrasmall tunnel junction. The evolution of the charge
probability, $\sigma(Q,t)$, as the charge builds up and tunnels is shown. A
water drop analogy is also sketched.

Conditions. Several conditions have to be fulfilled to allow the Coulomb blockade to
dominate and the correlation effects to be observable:

•   the charging energy must be larger than the thermal fluctuations, $e^2/2C > k_BT$

•   the capacitive energy must be larger than quantum fluctuation energies, $e^2/2C \gg h/\tau$,
where the characteristic time $\tau = R_TC$ or R_SC (whichever is smallest of the tunneling
and shunt resistances) leading to $R_T, R_S \gg R_Q = h/4e^2 \approx 6.5k\Omega$.

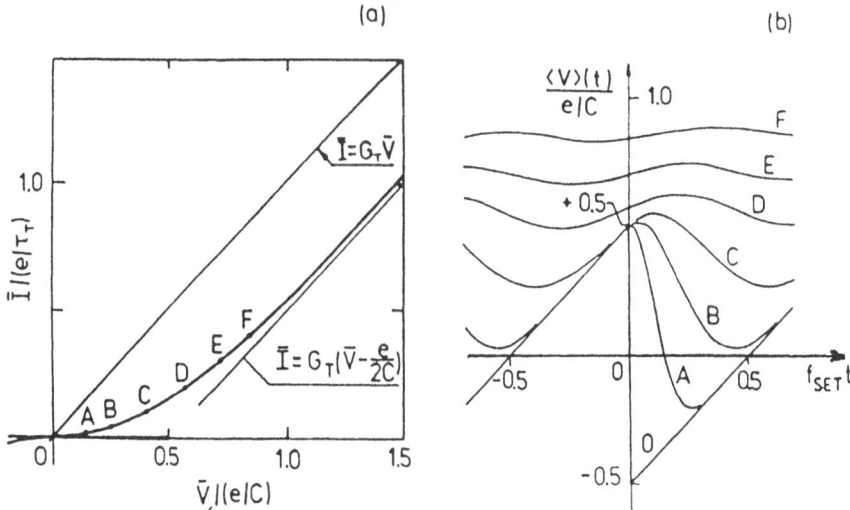

Fig. 4.   SET effects in an unshunted tunnel junction with non-superconduc-
ting electrodes in the low temperature limit: (a) dc I-V curve,
(b) oscillation waveforms at several biases as marked in (a).

Table 1.Single Electron Tunneling Characteristics

Properties:

Coulomb blockade, $eV > E_c = e^2/2C$

Offset voltage, $2V_{off} = e/C$

Oscillations, $f = <I>/e$

Conditions:

$E_c > k_B T$

$E_c > h/RC$ or $R_T > R_Q = h/4e^2$

$Z_{shunt} > R_T$

This means that the capacitance (and therefore the junction size) must be small, the temperature low and the tunneling resistance (as well as shunting impedances) higher than the quantum value. As we have already noted, the junction area must be less than about 0.1 μm^2 to allow an operating temperature of 1K.

Environment. Another limitation is set by the electro-magnetic environment. The Coulomb blockade can be completely wiped out by a low impedance environment. There will be stray capacitances to the tunnel electrodes. In fact, if no precautions are taken, the stray capacitance, C_s, from the leads and other sources will completely dominate over the junction capacitance ($C_{eff} = C + C_s$ for the two parallel capacitances). Several solutions to this problem have been suggested:

- connect very large resistances in series and in direct contact to the junction in order to cut off most of the stray capacitance;

- imbed the junction in an array of high resistance tunnel junctions that also will isolate the junction from the environment. (However, since the charging of one junction capacitor in the array will influence the others, there will be a transport of extended charge solitons in the array rather than single electron transfers in single junctions);

Bias. The single electron tunneling effects we consider are mainly manifested in variations in voltage over the junction. This means that we need a constant current source, e.g., by connecting large resistors in series with the junction, rather than a constant voltage source (which would be generated by a large stray capacitance in parallel with the junction). For two or more series coupled tunnel junctions in an array, there is an additional degree of freedom, namely the voltage of the intermediate electrode(s). This means that correlation effects can be present in arrays also using voltage bias sources.

Experiments

Distorted I-V curves and an off-set voltage at large bias have been seen in a number of experiments[11-14].

Thin-film junction. Two I-V curves of a single Al-AlOx-Al Junction are shown in fig. 5. The junction is connected through 30 μm long, thin film NiCr resistors of sheet resistance 700 Ω/sq, located very close to the junction. The junction resistance is 2.6 $k\Omega$, well below the value $R_Q = 6.5$ $k\Omega$ where quantum fluctuations should weaken the blockade. We see a distinct difference between the I-V curve in the superconducting state and the normal state which is achieved by supressing superconductivity with a magnetic field. This difference persists even out to voltages well above the gap voltage, $V_{2\Delta} = 200\mu V$ for this junction. We submit that his difference is due to the supression of quantum fluctuations at bias less than $V_{2\Delta}$, where the junction is sensitive to the quasiparticle resistance $R_{qp} >> R_Q$. As the voltage is increased beyond approximately $3V_{2\Delta}$, the effect of the quasiparticle resistance becomes negligible and the I-V curve returns to its normal state asymptotic value.

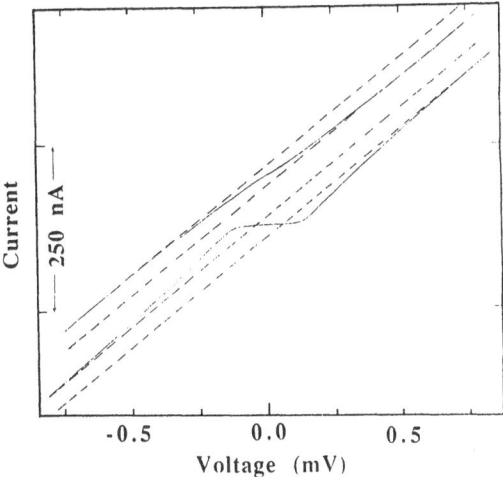

Fig. 5. The I-V curve of an Al-AlOx-Al junction in the superconducting state (lower curve) and the normal state (upper curve) which was measured in a magnetic field (H>H_{C1}). The dashed lines show the asymptotic resistance values, offset by 90μV due to the Coulomb blockade. Both curves were measured at T=55mK.

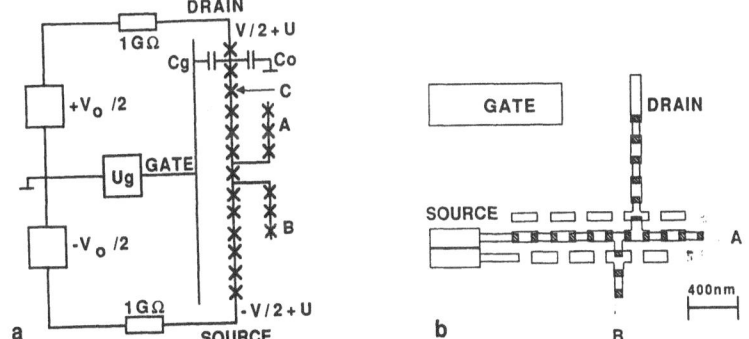

Fig. 6. The geometry of an one-dimensional array of ultrasmall tunnel junctions.
(a) Equivalent circuit of the array and the dc bias supply.
(b) The lay-out of the array. The junction electrodes are deposited using a form of a shadow evaporation technology. The latter uses self-aligned masks in an e-beam exposed double layer resist where the lower, supporting layer has been over-developed to give hanging bridges. Evaporation from separate angles gives the desired overlap and Al/insulator/Al tunnel junctions of the order of 80nmx80nm.
The additional metal islands parallel to and disconnected from the array are resulting as a side effect of the fabrication; they do not affect the array properties.

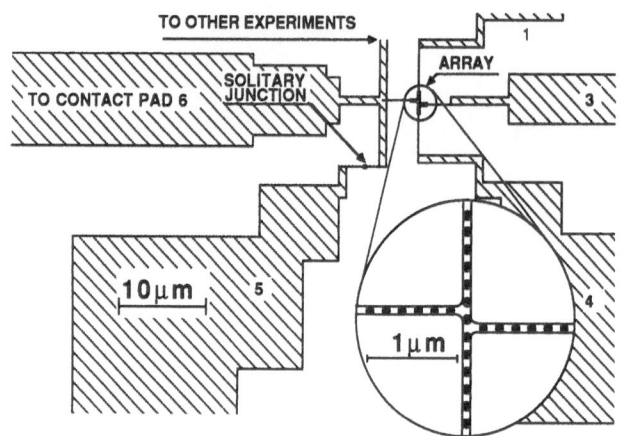

Fig. 7. The configuration of a thin film structure containing arrays of tunnel junctions and a single (solitary) tunnel junction. The array region is blown up.

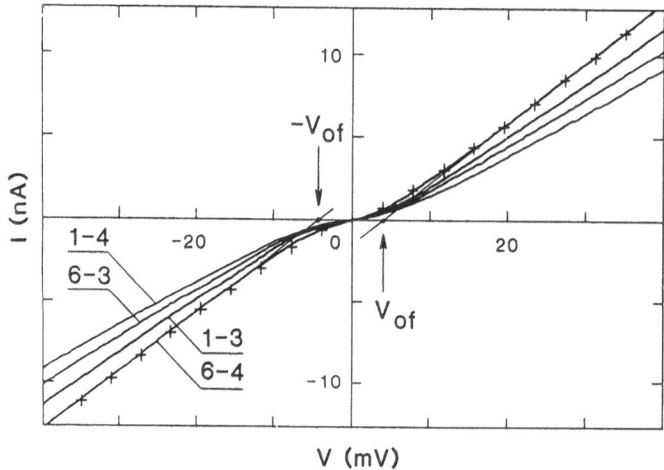

Fig. 8. Current and conductance as functions of voltage for the various branches of an array of the type shown in figure 7. Indices refer to the numbers of the contact pads given in figure 7. The solid lines are experimental curves. The variation in asymptotic resistance of the different branches does not exceed 20%. This indicates a good uniformity of the junctions. The crosses are results of a numerical calculation (ref.29) of the dynamics of a 13-junction 1-D array with parameters: $C_{i\neq7}=2.0 \times 10^{-16}$F, $C_7=2.8 \cdot 10^{-16}$F, $R_{i\neq7}=212$ kΩ, $R_7=150$ kΩ, $(C_0)_{i\neq6,7}=0.2 \cdot 10^{-16}$F, $(C_0)_{6,7}=0.6 \cdot 10^{-16}$F. T = 1.3 K.

<u>Junctions inside arrays</u>. A clear Coulomb blockade was seen[14] in a "conventional", planar, small tunnel junction that was "protected" from the low shunting impedance of the environment by arrays of similar tunnel junctions in the current and voltage leads. The geometry is shown in figs. 6 and 7 and the I-V curve in figs. 8 and 9. An offset voltage at large bias is clearly seen in fig. 9a. Also for an array of junctions as a whole, there is a similar, but enhanced, Coulomb blockade range. The interpretation of the experiment is complicated by the transport of solitons, as will be discussed below.

On the contrary, no, or a very small, Coulomb blockade was seen for a solitary, single junction that was not protected from the low load impedance of the surrounding circuit but was connected to it by low resistance leads. The junction was processed simultaneously with those in the array. The circuit geometry is shown in fig. 7. The small Coulomb blockade region indicated by the conductance, dI/dV vs V, curve of fig. 9b may be due to SET in the solitary junction but may as well, or rather, be due to a two step tunneling process via an intermediate state or inclusion.

<u>STM</u> experiments by van Bentum et al.[11] show a Coulomb blockade region when tunneling from a tungsten tip into pieces of stainless steel or a ceramic superconductor at room temperature as shown in fig. 10. At low bias, there is a quadratic dependence of the current on voltage. From the offset voltage one calculated a junction capacitance of $C \approx 10^{-18}F$, a very small value taking into account the tip geometry. Such a small value can only be obtained by assuming that the tunneling time, τ_T, is short. Charge is supplied to the tunnel electrodes from a region determined by the tunneling time and any contributions to the line impedance further away than $\tau_T c$ (c=the speed of light on the line) are then disregarded. The tunneling time in question would equal the traversal time of the electron across the tunnel barrier. We will argue against such a short τ_T later in this paper. There have also been suggestions that the observed STM tunneling process is a two-stage process and goes via an intermediate state[12], like a particle on the tip.

TUNNELING VIA A TWO-STEP PROCESS IN A DOUBLE JUNCTION

If two small tunnel junctions are coupled in series, the voltages across them will distribute according to tunnel impedances (capacitances). As long as the voltage on either of the junctions is not sufficient to overcome the Coulomb blockade, no current will flow. When the bias voltage is sufficiently large to allow a tunneling of an electron through one of the junctions, the middle electrode will charge up and there is a large probability for an electron to tunnel across the second junction. Thus a spatial correlation of tunneling events arises[1,17,18].

The tunneling will depend upon the charge on the middle electrode. This charge can be controlled via a gate electrode, or may result from a non-symmetric circuit. The Coulomb blockade may be completely suppressed by injecting a charge of e(n+1/2) onto the middle electrode. The tunneling dynamics is strongly effected even outside the Coulomb blocking range.

A characteristic staircase pattern in the I-V curve results. This is shown in fig.11. Each step corresponds to a change by e in the average charge of the middle electrode. Generally, there are two periods, $\Delta V_1 = e/C_1$ and $\Delta V_2 = e/C_2$, corresponding to the two junctions. If one of the junctions has a much larger resistance than the other one, $R_1 >> R_2$ (or a large capacitance $C_1 >> C_2$ if the resistances are comparable) only one period, $\Delta V = e/C_1$, can be distinguished. The threshold voltage where tunneling first occurs, i.e. the blockade voltage, is given by the smallest of $e/2C_1$ and $e/2C_2$ at zero charge on the middle electrode. The current jumps of the staircase are determined by the junction resistances and capacitances. For $R_1 >> R_2$ the current jump would be $\Delta I = e/R_1(C_1+C_2)$; half this value for the initial jump in current (at $V=e/2C_1$).

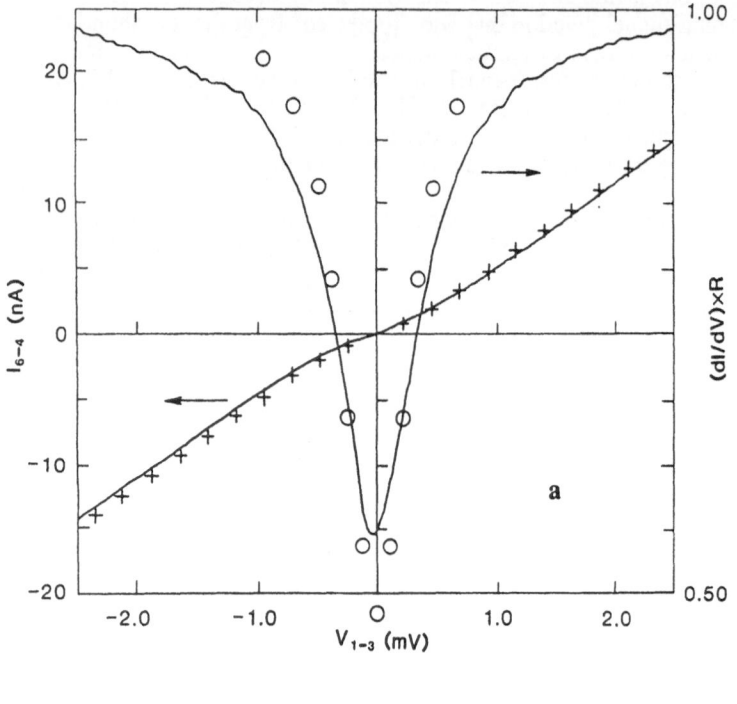

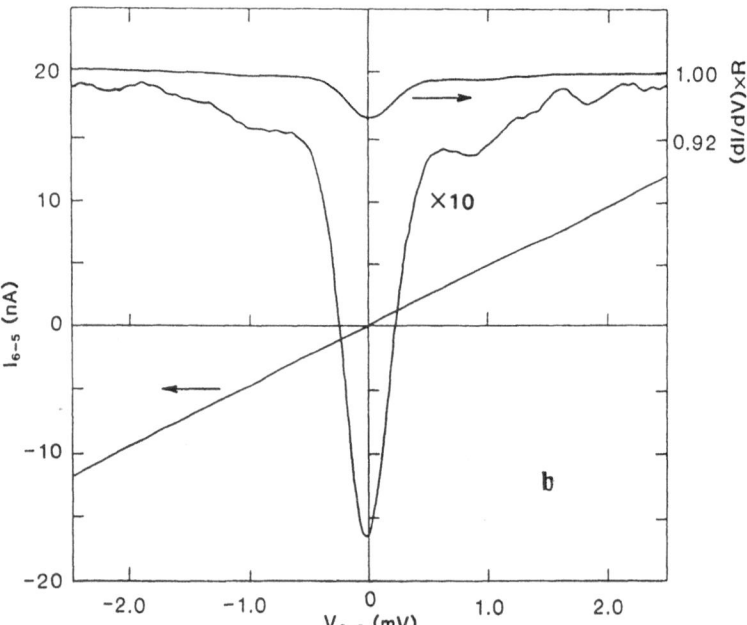

Fig. 9. dc current and differential conductance as functions of dc voltage for
(a) the middle (protected) junction of the array of figure 7 and
(b) the solitary (unprotected) junction of figure 7. Note the large
differences in the Coulomb blockade structure and the off-set voltage
between the two junctions. These features are barely distinguished for the
solitary junction. The ten times magnified conductance curve in (b) is
offset vertically in order to approach the same horisontal asymptote at
large voltage as the original dI/dV curve. The circles are results of a
numerical calculation (ref. 29) using the parameters given in figure 8.

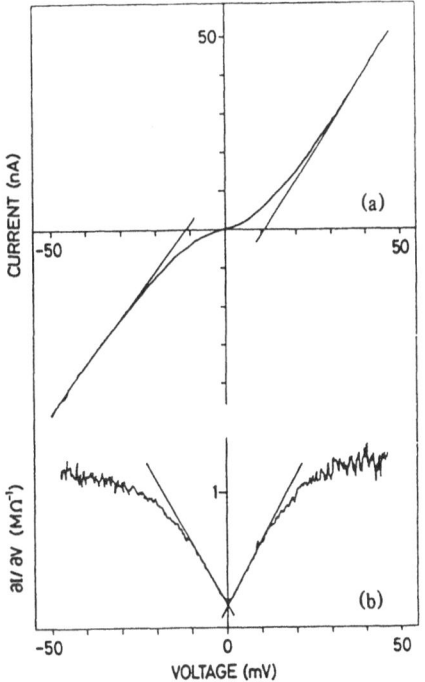

Fig.10. (a) The I(V) characteristic of a junction between a W tip and an YBa$_2$Cu$_3$O$_{7-\delta}$ surface.
(b) Corresponding dI/dV vs V. The data were taken with an STM immersed in a pumped liquid helium bath. From Ref.11.

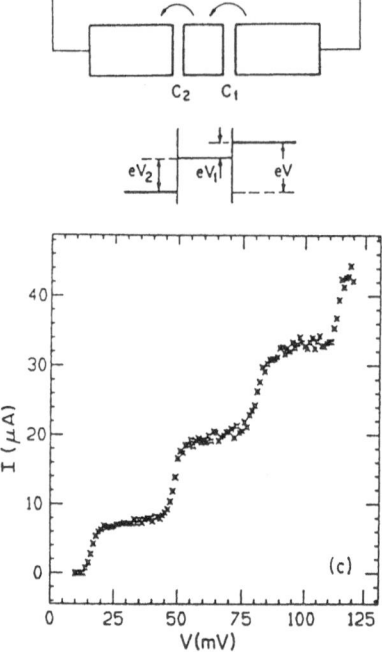

Fig.11. A schematic representation of a pair of ultrasmall capacitance tunnel junctions driven by an ideal voltage source is shown in the top part of the figure. The distribution of voltages between the two junctions is shown. The lower part of the curve gives the I-V curve correlated with the parameters: $R_1=25\Omega$, $R_2=2500\ \Omega$, $C_1=0.001$ fF, $C_2=0.005$ fF, $T=10K$. The steps are $e/C_2=32mV$ wide and have a height $e/R_2(C_1+C_2)=11\mu A$. Data from Ref.17.

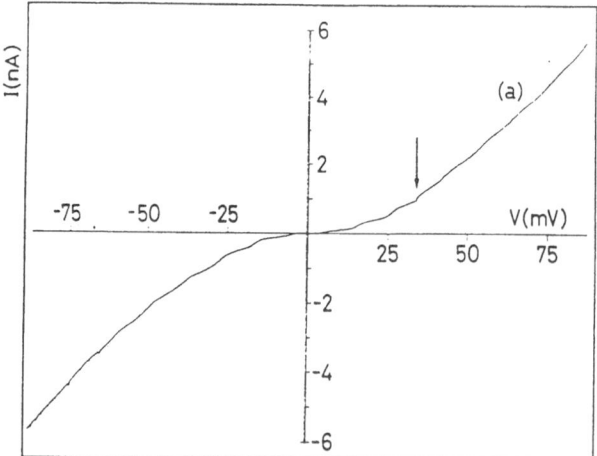

Fig.12. Staircase structure in the I-V curve from point contact tunneling, using an STM, into a high temperature superconductor, $YBa_2Cu_3O_{7-\delta}$. The data were taken at 1.2 K. The structure is supposed to be due to tunneling via a small grain in the superconductor. From Ref.25.

Most of the reported experimental results on single electron tunneling has been for more or less intentionally made double junction structures rather than for other types of structures. The first experiments[19-24], discovering the Coulomb blockade, were performed on grains embedded in an oxide with different thicknesses of the oxide on the top and the bottom. Tunneling mostly occurred over many grains in parallel.

Using a point contact, it has been possible to tunnel into isolated grains[25,26]. The intermediate particle may be charged and the charging can change during the experiment. Hence the Coulomb staircase can shift from experimental trace to trace. An example of an I-V curve taken with an STM point contact is shown in fig. 12.

Some point contact tunneling measurements on high T_c superconductors have shown extended regions of low current around zero bias and staircase like structure. Initially this was misinterpreted as an anomalously large superconducting energy gap and multiple gap structure[24,27]. However, it is more likely that the structure was due to single electron tunneling via small grains, i.e. a two step tunneling process. It has also been suggested that the point contact measurements that indicate a single step tunneling and an extremely small junction capacitance, in fact are due to a two step process via intermediate states, perhaps located on the tip.

A well controlled experiment was performed by Fulton and Dolan[28]. They used two series coupled ultrasmall tunnel junctions (about 0.001 μm^2 in size). The configuration is shown in fig. 13. The voltage of the middle electrode could be measured with the aid of a third, high resistance junction and the charge on the middle electrode could be controlled via a capacitively coupled gate. I-V curves for intermediate electrode charging of about 0, 1/6, 1/3, 1/2, and 2/3 electron charge are shown in fig.14. They can be compared with theoretical curves[1] given in fig. 15. These have been calculated using junction parameters corresponding to the experimental configuration.

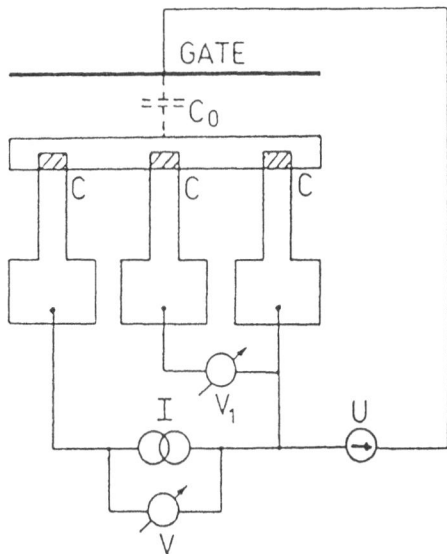

Fig.13. The layout of the experiment by Fulton and Dolan (Ref.28). It used two series coupled junctions to study the space correlated single electron tunneling. A third junction was used to allow the measurement of voltage on the middle electrode without loading it with an appreciable shunting capacitance. Charge could be capacitively coupled to the middle electrode via a gate on the back side of the substrate.

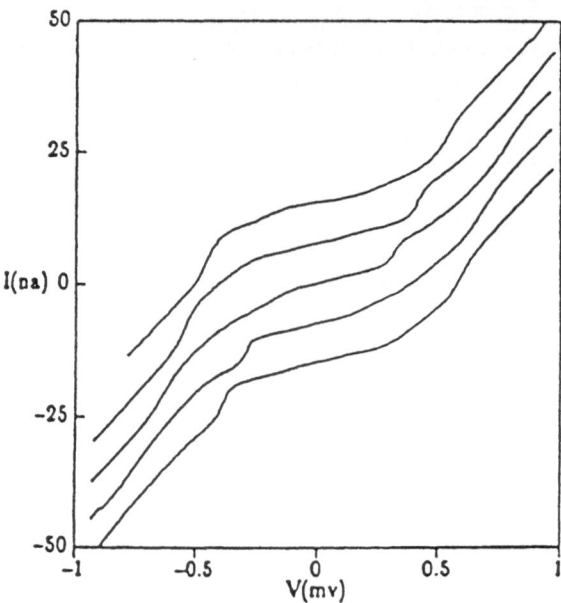

Fig.14. Experimental I-V curves from the experiment described in Fig. 13. The charge on the middle electrode is varied about e/6 from trace to trace. From Ref.28.

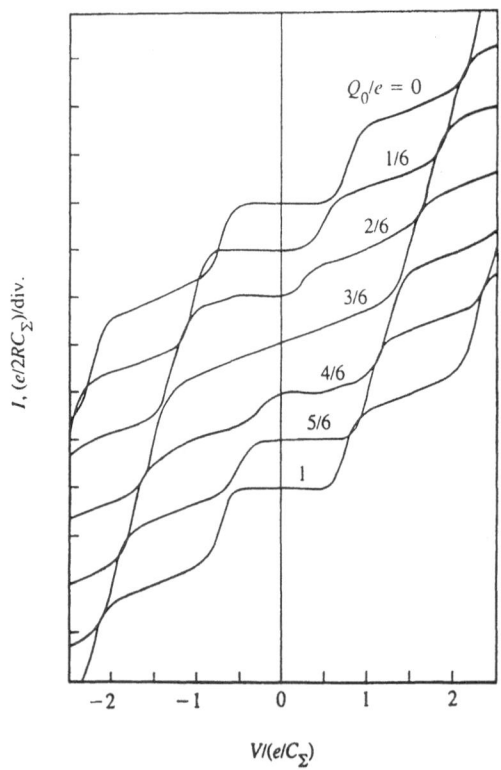

Fig.15. Theoretical I-V curves calculated using pertinent parameters corresponding to the experimental curves of the previous figure. From Ref.1.

Consider several tunnel junctions coupled in series. If one of the electrodes of a junction of the array is charged, the charge distributions on the surrounding junctions will also change, see fig. 16. With no stray capacitance, the charge on each junction capacitor is the same and the whole array can be considered as one extended capacitor. However, with stray capacitances C_0 to the environment, the charge on neighboring junctions will successively decay. If a charge carrier tunnels, the whole distribution will shift. It can be considered as a charge soliton (anti-soliton). Its extension (expressed in the number of junctions) is of the order of $2M=2[\text{ar cosh}(1+C_0/2C)]^{-1} \approx 2(C/C_0)^{1/2}$ (for $C_0<<C$), which typically is a few junctions[1,29].

If there is no voltage drop across the array, the charge soliton remains fixed. A threshold voltage $V_t = e/2C_0[1-(1+C_0/2C +\{C_0^2/4C^2+C_0/C\}^{1/2})^{-1}]$ is needed to inject and move a train of solitons, i.e. to allow current transport[1,29]. The off-set voltage at large bias is $V_{off} = (N-M)e/2C$ when the number of junctions, N, is considerably larger than the extension of a soliton, M.

The solitons are moving coherently in time and space according to theoretical models. They repel each other and form an ordered 1-D Wigner lattice. It moves with a frequency determined by the curent drawn through the array, $I=ef$. Corresponding entities in long Josephson junctions are fluxons (solitons). The movement of the fluxon lattice gives rise to a voltage $V=hf_J/2e$, i.e. the Josephson relation.

Fabrication of arrays

Modern electron beam lithography techniques allow the fabrication of sufficiently small tunnel junctions to allow a study of single electron tunneling phenomena. One can use a self aligned, hanging bridge technique developed by Dolan[30-32]. It is based upon two resist layers, see Fig. 17. The mask is formed in the upper layer which rests upon the lower, undercut layer. The mask is self-aligned as it is possible to evaporate under different angles in order to form varying amounts of overlaps of the base and top electrodes of the tunnel junction.

We have utilized a variant of that method to make arrays. The holes corresponding to bottom and top electrodes are formed as two parallel but shifted rows, see Fig. 6. The bottom aluminium electrodes are evaporated from an angle to the left and are oxidized to form a tunnel barrier. Then the upper electrodes are deposited through their masks by turning the substrate table and evaporating from an angle to the right. In that way, junction sizes down to about 50 nm x 50 nm can be made. The capacitance of such a junction is about 10^{-16} F. The resistance can be varied within a wide range, typically 30-1000 kΩ per junction in the experiments described here. The stray capacitance to the environment from each intermediate electrode is small - of the order of 10^{-17} F. The inductance of the intermediate strip is also small.

The number of junctions in an array can be varied over a large range. To avoid boundary effects, the length should be larger than 10-20 junctions. It is possible to couple to the array in different ways. The total voltage over all the junctions of the array can be determined. The voltage of a single junction in the array can be measured by leads consisting

Fig.16. A charge, e, on one junction, that is a member of a long array of series coupled junctions, affects the charge distributions on surrounding junctions. A charge soliton is formed. Its charge equals the supplied charge, e, and its extension is given by the junction and stray capacitances, which typically is of the order of a few junctions for common array parameters.

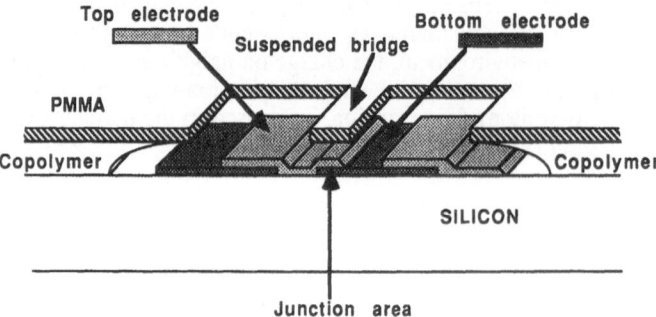

Fig. 17. The method of hanging resist bridges, in a two-layer resist technology, and evaportion from different angles is used to form overlap tunnel junctions as sketched above. The distances and angles are not to scale.

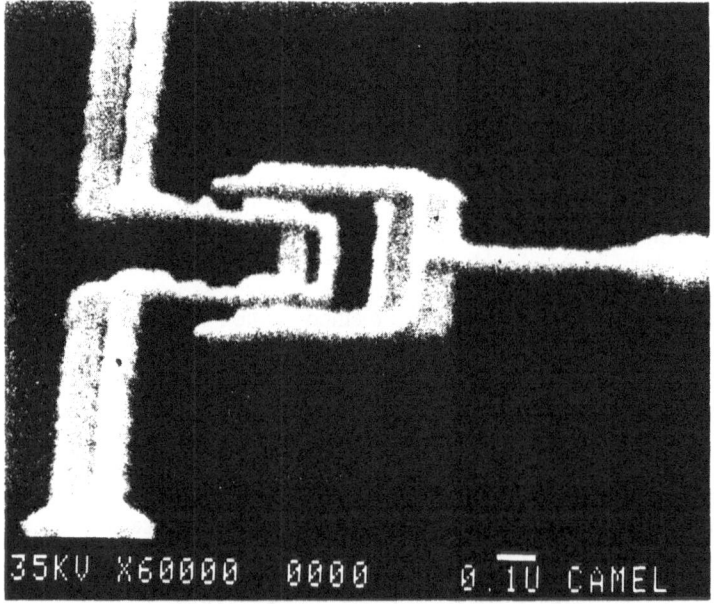

Fig. 18. A scanning electron microscope picture of one of the arrays of tunnel junctions studied. The two bent structures at the left are the bottom and top electrodes of two series coupled junctions (located in the middle of the horizontal parts. The fork like structure at the right is a capacitive gate used for the transistor experiments reported in this work. The double line (shadow like) structure is formed by the evaporation from diffeent angles, see the previous figure. Note the 0.1 μm marker.

of arrays of similar, high resistance tunnel junctions. Then the loading admittance is kept small. Charge can be coupled into intermediate electrodes capacitively (via a gate) or resistively (e.g., again via an array of junctions). A typical configuration is shown in fig. 7 . A picture taken by a scanning electron microscope is shown in fig. 18.

Coulomb blockade in 1-D arrays

An example[33] of the current-voltage curve for a nineteen-junction array is shown in fig. 19. There is no detectable current at low voltage bias. Note the 250 times magnification. The tunnel resistance is larger than 100 GΩ at zero bias. Current starts to flow at the threshold voltage, $V_t \approx 2$ mV. The offset voltage at large bias is $V_{off} \approx 7$ mV. The Coulomb blockade is very prominent at a temperature of 50 mK but is also discernible at 4.2 K. The I-V curve agrees well with a calculated curve using reasonable parameters[33], namely an asymptotic resistance of 210 kΩ per junction, a capacitance per junction, $C=2.4 \cdot 10^{-16}$ F, a stray capacitance per electrode, $C_0=1.2 \cdot 10^{-17}$ F, a resistance per junction at low bias of 1.4 MΩ if there were no charging effects, and a superconducting energy gap, $2\Delta(0)$, of 0.38 meV, see the bottom part of Fig. 19. The discrepancy between the experimental and theoretical curves at voltages just above the (Coulomb blockade) "supervoltage", where the experimental curve is not as steep as the theoretical one, may be due to noise smearing and to a crude sub-gap assumption.

The Coulomb blockade is smeared as the temperature is increased or as microwave power is applied. This is illustrated in Fig. 20 and Fig. 21.

Time correlation

The most basic relation[1] is the one between current and frequency (<I>=ef). Several attempts have been made to detect the time correlation[28,34-37]. There have even been doubts the effect ever will be detectable taking into account the effect of the environment.

The high frequency oscillation power is weak and will be difficult to detect. A way to verify the effect is to mix the oscillation with an external microwave signal. At zero intermediate frequency (which occurs when the internal frequency is a multiple of the external one, <I>/e=nf_{ext}) there may be induced voltage steps in the I-V curve. The corresponding effect in a Josephson junction is the Shapiro current steps at $f_J=2eV/h=nf_{ext}$.

Fig. 22 shows the tunneling resistance as a function of bias current (note, not voltage!) for the same nineteen-junction array as in fig. 19. When the array is irradiated by $f_{ext}=0.75$ GHz microwave radiation, pronounced peaks appear at bias currents corresponding to $I=nef_{ext}$. Very similar, but even more pronounced, resistance peaks show up in the calculated curves[33]. The calculations used the same parameters as those in the theoretical Coulomb blockade curves of fig. 19. Curves are given for three microwave powers (including no power) and the difference in the two microwave powers is the same as for the experimental traces. The additional broadening of the experimental curves, as compared to the theoretical ones, may be due to the ac signal used in the measurement, possible noise reaching the junctions, and inhomogeneities in junction parameters.

The peak magnitude depends, as expected, upon the microwave power. The peak location, however, remains fixed in bias current. The voltage location, on the other hand, depends on microwave power, temperature, etc. The peak structure is most easily seen at low temperature, 50 mK in fig. 22, but it can be distinguished for temperatures up towards 1 K. A good microwave coupling to the junction could only be obtained within a few frequency bands. The microwave induced structure decreased with increasing frequency and became difficult to distinguish above 5 GHz. This is close to the RC cut-off frequency of the junctions.

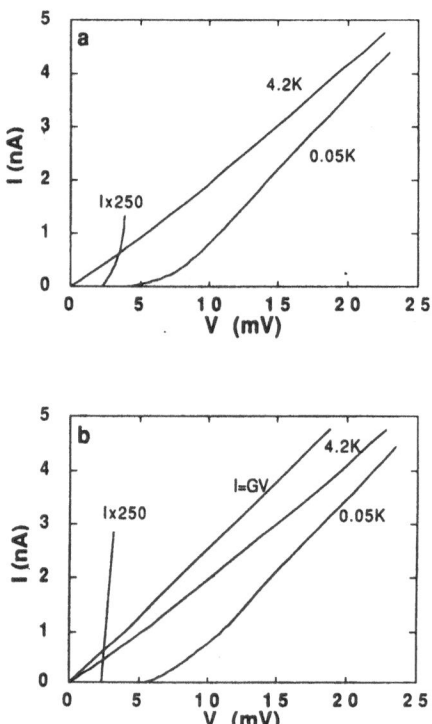

Fig.19. dc I-V curves from an array of nineteen tunnel junctions for two temperatures. (a) Experimental curves for 70nmx90nm Al/I/Al junctions. (b) Calculations using the theory of Ref.29. $C_0=1.2\cdot10^{-17}$ F, $C=2.4\cdot10^{-16}$ F, R=210 kΩ per junction. The superconducting gap has also been taken into account. A blow-up of the current scale for the 50 mK curve shows that the current is practically zero at low bias. From Ref.33.

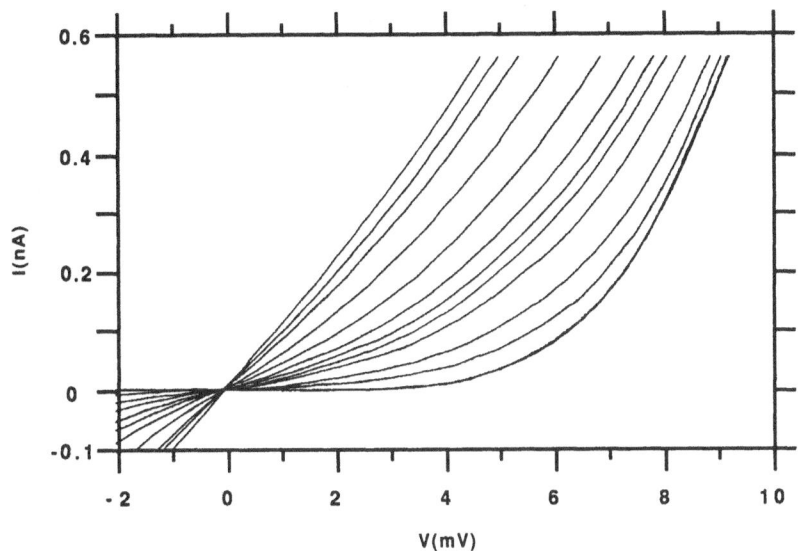

Fig.20. I-V curves at different temperatures for the same array as in fig.19.
Counted from the right (in the upper part of the figure) T= 0.06, 0.12,
0.25, 0.50 (these first four curves coincide), 0.71, 0.82, 0.94, 1.02,
1.07, 1.13, 1.22, 1.30, 1.43, 1.52, and 1.83 K. The Coulomb blockade
region is successively smeared as the temperature is increased.

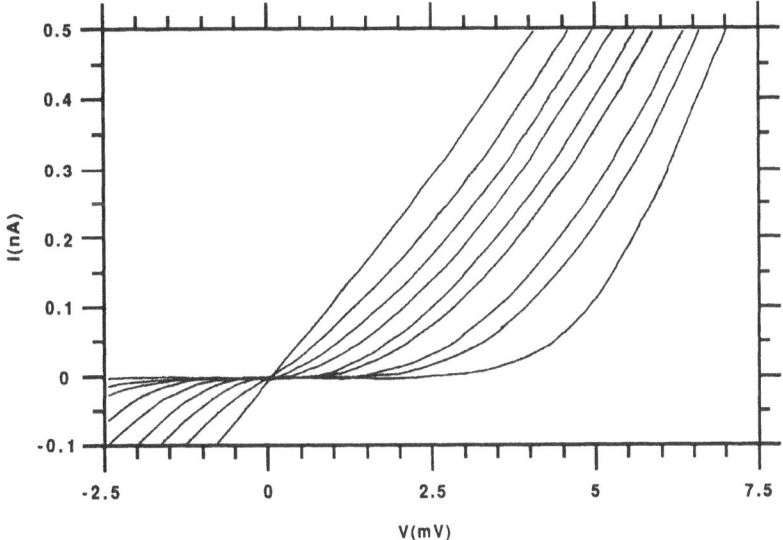

Fig.21. The smearing of the I-V curve of fig.19 with increasing microwave
power. Traces are given with no microwave power and with decreasing
damping of the microwave power: -20, -17, -15, -14, -13, -12, -11, and
-10 dBm (as refered to the room temperature end of the coaxial cable).
T=50 mK. f=687 MHz.

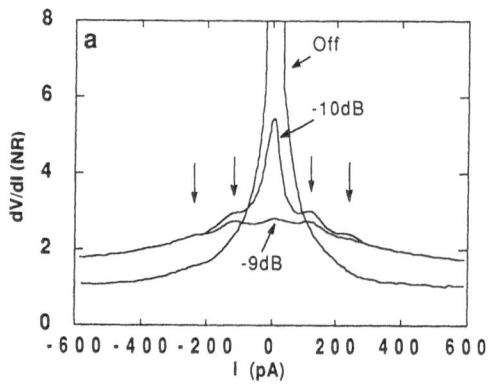

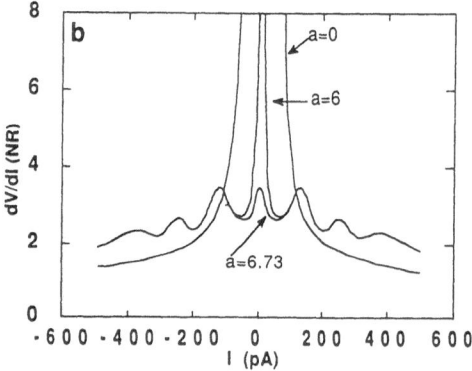

Fig.22. The dynamic resistance as a function of bias current for the same array as in Fig.19 at 50 mK for three values of the microwave power (f=0.75 GHz). It is normalized to the array resistance (NR). (a) experimental curves; (b) calculated ones. The difference in amplitude, a, between the two pumped curves in (b) corresponds to 1 dB in power, i.e a similar difference as in (a). Arrows show nominal positions of the resistance peaks as expected from I=nef. From Ref.33.

The relation between current and frequency is stressed in fig. 23. There, the locations in current of the resistance peaks are plotted as a function of microwave frequency. Not all wavelengths are easily coupled into the junctions, but there is a sufficient amount of resonance regions where there is a response in the form of discernible resistance structure. The photon induced structure is generally stronger the lower the frequency in accordance with theoretical expectations. The highest frequency of our investigation is approaching the RC cut-off frequency.

There is a good correspondence, see fig. 23, between bias current and frequency, $I=nef_{ext}$, $n=\pm1$, ±2. The slope of I vs. f_{ext} gives a multiple of the electron charge within a few percent. This leaves no doubt that the time correlation is present in the form of phase locked single electron (or soliton) tunneling events in arrays of high resistance junctions.

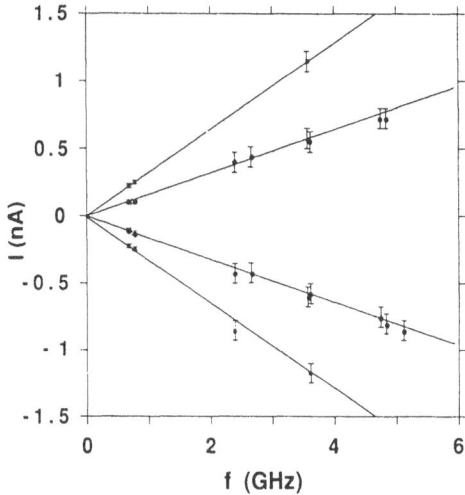

Fig.23. dc current positions of the $n=\pm1$ and $n=\pm2$ resistance peaks for two arrays. N=12 and N=15. The lines have slopes $\pm e$ and $\pm2e$. The data points fall close to these lines in good accord with the theory of time correlated tunneling. From Ref. 33.

Correlation in space

A tunneling event in one junction is expected to be shortly followed by an event in next junction, and so on. We have already noted that the current through a two-junction array can be controlled by inducing charge in the middle electrode. An example from the Delft group[38] is given in fig. 24. It shows I-V curves taken at very low temperature for two junctions in series. The maximum Coulomb blockade curve corresponds to the case of an integer charge on the middle electrode. The minimal blockade curve is presumably due to a half integer charge on the intermediate electrode. Fig. 25 displays the current through two junctions at a certain fixed voltage as a function of the gate electrode voltage. The period corresponds to an integer electron charge. With three junctions in series a more complicated periodic pattern can be seen. The result for a 13 junction array[39] is shown in fig. 26. That experiment was carried out at 1.3 K. The charge distribution on intermediate electrodes was varied with the aid of a gate, see fig. 6 for the lay-out. In this case, a dc signal was applied to the gate and the voltage response over the array, at a fixed current, was measured. The response is periodic and the period corresponds to a change in charge of an electron charge on an intermediate electrode. A change much less than that can be resolved, i.e. the correlated motion of single electrons (solitons) is controlled by sub-electron charges. The result can be reproduced numerically if we assume that one junction is particularly small.

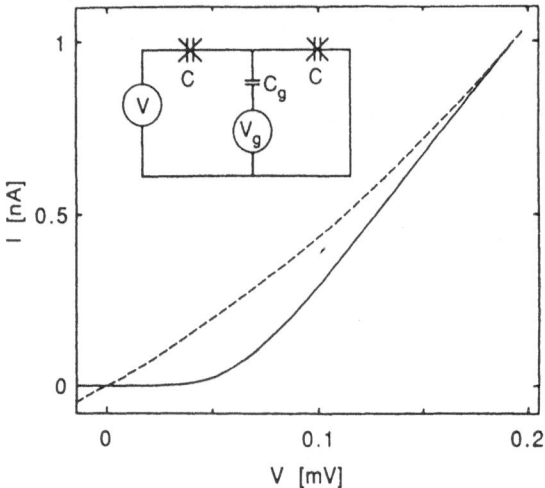

Fig.24. Modulation of the current through two junctions by a gate voltage. T= 10 mK. The curves shown display maximum and minimum Coulomb gaps. This presumably corresponds to an integer and half-integer additional charge on the middle electrode. From Ref.38.

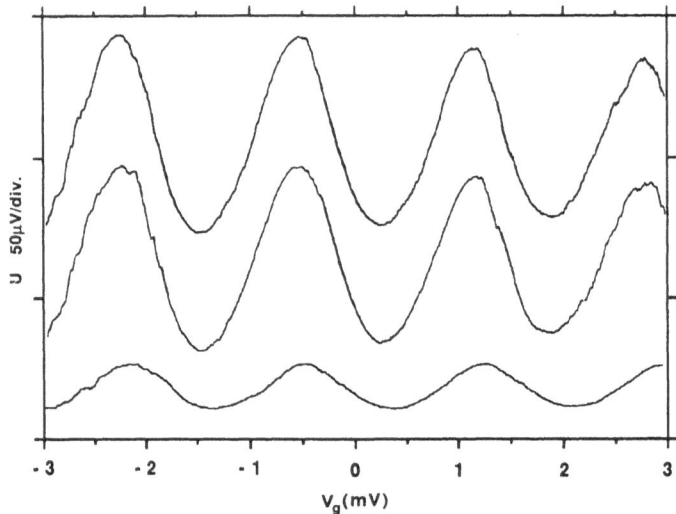

Fig.25. The modulation of the voltage (at a fixed bias current of 30 pA) of two 100 nm x 100 nm series coupled junctions as a function of gate voltage. T=40, 100, and 250 mK counted from the top. R_1+R_2=310 kΩ. $2V_{off}$= 425 mV. C_g=10^{-16} F is obtained from the period of modulation (ΔV_g= e/C_g). The superconductivity of the electrodes is suppressed by a magnetic field.

Beautiful experiments have been designed and carried out in a cooperation between the Saclay and Delft groups[40]. In the so called electron turnstile experiment (fig. 27), the middle electrode of a multi-junction array is charged negatively or positively by a sinusoidal signal of frequency f applied to a gate. During one half period an electron is transfered through the two junctions at the left of fig. 27 and charges up the middle electrode. During the next half period the gate voltage is switched and the intermediate electrode is discharged as the electron passes out through the two junctions at the right. Hence the electron transfer, and the current, through the array is controlled by the frequency of the gate signal: I=ef. The I-V curves for a number of rf frequencies, applied on the gate, are shown in fig. 28. Distinct plateaus are displayed and their current values are given by the relation above. f was varied in steps of 4 MHz. It may be noted that the frequency was considerably smaller than the one used to phase lock the current driven oscillations discussed in section 4.3; this may improve the sharpness of the induced structure.

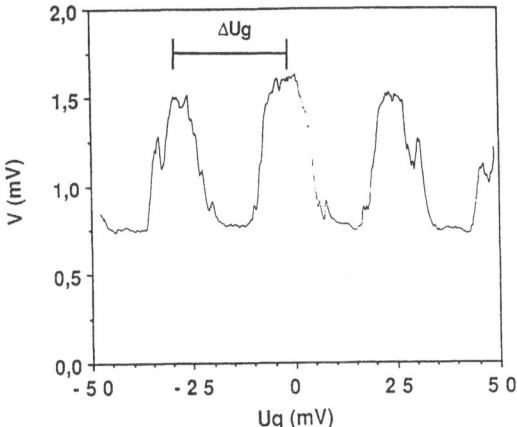

Fig.26. The response to a gate voltage is illustrated. The array configuration of
        Fig.6 was used. The array was biased with a constant current ($\approx$5 pA) at
        T=1.3 K and the voltage over the array is plotted as a function of applied
        gate voltage. A periodic response is seen - the period corresponds to a
        change of one electron charge on an intermediate electrode(s). By
        measuring the noise level at 10 Hz, biased at the steepest part of the
        response curve, we deduce a sensitivity for charge change of
        $2 \cdot 10^{-4} e / \sqrt{Hz}$. The transistor "gain", i.e. the relation between changes
        in the output and gate voltages, is of the order of 0.2 for this curve.

Averin et al.[41] have shown that a charge can traverse all the junctions in one step. This is an analog of the macroscopic quantum tunneling (MQT) of flux in a current biased Josephson junction giving rise to a voltage rounding of the I-V curve. It has been named a charge-MQT and it may limit the accuracy to which a step can be defined.

The so called charge pump[42] is a further development of the turnstile. However, it functions also when no bias is applied across the array. By introducing a phase shift between the signals applied to two succesive gates of an array, it is possible to pump electrons from a source via the two gates to a drain electrode without applying any bias between source and drain. Voltage steps cross the current axis (V=0) in the I-V curve. The phenomenon can be compared with the zero bias current steps induced in a Josephson junction by a microwave signal.

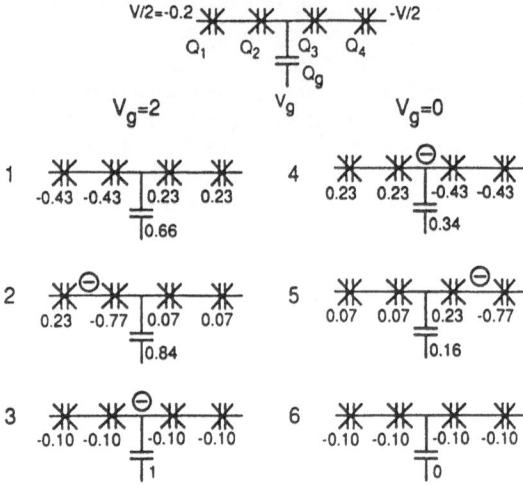

Fig.27. Principle of the so called turnstile for transfer of single electrons through a linear array of small tunnel junctions controlled by an rf-voltage applied to a gate. Junctions, with capacitance C, are denoted by crossed capacitor symbols. The gate voltage is applied via a capacitor C_g. If $C_g=C/2$, tunneling across any junction can only occur if for that junction $|Q| >e/3$. Voltages and charges are given in units of e/C and e. 1-6 indicates consecutive times in the cycle. Left: first half of the cycle, $V_g=2$. An elementary charge ends up trapped on the central electrode. Right: second half of the cycle, $V_g=0$. The charge can only leave on the right-hand side.

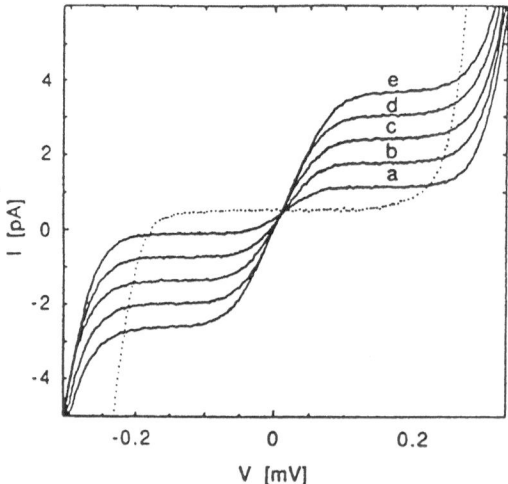

Fig.28. Results obtained with a turnstile with layout close to the one described in fig.27. Current-voltage curves are given for no ac gate voltage (dotted curve) and for ac gate voltage at frequencies f from 4 to 20 MHz in 4 MHz steps (a-e). Current plateaus are seen at I=ef. T=10 mK. The of the Al electrodes was quenched by a magnetic field.

OTHER WAYS TO MODULATE THE BARRIER. 1 DEG CHANNELS

Several other ways to control single tunnel events may be envisioned. Guinea and Garcia[43] have proposed the use of an STM where the tip is mechanically vibrated. For a short moment, the tip is close to a small particle or metal film island and one electron can tunnel resonantly to one of the discrete elelctron levels of the small particle. The tip is withdrawn and the particle discharges to the substrate via another tunneling barrier during a longer tip recess time and the process is repeated. The limited ability to rapidly move the tip mechanically sets an upper frequency limit.

A similar action might be obtained using a point contact and an optically active medium such that the barrier can be modulated by short (fs) laser pulses with a high repetition rate. Or electrodes connected by narrow photoconductive bridges. (Different antenna arrangements might be used to couple the optical signal to the high resistance junction.)

Another possibility, which has received attention[44], is to use narrow, effectively one dimensional channels connecting pockets of a two-dimensional electron gas. Such a device can be fabricated by using, e.g., a GaAs/AlGaAs heterojunction and defining gates on top of the structure. The application of a voltage to the gate may deplete the underlying region of the electron gas and an extremely narrow channel can be realized. The channel acts effectively as a junction and by closing and opening it, the transport of single electrons might be controlled. Several of the configurations described can be used as two, or more, channel devices by coupling gates in series and using a phase delay for the gate control signal.

THE EFFECT OF THE ENVIRONMENT. TUNNELING TIMES

The impedance of the environment has a profound influence on the properties of small tunnel junctions. Stray capacitances, or any low impedance shunting elements, rapidly reduce the single electron tunneling effects. The dc properties of ultrasmall tunnel junctions are strongly affected by their high frequency electrodynamical environment, which is characterized by the impedance $Z_e(\omega)$, where ω depends upon the bias; it is typically the frequency determined by the Coulomb blockade energy. We can compare with Josephson junctions where the dc I-V curve is determined by the correlation of (mixed down) tunneling events at the bias Josephson frequency.

Several authors have considered the effect of the imbedding circuit theoretically[45-50]. Nazarov[45] found that the most important contribution to the dc behavior comes from $Z_e(\omega)$ at a frequency determined by the bias voltage. A frequency dependent impedance was also treated by Devoret et al.[46] and Girwin et al.[47,48] Actually, it is possible to obtain the frequency dependence of the imbedding impedance network from the I-V curve and its derivatives. Compare, e.g., the phonon tunneling spectroscopy of normal or superconducting metals where d^2I/dV^2 vs. V informs about the frequency dependence of phonon excitations[51]. In our case of ultrasmall tunnel junctions, we study the excitations of the environmental circuitry, and the second derivative curve gives a qualitative picture of the excitation modes.

Discussing the point contact measurements[11], we remarked that if the Coulomb blockade in the I-V curve were due to single electron tunneling across one junction it would imply a very small capacitance of the order of $5 \cdot 10^{-18}$F. This would correspond to a semispherical tip of radius 200 Å and a distance to the sample of 10 Å. But the stray capacitance is expected to be much larger than this value as the STM structure is much larger than the utmost tip region. A way around the problem may be to state that the only capacitance that enters is the one corresponding to the region of the STM tip that the electron senses during the time of flight of the tunnel event itself.[52] If one assumes that a stray capacitance stems only from a short charge redistribution range equal to the speed of light along the transmission line (tip) structure times the tunneling time, one deduces a tunneling time of about $5 \cdot 10^{-16}$ s. (This would be roughly equal to the time needed for a Fermi velocity electron to traverse a separation distance of 10 Å.) However, there have been doubts that the tunneling time is merely the traversal time.

There are several different times that are connected with the tunneling process. The time between tunneling events (determined by the current) has to be long enough to allow the charging and discharging of the tunnel junction, a time given by the so called RC-time. The sudden tunnel event leads to the generation of a number of electron-hole pairs at the Fermi energy, the non-equilibrium charge distribution is screened during a time characteristic of the inverse plasma frequency. The time under the tunneling barrier, the traversal time[52] that often is considered as the tunneling time, can be large, as for macroscopic quantum tunneling in a small gap superconducting junction (typically of the order of meV), while it is small for a large barrier as in a regular oxide tunnel junction (typically eV). In that case, the dominant time scale is rather set by the "probing time" (or "hesitation" time given by the Heisenberg uncertainty relation). Recent discussions of the tunneling time concept are given in Refs. 53 and 54.

If we apply the horizon argument, i.e. that the electron samples the impedance of the environment up to a distance determined by the product of the speed of the electromagnetic wave and the tunneling time, we can estimate the latter from tunneling data.

We can compare similar tunnel junctions, fabricated in the same process but embedded in different environments[14]. The configuration of the experiment is shown in fig. 7. One junction sits in the middle of an array of high resistance tunnel junctions that protects it from loading effects by the environment. Both the current and voltage leads consist of arrays of junctions, 6 in each branch. The other, "solitary" junction feels the load impedance: the capacitance, inductance, and resistance of the transmission lines that connect it to the probes. That is, if a short tunneling time does not prevent it from feeling the influence of a shunting admittance.

Experimental I-V and dI/dV-V curves are shown in figs. 8 and 9a. The Coulomb blockade is pronounced for the single junction in the middle of the array. The tunneling curves agree very well with calculated ones. (No fitting parameters were used in the calculations, merely values deduced from separate measurements and a calculated value of the stray capacitance.)

The tunneling structure for the "solitary" junction (fig. 9b), on the other hand, is weak. The conductance dip is an order of magnitude smaller than for the "protected" junction and the offset voltage about 20 times smaller. If the tunneling time were less than or about 10^{-14}s (i.e. the traversal time), one would expect a voltage offset of the order of 400 μV (instead of the measured value of 27±3 μV) and a conductance dip much wider than the observed one. In order to describe the observed behavior, one would need a load capacitance $C_L > 3 \cdot 10^{-15}$F, which is much larger than would be expected from the probing range corresponding to merely a short traversal time. We hence conclude that the time during which the tunneling electron probes the environment (or during which the tunneling degrees of freedom adjust) in our case is much longer than the traversal time.

A similar conclusion was reached theoretically by Nazarov[45] who found that the most important contribution to the dc behavior at small bias comes from $Z_e(\omega)$ at much lower frequencies than the one corresponding to the traversal time, $\hbar\omega \sim \Delta E \approx \max(eV, k_B T)$. At large bias voltage, the traversal time may be the dominant one.

With our voltage and temperature scales, this would correspond to $\omega \approx 10^{12}$ s^{-1} which in turn would imply a relativistic radius of the order of 0.3 mm and $C_L \approx 3 \cdot 10^{-15}$F. The experimental conductance dip (fig. 9) can then be reproduced.

COMPARISONS WITH JOSEPHSON EFFECTS

We have already remarked that there is a very strong (although incomplete) analogy between the coherent single electron tunneling in small junctions and macroscopic quantum tunnel effects in large superconducting junctions. Corresponding entities are:

SET		JOSEPHSON
q	$\leftrightarrow$	Φ
e	$\leftrightarrow$	$\Phi_0 = h/2e$
$E_C = e^2/2C$	$\leftrightarrow$	$E_J = hI_J/2e$
I	$\leftrightarrow$	V
C	$\leftrightarrow$	L
R	$\leftrightarrow$	G
$R_Q = h/4e^2$	$\leftrightarrow$	R_Q^{-1}
Series	$\leftrightarrow$	Parallel

The Josephson effects have been utilized in a number of realized or prospective applications. Analog uses can be proposed for SET based devices. For example a current standard analog to the Josephson voltage standard; a sensitive electrometer based on two junctions in series corresponding to the two junctions in parallel in the dc-SQUID; a charge soliton oscillator based on an array of series coupled ultrasmall tunnel junctions and a flux flow oscillator using a long Josephson junction; amplifiers based on parametric capacitances and inductances, respectively; digital elements based on charge or flux solitons.

BLOCH OSCILLATIONS

We have considered the tunneling of single electrons. In superconductor/insulator/superconductor junctions, where $E_J = hI_J/2e = \Phi_0 I_J/2\pi$ is comparable to E_c, pairs of electrons will tunnel. Interesting phenomena will occur when the charging energy ($Q^2/2C$; $Q=2e$) becomes of the same order as the Josephson coupling energy $E_J\cos\phi$. We can compare the situation with that of a crystal when the kinetic energy of an electron becomes of the same order as the periodic crystal potential energy. In that case we have standing Bloch waves, Bragg reflections, energy bands separated by energy gaps, effective electron masses, quasi-momentum, etc. Similarly, there may be oscillations of Cooper pair tunneling through the periodic "wash-board" potential in phase space. These are named Bloch oscillations. Analogously, we also have energy bands and energy gaps, quasi-charge, effective capacitance, etc. Single electron ($f=I/e$) and pair ($f_B=I/2e$) oscillations may occur together. At low bias, the sharp linewidth SET oscillations will dominate, but at higher current bias, the broader linewidth Bloch oscillations will grow in amplitude. In the weak coupling ($E_J/E_C < 1$) limit, the SET oscillations dominate and are of comparable strength to the Bloch oscillations, even at high current bias. In the strong coupling limit ($E_J/E_C > 1$) the Bloch oscillations prevail at high bias.

The time correlated pair oscillations have not been observed up till now, but other aspects of pair tunneling in low capacitance junctions have been studied. The subject is worth a review in itself. For reviews of the properties, see Ref. 1.

APPLICATIONS OF SINGLE ELECTRON TUNNELING

Current standard

Many national standards laboratories maintain the volt with the aid of the Josephson ac effect. Current steps are induced with a voltage separation determined by the frequency of the applied microwaves, $\Delta V=(h/2e)f$. Frequency can be determined with extremely high accuracy. By using an accepted (long term agreed) value of $h/2e$, it is presently possible to determine the voltage unit with an accuracy of considerably less than one ppm.

Likewise, it may be possible to use the microwave induced SET voltage steps at $I=(n/m)ef$ (n and m integers, in the experiments described above m=1) to maintain a current standard. Sharp steps, much sharper than those observed up till now, are needed to give well defined current values. The initial measurements gave a photon induced structure that was smeared to become indistinguishable as the RC-frequency was approached.

Much sharper steps were realized in the MHz range using the "turnstile"[40]. The value of e that was extracted agreed within 10^{-3} with the accepted value already in the first experiment. A problem hampering attempts to better the accuracy may be the macroscopic tunneling of charge directly through the array without successively charging the intermediate electrodes as pointed out by Averin[41,55].

Low currents can be determined accurately by counting electrons one by one. However, going up to a calibration current of 1 A, as needed for standards, difficulties have to be overcome. One needs a high frequency to get large current steps and we already noted that the steps are broadened at high frequency unless the junctions are made still smaller. A large number of arrays have to be coupled in parallel to increase the current level and the microwave radiation must be efficiently coupled to each of them. The last orders of magnitudes may be achieved by using superconducting current comparators. We can compare with Josephson voltage standards that now, after 25 years of development, operate with thousands of junctions connected in series.

If a current standard may be accomplished with a sufficiently high accuracy, say at least $1:10^8$, it may even be possible to test basic relations by closing the so called quantum metrology triangle, see fig. 29. It combines the SET (or Bloch) oscillations, that give a relation between current and frequency, with the Josephson voltage standard, converting frequency to voltage, and the quantum Hall resistance, that gives quantized resistance, $R=h/ne^2$.

Sensitive electrometer

The I-V curve of two, or more, series coupled junctions may be shifted by changes in charge on the middle electrode(s) of much less than an electron charge. The experiments[39] described previously (see fig. 26) gave a charge sensitivity of about $2 \cdot 10^{-4}$ $e/(Hz)^{1/2}$ at 10 Hz.

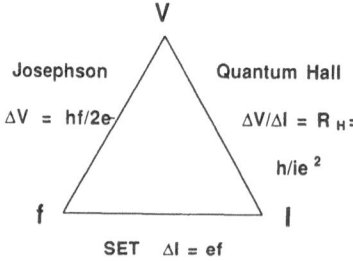

Fig.29. The so called quantum metrology triangle illustrates that single electron tunneling (SET) oscillations relate current and frequency; the ac Josephson effect gives a similar relation between voltage and frequency; while the quantum Hall effect and its quantized resistance value relates voltage and current.

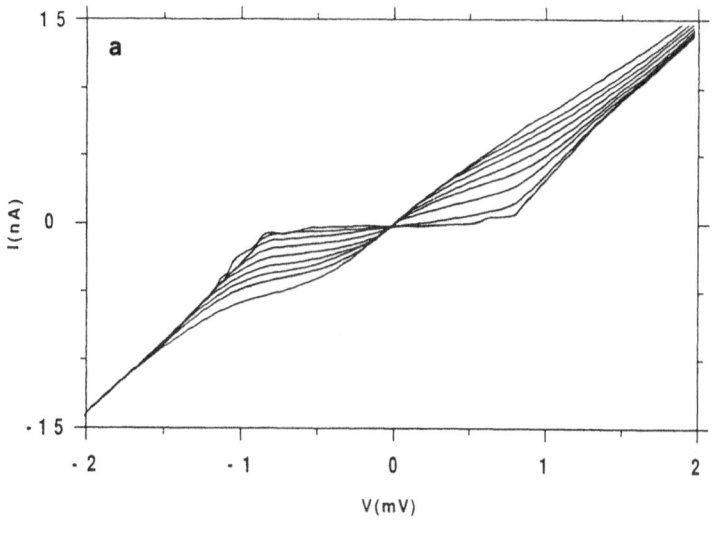

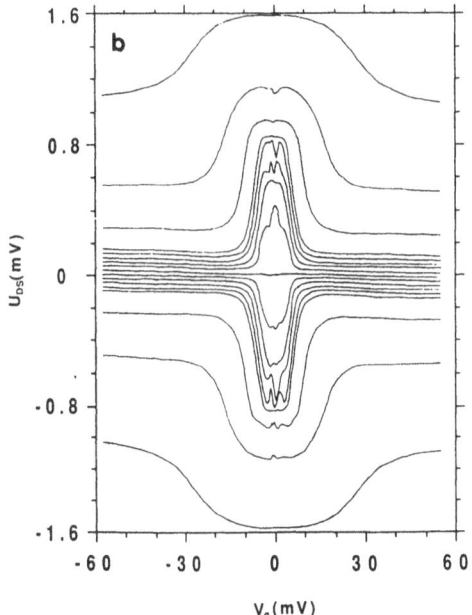

Fig.30. A resistively coupled transistor where the resistor consists of 13 Al/I/Al tunnel junctions fabricated in the same process as the source (S) and drain (D) junctions. (a) Current vs source-to-drain voltage for gate voltage increasing in 2.5 mV steps. (b) Source-to-drain voltage as a function of gate voltage for different bias currents. T=50 mK.

$R_D+R_S=0.11$ MΩ. $R_g=3.0$ MΩ. A maximum voltage gain of 0.15 and power gain of 4.1 were estimated for this device.

Obviously, the sensitivity is extremely high but the difficulty is to couple charge to the middle electrode in an efficient way. The object to be measured will have a much larger capacitance than the one coupling to the middle electrode. Hence a large portion of the sensitivity will be lost.

The corresponding Josephson device is a dc-SQUID consisting of two junctions connected in parallel. It has a flux sensitivity much less than the quantized flux, $\Phi_0 = h/2e$.

Transistors

If the middle electrode is sensitive to charge changes, the voltage over the junctions, at a given current bias, will also respond to the voltage applied to a gate. This means that it is possible to construct a three terminal transistor. The gate can be coupled capacitively or resistively, in the latter case via a large resistance to an intermediate electrode.

The "gain", i.e. the ratio of the output voltage at constant current to the input gate voltage, has hithertoo been less than one. The experiment of Ref.28 used two junctions and a capacitive coupling. It gave a "gain" of about 0.01. The configuration described previously in this review was based on an array (probably somewhat inhomogenous) and, again, a capacitive coupling. It gave a considerably higher "gain", namely 0.2, at the optimal response region. Somewhat higher values have later been obtained[56], the unexpectedly high values are maybe partly due to a coupling via the leakage resistance of the substrate.

The intermediate electrode can also be resistively coupled to the external gate and theoretical calculations give a higher attainable gain for such a transistor. Using a large resistance array of tunnel junctions as a resistive gate, it was possible to obtain a voltage gain of close to unity[56,57]. The response of the device, as shown in fig. 30, is complicated by the resistor being an array of ultrasmall junctions. (It is actually easier to fabricate arrays of junctions in the same process as the junctions under investigation than to deposit a high resistance lead with sufficiently small capacitance to ground.) No charge is transfered through the gate array until a Coulomb blockade threshold voltage is exceeded. Then there is a sharp response.

The current gains of the transistors can be considerably larger than one as the gate impedance can be made larger than the resistance between the source and drain electrodes. Power gains of the order of 4 were registered for both the capacitively and the resistively coupled transistors[56,57].

Computer elements

A transistor with true gain can be used as a logic element. The possibility to control the shuffling of charge solitons along arrays may be utilized in shift registers. A number of possible digital applications can be foreseen[1,58,59].

CRITERIA FOR OBSERVATION

The single electron tunneling effects have been observed in ultrasmall junctions at low temperature. The phase locking of SET oscillations were clearly seen only for T<1K. The smaller the junction, the higher the temperature of operation as listed in Table 2. If we assume that the state-of-the-art lithography technology is capable of producing junctions as small as 10 nm x 10 nm (within some distant future), it should be possible to operate SET components up towards 300 K. Coulomb blockade features have been observed even at room temperature by point contact measurements on, e.g., a high T_c superconductor[60] or ferritin, that is an iron atom cluster surrounded by a protein shell[61].

It should be noted from the table, though, that the voltage level increases as the junction size is decreased. This is good as it discriminates against voltage fluctuations, but it also means that the power level of the device increases. This may be an obstacle in attempting to realize an extremely high packing density in logic circuits.

Table 1   Estimates of parameters for single electron tunneling junctions (ref. 1)

Level of fabrication technology	Junction area (nm²)	Junction capacitance (aF)	Temperature limit* $e^2/2C$ (K)	Voltage scale $V_t = e/2C$ (mV)	Power scale** V_t^2/R_T (W)	Time scale** $\tau = R_T C$ (ps)
Present technology junctions	80x80	200	4	0.4	10^{-12}	20
State-of-the-art junctions	30x30	30	30	2.5	10^{-10}	3
Apparent limit of resist lithography	10x10	3	300	25	10^{-8}	0.3

   * defined as $e^2/2C$; any practical applications taking advantage of SET should be operated well below this temperature

** R_T is assumed to be of the order of 100 kΩ

CONCLUSIONS

The single electron tunneling effects have been conclusively proven during the last couple of years. It has been possible to realize ultrasmall junctions in a controlled way. Their capacitances have been sufficiently small to make the charging energy of a tunneling event larger than the energies corresponding to thermal or quantum fluctuations.

The Coulomb blockade has been observed in a number of experiments - lately also in a single junction shielded by high resistance leads in the near environment. Most of the observations have been made in double and multi-junction structures. Theoretical fits give very reasonable tunnel parameter values. Stray capacitances to the environment usually prevent SET in single junctions. However, when they are sufficiently protected from the high frequency shunting admittance of the environment, the effects of space and time correlated tunnel events can be distinguished.

The sensitivity of the I-V curve to changes in the charge distribution on intermediate electrodes in multi-junction configurations may introduce switching between states bounded by a change in half an electron charge. This may be a hindrance in observing clearcut I-V curves (due to, e.g., an unsymmetric bias, mechanical vibrations, or intermediate traps of different origin). But it also gives a possibility to control the single electron tunneling by sub-electron charges, which might be used in extremely sensitive Coulomb meters. The space correlation by successive tunneling events in an array of junctions is well established. Resistively and capacitively coupled transistors have been realized, moderate power gains have been obtained in the first devices.

The time correlation of tunneling events has also been shown. An applied microwave signal phase locks the tunneling events, and microwave induced steps in the I-V characteristic follow the fundamental relation $I=nef_{ext}$. This may give rise to a quantum current standard in the future. Tunnel events have also been induced by applying a periodic potential to a gate electrode. A free running oscillation has not yet been proven experimentally, but all observations are fully in line with theoretical concepts that include SET oscillations.

Another milestone will be the observation of Cooper pair, or Bloch oscillations in small superconducting junctions. The transition between single electron and pair tunneling as E_c/E_J is varied remains to be mapped out.

Many of the effects and applications that have been discussed have there counterparts in superconducting Josephson junctions.

REFERENCES

1. For reviews, see K.K. Likharev, "Correlated Discrete Transfer of Single Electrons in Ultrasmall Tunnel Junctions", *IBM J. Res. Dev.* **32**, 144 (1988); D.V. Averin and K.K. Likharev, "Single-Electronics: A Correlated Transfer of Single Electrons and Cooper Pairs in Systems of Small Tunnel Junctions", in *Quantum Effects in Small Disordered Systems*, B.L. Altshuler, P.A. Lee, and R.A. Webb, eds., Elsevier, to be publ.; K.K. Likharev, *Dynamics of Josephson Jucntions and Circuits*, Gordon and Breach Sci. Publ., N.Y., 1986; G. Schön and A.D. Zaikin, "Quantum Coherent Effects, Phase Transitions, and the Dissipative Dynamics of Ultra Small Tunnel Junctions", *Physics Reports*, to be publ.
2. D.V. Averin and K.K. Likharev, "Possible Coherent Oscillations at Single-Electron Tunneling," in *SQUID'85*, H. Lübbig and H.-D. Hahlbohm, Eds., W. de Gruyter, Berlin 1985, p. 197.
3. D.V. Averin and K.K. Likharev, "Coulomb Blockade of the Single- Electron Tunneling, and Coherent Oscillations in Small Tunnel Junctions", *J. Low. Temp. Phys.* **62**, 345 (1986); *Zh. Eksp. Teor. Fiz.* **90**, 733 (1986) [*Sov. Phys.-JETP* **83**, 427 (1986)] .
4. E. Ben-Jacob and Y. Gefen, "New Quantum Oscillations in Current Driven Small Junction," *Phys. Lett.***108A**, 289 (1985).
5. R. Brown and E. Simanek, "Transition to Ohmic Conduction in Ultrasmall Tunnel Junctions", *Phys. Rev. B* **34**, 2957 (1986).
6. F. Guinea and G. Schön, "Coherent Charge Oscillations in Tunnel Junctions", *Europhys. Lett.* **1**, 585 (1986).
7. D.V. Averin, "Characteristics of Small Tunnel Junctions in the Zero Bias Current Limit", *Zh. Eksp. Teor. Fiz.* **90**, 2226 (1986) [*Sov. Phys.-JETP*].
8. D.V. Averin, "Effects of Temperature on Single-Electron and Bloch Oscillations in Tunnel Junctions", *Fiz. Nizk. Temp.* **13**, 364 (1987) [*Sov. J. Low Temp. Phys.* **13**, 209 (1987)].
9. D.V. Averin and K.K. Likharev, "New Results in the Theory of SET and Bloch Oscillations in Small Tunnel Junctions", *IEEE Trans. Magnetics* **23**, 1138 (1987).
10. U. Geigenmüller and G. Schön, "Quantum Dynamics and Single Electron Effects in Networks of Josephson Junctions", *Physica B* **152**, 186 (1988).
11. P.J.M. van Bentum, H.van Kempen, L.F.C. van de Leempert, and P.A.A. Teunissen, "Single-Electron Tunneling Observed with Point-Contact Tunnel Junctions", *Phys. Rev. Lett.* **60**, 369 (1988).
12. U. Hartmann, R. Berthe, and C. Heiden, "Temperature Dependence of Correlated Single Electron Tunneling Observed with a Scanning Tunneling Microscope",
13. L.J. Geerligs, V.F. Anderegg, C.A. van der Jeugd, J. Romijn, and J.E. Mooij, "Influence of Dissipation on the Coulomb Blockade in Small Tunnel Junctions", *Europhys. Lett.* **10**, 79 (1989).
14. P. Delsing, K.K. Likharev, L.S. Kuzmin, and T. Claeson "Effect of High-Frequency Electrodynamic Environment on the Single-Electron Tunneling in Ultrasmall Junctions", *Phys. Rev. Lett.* **63**, 1180 (1989).
15. D. Haviland et al, to be published.
16. R. Berthe, U. Hartmann, C. Heiden, and J. Halbritter, "Temperature and Adsorbate Dependence of the Coulomb Barrier Observed with a Scanning Tunneling Microscope", to be published.
17. K. Mullen, E. Ben-Jacob, R.C. Jaklevic, and Z. Schuss, "I-V Characteristics of Coupled Ultrasmall-Capacitance Normal Tunnel Junctions", *Phys. Rev. B* **37**, 98 (1988).

18. I.O. Kulik and R.I. Shekhter, "Kinetic Phenomena and Charge Discreteness in Granular Media", *Zh. Eksp. Teor.* **68**, 623 (1975) [*Sov. Phys.-JETP,* **41**, 308 (1975)].

19. C.A. Neugebauer and M.B. Webb, "Electrical Conduction Mechanism in Ultrathin Evaporated Metal Films", *J. Appl. Phys.* **33**, 74 (1962).

20. H.R. Zeller and I. Giaever, "Tunneling, Zero-Bias Anomalies, and Small Super-conductors", *Phys. Rev.* **181**, 789 (1969).

21. J. Lambe and R.C. Jaklevic, "Charge Quantization Studies Using a Tunnel Capacitor", *Phys. Rev. Lett.* **22**, 1371 (1969).

22. R.E. Cavicchi and R.H. Silsbee, "Coulomb Suppression of Tunneling Rate from Small Metal Particles", *Phys. Rev. Lett.* **52**, 1453 (1984).

23. L.S. Kuzmin and K.K. Likharev, "Direct Observation of the Discrete Correlated Single Electron Tunneling", *Pis'ma v. Zh. Teor. Eksp. Fiz.* **45**, 389 (1987) [*JETP Lett.* **45**, 495 (1987)].

24. J.B. Barner and S.T. Ruggiero, "Observation of Incremental Charging of Ag Particles by Single Electrons", *Phys. Rev. Lett.* **59**, 807 (1987); S.T. Ruggerio and J.B. Barner, "Multiple-Gap Tunneling Structure Observed for the High-T_c Superconductors: Charging Effects as Possible Cause", *Phys. Rev. B* **36**, 8870 (1987).

25. P.J. M. van Bentum, R.T. M. Smokers, and H. van Kempen, "Incremental Charging of Single Small Particles", *Phys. Rev. Lett.* **60**, 2543 (1988).

26. R. Wilkins, E. Ben-Jacob, and R.C. Jaklevic, "Scanning-Tunneling-Microscope Observations of Coulomb Blockade and Oxide Polarization in Small Metal Droplets", *Phys. Rev. Lett.* **63**, 801 (1989).

27. M. Naito, D.P.E. Smith, M.D. Kirk, B. Oh, M.R. Hahn, K. Char, D.B. Mitzi, J.Z. Sun, D.J. Webb, M.R. Beasley, Ø. Fisher, T.H. Geballe, R.H. Hammond, A. Kapiltunik, and C.F. Quate, "Electron-Tunneling Studies of Thin Films of High-T_c-Superconducting La-Sr-Cu-O", *Phys. Rev. B* **35**, 7228 (1987); J.R. Kirtley, R.T. Collins, Z. Schlesinger, W.J. Gallagher, R.L. Sandstrom, T.R. Dinger, and D.H. Chance, "Tunneling and Infrared Measurements of the Energy Gap in the High-Critical-Temperature Superconductor Y-Ba-Cu-O", *Phys. Rev. B* **35**, 8846 (1987); I. Iguchi, H. Watanabe, Y. Kasai, T. Mochiku, A. Sugishita, and F. Yamaha, *Jpn. J. Appl. Phys.* **26**, L645 (1987); J. Moreland, L.F. Goodrich, J.W. Elkin, T.E. Capobianco, and A.F. Clark, "Electron Tunneling Measurements of High T_c Compounds Using Break Junctions", *Jpn. J. Appl. Phys.* **26** (suppl.3), 899 (1987); M.E. Hawley, K.E. Gray, D.W. Capone II, and P.G. Hinks, "Energy Gap in La$_{1.85}$Sr$_{0.15}$CuO$_{4-y}$ from Point-Contact Tunneling", *Phys. Rev. B* **35**, 7224 (1987).

28. T.A. Fulton and G.J. Dolan, "Observation of Single Electron Charging Effects in Small Tunnel Junctions," *Phys. Rev. Lett.* **59**, 109 (1987).

29. K.K. Likharev, N.S. Bakhvalov, G.S. Kazacha, and S.I. Serdukova, "Single-Electron Tunnel Junction Array: An Electrostatic Analog of the Josephson Transmission Line", *IEEE Trans. Magn.* **MAG-25**, 1436 (1989).

30. G.J. Dolan, "Offset Masks for Lift-off Photoprocessing", *Appl. Phys. Lett.* **31**, 337 (1977).

31. A variant of the technology, but with a supported fiber above the substrate, was used by J. Niemeyer, *P.T.B. Mitt.* **84**, 251 (1974).

32. G.J. Dolan and J.H. Dunsmuir, "Very Small ($\geq$ 20 nm) Lithographic Wires, Dots, Rings, and Tunnel Junctions", *Physica B* **152**, 7 (1988).

33. P. Delsing, K.K. Likharev, L.S. Kuzmin, and T. Claeson, "Time-Correlated Single-Electron Tunneling in One-Dimensional Arrays of Ultrasmall Tunnel Junctions", *Phys. Rev. Lett.* **63**, 1861 (1989).

34. R. Ono, M.W. Cromar, R.L. Kautz, R.J. Soulen, J.H. Colwell, and W.E. Fogle, "Current-Voltage Characteristics of Nano-ampere Josephson Junctions", *IEEE Trans. Magnetics* **23**, 1670 (1987).

35. P. Delsing, MQT´86, Imperial College, London, 1986.

36. P.J. M. van Bentum, private comm.

37. L.J. Geerligs, private comm.

38. L.J. Geerligs, V.F. Anderegg, J. Romijn, and J.E. Mooij, "Single Cooper-Pair unneling in Small-Capacitance Junctions", *Phys. Rev. Lett.* **65**, 377 (1990).

39. L.S. Kuzmin, P. Delsing, T. Claeson, and K.K. Likharev, "Single-Electron

Charging Effects in One-Dimensional Arrays of Ultra-Small Tunnel Junctions", *Phys. Rev. Lett.* **62**, 2539 (1989).

40. L.J. Geerlings, V.F. Anderegg, P.A.M. Holweg, J.E. Mooij, H. Pothier, D. Estève, C. Urbina, and M.H. Devoret, "Frequency-Locked Turnstile Device for Single Electrons" *Phys. Rev. Lett.* **64**, 2691 (1990).

41. D.V Averin and A.A. Odintsov, "Macroscopic Quantum Tunneling of the Electric Charge in Small Tunnel Junctions", *Phys. Lett. A* **140**,?(1989); L.J. Geerligs, D.V. Averin, and J.E. Mooij, "Observation of Macroscopic Quantum Tunneling of the Electric Charge", to be publ.

42. H. Pothier, P. Lafarge, P.F. Orfila, C. Urbina, D. Esteve, and M.H. Devoret, "Single Electron Pump Fabricated with Ultrasmall Normal Tunnel Junctions", to be publ.

43. F. Guinea and N. Garcia, "Scanning Tunneling Microscopy, Resonant Tunneling, and Counting Electrons: A Quantum Standard of Current", *Phys. Rev. Lett.* **65**, 281 (1990).

44. A.A. Odintsov, "Single Electron Transport in 2DEG System with Modulated Transparencies of Gates: Possible Basis of the Standard of Current", to be publ.; M Persson, private comm.

45. Yu. V. Nazarov, "Anomalous Current-Voltage Characteristics of Tunnel Junctions", *Zh. Eksp. Teor. Fiz.* **95**, 975 (1989) [*Sov. Phys. JETP* **68**, 561 (1989)]; D.V. Averin and Yu.V. Nazarov, "Coulomb Fingerprints on the I-V Curves of the Normal Tunnel Junctions", to be publ.

46. M.H. Devoret, D. Estève, H. Grabert, G.-L. Ingold, H. Pothier, and C. Urbina, "Effect of the Electromagnetic Environment on the Coulomb Blockade in Ultrasmall Tunnel Junctions", *Phys. Rev. Lett.* **64**, 1824 (1990).

47. S.M. Girvin, L.I. Glazman, M. Jonson, D.R. Penn, and M.D. Stiles, "Quantum Fluctuations and the Single-Junction Coulomb Blockade, *Phys. Rev. Lett.* **64**, 3183 (1990); K. Flensberg, S.M. Girvin, and M. Jonson, "Quantum Fluctuations and Charging Effects in Small Tunnel Junctions", to be publ.

48. M. Ueda and S. Kurihara, "Quantum Theory of Ultrasmall Tunnel Junctions", *Proc. LT-19 Satellite Workshop on Macroscopic Quantum Phenomena* (ed. T.D. Clark), World Sci., Singapore, 1990, to be publ.

49. G.-L. Ingold, "Influence of the Electrodynamical Environment on Tunnel Junctions, to be publ.

50. A.N. Cleland, J.M. Schmidt, and J. Clarke, "Charge Fluctuations in Small-Capacitance Junctions", *Phys. Rev. Lett.* **64**, 1565 (1990).

51. E.L. Wolff, *Principles of Electron Tunneling Spectroscopy*, Oxford Univ. Press, N.Y. 1985.

52. M. Büttiker and R. Landauer, "Traversal Time for Tunneling," *IBM J. Res. Develop.* **30**, 451 (1986).

53. R. Landauer, "Barrier traversal time", *Nature* **341**, 567 (1989).

54. E.H. Hauge and J.A. Støvneng, "Tunneling Times: A Critical Review", *Rev. Mod. Phys.* **61**, 917 (1989).

55. D.V. Averin, "Quantum Tunneling of the Electric Charge", *NATO Adv. Res. Workshop: Rate Processes in Dissipative Systems: 50 Years After Kramers*, to be publ.

56. P. Delsing, "Single Electron Tunneling in Ultrasmall Tunnel Junctions", thesis, Chalmers Univ. Techn., Göteborg, 1990.

57. P. Delsing, T. Claeson, G.S. Kazacha, L.S. Kuzmin, and K.K. Likharev, "1-D Array Implementation of the Resistively-Coupled Single Electron Transistor", *IEEE Trans. Magn.*, to be publ.

58. K.K. Likharev, "A Possibility to Design Analog and Digital Integrated Circuits Based on Discrete Single-Electron Tunneling", *Mikro-elektronika* **16**, 195 (1987) [*Sov. Micro-electron.*].

59. K.K. Likharev and V.K. Semenov, "Possible Logic Circuits Based on the Correlated Single-Electron Tunneling in Ultrasmall Junctions," *Extended Abstracts, International Superconducting Electronics Conference*, Tokyo, 1987, p. 182.

60. P. Davidsson, private communication.

61. H. Olin, private communication.

COOPER PAIR TRANSPORT IN SYSTEMS OF ULTRASMALL JOSEPHSON JUNCTIONS

S. A. Hattel [a], S. P. Plooij [b], and A. B. Zorin [c]

[a] Physics Laboratory I, The Technical University of Denmark
DK-2800 Lyngby, Denmark
[b] Department of Applied Physics, Delft University of
Technology, 2600 GA Delft, The Netherlands
[c] Department of Physics, Moscow State University
Moscow 119899 GSP, U S S R

INTRODUCTION

In the recent years the tunnel junctions of ultrasmall size have drawn attention of many researchers. This is due to new effects of correlated tunneling of charged particles (single electrons and Cooper pairs) in the junctions and their very alluring applications (see, e.g., Refs. 1-3). The physics of correlated transfer of single electrons in ultrasmall tunnel junctions and complex systems of such junctions is well investigated both theoretically and experimentally.[3-5] Unlike this single electron tunneling (SET) domain the study of Cooper pair tunneling (CPT) in ultrasmall objects has been basically restricted to the simplest system of a single junction and the energy source.

The ultrasmall tunnel junction with finite Josephson coupling between superconducting electrodes or, in other words, the ulrasmall Josephson junction is a very interesting mesoscopic object which can reveal both charging effects and quantum behavior of the Josephson phase difference. The most important result of it consists in generation of ac voltage with frequency $f = I/2e$ across the junction at dc current bias I . For proper parameters of the junction and under suitable experimental conditions these so-called Bloch oscillations caused by coherent tunneling of Cooper pairs can be characterized by relatively narrow linewidth.[6] However, the unsolved problem of high-impedance energy source remains the serious obstacle for direct reliable observation of the Bloch oscillations. Actually, here as in the SET domain, the parasitic role is played by the stray capacitance of the leads. Therefore high-Ohmic ultrasmall resistors should be inserted into the current leads in the very vicinity of the junction,which is not easy even for the present-day submicron technology. Another approach to this problem consists in fabrication of more complex systems of ultrasmall tunnel junctions where correlated tunneling of Cooper pairs can occur. The first experimental steps in this direction were recently made by Fulton et al.[7] and Geerligs et al.[8] who investigated the simple few-junction systems.

The goal of this work is to analyse dynamics of a few multy-junction systems of ultrasmall tunnel junctions with weak Josephson coupling. We shall also focus on peculiarities in behavior resulted from CPT. Among the systems under consideration are the double-junction structure (or

Nonlinear Superconductive Electronics and Josephson Devices
Edited by G. Costabile *et al.*, Plenum Press, New York, 1991

transistor), the combination of a 1-D normal-junction array with a Josephson junction, and a 1-D array of ultrasmall Josephson junctions.

BACKGROUND

The Object

We shall deal with the ultrasmall Josephson junctions whose self-capacitance C is so small (in practice of femto-Farade scale) that the elementary charging energy E_C of the junction as capacitor

$$E_C \equiv e^2/2C \gg E_J \,, \tag{1}$$

where $E_J \equiv (\hbar/2e)I_c$ is the magnitude of the Josephson coupling energy which is proportional to the critical current I_c. The temperature of the thermal bath will be assumed low (practically in mK range):

$$k_B T \ll E_J \tag{2}$$

which provides low level of thermal fluctuations. In order to provide low level of quantum fluctuations we also restrict ourselves by the important case of very weak damping in the system

$$\alpha \equiv G\, R_Q \ll 1 \quad (\text{or rather} \ll (E_J/E_C)^2) \,, \tag{3}$$

where the fundamental resistance unit $R_Q = h/4e^2 \cong 6.5$ kOhm and G is the tunnel conductance of the junction measured below the gap voltage $V_\Delta = 2\Delta/e$ (practically, say, in the nano-Siemens range).

Hamiltonian of a System

The Hamiltonian of arbitrary system of ultrasmall tunnel junctions and energy sources can be symbolically presented as

$$H = H_{sources} + H_{charge} + H_{pair} + H_{qp} + H_{environment} \tag{4}$$

where the first term on the r.h.s. of Eq.(4) is simply the work function of the system sources + charges, the second one represents the electrostatic energy of the system or the sum of energies of the capacitors, the third is the sum of Josephson coupling energies over all junctions, the fourth is the sum of the quasiparticle (single electron) tunnel Hamiltonians, and the last term is Hamiltonian of the environment, which in our case is presented by electron systems of all electrodes forming the structure. With assuptions Eqs.(1)-(3) the leading role in Hamiltonian Eq.(4) is played by the first two terms while other terms can be regarded as perturbation. It means that for our analysis we can mostly apply the approach developed in Ref.9 for systems of ultrasmall junctions with single-electron tunneling. This approach is based on the equation for the density matrix and the quantum-mechanical Golden Rule and described in Refs.2,9.

Cooper Pair Tunneling

The finite Josephson coupling in the system introduces some changes in the behavior. In fact, the simple picture of electrostatic eigenstates with low rate of decay of a charge configuration [9] breaks when Josephson interaction is turned on. However this interaction is essential only if the energy difference $\Delta E = |E_1 - E_2|$ between the states E_1 and E_2 of the system is of the order of E_J, where E_1 is the energy of initial state while a charge configuration of the state E_2 is characterized by a transfer of one Cooper pair in one of the junctions. In this case the density matrix of the system is nondiagonal and the eigenvalues of two interacting levels should be calculated in accordance with the quantum-mechanical rules.

Nevertheless, under condition Eq.(1) the Josephson interaction in an arbitrary junction will be revealed relatively seldom. On the other hand, weak coupling guarantees against multiple Cooper pair tunneling (whose probability $\propto (E_J/E_C)^2$). This gives the possibility to describe the dynamics of a system taking into account only the Josephson interaction when level E_1 crosses the position of level E_2 .

We will now regard adiabatic and non-adiabatic level crossing. The first one takes place when levels approach each other continuously which is realized by moving the level with an external ac bias. In this case the probability of CPT can be expressed in terms of Zener tunneling (ZT)

$$P_{CPT} = 1 - P_{ZT} , \tag{5}$$

$$P_{ZT} = \exp\left[-\pi E_J^2 / (\hbar \dot{\Delta E})\right] , \tag{6}$$

where Eq.(6) is a generalization of the well-known formula (see, e.g., Ref.10) for the probability of Zener tunneling in a single junction. Here $\dot{\Delta E}$ is the absolute value of relative velocity of one level with respect to another one.

The second case of a sudden change of the level position takes place if a random tunneling of single electron occurs in the same or a close junction. The duration of this tunneling is extremely short on the scale of the Josephson interaction time $\tau_J = (\hbar/E_J)$. Moreover, because of the stochastic nature of seldom (on the same time scale τ_J) SET events[2] the phase difference between two states becomes a random variable and the coherence breaks. It results in averaging out of quantum-interference terms on the time-scale of the measurement. Hence, the level crossing (or CPT) here is described by a formula which generalizes the result of the paper by Shimshoni et al.[11] for the incoherent case

$$P_{CPT} = 1 - \frac{(E_J^2 + f\,f^{*})^2}{(E_J^2 + f^2)(E_J^2 + f^{*2})} , \tag{7}$$

$$f = (E_1 - E_2) + \text{sign}(E_1 - E_2)\,[(E_1 - E_2)^2 + E_J^2]^{1/2} , \tag{8}$$

$$f^{*} = (E_1^{*} - E_2^{*}) + \text{sign}(E_1^{*} - E_2^{*})\,[(E_1^{*} - E_2^{*})^2 + E_J^2]^{1/2} , \tag{9}$$

where E_1^{*} and E_2^{*} are the new values of the energies E_1 and E_2 after a SET event.

Finally, note that we neglected the influence of the environment on the quantum-mechanical process of Cooper pair tunneling. This assumption is quite reasonable if conditions Eqs.(1)-(3) are satisfied.[12] However, in practice this quantum dynamics of CPT is very sensitive to the environment admittance Y_e at high frequencies (of the order of $\omega_J = (E_J/\hbar)$). In particular, as can be shown from the master equation,[1] the finite Re $Y_e(\omega_J)$ could make transitions to lower level more probable. Here this problem which needs analysis of a concrete experimental setup, is beyond our scope.

MODELLING OF DYNAMICS OF SYSTEMS

Using described above approach we have modelled the behavior of systems with ultrasmall Josephson junctions in the time domain. The computer program combined both the early developed Monte Carlo technique for simulations of SET events[9] and a probabilistic algorithm based on Eqs.(5)-(9) for CPT. Below we present the results obtained.

Superconducting Transistor

We start with the simple system (see Fig.1a) comprising two ultrasmall Josephson junctions and ultrasmall gate-like capacitor C_g, biased by two dc voltage sources. As was demonstrated in experiments by Fulton et al.,[7] at finite dc voltage bias the electric charge transport in the system is due to both SET and CPT. For weak Josephson coupling Eq.(1) the former gives the background quasiparticle dc I-V curve while the latter is revealed by a number of dc current peaks in it. The position of these resonant peaks depends on capacitance values and a gate voltage U_g which induces on the central electrode (island) charge Q_0.[13] Each peak reflects the process of tunneling of one Cooper pair in one junction and two electrons in the other junction.

Fig.1b shows the simulated I-V curves of such a system whose parameters were taken close to a recent experiment by Geerligs et al.[8] The quasiparticle nonlinearity was modelled by the broken line with ratio G/G_n = 0.1, where G_n is the normal-state conductance of a junction. The capacitances of the junctions C_1 and C_2 were taken slightly different in order to remove degeneracy in the peak positions. Comparing the obtained result with the experimental I-V curves presented in Fig.1b from Ref.8 one can see that the results are in a reasonable agreement. In particular, our model adequately describes the Coulomb blockade range and the voltage gate control of the device. However, a number of low-voltage peaks in the experimental curves can hardly be explained in our model, because they are presumably due to high frequency properties of the environment.[8]

1-D Normal Junction Array with Inserted Josephson Junction

The previous example just demonstrated Cooper pair transport in the two-junction voltage biased system, where CPT was controlled by stochastic SET in the other junction. However, the realization of time-correlated CPT in a voltage biased system seems much more attractive. As was mentioned in the Introduction, a natural solution of this problem is inserting high-Ohmic resistors in the vicinity of the junction. These resistors being the electron systems with many degrees of freedom provide continuous change of the charge on the junction,[2] in other words they make the junction be dc current biased. Here we try to solve this problem using

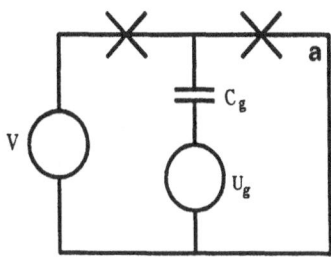

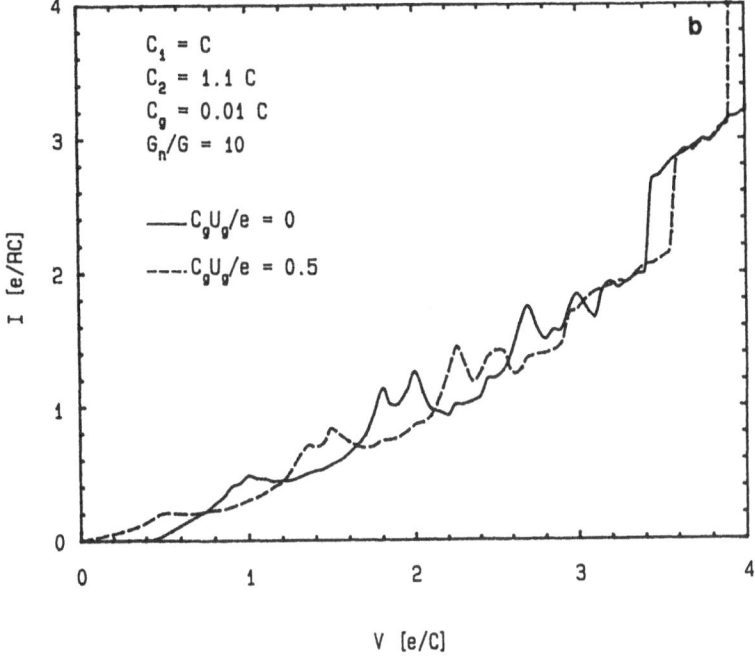

Fig. 1. The two-junction system. (a) The equivalent circuit where
crossed symbols denote the Josephson junctions; (b) the
I-V curves for two values of the gate voltage Ug.

a one-dimensional array of ultrasmall normal junctions instead of
resistors.[6]

The equivalent circuit of the system depicted in Fig.2a comprises the
chain of normal-metal tunnel junctions whose electrodes form stray
capacitances C_0 with the ground plane. The Josephson junction is placed in
the centre of the chain and the voltages are applied to the ends in a
symmetric way. As we are interested in measuring the dc voltage drop across
the Josephson junction, we assume that similar normal tunnel junction arrays
(not shown in Fig.2a) connect it to a registration apparatus. Hence, we
take into account increased values of the two stray capacitancies closest to
the Josephson junction.

Figure 2b shows the calculated dc I-V curves of the Josephson junction
inserted in the 14-junction array for several amplitudes A of ac
component of the biased voltage

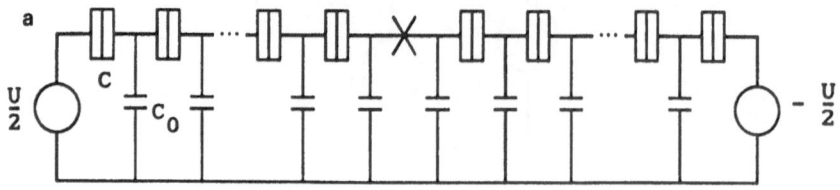

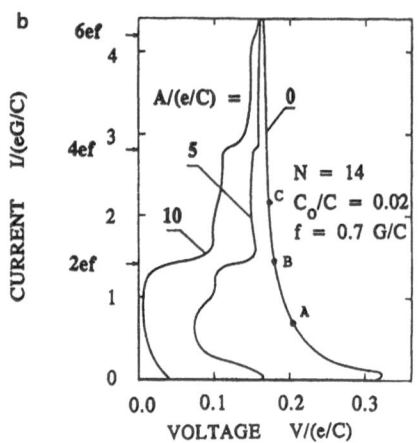

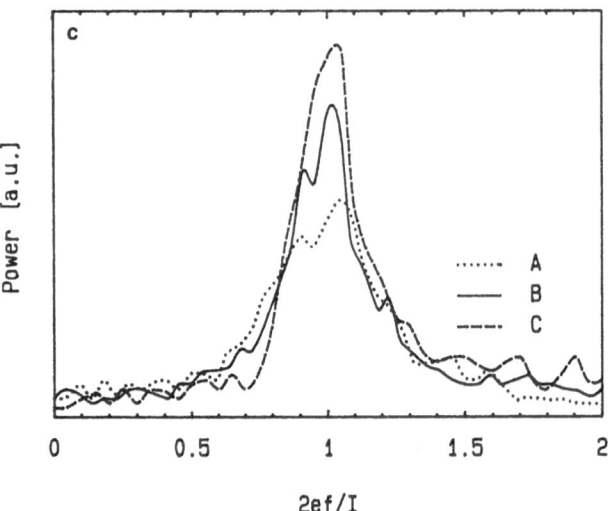

Fig. 2. (a) The equivalent circuit of the normal junction array with inserted Josephson junction. Symbol ⊐⊏ denotes a normal tunnel junction; (b) the average current flowing throw the system vs average voltage across the Josephson junction for the several amplitudes of external ac drive. (c) The power spectrum of the Bloch oscillations in the points A, B, and C of the autonomous I-V curve.

$$U(t) = U_{dc} + A \sin(2\pi f t) \ . \tag{10}$$

The plot represents the average current I   flowing throw the system vs the average voltage V across the Josephson junction. The parameters of the system were not specially chosen in the optimum way, in particular the capacitance and leakage (i.e. C and G) of the Josephson junction were the same as those of the array. Nevertheless, the curves distinctively show staircase structure when ac voltage signal is applied. These dc voltage steps correspond to quantized values of current   I = 2nef (n is integer) and reflect phase- locking of the Bloch oscillations by an external high frequency signal.

Figure 2c shows the power spectrum of the voltage oscillations across the junction in a few points of the autonomous I-V curve. One can see that although the linewidth is relatively large it clearly indicates the time coherence in the system. The origin of such a wide line (and, as result, the smoothed steps) is essential leakage in the Josephson junction and the imperfection of the array with so large a ratio $C_0/C = 0.02$ as resistor.

The work of choosing the optimum parameters of the system both from the point of view of theory and experiment is in progress now. Below we present the result of a study of the behavior of the ultrasmall Josephson junction array as a possible substitute for a single junction in a normal array.

Josephson Junction Array

In this paragraph we consider the superconducting tunnel junction array supplied by the dc current source (see Fig.3a). The Josephson coupling energy and damping are assumed to be small in the sense of Eqs.(1) and (3). In this case the behavior of the system can be adequately described as dynamics of the Cooper pair solitons.[14] Each soliton represents a 2e-charged island with polarized surrounding electrodes similar to the single-electron (or 1e-) soliton.[9] Owing to an applied dc current the 2e-solitons move from one end of the array to another one (opposite polarity in opposite direction). In absence of quasiparticle damping (or SET events) they carry electric charge coherently along the axis of the array, because of the deterministic nature of the equations for relatively low current (lower than the Zener current threshold). In the general case of finite α the stochastic SET events destroy the coherent pattern of 2e-solitons. In other words, there are two coexisting types of solitons which move in the system.

Figure 3b shows the calculated I-V curve (I vs average voltage V) of 5-junction uniform superconducting array with relatively low stray capacities. The comparison of it with the I-V curve of a single junction (see curve III in Fig.9 from paper by Geigenmüller and Schön[10]) confirms the trivial fact that the resulting voltage is the sum of voltage drops across the junctions. Nevertheless the dynamics of our system differs from that of a single junction in the same way as the dynamics of a long Josephson junction differs from that of a lumped Josephson element.

The calculation demonstrates that such an electrostatically coupled array behaves as a single junction with effective parameters

$$C_{eff} = C / N , \qquad G_{eff} = G / N , \tag{11}$$

where N is the number of junctions. This can be useful in making the systems of normal and/or superconducting junctions, because it makes it possible to reduce effective damping and capacitance.

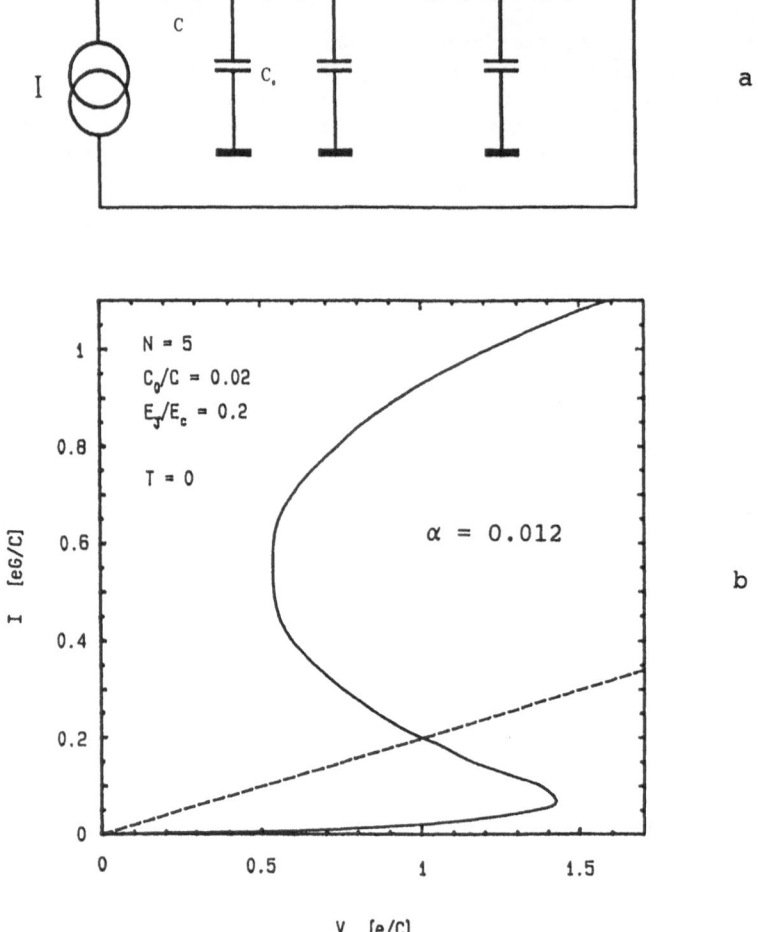

Fig. 3. The equivalent circuit (a) and the calculated dc I-V curve
(solid line) of the 5-junction array of ultrasmall
Josephson junctions (b). Dashed line is the high-voltage
asymptote of the I-V curve.

CONCLUSION

      We have analyzed a few systems with weak Josephson coupling where the
CPT provides essential electric charge transport. CPT is a quantum-coherent
phenomena and it can give rise to time-coherent oscillations if the
ultrasmall Josephson junction or array sees a dc current source rather than
a voltage source. This can hardly be met in practice, therefore the
reasonable solution of this problem seems in using 1-D or 2-D arrays of SET
junctions instead of high-Ohmic resistors.

      Finally we note that realization of the time-correlated CPT is possible
in systems of ultrasmall junctions with weak Josephson coupling biased by ac
voltages. Although the systems in this case do not reveal the time
coherence the external ac drive can control tunneling of Cooper pairs making
this tunneling periodic in time. The performance of such a device based on
the stimulated CPT would be close to that of SET turnstile device.[5] A

principal difference consists in that in the former the SET events play a parasitic role shunting the chanel of Cooper pairs flow.

ACKNOWLEDGEMENTS

For useful discussions and assistance we thank L.J. Geerligs, U. Geigenmüller, K.K. Likharev, J.E. Mooij, and K.Yu. Platov.

REFERENCES

1. K.K. Likharev and A.B. Zorin, Theory of the Bloch-Wave Oscillations in Small Josephson Junctions, J. Low Temp. Phys. 59:347 (1985).
2. K.K. Likharev, Correlated Discrete Transfer of Single Electrons in Ultrasmall Tunnel Junctions, IBM J. Res. and Develop. 32:144 (1988).
3. D.V. Averin and K.K. Likharev, Single-Electronics, in: "Quantum Effects in Small Disordered Systems," B. Al'tshuler et al. eds., Elsevier, Amsterdam, to be published.
4. P. Delsing, K.K. Likharev, L.S. Kuzmin, and T. Claeson, Observation of the Time-Correlated Single-Electron Tunneling, Phys. Rev. Lett. 63:1861 (1989).
5. V.F. Anderegg, L.J. Geerligs, J.E. Mooij, H. Pothier, D. Esteve, C. Urbina, and M.H. Devoret, Frequency-Determined Current with a Turnstile Device for Single Electrons, Physica B 165&166:61 (1990).
6. A.B. Zorin, L.S. Kuzmin, and K.K. Likharev, Two Ways Toward Experimental Observation of the Bloch Oscillations in Ultrasmall Josephson Junctions, Physica B 165&166:933 (1990).
7. T.A. Fulton, P.L. Gammel, D.J. Bishop, L.N. Dunkleberger, and G.J. Dolan, Observation of Combined Josephson and Charging Effects in Small Tunnel Junction Circuits, Phys. Rev. Lett. 63:1307 (1989).
8. L.J. Geerligs, V.F. Anderegg, J. Romijn, and J.E. Mooij, Single Cooper Pair Tunneling in Small Capacitance Junctions, Phys. Rev. Lett. 65:377 (1990).
9. K.K. Likharev, N.S. Bakhvalov, G.S. Kazacha, and S.I. Serdyukova, Single-Electron Tunnel Junction Array: an Electrostatic Analog of the Josephson Transmission Line, IEEE Trans. Magn. 25:1436 (1989).
10. U. Geigenmüller and G. Schön, Single Electron Effects and Bloch Oscillations in Normal and Superconducting Tunnel Junctions, Physica B 152:186 (1988).
11. E. Shimshoni, D.V. Averin, and G. Schön, Zener Tunneling for a Spiky Drive, Physica B 165&166:943 (1990).
12. P. Ao and J. Rammer, Level Crossing in a Dissipative Environment, Physica B 165&166:953 (1990).
13. V.Ya. Aleshkin and D.V. Averin, Resonant Tunneling of Cooper Pairs in a Double Josephson Junction System, Physica B 165&166:949 (1990).
14. S.A. Hattel and A.B. Zorin, work in progress.

DYNAMICAL SIMULATIONS OF JOSEPHSON JUNCTIONS

USING SCHRÖDINGER'S EQUATION WITH DISSIPATION

A. Davidson and P. Santhanam

IBM Research Division, T.J. Watson Research Center
Yorktown Heights, NY 10598, U.S.A.

ABSTRACT

We use a rotational canonical coordinate to describe the physics of small Josephson tunnel junctions. This formulation is combined with a phenomenological damping term to allow dynamical simulations of quantum limited junctions. Our model incorporates the Coulomb blockade as well as the Josephson effects, and both single electron and pair tunneling events are accounted for. The simulations show the possible existence of chaotic dynamics due to competition between tunneling of single particles and pairs.

INTRODUCTION

The dynamics of low current density tunnel junctions not much smaller than the Josephson penetration depth have been well modeled as depending on a single macroscopic dynamical variable, the Josephson phase. This is the famous resistance-capacitance shunted junction (RCSJ) model[1]. For much smaller junctions, the order of 0.01 μm^2 or less, quantum effects become important, and many authors[2,3,4] have endeavored to quantize the RCSJ model. Generally, the quantization runs into two problems: the washboard potential seems to spoil the symmetry of the Hamiltonian; and there has been no manageable way of handling the dissipation that is always there in macroscopic systems.

In this paper we will present a time dependent Schrödinger equation for the phase, including the tilted washboard potential, and a dissipation term. We will show that this equation accounts for single electron tunneling, as well as pair tunneling; and, that the Coulomb blockade emerges naturally in the limit where the charging energy dominates the Josephson coupling. Explicit results for the wave function and energy band pictures will be presented.

Nonlinear Superconductive Electronics and Josephson Devices
Edited by G. Costabile *et al.*, Plenum Press, New York, 1991

THE SCHRÖDINGER EQUATION

Consider a small area Josephson tunnel junction, of capacitance C, critical current I_0, and resistance R, and biased at current I. The normalized Schrödinger equation we use is

$$-\frac{1}{2}\frac{\partial^2}{\partial\theta^2}\psi - \Gamma\cos(2\theta)\psi - \eta\theta\psi - \alpha(\xi^2\frac{J_\theta}{\rho})\psi = i\frac{\partial}{\partial t}\psi . \qquad (1)$$

The first term on the left is usually referred to as the charging energy. The second term is the normalized Josephson potential energy, where $\Gamma = \hbar CI_0/2e^3$ is essentially the ratio of Josephson to charging energy. $\eta = \hbar CI/e^3$ is the normalized coefficient for the external bias term. The last term on the left hand side is for dissipation[5,6,7], and will be discussed later, but $\alpha = \hbar/e^2R$ is the normalized junction conductance. θ is half of the usual Josephson phase, $\theta = \theta_j/2$. Time is in units of $C\hbar/e^2$, and e is the absolute value of electron charge.

There are several features of this equation which are not obvious, and which will be discussed, but this is a good place to show some current-voltage curves generated by integrating Eq. 1. In Fig. 1a these curves are for relatively large values of Γ, and tend to look more and more RCSJ-like as Γ is increased. The

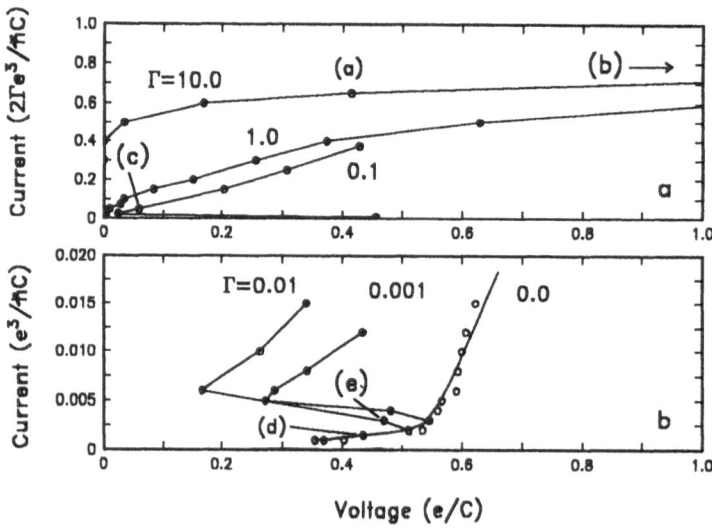

Fig. 1. Current-voltage characteristics generated by Eq. 1. In Fig. 1a the current has been normalized to I_0, the critical current, which is proportional to Γ. As Γ goes from 10 to 0, the curves go from Josephson behavior to Coulomb blockade behavior. The dynamics for points (a) through (e) are shown in subsequent figures, and discussed in the text. As indicated by the arrow, point (b) is slightly off the graph.

current scale in Fig. 1a is normalized to I_0. Fig. 1b is for much smaller Γ values, and in the limit shows a very clear Coulomb blockade characteristic. In between there is reentrant behavior, producing a feature sometimes called Bloch's nose.

In the next section, we address questions concerning the formulation of Eq. 1. Then the I-V curves will again be discussed, and some of the dynamics will be shown from the wave function and energy band perspectives.

Choosing the Coordinate

The canonical coordinate for this problem can be chosen arbitrarily, so long as it represents a rotation. The conjugate momentum is then quantized, and can be used to represent the charge on a capacitor. The only condition[3] is that the product of the angle required for a full rotation and the quantum number are such that the charge quantum is of the experimentally measured size (namely, 1 electron). To understand this, consider Eq. 1 with $\eta = \Gamma = \alpha = 0$. This is the equation for an isolated capacitor, and the eigenstates are simple exponentials, $\psi \propto e^{in\theta}$.

If θ is a rotational coordinate, and the Hamiltonian has rotational symmetry, then it is sufficient that ψ is single valued. For ψ to be single valued, the phase can change by only an integer multiple of 2π along a path that makes a complete revolution. So, if $\Delta\theta$ is the range of angle going once around, and n' is the smallest non-zero quantum number, then we have $\Delta\theta = 2\pi/n'$. If we choose to count electrons as integers, then $\Delta\theta = 2\pi$, and we have the normalization of Eq. 1. Some self consistent consequences of this choice are that the expressions for the voltage (in the classical limit) and the critical current differ by a factor of 2 from those that apply for the more usual choice of phase. That is $V = \frac{\hbar}{e}\dot{\theta}$, and the minima in the washboard potential disappear for $\eta = 2\Gamma$. We do not consider the case where only pair tunneling is allowed, since in that case there would be no tunneling losses.

The External Bias

If $\eta > 0$, a single valued wave function is no longer sufficient, since the potential energy is multivalued. The wave function must then also be multivalued, and it can be shown[8] that it must have the Bloch form:

$$\psi = u(\theta)e^{ik\theta}, \tag{2}$$

where $u(\theta)$ is complex and rotational, and k is some real number. Putting Eq. 2 into the Schrödinger equation, Eq. 1, yields

$$\frac{1}{2}\left(-i\frac{\partial}{\partial\theta} + k\right)^2 u - \Gamma\cos(2\theta)u - \eta\theta u - \alpha\left(\xi^2\frac{J_0}{\rho}\right)u \tag{3}$$

$$= i\frac{\partial u}{\partial t} - \frac{\partial k}{\partial t}\theta u .$$

All terms here have rotational symmetry, except the two explicitly proportional to θ. Equating terms of equivalent symmetry yields two equations:

$$\frac{1}{2}\left(-i\frac{\partial}{\partial\theta}+k\right)^2 u - \Gamma\cos(2\theta)u - \alpha(\xi^2\frac{J_\theta}{\rho})u = i\frac{\partial}{\partial t}u \qquad (4a)$$

and

$$\eta = \frac{\partial}{\partial t}k. \qquad (4b)$$

Equations 1 and 4 are related by a gauge transformation[3,4,8], in which the number k appears explicitly either as an extra phase in the wave function, or as a time dependent "vector potential" in the Hamiltonian.

The main result is that in the gauge of Eq. 1, the wave function is multiple-valued and of the Bloch form. <u>All</u> wave functions that solve Eq. 1 for $0 < \theta \le 2\pi$, both time dependent and stationary, are of the Bloch form. (For $-\infty < \theta < \infty$ only the eigenstates would have to be of the Bloch form.)

Damping

It has been known for many years[5,6] that there exist Schrödinger operators to remove energy. One such operator has recently[7] been shown to give results in reasonable agreement with those predicted by path integral analysis. This term has the following form:

$$S(\theta) = -\alpha\xi^2\frac{J_\theta}{\rho} \qquad (2)$$

where α is a damping coefficient, J_θ is the divergence of the probability current, ρ is the probability density, and ξ is the length scale of the wave function. For a Gaussian wave packet, in which $\xi^2 = <\theta^2>$, $S(\theta)$ agrees remarkably well[7] with tunneling out of a cubic potential[9], oscillating in an harmonic potential, and spreading in a flat potential[10].

It should be noted that this loss operator has rotational symmetry. Therefore, in the gauge transformation that leads to Eq. 4, it is most convenient to leave it in the Hamiltonian, so it appears as a term on the left side of Eq. 4a. Physically, this means it will have no effect on the external charge delivered to the junction, represented by k and determined by Eq. 4b. This is an important property of tunneling losses. Other possible terms, and particularly that of Kostin[5,6], would not have this property.

An example of the damping term relevant to the problem at hand is shown in Fig. 2. This graph is a perspective view of the dynamic evolution of a wave function in the $\Gamma\cos(2\theta)$ potential, for $\Gamma = 1$, $\alpha = 0.015$, and $\eta = 0$. The inital wave function was taken to be real with value $1/\sqrt{2\pi}$. The damped response is obvious in the figure, and the final shape, in the upper right, is a close approximation to the ground state solution. For the case of the Josephson potential, the wave function is clearly not Gaussian, and as a practical matter we have made $\xi^2 = \text{constant} = 1.8$, which is the appropriate width for a wave function spread uniformly over 2π.

This graph illustrates several interesting things about this kind of damping term. First, for Fig.2, there is no analogous classical damping. That is, there is no change in the expectation values of position or momentum. Second, this damping term does not disturb any of the eigenstates. That is, $S(\theta)$ is zero when u is an exact eigenstate. Any fluctuation from an exact eigenstate, however, will result in loss of energy, and so the only stable equilibrium state is the ground state.

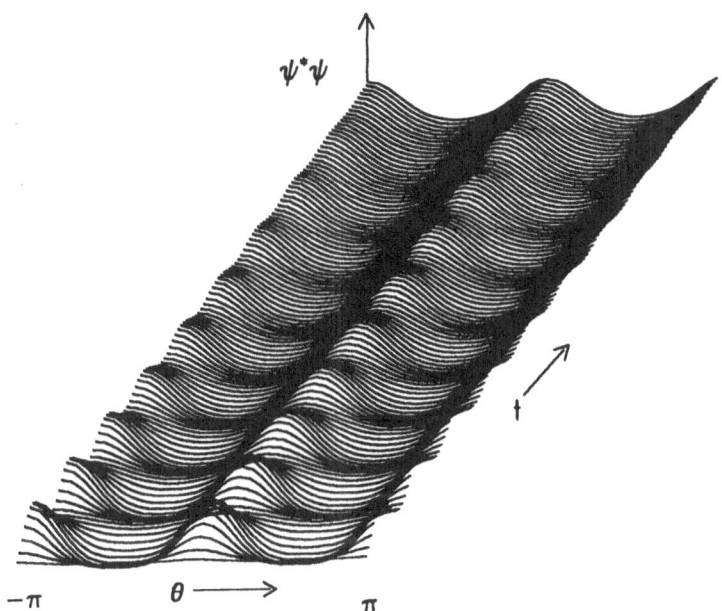

Fig. 2. A demonstration of the damping term, Eq. 2. The coordinate θ runs horizontally, the probability density runs vertically, and succeeding curves are offset toward the upper left to give a perspective view. The initial wave function was a constant, and after several damped oscillations, the ground state emerges at the upper right. Parameters: $\Gamma = 1$, $\eta = 0$, and $\alpha = 0.015$.

This damping term is entirely phenomenological, and its only justification is its ability to approximate reality. The I-V curves of Fig. 1, the damped wave functions of Fig. 2, and Ref.(7) suggest that a reasonable degree of approximation has been attained.

Numerical Techniques

We numerically integrate Schrödinger's equation in the form of Eq. 4, since the wave function u in this gauge obeys periodic boundary conditions. The method of lines was used, with a fourth order Runge-Kutta algorithm for the time step, and five point numerical derivatives with respect to θ. Since this is a diffusion type equation, there is a large penalty for choosing a small discretization grid for θ: the time step scales as the square of the step in θ. Where precision was most needed, we used a time step of 0.002, and 80 coordinate segments between $-\pi$ and π. This was necessary at high values of current where k evolves rapidly, and at high values of Γ, where the wave functions begin to get steep. In other areas, it was sufficient to use a time step of 0.02, and 40 segments.

A disadvantage of using Eq. 4, is that u always develops a progressively tighter twist for finite η, even in the steady state. If nothing were done, the gradients would become steeper than the discretization grid allows. However, the Bloch factor, $e^{ik\theta}$, develops the opposite twist. Therefore whenever k = 0.5, u can be untwisted by an umklapp: $u \rightarrow ue^{2ik\theta}$, and $k \rightarrow -k$. Such a process has no effect on the dynamics or the boundary conditions, provided it is done when k is sufficiently close to 0.5. We use an algorithm to modify the time step to insure that k approaches 0.5 to 8 decimal places, to allow the umklapp.

The damping term has a logarithmic singularity at $\rho = 0$, and so requires some care in the simulations. We coped with this problem by adding a small real number to ρ in the denominator of Eq. 2. We found that 10^{-5} was sufficiently small.

RESULTS

Let us look at the dynamics, as found by integrating Eq. 4, for several points on the I-V curves in Fig. 1. We start with the $\Gamma = 10$ curve, where pair transitions dominate. Then we work down to $\Gamma = 0$, where single electron transitions dominate and there is a Coulomb blockade. At the end there will be a step back to look at the dynamics on the reentrant branch, where there is a transition between one and two electron events. In all these curves, the damping coefficient has been kept constant at $\alpha = 0.1$. (Only Fig. 2 is for a different value, $\alpha = 0.015$.)

Begin at the point marked (a) in Fig. 1, on the $\Gamma = 10$ curve. The time evolution of the wave function is shown in Fig. 3. This figure is basically the response of the probability density to a step from $\eta = 13$ to 15, starting at the lower left. Generally speaking, the wave function moves a bit to the right, corresponding to the increased tilt of the washboard, and at first remains smooth, like the untilted response of Fig. 2, which was for a smaller Γ. As time goes on, the probability density in Fig. 3 begins to build up an oscillation, and starting about half way up,

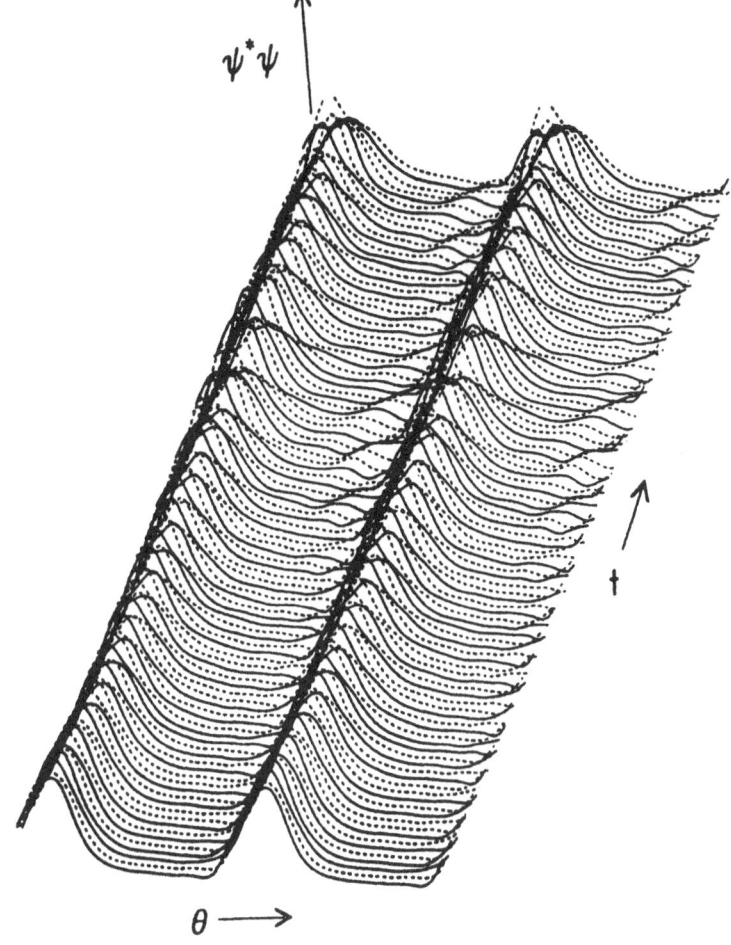

$\psi^* \psi$

t

$\theta \longrightarrow$

Fig. 3. A perspective view similar to Fig. 2, but now with bias applied, and with different parameters corresponding to (a) in Fig. 1: $\Gamma = 10$, $\eta = 15$, and $\alpha = 0.1$.

small amounts of probability can be seen leaking from left to right. The leaking over is shown a bit more graphically in the contour plot of the same data in Fig. 4.

Although it is tempting to think of this leakage as marking the beginning of tunneling, this is not so. There is a probability current flowing from left to right for the entire time interval, and for all finite bias currents, not just when the probability density begins to visibly tunnel over. This is due to the Bloch form of the wave function.

If η is increased slightly to 16, point (b) in Fig. 1, the wave function behaves as shown in Fig. 5. This contour plot shows more vigorous leakage, but no great regularity. That is, there is no obvious periodicity. A plot of the expectation values of the energy against the total charge on the capacitor is shown in Fig. 6, for

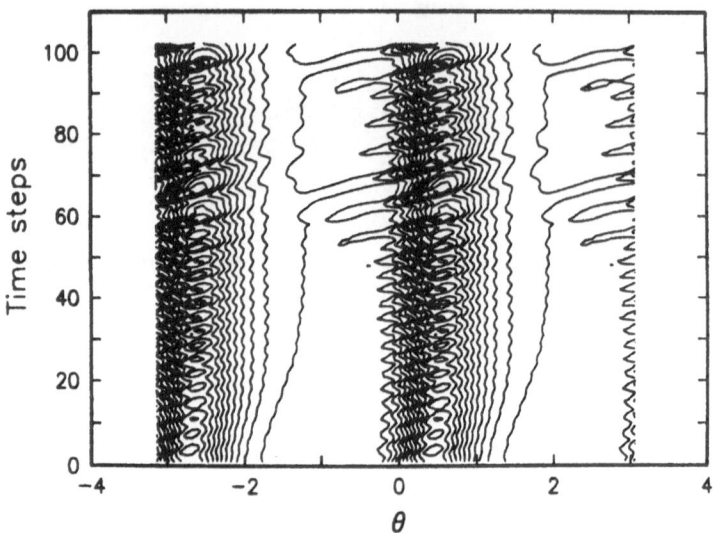

Fig. 4. The same data as Fig. 3, but shown as a contour plot. Discrete leakage of the probability density to the right has just begun.

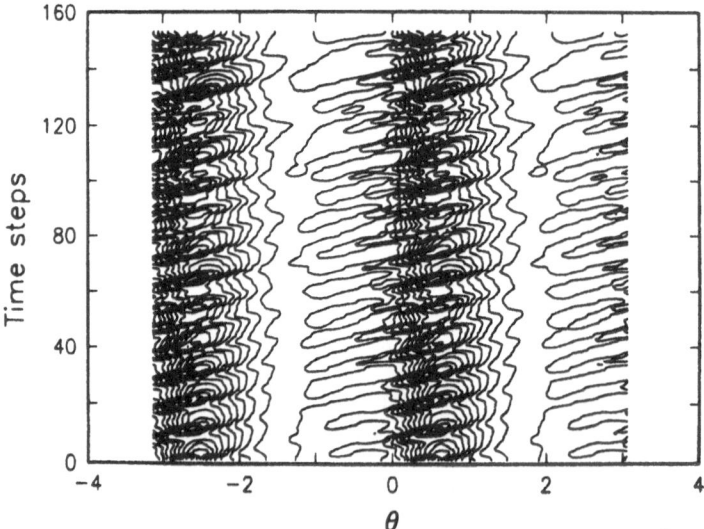

Fig. 5. Same as Fig. 4, except that η has been stepped from 15 to 16 at t = 0. This corresponds to point (b) in Fig. 1. The leakage is much more vigorous than that in Fig. 4.

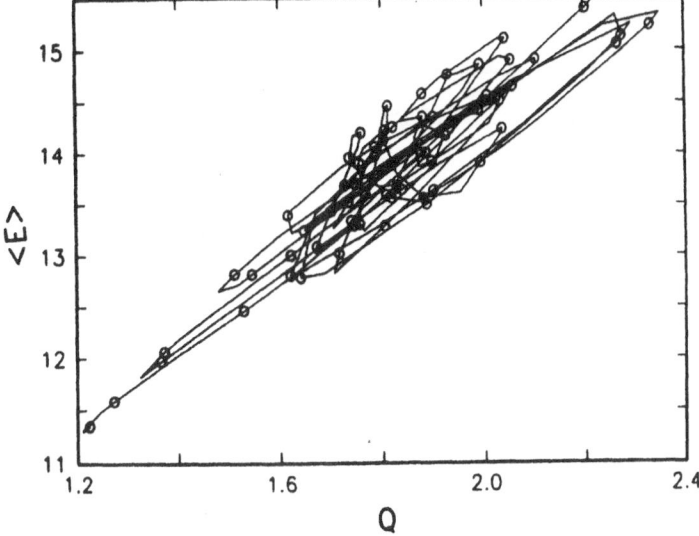

Fig. 6. The energy expectation value versus total charge on the
capacitor for the same parameters as Fig. 5, corresponding
to point (b) on Fig. 1. This complex non-repeating curve
could be chaotic.

$\eta = 16$, and again, there is no regularity. Only much longer simulation times will confirm that this behavior is chaotic.

Now let us move on to point (c) in Fig. 1. For this point there is a comparison between the simulation and the predictions of the energy band picture. This is done in Fig. 7. The solid lines here are the parabolic bands expected for the free particle picture, and the circles are the results of our simulation. If we start at $k = 0$ on the lower branch, we see that the parabolic prediction is followed very well, including the umklapp at $k = 0.5$, to the left hand side. The energy then moves along the higher branch of the parabola, and near the peak makes a pair transition, which umklapps again at $k = 0.5$, to the lower branch, and the cycle is repeated. This is a pair transition because it is between the upper branches, and takes two umklapps to complete a cycle. (A single particle transition would have gone directly from the left upper branch to the lower branch.) Pair transitions occur for the curves in Fig. 1 for $\Gamma = 10$ and 1, and for the upper branches of the curves with $\Gamma = 0.1, 0.01$, and 0.001. It was not observed for $\Gamma = 0$.

For $\Gamma = 0$, and for the lower branches of $\Gamma = 0.1, 0.01$, and 0.001, band diagrams like Fig. 8 were observed, showing clear single particle transitions. This plot was taken specifically for point (d) in Fig. 1, corresponding to $\Gamma = 0.1$, and $\eta = 0.002$. This single electron behavior persists to higher current for the $\Gamma = 0$ curve, which has the shape expected for the Coulomb blockade. Notice also that

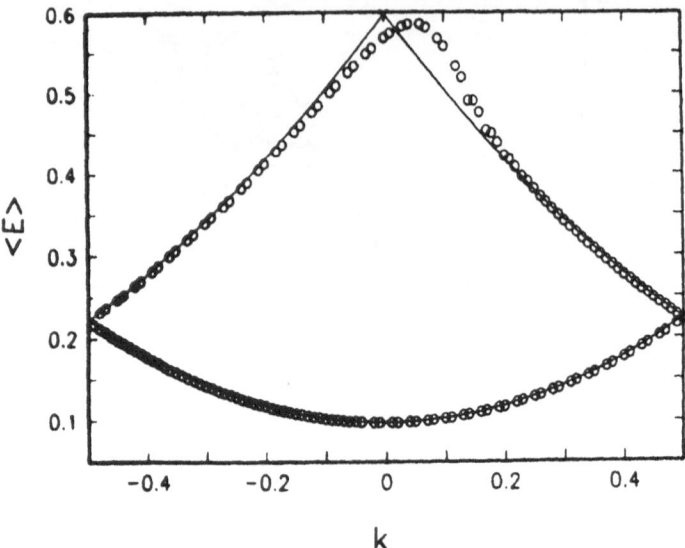

Fig. 7. Energy expectation value versus charge from the source for point (c) in Fig. 1, corresponding to $\Gamma = 0.1$, $\eta = 0.01$, and $\alpha = 0.1$. The solid lines represent the parabolic bands expected in the free particle model, and the circles are from integration of Eq. 4. The transition from the upper left branch to the upper right branch is unambiguously a two electron (pair) process.

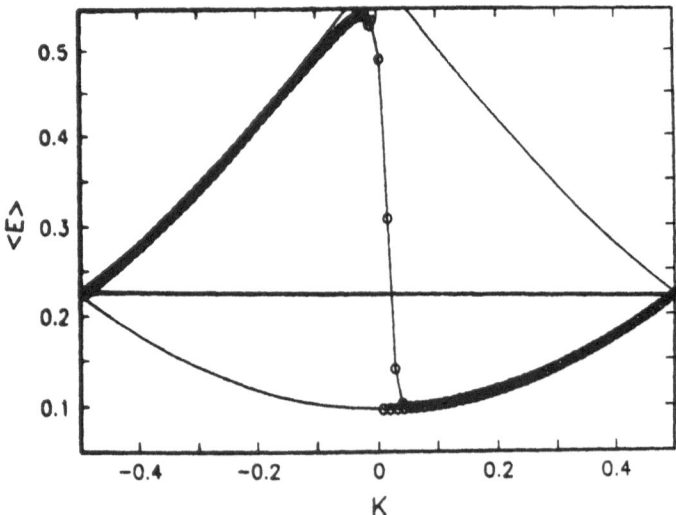

Fig. 8. Energy expectation value versus charge from the source for point (d) in Fig. 1, for $\Gamma = 0.01$, $\eta = 0.002$, and $\alpha = 0.1$. This curve is characteristic of single electron transitions.

248

this curve bends at a voltage of e/2C, also as expected for the Coulomb blockade. It is worth noting that we have gone from RCSJ behavior to Coulomb blockade behavior with the same equation, Eq. 1 or 4, by changing only one parameter, the coefficient of the Josephson potential. The question is, for finite Γ, how do we go from the one electron transition to the pair transitions? Chaotically, perhaps.

If we look at point (e) in Fig. 1, corresponding to a reentrant portion of the I-V curve, we see the behavior shown in Figs. 9 and 10. Fig. 9 plots the expectation value of the energy verse the total charge, showing showing some kind of complicated behavior. It is not obvious whether a limit cycle will emerge, or if chaos will prevail, but certainly nonlinear behavior is evident. Here two transitions are seen almost all the way across the parabola, corresponding to two pair transitions. The three branches on the lower right all clearly correspond to single electron events, and the three branches in the middle are ambiguous.

The same data are replotted in Fig. 10, which shows the expectation value of the energy versus the applied charge, k. Now we see what is going on. The large momentum transitions correspond to the right hand side of this figure, where the junction clearly goes form the left hand upper branch of the parabolic band to the right hand upper branch, a pair transition. But it stays on that branch only briefly, and then falls down to the lower branch before the umklapp. Clearly, a pair tunneled across, and a single electron tunneled back. This process was not repeated exactly, suggesting the possibility of a chaotic transfer of pairs and single electrons on this negative resistance branch.

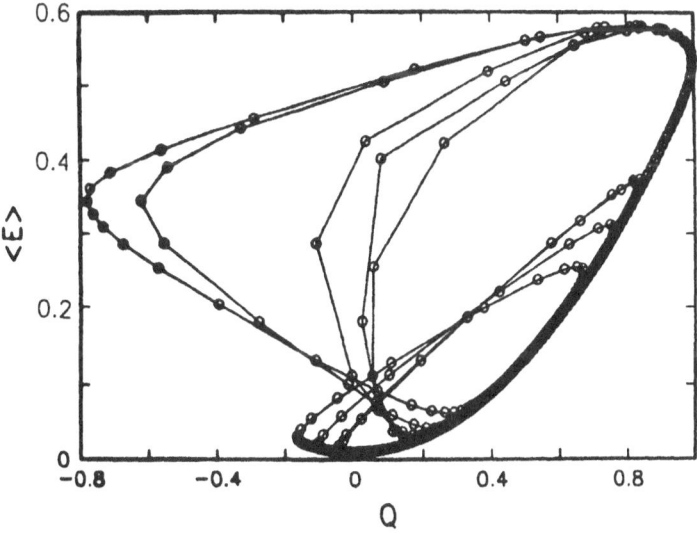

Fig. 9. Energy expectation value versus total charge on the capacitor for point (e) in Fig. 1, for $\Gamma = 0.01$, $\eta = 0.003$, and $\alpha = 0.1$. This chaotic-like curve seems to show mixed pair and single electron transitions.

It is interesting to speculate that there may be a correspondence between period doubling in ordinary differential equations, and quantum doubling in Schrödinger's equation. In any event, when the Josephson cosine potential disappears, the two electron process seems to go with it. Apparently, the presence of even a very small Josephson potential influences the symmetry of the wave function, such that the two electron process is more likely.

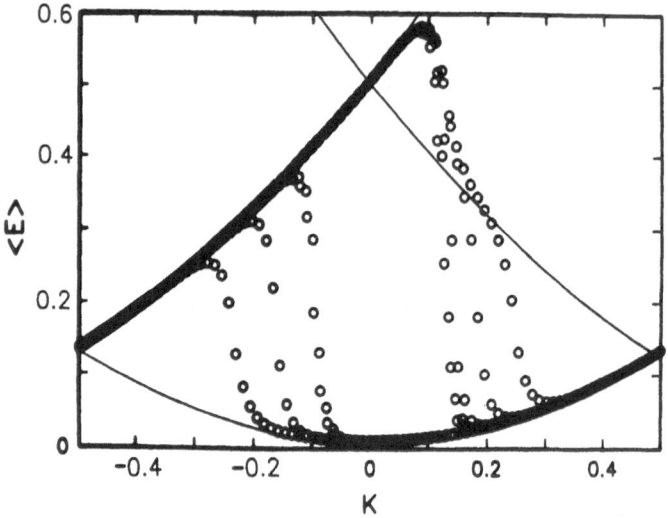

Fig. 10. The same data as Fig. 9, (corresponding to point (c) in Fig. 1) but now the energy expectation value is plotted against charge from the source. The three branches on the left are clearly single electron events. The pair events on the right dwell for some time on the free particle parabola before making a second transition. This behavior corresponds to two electrons tunneling over and one tunneling back.

CONCLUSIONS

We have treated the time dependent Schrödinger equation for a small area Josephson junction, in such a way that the rotational symmetry of the coordinate is preserved, thus insuring quantized behavior. By incorporating a phenomenological damping term, we have been able to go smoothly from the Josephson regime to the Coulomb blockade by changing a single parameter, the normalized Josephson energy. For large Josephson coupling and high current bias, we observe only pair tunneling. At low currents and for very small Josephson coupling, we observe single electron events. In between there were complicated dynamics involving both pairs and single electrons, which may be chaotic.

REFERENCES

1. W.C. Stewart, Appl. Phys. Lett. $\underline{12}$, 277 (1968), and D.E. McCumber, J. Appl. Phys. $\underline{39}$, 3113 (1968).
2. K.K Likharev, IBM J. Res. Develop. $\underline{32}$, 144 (1988).
3. Gerd Schön and A. D. Zaikin, Physica B, $\underline{152}$, 203 (1988).
4. A. Davidson and P. Santhanam, "Quantum Rotors, Phase Slip Centers, and the Coulomb Blockade," Physics Letters A, in press.
5. M.D. Kostin, J. Chem. Phys. $\underline{57}$, 3589 (1972).
6. M.D. Kostin, J. Stat. Phys. $\underline{12}$, 145 (1975).
7. A. Davidson, Phys. Rev. A $\underline{41}$, 3395 (1990).
8. N. Byers and C.N. Yang, Phys. Rev. Lett. $\underline{7}$, 46 (1961).
9. A.O. Caldeira and A.J. Leggett, Ann. Phys. (NY) $\underline{149}$, 374 (1983).
10. Vincent Hakim and Vinay Ambegaokar, Phys. Rev. A $\underline{32}$, 423 (1985).

PHASE LOCKING OF FLUXON OSCILLATIONS IN LONG JOSEPHSON JUNCTIONS

G. Filatrella[a], N. Grønbech-Jensen[b], R. Monaco[c], S. Pagano[c],
R. D. Parmentier[a], N. F. Pedersen[b], G. Rotoli[a], M. Salerno[d], and
M. R. Samuelsen[b]

[a]Dept. Physics, Univ. Salerno, I-84081 Baronissi (SA)
[b]Physics Lab. I, Tech. Univ. Denmark, DK-2800 Lyngby
[c]Inst. Cybernetics of the CNR, I-80072 Arco Felice (NA)
[d]Dept. Theoretical Physics, Univ. Salerno, I-84081 Baronissi (SA)

INTRODUCTION

The study of phase locking of fluxon oscillations in long Josephson tunnel junctions has been stimulated by potential electronic applications of these devices. Single-junction fluxon oscillators subjected to an external microwave field can generate constant-voltage steps that cross over the zero-current axis, a phenomenon of some interest for voltage standard applications.[1] Series-biased arrays of long junctions operating in the resonant fluxon oscillation mode appear to be serious candidates for employment as local oscillators in integrated superconducting heterodyne receivers for radio astronomy applications[2]; operating such an array in a coherent, phase-locked mode offers increased output power and reduced oscillator linewidth with respect to the figures attainable from a single-junction oscillator.[3]

In order to design a practical device, we need an accurate model of how a junction couples to a microwave field, or how the individual junctions in an array couple between themselves, and we need an efficient computational scheme to calculate the dynamics of the model. The derivation of an appropriate model evidently depends in a fundamental way on the geometrical configuration of the device under study, i.e., the physical layout of the junction(s), dc biasing leads, associated microwave circuit elements, and, more generally, the overall microwave environment in which the device is placed. This is a difficult problem, and most often we must be content with fairly drastic simplifying approximations. On the other hand, a computational scheme that has found a rather broad range of applicability is provided by the classic McLaughlin-Scott soliton perturbation theory[4], which reduces the description of a fluxon in a long junction to that of a relativistic particle in a box, subjected to energy input and dissipation in a form depending on the details of the model under study. In the simplest cases the resulting model can be reduced explicitly to a two-dimensional functional map, in which phase-locked states of the fluxon dynamics correspond to fixed points of the map, the existence and stability of which can be studied analytically.[5] More complicated cases may require a repeated numerical inversion of a functional equation, or even a repeated numerical

Nonlinear Superconductive Electronics and Josephson Devices
Edited by G. Costabile *et al.*, Plenum Press, New York, 1991

253

integration of an ordinary differential equation; however, even in the most complicated cases, the computational labor involved is enormously reduced with respect
to a full simulation of the original partial differential equation model.

In the present chapter we outline the basic ideas of the perturbation theory
analysis, and we apply these ideas to several, progressively more complicated,
specific cases (with only a cursory justification for the form of the models
employed). Where possible, we compare the results with those obtained from p.d.e.
simulations and from experimental measurements. More detailed studies may be
found in several other chapters of this volume.

IN-LINE GEOMETRY JUNCTION IN FIXED EXTERNAL FIELD

The simplest case to study is that of a single in-line geometry junction[6] in
an external microwave field. The appropriate normalized p.d.e. model in this case
is

$$\varphi_{xx} - \varphi_{tt} - \sin\varphi = \alpha\varphi_t - \beta\varphi_{xxt} , \qquad (1a)$$

with boundary conditions

$$\varphi_x(0,t) + \beta\varphi_{xt}(0,t) = -x + \eta ; \quad \varphi_x(L,t) + \beta\varphi_{xt}(L,t) = +x + \eta . \qquad (1b,c)$$

The terms in Eqs. (1) have their usual meanings: φ is the phase difference between
the junction electrodes, x is the spatial coordinate normalized to the Josephson
penetration length λ_J, t is time normalized to the inverse of the Josephson angular plasma frequency ω_0, the term in α represents shunt dissipation caused by
quasiparticle tunneling (here, as usual, assumed ohmic), the term in β represents
dissipation caused by the surface resistance of the junction electrodes, x is the
normalized bias current, η is the normalized magnetic field in the plane of the junction and perpendicular to its long dimension, L is the normalized junction length,
and subscripts denote partial derivatives.

For analytical convenience we now make three further simplifying assump-
tions: i) $\beta = 0$; ii) x = constant; iii) $\eta = \eta_0\sin(\omega t + \vartheta)$, where ω is the angular frequency of the applied microwave field, and ϑ is an arbitrary phase angle, which,
in the analysis that follows, will be seen to be just the steady-state phase shift
between the fluxon and the external field in the locked state. Assumption i) can
be eliminated at the cost only of a slightly more complicated notation in the analysis.[7] Assumptions ii) and iii) imply that we are supposing that the external field
interacts with the junction only through the magnetic field term; other cases can
also be readily handled.[5,8]

The basic idea of the perturbation theory analysis is to consider that a fluxon
in a long junction can be described as a relativistic Newtonian particle, characterized by the position, X(t), and the velocity, u(t), of its center of mass. Assuming
that the junction length L >> 1, so that results for the infinite-length limit are
appropriate, we may define the momentum, P(t), of a fluxon as

$$P(t) = - \int_{-\infty}^{+\infty} \varphi_x\varphi_t \, dx . \qquad (2)$$

It is easy to demonstrate that the insertion of Eq. (2) into Eq. (1a) gives that the
fluxon momentum is described (in the infinite-length limit, for $\beta = 0$) by the simple, linear o.d.e.

$$\dot{P} = -\alpha P , \qquad (3)$$

where the overdot denotes a time derivative. Eq. (3), obviously, may be integrated explicitly. For the pure sine-Gordon equation (i.e., with also $\alpha = 0$) in the infinite-length limit the fluxon momentum is related to its velocity by

$$P = 8u/(1-u^2)^{1/2} . \tag{4}$$

Assuming that Eq. (4) continues to hold approximately also for $\alpha > 0$, a given solution of Eq. (3) for $P(t)$ furnishes a corresponding solution for $u(t)$. From this, in turn, the fluxon position is given simply by

$$X(t) = x_0 + \int_0^t u(\tau)\, d\tau . \tag{5}$$

The other major ingredient of the analysis is the treatment of the boundary conditions, Eqs. (1b,c). Following Olsen et al.[9], we observe that during a reflection from a boundary, a fluxon undergoes an energy variation ΔH, given by

$$\Delta H = 4\pi(x \pm \eta) , \tag{6}$$

where the plus sign is taken at (say) the left-hand boundary and the minus sign at (say) the right-hand boundary. To relate this energy variation to the fluxon trajectory, $X(t)$, we recall that the energy of a single-fluxon solution of the pure sine-Gordon equation in the infinite-length limit is given simply by

$$H = P/u . \tag{7}$$

Assuming again that Eq. (7) continues to hold approximately also for $\alpha > 0$, an energy variation given by Eq. (6), assumed to occur instantaneously, may be related to a velocity variation through Eq. (7), whereupon the calculation of $X(t)$ proceeds as before.

At this point, we may combine the boundary condition analysis with the basic McLaughlin-Scott formalism by adopting the following simple mathematical artifice: we consider a periodically extended junction structure of length L lying along the positive x axis, between 0 and $+\infty$, in such a way that the back and forth shuttling motion of the fluxon in the physical junction is transformed into a unidirectional, left-to-right motion on the extended structure, with boundary reflection effects taking place at spatial points equal to integer multiples of L. In this way, Eq. (3) is generalized to

$$\dot{P} = -\alpha P + 4\pi \sum_{k=0}^{\infty} \left[x + (-1)^{k+1} \eta_0 \sin(\omega t + \vartheta) \right] \delta\big(X(t) - kL\big) , \tag{8}$$

in which $\delta()$ is the Dirac delta, $X(t)$ is given by Eq. (5), and the factor $(-1)^{k+1}$ accounts for the alternating sign of the energy variation, due to the magnetic field, felt by the fluxon during successive reflections.

As shown in detail in Ref. 5, it is possible to manipulate Eq. (8) into the form of an explicit, two-dimensional functional map in terms of the variables t_k, the time, modulo the period of the applied microwave field, of the k'th boundary reflection, and E_k, the fluxon energy at the k'th boundary. Reduction to a map is particularly convenient inasmuch we can use standard techniques to study the existence and stability of fixed points, which correspond to phase-locked states of the fluxon dynamics. Defining the time of flight, i.e., the time employed by the fluxon to traverse the junction, as $T_k \equiv t_k - t_{k-1}$, we may express the general condition for the existence of a phase-locked fixed point as

$$T_k + T_{k+1} = 2\pi m/\omega n ; \quad E_{k+\mu} = E_k . \tag{9a,b}$$

255

where m, n, and p are integers. To analyze the stability of such a fixed point we linearize the map around the fixed point, and we calculate the Jacobian matrix of this linearized map. The condition for stability is that the eigenvalues of this matrix lie within the unit circle in the complex plane.

The simplest class of phase-locked fixed points of the map corresponds to single-time-of-flight solutions, i.e., with $p = 1$ in Eq. (9b). We may infer from the symmetry of the map that such solutions may exist (for the magnetic coupling case) only for locking at the fundamental frequency and all odd-order subharmonics, i.e., with $n = 1$ and $m = 1, 3, 5, \ldots$, in Eq. (9a). A detailed analysis[5] shows that these solutions exist for bias currents in the range $\Delta x = 2\eta_0$, centered about the current value of the unperturbed zero-field step at the voltage corresponding to the frequency ω ($V = 2\omega/m$ in this case). The condition for stability is

$$0 < \cos\vartheta < \Lambda/\eta_0 \,, \tag{10a}$$

where

$$\Lambda \equiv \frac{4\alpha E(1 + E)\sinh^2(\alpha L)}{\pi\omega[1 - 2E\cosh(\alpha L) + E^2]^{3/2}} \quad , \text{ with } E \equiv \exp(-m\pi\alpha/\omega) \,. \tag{10b,c}$$

Condition (10a) implies that instability occurs first at the center of a step, inasmuch as $\cos\vartheta = 1$ there, rather than at the extremes of the step, as one might intuitively expect. Moreover, we note from Eqs. (10b,c) that Λ decreases rapidly with increasing m, which means that the stability range for subharmonic steps decreases rapidly with increasing subharmonic order.

Fig. 1 shows a typical current-voltage characteristic, calculated by iterating the map numerically, for the case of locking at the fundamental frequency. Details of the numerical procedure may be found in Ref. 5. The junction voltage is given by $V = 2\pi/T_{av}$, where T_{av} is the average time of flight. In this figure the smooth curve is the profile of the zero-field step in the absence of a microwave field, and the discontinuous curve is the step induced by a field of amplitude $\eta_0 = 0.4$. The observed step amplitude agrees well with the analytic prediction, $\Delta x = 2\eta_0$. From condition (10) the fixed point corresponding to Fig. 1 should be stable for $\eta_0 \leq 4.824$, which is much larger than the value used.

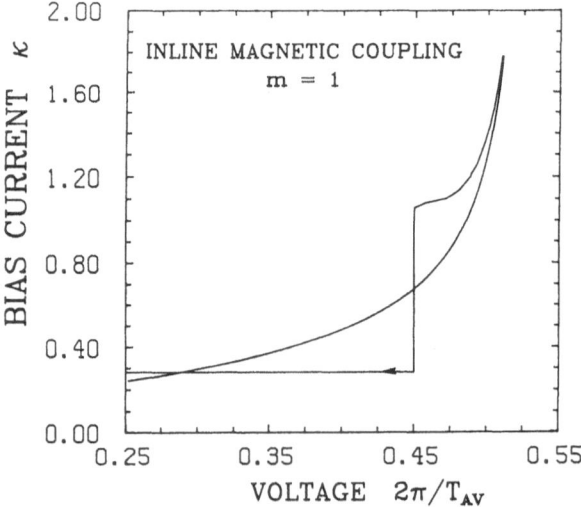

Fig. 1. Current-voltage characteristic of in-line junction with magnetic coupling. Smooth curve: no field; discontinuous curve: with field. Parameters: m = n = p = 1; $\omega = 0.225$; $\alpha = 0.05$; $\beta = 0$; $L = 12$; $\eta_0 = 0.4$. (From Ref. 5.)

In order to check the validity of the map formalism, it is useful to compare results with those obtained from a full simulation of the original p.d.e. model, given by Eqs. (1). Such a comparison shows that the strongest limitation of the map approach derives from the assumption of single fluxon dynamics. Physically, sufficiently large or sufficiently small energy exchange terms can give rise to either the creation or the annihilation of fluxons; such effects are clearly manifested in p.d.e. simulations,[10] but they are not contained in the map model. However, within its range of validity, the map predicts results to a surprising level of detail and accuracy. Moreover, small quantitative discrepancies with p.d.e. results normally can be reduced substantially by incorporating several simple corrections, e.g., those due to phase shift and energy dissipation during reflections, into the map formalism.[5]

A question that naturally arises is what happens to the fluxon dynamics when the stability limit given by condition (10) is exceeded. Fig. 2 suggests the answer to this question. For the parameter values used in this figure, condition (10) gives that $\cos\vartheta = 1$ for $\eta_0 = 0.03324$. For just this value of η_0, we see from Fig. 2 that the single-time-of-flight solution bifurcates into a two-times-of-flight solution. The average value of the two times of flight after the bifurcation is just the same as the single time of flight before the bifurcation; this means that the bifurcation phenomenon leaves no signature in the current-voltage characteristic of the junction, since the junction voltage depends only on the average time of flight, T_{av}. In addition, Fig. 2 shows a second bifurcation at $\eta_0 = 0.0403$ and a third at $\eta_0 = 0.0419$. The appearance of Fig. 2 strongly suggests the existence of a Feigenbaum cascade.[11] Assuming this to be the case, we find from these first three bifurcations an estimate of 4.44 for the Feigenbaum ratio, as compared with the universal asymptotic value of 4.669.... .

As is well known, the Feigenbaum bifurcation cascade leads to a state of chaotic dynamics.[11] Within the bifurcation tree, i.e., before the accumulation point of the cascade, the map predicts that T_{av} remains equal to its value before the

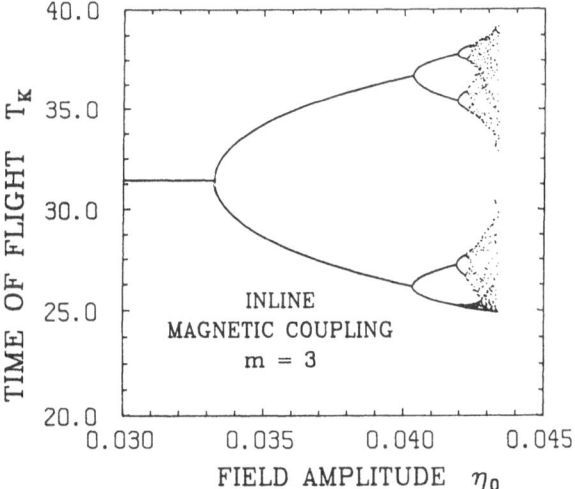

Fig. 2. Dependence of time of flight on field amplitude for in-line junction with magnetic coupling at third subharmonic. Parameters: m = 3; n = 1; p = variable; ω = 0.3; α = 0.05; β = 0; L = 10; x = 0.1295 (center of step). (From Ref. 5.)

first bifurcation. Recent map calculations[12] indicate that this situation continues to hold also for some range of η_0 values after the accumulation point, i.e., for some distance into the chaotic region. This fact is strongly reminiscent of results reported by Kautz[13] for a model of a small Josephson junction with an external rf drive, even though the details of the two models are quite different. Since, as mentioned above, constant T_{av} implies constant junction voltage, it is desirable to find some other signature of the underlying dynamics in order compare model predictions with experimental measurements. A signature that is, at least in principle, reasonably convenient for the experimentalist is the power spectrum of the microwave radiation emitted by the junction. Salerno[14] has shown that the transition to chaos in the map model is accompanied by a transition from a sharp line spectrum to a noisy broadband spectrum, in much the same way as occurs for the small junction.[13]

It is particularly important to compare map predictions with p.d.e. simulations in the case of chaotic phenomena because such phenomena are inherently rather complicated. P.d.e. simulations of Eqs. (1) performed by Rotoli and Filatrella[12] have shown the existence of a bifurcation cascade leading to chaos having a form qualitatively very similar to, although quantitatively somewhat different from, the cascade calculated from the map model. In particular, the value of η_0 at the first bifurcation point is approximately 30% higher in the p.d.e. than in the map, and, consequently, the whole p.d.e. bifurcation tree is shifted to higher field values. This discrepancy may be attributed essentially to the fact that a fluxon in the map formalism is a point particle with no spatial extension, whereas a p.d.e. fluxon has a non-zero width. Consequently, whereas energy exchange at the junction boundaries occurs instantaneously in the map, the exchange interaction is smeared over a certain time interval in the p.d.e. Nonetheless, the qualitative nature of the two bifurcation cascades is quite similar. Fig. 3 shows the appearance of the strange attractors, displayed in terms of the relative phase of the microwave field vs. the time of flight of the fluxon, for the two cases; Fig. 3(a) refers to the p.d.e. and Fig. 3(b) to the map. Both attractors have been calculated for η_0 values just slightly into the chaotic region, i.e., slightly beyond the values corresponding to the accumulation points of the respective bifurcation cascades.

The reader may notice that the value of the bias current x used in Fig. 3(a) is different from that used in Fig. 3(b). The reason for this difference is that the map, particularly when effects due to phase shift and energy loss during reflections[5] are not taken into account (as was the case for the calculations represented in Fig. 3), gives an unperturbed zero-field step profile that is somewhat lower in bias current than the p.d.e step profile. Consequently, the x value at the center of the p.d.e phase-locking step (where, from condition (10), we expect first to see the onset of bifurcations) is somewhat higher than the x value at the center of the map phase-locking step. However, even when the phase shift and energy loss corrections are taken into account, in such a way as to bring the x values at the centers of the two steps into coincidence, the discrepancy in the threshold value of η_0 persists.

For values of η_0 larger than those indicated in Fig. 3 the similarity between the map and the p.d.e. descriptions of chaos begins to disappear. In particular, times of flight that are unphysically long begin to appear in the map attractor, whereas the p.d.e. attractor undergoes a distinct change of appearance, filling much more densely and irregularly the relative phase vs. time of flight plane.[12] It should be mentioned also that some cases of qualitative disagreement between map and p.d.e. results have been found: in particular, for the parameters of Fig. 2, where chaos is clearly evident in the map, no corresponding chaotic behavior has been found in the p.d.e.

Another standard geometry for long Josephson junctions is the overlap geometry.[6] The normalized p.d.e. model in this case is

$$\varphi_{xx} - \varphi_{tt} - \sin\varphi = \alpha\varphi_t - \beta\varphi_{xxt} - \gamma \ , \tag{11a}$$

with boundary conditions

$$\varphi_x(0,t) + \beta\varphi_{xt}(0,t) = \varphi_x(L,t) + \beta\varphi_{xt}(L,t) = \eta \ . \tag{11b}$$

The essential difference with respect to the in-line case is that the bias current, γ, now enters the problem through the p.d.e. rather than through the boundary conditions. Two limiting cases of Eqs. (11) have received attention in the literature: i) the M-model, in which γ = constant and $\eta = \eta_0 \sin(\omega t + \vartheta)$, and ii) the U-model, in which $\gamma = \gamma_0 + \gamma_1 \sin(\omega t + \vartheta)$ and $\eta = 0$. The M-model is evidently fairly similar

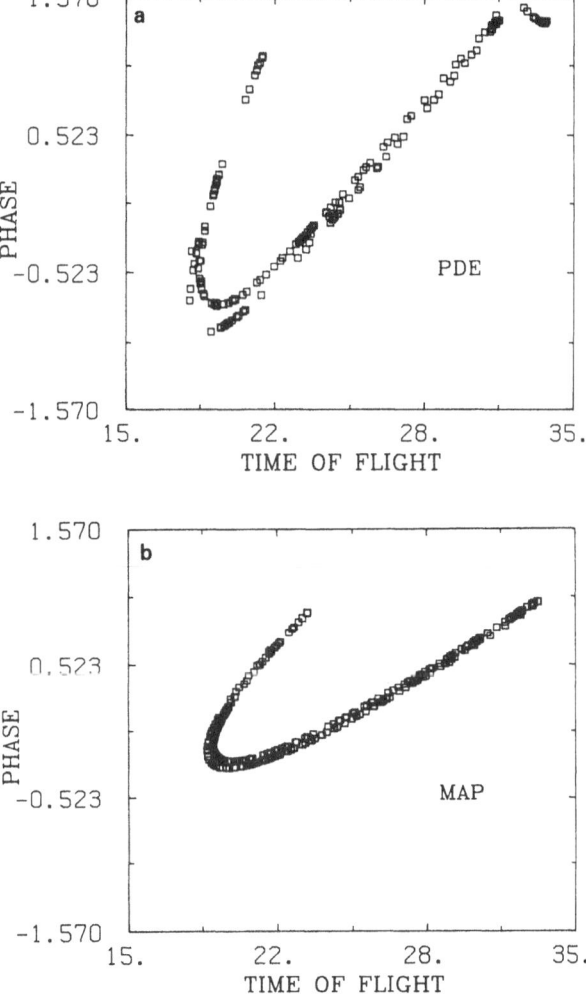

Fig. 3. Chaotic strange attractors for η_0 values slightly larger than accumulation points of bifurcation cascades. (a) P.d.e., Eqs. (1). Parameters: m = 3; n = 1; p not defined; ω = 0.4; α = 0.1; β = 0; L = 10; x = 0.493 (center of step); η_0 = 0.185. (b) Map. Parameters: m = 3; n = 1; p not defined; ω = 0.4; α = 0.1; β = 0; L = 10; x = 0.447 (center of step); η_0 = 0.106.

to the in-line case discussed in the previous section, inasmuch as the interaction with the microwave field takes place only through magnetic (hence M-model) field boundary conditions. The U-model, on the other hand, supposes that the effect of the microwave field is to generate a spatially uniform (hence U-model) rf current along the length of the junction.

M-model

The perturbation theory formalism for the M-model is quite similar to that described above for the in-line geometry junction with magnetic coupling. In particular, with $\beta = 0$, Eq. (8) is replaced by[5]

$$\dot{P} = -\alpha P + 2\pi\gamma + 4\pi \sum_{k=0}^{\infty} (-1)^{k+1} \eta_0 \sin(\omega t + \vartheta) \, \delta\big(X(t) - kL\big) . \tag{12}$$

The difference between Eqs. (8) and (12) may appear to be trivial, but it has one important consequence: it is no longer possible to manipulate Eq. (12) into a functional map in explicit form. The basic character of a map is still implicitly present, but it is now necessary to perform a numerical inversion of a functional equation at each iteration step in order to find the successive times of flight. From a practical point of view this difference is perhaps not highly significant, since the numerical iteration procedure used to obtain results like those shown in Fig. 1 becomes now only slightly more complicated,[5] but it does mean that we no longer have available simple, exact analytic expressions for the ranges of existence and stability of the fixed points.

In spite of the lack of an explicit representation, it is nonetheless possible to obtain some approximate analytic results from the M-model map. For example, we can show[5] that the range in bias current for the existence of a phase-locked state at the fundamental frequency or an odd subharmonic is given approximately by $\Delta\gamma = 4\eta_0/L$, once again centered about the unperturbed zero-field step, as compared with $\Delta x = 2\eta_0$ for the in-line case. Direct numerical iteration of the maps gives results that are qualitatively very similar in the two cases.[5]

The phase-locking conditions of Eqs. (9) may be applied in two different ways: i) fix ω and sweep ϑ over a certain range (in the in-line case), or simply iterate the map (in both cases), to find the range of phase locking in bias current, Δx or $\Delta\gamma$; ii) fix x or γ and vary ϑ (or iterate the map) to find the range of phase locking in frequency, $\Delta\omega$. The first procedure has been described above; a typical result is illustrated in Fig. 1. The second procedure corresponds to what the experimentalist does when he observes the microwave radiation emitted by the junction using a spectrum analyzer; a typical experimental result is shown in Fig. 3 of Monaco et al.[15] The phase-locking map can be used to model qualitatively this aspect of fluxon dynamics by representing the time sequence $\{T_k + T_{k+1}\}$ as a sequence of unit δ functions, and by calculating the Fourier transform of this sequence. Fig. 4 shows an example of the dynamics of the phase-locking process in the frequency domain for the M-model, calculated in this way. In this figure the fluxon frequency is defined as $2\pi/(T_k + T_{k+1})$, i.e., we consider locking at the fundamental frequency. The drive frequency ω is 0.230 in the lowest trace; it increases in increments of 0.003 in the higher traces. Clearly evident in Fig. 4 is the fact that as the drive frequency approaches the unlocked fluxon frequency, frequency pulling is observed, and mixing products (the small peaks near the edges of the figure) begin to appear. At a certain point, the fluxon frequency is pulled into synchronism with the driver, where it remains locked for a certain interval (the central region of the figure). Beyond this region the fluxon frequency unlocks, and mixing products are once again seen. The resemblance between Fig. 4 and Fig. 3 of Ref. 15 is evident.

U-model

The perturbation theory analysis for the U-model has not yet been developed and tested to the same level of detail as it has for the M-model. Fernandez et al.[16] have taken an important step in this direction by reducing Eqs. (11) to an ordinary differential equation via the McLaughlin-Scott formalism and integrating this o.d.e., with appropriate boundary conditions, numerically; Fernandez et al.[17] also consider a generalized case, consisting of a combination of the M- and U-models. On the other hand, Cirillo[18,19] has reported several detailed studies of the U-model based on p.d.e simulations of Eqs. (11); however, with only few exceptions[16], perturbation theory results and p.d.e. simulation results have not yet been subjected to a detailed comparison.

A somewhat different approach to the perturbation theory analysis of the U-model may be formulated as follows:[20] For $\beta = 0$, we replace Eq. (12) by

$$\dot{P} = -\alpha P + 2\pi\left(\gamma_0 + \gamma_1 \sin(\omega t)\right) + \sum_{k=0}^{\infty} \Delta E_k \, \delta\left(X(t) - kL\right) , \qquad (13)$$

where ΔE_k is the energy variation due to phase shift and energy dissipation[5] during the k'th reflection, and the phase angle ϑ in the external field has been set to zero. For later computational convenience we now redefine the time scale so that t is set to zero at each boundary reflection. With the use of this artifice we may write the general solution of Eq. (13) as

$$P(t) = P_\infty + P_{rf} \sin\left(\omega(S+t) - \Theta\right) + \left[P_0 - P_\infty - P_{rf}\sin(\omega S - \Theta)\right]\exp(-\alpha t) , \qquad (14)$$

where $P_\infty \equiv 2\pi\gamma_0/\alpha$, $P_{rf} \equiv 2\pi\gamma_1/(\alpha^2 + \omega^2)^{1/2}$, $\Theta \equiv \arctan(\omega/\alpha)$, t ≡ time elapsed since last reflection, S ≡ sum of previous times of flight, and $P_0 \equiv P(0)$. We now assume that the phase function $\varphi(x,t)$ in Eqs. (11) can be separated into a fluxon part and a spatially uniform background part, i.e.,

$$\varphi(x,t) = \varphi_{fl}(x,t) + \varphi_{bk}(t) . \qquad (15)$$

The utility of this idea was demonstrated by If et al.[21] Inserting the ansatz of

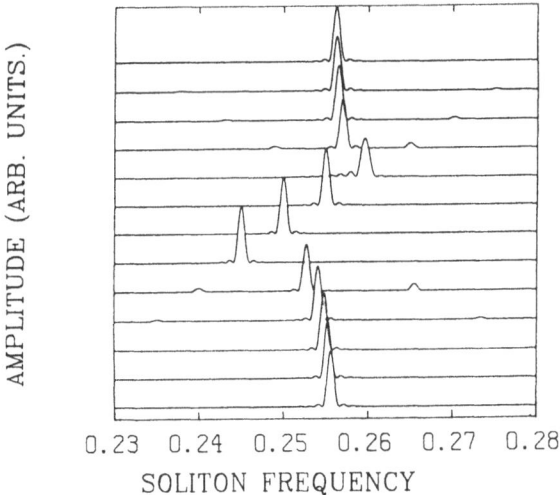

Fig. 4. Frequency-domain representation of phase-locking process for M-model at fundamental frequency. See text for details. Parameters: m = n = p = 1; $\alpha = 0.05$; $\beta = 0$; L = 12; $\gamma = 0.34$; $\eta_0 = 1.0$. (From Ref. 5.)

Eq. (15) into Eq. (2), we find that the total momentum may be divided into a fluxon part and a background part, given by

$$P(t) = P_{fl}(t) + 2\pi\dot{\varphi}_{bk}(t) \; . \tag{16}$$

We now assume that $\varphi_{bk}(t)$ is governed by the o.d.e.

$$\ddot{\varphi}_{bk} + \alpha\dot{\varphi}_{bk} + \sin\varphi_{bk} = \gamma_0 + \gamma_1\sin(\omega t) \; . \tag{17}$$

Eq. (17) in general requires numerical solution. To proceed analytically, we now assume that γ_1 is sufficiently small that we can linearize Eq. (17) around the static solution, i.e., we write

$$\varphi_{bk}(t) = \arcsin(\gamma_0) + \Phi(t) \; , \tag{18}$$

with $\Phi(t)$ governed by the linear o.d.e.

$$\ddot{\Phi} + \alpha\dot{\Phi} + (1-\gamma_0^2)^{1/2}\Phi = \gamma_1\sin(\omega t) \; . \tag{19}$$

The steady-state solution of Eq. (19) may easily be found; from this solution we find that

$$2\pi\dot{\varphi}_{bk}(t) = 2\pi\dot{\Phi}(t) = P_{bk}\cos\big(\omega(S+t) - \Psi\big) \; , \tag{20a}$$

with

$$P_{bk} \equiv 2\pi\omega\gamma_1\Big/\Big[\big((1-\gamma_0^2)^{1/2} - \omega^2\big)^2 + \big(\alpha\omega\big)^2\Big]^{1/2} \; , \text{ and} \tag{20b}$$

$$\Psi \equiv \arctan\Big[\alpha\omega\big/\big((1-\gamma_0^2)^{1/2} - \omega^2\big)\Big] \; . \tag{20c}$$

Substituting for P(t) from Eq. (14) and for $2\pi\dot{\varphi}_{bk}(t)$ from Eqs. (20) we may solve Eq. (16) explicitly for $P_{fl}(t)$. Assuming now that Eq. (4) holds for $P_{fl}(t)$ we may write the fluxon velocity as

$$u(t) = \big(P_{fl}(t)/8\big)\Big/\Big[1 + \big(P_{fl}(t)/8\big)^2\Big]^{1/2} \; . \tag{21}$$

A generic time of flight of the fluxon, T, may then be calculated by solving the equation

$$L = \int_{\tau=0}^{T} u(\tau)\, d\tau \; , \tag{22}$$

where L is the junction length. Having solved for the k'th time of flight, T_k, from Eq. (22), we calculate the initial condition for the next iteration from

$$P_0^{(k+1)} = 8\Big[\big\{\big(1 + [P_{fl}(T_k)/8]^2\big)^{1/2} - \big(\Delta E_k/8\big)\big\}^2 - 1\Big]^{1/2} \; . \tag{23}$$

Phase locking once again is obtained by imposing the conditions of Eqs. (9).

A general feature of the U-model that emerges both from the perturbation theory approach outlined above and from that employed by Fernandez et al.[17] is that the locking range in bias current is much smaller than for the M-model and that the induced current steps are no longer symmetric about the unperturbed zero-field step, but are shifted upwards in current.

Phase locking, denoted also frequency entrainment, of coupled nonlinear oscillators is by no means a new phenomenon: published reports go back at least three centuries, to Huygens, who observed that two mechanical clocks became locked into a synchronous state when weakly coupled by being mounted on a common support.[22] The fact that the phenomenon should also manifest itself in arrays of long Josephson junctions is therefore not particularly surprising. Interest in studying phase locking in arrays of long junctions stems essentially from the practical considerations mentioned above in the Introduction; the information that one seeks to obtain from model studies is how to design the array to optimize the desired practical performance features.

Experimental studies of phase locking in long-junction arrays go back several years. The pioneering work in this direction was that of Finnegan and Wahlsten[23], who considered locking phenomena in a system of two coupled junctions. Cirillo and Lloyd[24] reported further experimental results on a two-junction array having a different geometry. More recently, Pagano et al.[25] have described measurements on arrays of 8, 20, and 100 junctions, and Holst et al.[26] have reported on experimental and numerical results obtained with a two-junction array having a quite different coupling mechanism.

The 20-junction device of Pagano et al.[25] consists of a series-biased array of overlap-geometry junctions loosely coupled to a linear distributed resonator that is one half-wavelength long at the operating frequency ($\sim$ 10 GHz), which in turn is capacitively coupled to a 50 ohm microstrip transmission line to carry the generated signals to a spectrum analyzer. A first attempt to model the behavior of this device was described by Grønbech-Jensen et al.[27] In their analysis the distributed resonator is modelled as a lumped LC tank, which gives a valid approximation if attention is limited to the half-wavelength mode of the physical resonator. The motion of the fluxon in each junction of the array is described using the map formalism appropriate to junctions of overlap geometry, as outlined above, and each junction is coupled to the tank through a coupling resistance. The basic idea

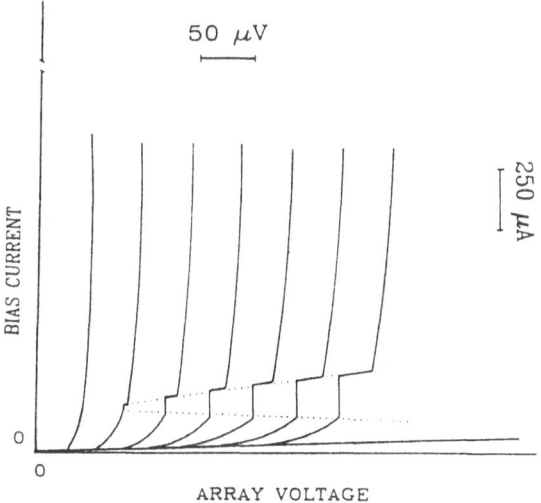

Fig. 5. Multiple-exposure oscilloscope tracing of the current-voltage characteristic of an experimental array having Q $\approx$ 1000, with $1 \leq n \leq 7$ junctions biased on the first zero-field step. The dotted lines indicate a linear dependence of the locking range in bias current on n. (From Ref. 3.)

of the analysis is thus that each junction pumps the tank through a sequence of delta-function voltage pulses, due to the successive reflections of the fluxon from the tank boundary, and that the tank furnishes to each junction, at that boundary, a current proportional to the tank voltage. This current is proportional to the term η in Eq. (11b); however, it is applied at only one end of the junction, the other end being assumed open-circuited.

The result of this analysis[27] is that when a single junction is connected to the tank, the fluxon frequency is pulled by the tank toward its resonant frequency, causing a deformation of the current-voltage characteristic of the junction. When two junctions having slightly different normalized lengths, and hence slightly different natural frequencies, are connected to the tank, phase locking of the two junctions is observed over a significant range of the common bias current γ. The phase locking is evidenced by a coincident segment of the current-voltage characteristics of the two junctions and by the fusion of the two lines relative to the two junctions in the frequency spectrum of the system into a single line of larger amplitude. In the locking region the relative phase with respect to the tank voltage of the fluxons in the two junctions adjusts itself so that the longer junction receives a larger energy input from the tank, thus offsetting its lower free-running frequency. A p.d.e. simulation study of this very same model was reported by Grønbech-Jensen et al.[28] Also in this case excellent qualitative and semi-quantitative agreement between map and p.d.e. results was found, with discrepancies again attributable to the fact that the map fluxon is a point particle whereas the p.d.e fluxon has a non-zero width.

A more detailed study of essentially the same 20-junction device has recently been described by Monaco et al.[3] With respect to the previous experimental configuration, this new device is characterized by the fact that it has been constructed using an improved $Nb/Al-Al_2O_3/Nb$ fabrication technology, thus yielding a significantly higher degree of uniformity of the individual junctions in the array, and by the addition of a superconducting ground plane on the bottom side of the glass substrate, thus yielding a higher quality factor of the linear resonator ($Q \approx 1000$). With these modifications, spontaneous phase locking of a controlled, variable number of junctions is observed in the array. The locking is sufficiently strong to

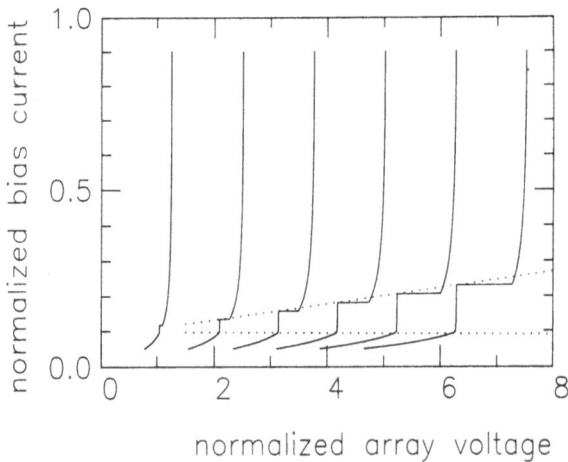

Fig. 6. Current-voltage characteristic of the model array, with $1 \le n \le 6$ junctions biased on the first zero-field step, computed using parameters selected to match (approximately) the experimental result shown in Fig. 5. The dotted lines indicate a linear dependence of the locking range in bias current on n. (From Ref. 3.)

manifest itself also in the current-voltage characteristic of the array, rather than just in the frequency spectrum. Fig. 5 shows a multiple-exposure oscilloscope tracing of the current-voltage characteristic of the array, with a variable number, from one to seven, of junctions biased on the first zero-field step; for each trace in the figure, all of the non-active junctions are biased in the zero-voltage state. For this device, the emission frequency associated with the asymptotic limiting voltage of the first zero-field step of each junction is about 12 GHz, and the resonant frequency of the linear resonator is about 10 GHz.

With a bias point chosen near the center of the near-vertical portion of the steps in Fig. 5, the integrated power emitted by the array is observed to follow a quadratic dependence on the number of active junctions. For ten participating junctions, a power level in a 50 ohm external load of approximately 10 nW has been measured.[3] Although still not quite large enough to be useful for practical applications of the sort described in Ref. 2, such a power level, together with the observed dependence on the square of the number of active junctions, certainly indicates that the approach of coupling the junctions in an array through a linear resonator is a potentially promising one.

In order to model the experimental device, Monaco et al.[3] have employed an approach very similar to that reported by Grønbech-Jensen et al.[27], i.e., overlap junctions described by the map formalism coupled to a lumped linear tank circuit. The most important difference between the two models is that Monaco et al.[3] take a capacitance as the coupling element in place of the resistance employed by Grønbech-Jensen et al.[27] The justification for this choice is the observation that locking phenomena in the experimental device occur at frequencies just below the resonant frequency of the resonator[3], whereas the resistance-coupled tank model gives rise to locking phenomena at frequencies just above the resonant frequency[27]; moreover, a capacitive coupling is intuitively more plausible physically than is a resistive coupling.

Fig. 6 shows the resulting current-voltage characteristic computed for the model array. Details of the computational procedure may be found in Ref. 3. The resemblance to the experimental result of Fig. 5 is evident; as in Fig. 5, the voltage

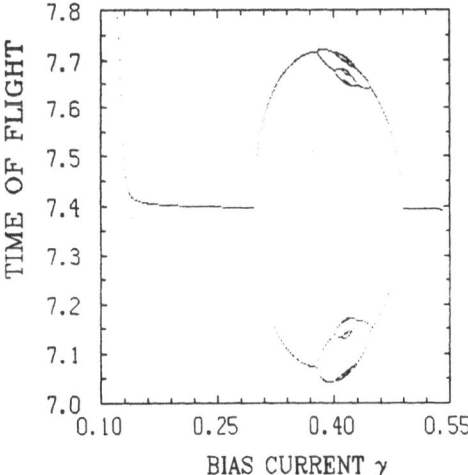

Fig. 7. Dependence of fluxon time of flight on junction bias current for overlap junction capacitively coupled to a lumped tank, at second subharmonic of the tank resonant frequency. Parameters: $\alpha = 0.05$; $\beta = 0$; $L = 3$; tank resonant frequency = 1.7; tank quality factor = 1000; coupling capacitance = 0.0002

indicated is the total voltage across the array, i.e., the sum of the individual junction voltages. A salient feature of this system, seen both in the experimental characteristic of Fig. 5 and in the model characteristic of Fig. 6, is that the locking range in bias current, i.e., the height of the near-vertical portion of the steps, varies linearly with the number of active junctions.

Very recent calculations[29] with this model have predicted several dynamic phenomena that have not yet been observed experimentally: locking at subharmonics of the tank resonant frequency, and a bifurcation cascade leading to chaos in the times of flight of a fluxon across the junction. In particular, with a single junction connected to the tank, locking can be observed not only when the fluxon frequency is close to the resonant frequency of the tank, but also at subharmonics of all integer orders, up to a certain maximum, of the tank resonant frequency. In the cases studied, the height in current of the subharmonic steps varies approximately inversely with the subharmonic order number. Moreover, for appropriate system parameters, bifurcations and chaos can be observed on subharmonic steps. Fig. 7 shows a bifurcation "egg" in the fluxon times of flight, associated with a second-subharmonic step. For sufficiently small values of the junction bias current the fluxon time of flight is just twice the period of the resonant oscillation of the tank. At a certain threshold value of the current this single-time-of-flight solution bifurcates into a two-times-of-flight solution, in a manner similar to that observed above in Fig. 2. Depending on the system parameters, the bifurcation cascade may or may not proceed all the way to chaos. In either case, with a further increase of the bias current, we enter an inverse bifurcation cascade, and, at a second threshold value, the "egg" closes, and we obtain again a single-time-of-flight solution. This situation is strongly reminiscent of that depicted by Salerno[14] in his Figs. 2 and 3 for a single junction in a fixed external field.

Although it is not readily apparent on the scale shown, the bifurcation egg of Fig. 7 does in fact contain a small chaotic region, of width $\Delta\gamma = 0.013$, centered near $\gamma = 0.43$. This fact may be evidenced by observing Fig. 7 on a greatly magnified scale, using a slower sweep rate of the bias current, or, alternatively, by calculating the resulting frequency spectrum of the tank voltage. For a value of the bias current within the single-time-of-flight region of Fig. 7, for example at $\gamma = 0.31$, this spectrum shows a strongly dominant line at $\omega = 1.7$, the tank resonant frequency, and a much weaker, but still distinct, line at $\omega = 0.85$, the fluxon propaga-

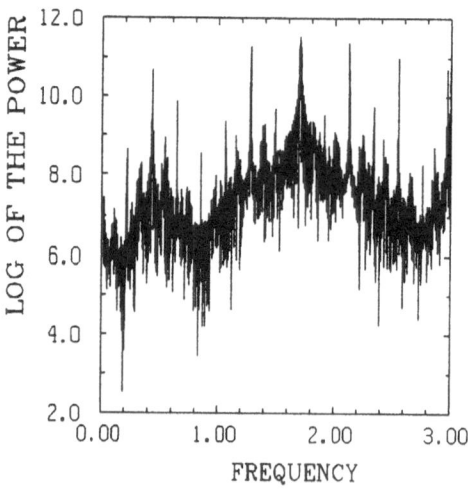

Fig. 8. Power spectrum of the tank voltage for the situation depicted in Fig. 7, for bias current $\gamma = 0.43$.

tion frequency. Fig. 8, instead, shows the spectrum at $\gamma = 0.43$, within the chaotic region. The noisy, broadband nature of this spectrum typifies a chaotic underlying dynamics. As was the case in the calculations reported by Rotoli and Filatrella[12] for a single junction in a fixed external field, the complicated phenomena depicted in Figs. 7 and 8 leave no signature in the current-voltage characteristic of the junction.[29]

The experimental device described by Holst et al.[26] consists of two long $Nb/NbO_x/Pb$ junctions lying parallel to one another in close proximity (35-75 μm) and connected by a common top (Pb) electrode. The device permits the employment of two independent bias current sources, one providing a common dc bias current, and the other allowing for adjustment of the current through one of the junctions. The range of adjustment of the second source was such as to permit operation of the two junctions in both "series-aiding" and "series-opposed" configurations.[26] Phase locking was monitored both by measuring the current-voltage characteristics of the junctions and by observing the microwave radiation emitted by the system using a spectrum analyzer. The model proposed by Holst et al.[26] to account for their experimental observations consists of two p.d.e.'s of the form of Eqs. (11) which are coupled along their entire lengths by a uniformly distributed mutual inductance. A numerical simulation of this model shows good qualitative agreement with the experimental data.

A more detailed study of models with distributed longitudinal coupling[30], employing both perturbation theory and numerical simulation, casts further light on the experimental results obtained by Holst et al.[26] Moreover, very recent measurements by Monaco and Costabile[31] provide additional experimental support for the proposed mechanism of distributed longitudinal coupling: using a device with the 100-junction geometry of Pagano et al.[25], but with the distributed resonator removed, they observe current-voltage characteristics indicative of the preferential operation of variable numbers of pairs of junctions, i.e., the voltage spacing between successive steps in the characteristic is twice that of the first zero-field step of each junction. Presumably then, both distributed longitudinal coupling and coupling through the linear distributed resonator are present in arrays of the type described by Monaco et al.[3] A detailed analysis of such devices will present a considerable challenge to system modellers.

ACKNOWLEDGEMENTS

This work profitted from many fruitful discussions with G. Costabile and A. Davidson. Financial support from the EEC Stimulation Program, under contract no. St-2-0267-J-C(A), and from the Progetto Finalizzato "Tecnologie Superconduttive e Criogeniche" of the Consiglio Nazionale delle Ricerche, the Ministero dell'Università e della Ricerca Scientifica e Tecnologica, and the Fondazione Angelo della Riccia (Italy) is gratefully acknowledged.

REFERENCES

1. G. Costabile, R. Monaco, S. Pagano, and G. Rotoli, "Inverse ac Josephson effect in long junctions", Phys. Rev. B 42, 2651 (1990).
2. D.-G. Crete, P. Feautrier, M. Hanus, R. Monaco, and P. J. Encrenaz, "Nb/AlO$_x$/Nb junctions for integrated superconducting heterodyne receiver", in: Stimulated Effects in Josephson Devices, M. Russo and G. Costabile, eds. (World Scientific, Singapore, 1990), pp. 87-98.
3. R. Monaco, N. Grønbech-Jensen, and R. D. Parmentier, "Self locking of fluxon

oscillations in series arrays of niobium Josephson tunnel junctions coupled to a linear resonator", Phys. Lett. A (in press).

4. D. W. McLaughlin and A. C. Scott, "A multisoliton perturbation theory", in: Solitons in Action, K. Lonngren and A. Scott, eds. (Academic Press, New York, 1978), pp. 201-256.

5. M. Salerno, M. R. Samuelsen, G. Filatrella, S. Pagano, and R. D. Parmentier, "Microwave phase locking of Josephson-junction fluxon oscillators", Phys. Rev. B $\underline{41}$, 6641 (1990), and references therein.

6. A. Barone and G. Paternò, Physics and Applications of the Josephson Effect (Wiley, New York, 1982), Chap. 5.

7. G. Filatrella, G. Rotoli, and R. D. Parmentier, "Phase locking of fluxon oscillations in long Josephson tunnel junctions with surface losses", Phys. Lett. A $\underline{148}$, 122 (1990).

8. M. Salerno and M. R. Samuelsen, "Different microwave couplings to long Josephson junctions phase locked at a zero field step", this volume.

9. O. H. Olsen, N. F. Pedersen, M. R. Samuelsen, H. Svensmark, and D. Welner, "Perturbation treatment of boundary conditions for fluxon motion in long Josephson junctions", Phys. Rev. B $\underline{33}$, 168 (1986).

10. N. F. Pedersen and A. Davidson, "Phase locking of long Josephson junctions", Phys. Rev. B $\underline{41}$, 178 (1990).

11. H. G. Schuster, Deterministic Chaos (Physik-Verlag, Weinheim, 1984), Chap. 3.

12. G. Rotoli and G. Filatrella, "Chaotic dynamics in the map model of fluxon propagation in long Josephson junctions" (preprint, 1990).

13. R. L. Kautz, "Chaos in Josephson circuits", IEEE Trans. Magnetics $\underline{19}$, 465 (1983).

14. M. Salerno, "Phase-locking chaos in long Josephson junctions", Phys. Lett. A $\underline{144}$, 453 (1990).

15. R. Monaco, S. Pagano, and G. Costabile, "Superradiant emission from an array of long Josephson junctions", Phys. Lett. A $\underline{131}$, 122 (1988).

16. J. C. Fernandez, R. Grauer, K. Pinnow, and G. Reinisch, "Phase-locked dynamical regimes to an external microwave field in a long, unbiased Josephson junction", Phys. Lett. A $\underline{145}$, 333 (1990).

17. J. C. Fernandez, G. Filatrella, and G. Reinisch, "Comparison between electric and magnetic rf drive in long Josephson junctions", Phys. Lett. A (in press).

18. M. Cirillo, "Chaos and phase lock in long Josephson junctions", J. Appl. Phys. $\underline{60}$, 338 (1986).

19. M. Cirillo, "Dynamics of a non-autonomous sine-Gordon system modeling long Josephson junction", J. Appl. Phys. (in press).

20. R. D. Parmentier (unpublished notes, 1990).

21. F. If, P. L. Christiansen, R. D. Parmentier, O. Skovgaard, and M. P. Soerensen, "Simulation studies of radiation linewidth in circular Josephson-junction oscillators", Phys. Rev. B $\underline{32}$, 1512 (1985).

22. N. Minorsky, Nonlinear Oscillations (Van Nostrand, New York, 1962), Chap. 18.

23. T. F. Finnegan and S. Wahlsten, "Microwave emission from coupled Josephson junctions", in: Low Temperature Physics - LT-13, Vol. 3, K. D. Timmerhaus, W. J. O'Sullivan, and E. F. Hammel, eds. (Plenum, New York, 1974), pp. 272-275.

24. M. Cirillo and F. L. Lloyd, "Phase lock of a long Josephson junction to an external microwave source", J. Appl. Phys. $\underline{61}$, 2581 (1987).

25. S. Pagano, R. Monaco, and G. Costabile, "Microwave oscillator using arrays of long Josephson junctions", IEEE Trans. Magn. $\underline{25}$, 1080 (1989).

26. T. Holst, J. Bindslev Hansen, N. Grønbech-Jensen, and J. A. Blackburn, "Phase locking between Josephson soliton oscillators", Phys. Rev. B $\underline{42}$, 127 (1990).

27. N. Grønbech-Jensen, R. D. Parmentier, and N. F. Pedersen, "Phase-locking of fluxon oscillations in long Josephson junctions coupled through a resonator", Phys. Lett. A 142, 427 (1989).

28. N. Grønbech-Jensen, N. F. Pedersen, A. Davidson, and R. D. Parmentier, "Numerical study of long Josephson junctions coupled to a high-Q cavity", Phys. Rev. B 42, 6035 (1990).

29. G. Filatrella, N. Grønbech-Jensen, M. Salerno, and N. F. Pedersen, (unpublished notes, 1990).

30. N. Grønbech-Jensen, M. R. Samuelsen, P. S. Lomdahl, and J. A. Blackburn, "Bunched soliton states in weakly coupled sine-Gordon systems", Phys. Rev. B 42, 3976 (1990).

31. R. Monaco and G. Costabile (unpublished notes, 1990).

DIFFERENT MICROWAVE COUPLINGS TO LONG JOSEPHSON

JUNCTIONS PHASE LOCKED AT A ZERO FIELD STEP

M. Salerno

Department of Theoretical Physics
University of Salerno
I-84100 Salerno, Italy

M.R. Samuelsen

Physics Laboratory I
The Technical University of Denmark
DK-2800 Lyngby, Denmark

INTRODUCTION

When a long Josephson junction oscillator is phase locked to an external microwave field, vertical steps appear in the IV-curve at voltages corresponding to fluxon frequencies ω_F which are rationally related to the external driving frequency ω by

$$\omega_F = \frac{\omega}{M} \tag{1}$$

where M is an integer. This phenomena has been observed experimentally[1,2] and analysed as fluxon motion perturbed by the applied microwave field[3,4]. The main result is that the step hight (the locking range) increases linear with the microwave field amplitude as also more detailed calculations from the model shows[5].

An important question is how the microwaves couples into the Josephson junction. There are two distinct extreme ways. One is that the microwaves enters the junction through the wires. This is called electric coupling[6]. The other is that the microwave magnetic field acts as an external field. This is called magnetic coupling[6]. The difference between the two types of coupling is that in the first (electric) the fluxon at a given time would get the same energy input from the microwave field during the reflection in the two ends of the Josephson junction, while in the other (magnetic) the fluxon at a given instant would receive opposite energy input from the microwave field in the two ends. Another coupling which is in the middle of the electric and magnetic coupling is one in which the microwave field only enters through one end of the junction. That could be the case if the microwaves are lead to one end of an overlap tunnel junction by a microwave strip line.

In order to cope with all these possibilities we investigate a general coupling model allowing different amplitudes of the microwave field in the two ends of the junction together with a phase difference between the fields.

The result of this investigation is first of all the linear relation between locking

Nonlinear Superconductive Electronics and Josephson Devices
Edited by G. Costabile *et al.*, Plenum Press, New York, 1991

range and microwave field amplitude but also that all even M steps have the same size as will all the odd M steps. The ratio between the even and odd step size depends on the coupling.

THEORY

The dynamics of a long Josephson tunnel junction is assumed to be governed by the perturbed sine-Gordon equation[7]

$$\Phi_{xx} - \Phi_{tt} = \sin\Phi + \eta + \alpha\Phi_t. \tag{2}$$

Here Φ is the quantum mechanical phase difference between the two films. The spatial variable x is measured in units of the Josephson penetration depth λ_J and the time t in units of the reciprocal plasma frequency ω_p^{-1}. α is the loss parameter and η is the normalized uniform part of the bias current perpendicular to the length of the junction (the overlap part). The bias current through the ends of the junction (the inline part) enters through the boundary conditions[8]

$$-\Phi_x(0,t) = \kappa_1 \quad \text{and} \quad \Phi_x(l,t) = \kappa_2, \tag{3}$$

where κ_1 and κ_2 are the normalized bias current through the ends of the junction. l is the normalized length of the junction.

The normalized current through the tunnel junction is given by[8]

$$i = \kappa_1 + \kappa_2 + \eta l. \tag{4}$$

On a zerofield step the solution to the problem will be fluxons moving forth and back along the junction. On the first zero field step a single fluxon is shuttling. For the single fluxon the motion can be determined from a relativistic equation of motion with the current acting as a force. The average velocity u can be determined by a balance between the work done per period and the energy loss per period:

$$\frac{\kappa_1 + \kappa_2 + \eta l}{4\alpha l}\pi = u\gamma(u), \quad \text{where} \quad \gamma(u) = \frac{1}{\sqrt{1-u^2}}. \tag{5}$$

Here η is assumed to be constant ($\eta^{ac} = 0$) and κ_1 and κ_2 are the values at the times when the fluxon is reflected at the ends. An ac part of η would only give a second order contribution to the phase locking range. The average velocity is related to the fluxon frequency ω_F through

$$u = \frac{2l\omega_F}{2\pi}. \tag{6}$$

The normalized dc voltage v_{dc} is related to ω_F through

$$v_{dc} = 2\omega_F. \tag{7}$$

The voltage v_{dc} is also the normalized Josephson frequency.

If the junction is biased on a phase locked step on the zero field step, the average velocity is fixed to a value given by Eqs. (6) and (1). The reason for the phase locking is that Eq. (5) with a fixed u can be fulfilled for different dc current i_{dc} because of the ac part of $\kappa_{1,2}$. Assuming that the fluxon is reflected from end 1 at the time t_1 and from end 2 at the time t_2 we will with different ac amplitudes in the two ends and a phase difference ϕ have

$$\kappa_1 = \kappa^{dc} + \kappa_1^{ac} \cos(\omega t_1) \tag{8}$$

and

$$\kappa_2 = \kappa^{dc} + \kappa_2^{ac} \cos(\omega t_2 + \phi). \tag{9}$$

The phase locking condition can now be expressed as

$$t_2 = t_1 + M \frac{\pi}{\omega} + \Delta T. \tag{10}$$

The difference in the time of flight for the fluxon in the two directions $2\Delta T$ will be first order in $\kappa_{1,2}^{ac}$ and therefore yields second order contributions which we neglect. Collecting all these conditions into Eq. (4) gives

$$\Delta i = \sqrt{(\kappa_1^{ac})^2 + (\kappa_2^{ac})^2 + (-1)^M 2\kappa_1^{ac}\kappa_2^{ac}\cos\phi} \cdot \cos(\omega t_1 + \theta) \tag{11}$$

with

$$\tan\theta = \frac{(-1)^M \kappa_2^{ac}\sin\phi}{\kappa_1^{ac} + (-1)^M \kappa_2^{ac}\cos\phi}. \tag{12}$$

Due to the arbitrary phase ωt_1 in Eq. (11) the step hight (locking range) is given by

$$\Delta i_{max} = \sqrt{(\kappa_1^{ac})^2 + (\kappa_2^{ac})^2 + (-1)^M 2\kappa_1^{ac}\kappa_2^{ac}\cos\phi}. \tag{13}$$

We note first that the step hights are proportional to the microwave field amplitude and that all even M steps and all odd M steps have the same magnitude.

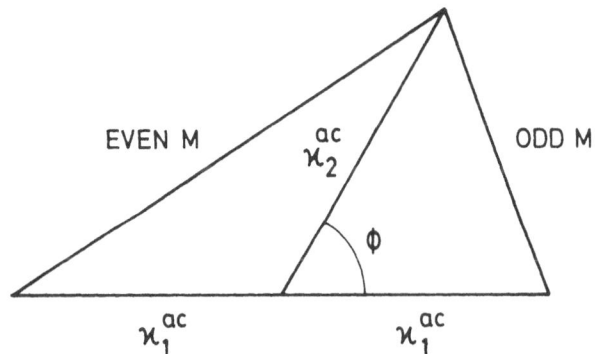

FIG. 1 The simple geometrical interpretation of Eq. (13). If κ_1^{ac} and κ_2^{ac} are two sides of a triangle then the third will be the size of the steps.

The expression for the locking range Eq. (13) can be given a simple geometrical interpretation as shown in Fig. 1. κ_1^{ac} and κ_2^{ac} are two sides of a triangle. If the angle between the two sides is ϕ the third side will be the size of the odd M steps. If the angle is $\pi - \phi$ the third side will be the size of the even M steps. When the phase angle ϕ is 0 or π the step sizes will be $|\kappa_1 \pm \kappa_2|$.

NUMERICAL EXPERIMENTS

In order 'to check the approximations made we have performed numerical experiments solving the relativistic equation of motion for the fluxon resonating in the Josephson tunnel junction by the same method as used in Refs. 3 and 4. We expect the deviation between our results and a numerical experiment to be largest for an inline junction. Therefore all the results reported on here are on inline junctions ($\eta=0$). The parameters are: length $l=16$, loss parameter $\alpha=0.05$, and applied frequency $\omega=1.5$.

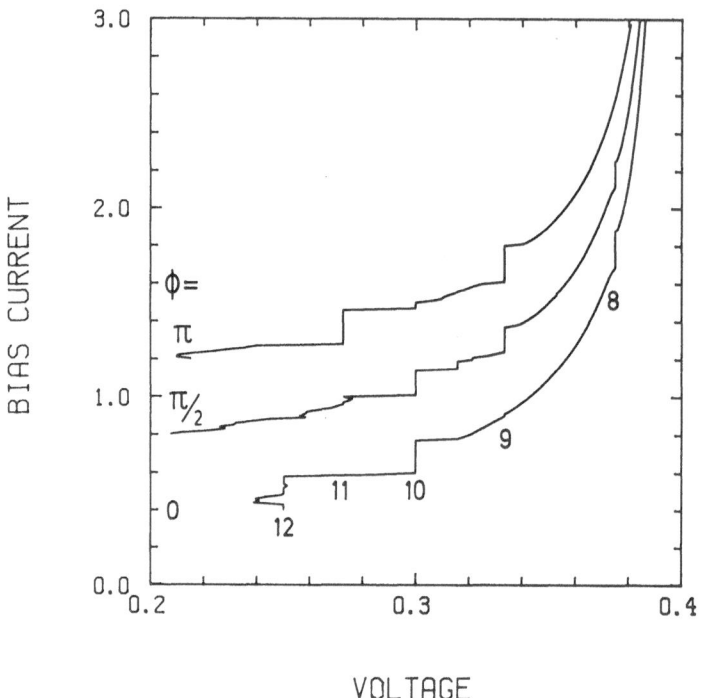

FIG. 2 Current-voltage characteristic of a inline junction with the same amplitude of the ac field in the two ends. $\kappa_1^{ac}=\kappa_2^{ac}=0.1$. Shown is κ^{dc} versus $v_{dc}=2\omega_F$ for different values of the phase angle ϕ. $\phi=0$ corresponds to electric coupling and $\phi=\pi$ to magnetic coupling. The integers indicate the number of the substep M. The curves are displaced 0.4 vertically for clarity.

First we calculate the IV-curve with the same amplitudes of the microwave field in the two ends of the junction and varies the phase difference ϕ. The result is shown in Fig. 2 for $\kappa_1^{ac}=\kappa_2^{ac}=0.1$. We notice that for $\phi=0$, the electric coupling, we have only even M steps all with the same size and for $\phi=\pi$, the magnetic coupling, we have only odd M steps also all with the same size. For $\phi=\frac{\pi}{2}$ all steps have the same size. All in accordance with Eq. (13).

Second we calculate the IV-curve if the amplitudes of the microwave fields are different in the two ends of the junction and varies the phase difference ϕ. The result is shown in Fig. 3 for $\kappa_1^{ac}=0.15$ and $\kappa_2^{ac}=0.05$. We notice that for $\phi=0$ the even M steps are larger than the odd M steps and for $\phi=\pi$ it is vice versa. The step sizes are $\kappa_1^{ac} \pm \kappa_2^{ac}$. For $\phi=\frac{\pi}{2}$ again all steps have the same size

$$\sqrt{(\kappa_1^{ac})^2 + (\kappa_2^{ac})^2}.$$

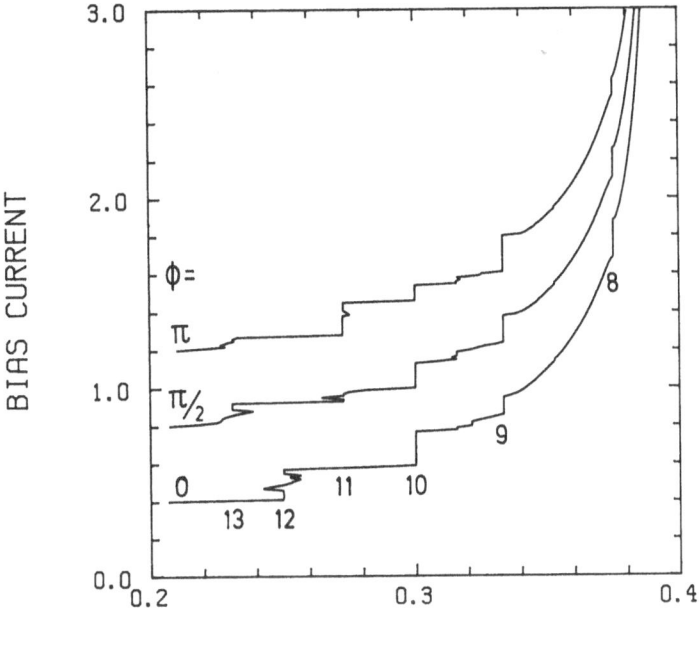

FIG. 3 Current-voltage characteristic of an inline junction with different amplitudes of the ac field in the two ends. $\kappa_1^{ac}=0.15$ and $\kappa_2^{ac}=0.05$. Shown is κ^{dc} versus $v_{dc} = 2\omega_F$ for different values of the phase angle ϕ. The integers indicate the number of the substep M. The curves are displaced 0.4 vertically for clarity.

Both in Fig. 2 and Fig. 3 we note small second order steps of the kind $\frac{p}{q} \cdot M$ where p and q are integers and we note instabilities especially for small currents where the steps overlap. Also it should be noted that instabilities apparently do not influence the size of the step.

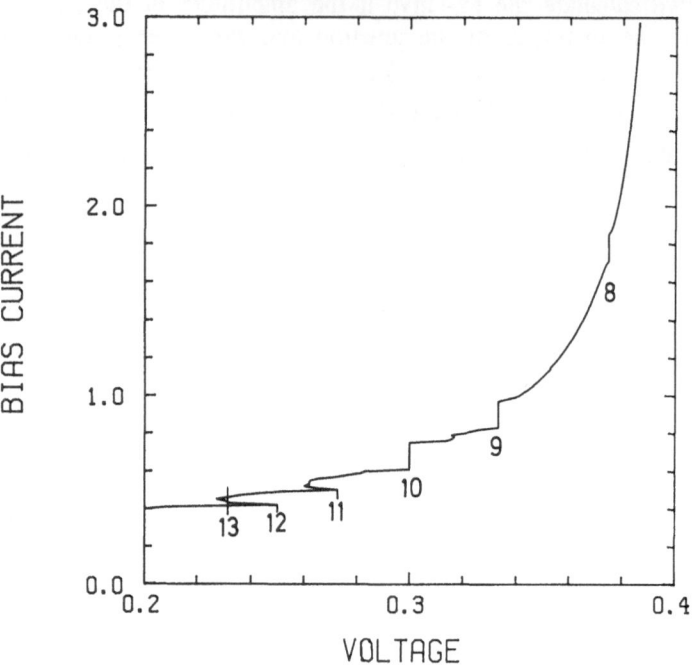

FIG. 4 Current-voltage characteristic of an inline junction with ac field only in one end. $\kappa_1^{ac}=0.15$ and $\kappa_2^{ac}=0$. Shown is κ^{dc} versus $v_{dc}=2\omega_F$ for different values of the phase angle ϕ. The integers indicate the number of the substep M.

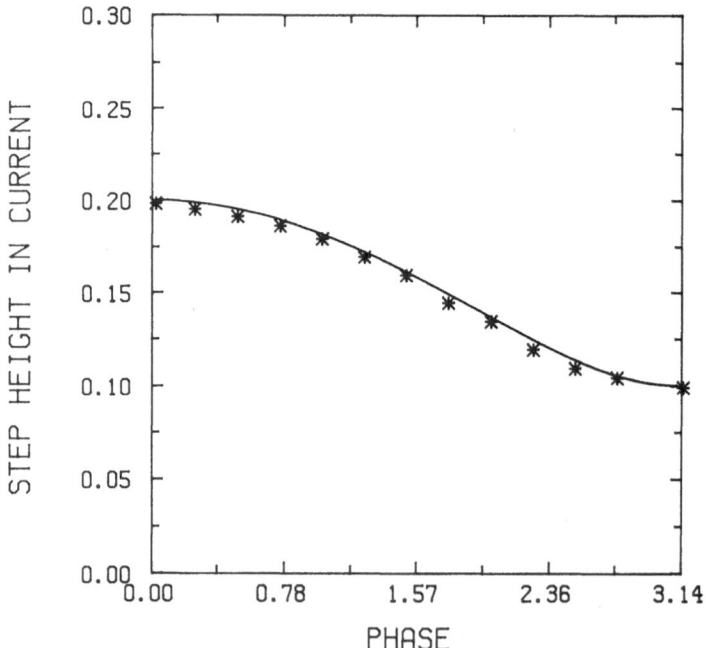

FIG. 5 The current step heights versus phase angle ϕ from Eq. (13) for $\kappa_1^{ac}=0.15$, $\kappa_2^{ac}=0.05$. Shown is Δi_{max} versus ϕ for the M=8 step of Fig. 3. The curve is the expression Eq. (13) while the points are the results of the numerical experiment.

Third we calculate the IV-curve with microwave field only in one end of the junction. The result is shown in Fig. 4 for $\kappa_1^{ac}=0.15$ and $\kappa_2^{ac}=0$. We notice that all steps have the same size.

This case which is right between the electric coupling and the magnetic coupling has the same qualitative appearance as $\phi = \frac{\pi}{2}$ independent of the microwave field amplitudes in the two ends. Instabilities appears to be more pronounced in these cases.

We have tried to check the linear dependence of the locking range on the microwave amplitude of Eq. (13) but not been able to detect any deviations. Also the phase dependence of the step hight has been checked. An example is shown in Fig. 5 where we show the phase dependence of the $M=8$ step of Fig. 3. We notice again the full agreement between Eq. (13) and the numerical experiment.

DISCUSSION

We have investigated different mechanisms for microwave coupling into a Josephson tunnel junction theoretically. First we note that an overlap ac current alone ($\eta^{ac} \neq 0$) will yield no phase locking at all, only a shift proportional to $(\eta^{ac})^2$ of the IV-curve towards lower voltages. This would be an electric coupling.

Only inline type ac current fed of microwaves into the junction leads to a locking range which is linear in the microwave field amplitude. In order to cover all possible coupling mechanisms we have investigated a general inline type of ac current fed into the junction.

The result is given by Eq. (13) and Fig. 1 and is a linear step hight in microwave field amplitude and that all even M steps and all odd M steps have the same magnitude. The even M step hight relative to the odd M step hight depends on the relative microwave amplitudes in the two ends of the Josephson junction and the phase difference according to Eq. (13). These quantities depends on the geometry of the junction and microwave setup. Normally $\phi = 0$ or $\phi = \pi$ due to the large wave length of the microwaves.

In the numerical experiments we have not been able to find any deviation from the linear relationship of Eq. (13) between locking range and microwave field. We observe that the steps get unstable probably due to overlap of the steps and the instability always begins in the middle of the step[9]. This instability did not influence the step size. If the step exists the size were given by Eq. (13) even if instabilities were observed on the step.

REFERENCES

1   M. Cirillo and Frances L. Lloyd, Phase lock of a long Josephson junction to an external microwave source, J. Appl. Phys. 61:2581 (1987).

2   G. Costabile, R. Monaco, and S. Pagano, Coherent oscillations of fluxons in arrays of long Josephson Junctions, p. 71 in "Stimulated Effects in Josephson Devices" M. Russo and G. Costabile, eds., World Scientific, Singapore (1988).

3   M. Salerno, M.S. Samuelsen, G. Filatrella, S. Pagano, and R.D. Parmentier, A simple map describing phase-locking of fluxon oscillations in long Josephson tunnel junctions, Phys. Lett. 137A:75 (1989).

4   M. Salerno, M.S. Samuelsen, G. Filatrella, S. Pagano, and R.D. Parmentier, Microwave phase locking of Josephson-junction fluxon oscillators, Phys. Rev. B41:6641 (1900).

5   N.F. Pedersen and A. Davidson, Phase locking of long Josephson junctions, Phys. Rev. B41:178 (1990).

6   Jhy-Jiun Chang, Theory of the microwave-soliton or antisoliton interaction in Josephson tunnel junctions, Phys. Rev. **B34**:6137 (1986).

7   D.W. McLaughlin and A.C. Scott, Perturbation analysis of fluxon dynamics, Phys. Rev. **A18**:1652 (1978).

8   O.H. Olsen, N.F. Pedersen, M.R. Samuelsen, H. Svensmark, and D. Welner, Perturbation treatment of boundary conditions for fluxon motion in long Josephson junctions, Phys. Rev. **B33**:168 (1986).

9   Mario Salerno, Phase-locking chaos in long Josephson junctions, Phys. Lett. **144A**:453 (1990).

SMALL JOSEPHSON JUNCTIONS STRONGLY
COUPLED TO MICROSTRIP RESONATORS

H. Dalsgaard Jensen

Danish Institute of Fundamental Metrology,
DK-2800 Lyngby, Denmark

A. Larsen*

Physikalisch-Technische Bundesanstalt,
D-3300 Braunschweig, Federal Republic of Germany

J. Mygind

Physics Laboratory I, Technical University of Denmark,
DK-2800 Lyngby, Denmark

INTRODUCTION

The Josephson junction is a highly nonlinear element and the interaction with its embedding circuitry have been studied extensively over more than two decades; see Ref. 1 and references therein.

High quality Josephson tunnel junctions consisting of two superconducting films separated by a few nanometer thick tunneling barrier are now routinely being fabricated using modern thin-film technology. The understanding of the nature of a close coupling to such elements inherently shunted by the self-capacitance of the tunneling structure is imperative for their use in fast electronics. In small tunnel junctions (i.e. with spatial dimensions shorter than λ_J, the Josephson penetration depth) narrow-band coupling can be achieved by resonating out the capacitance by an external low-loss inductive load or, at the so-called Josephson plasma frequency, by the inductance of the tunneling pair current. The internal electromagnetic (Fiske) resonances allow for a magnetically tuneable narrow-band coupling. In junctions with dimensions larger than λ_J also other internal (soliton/fluxon) resonances may be used.

In this paper we consider the interaction of a Josephson junction with a relatively long section of a low-loss superconducting transmission line. Due to its low impedance compared to free space the line is electrically open-ended along the periphery. Such a line may at certain frequencies provide an inductive load able to compensate the junction capacitance, and loaded by the full admittance of the junction it constitutes a multi-mode (nearly full-wave) resonant system which allows us to achieve an exceptionally close coupling.

Nonlinear Superconductive Electronics and Josephson Devices
Edited by G. Costabile *et al.*, Plenum Press, New York, 1991

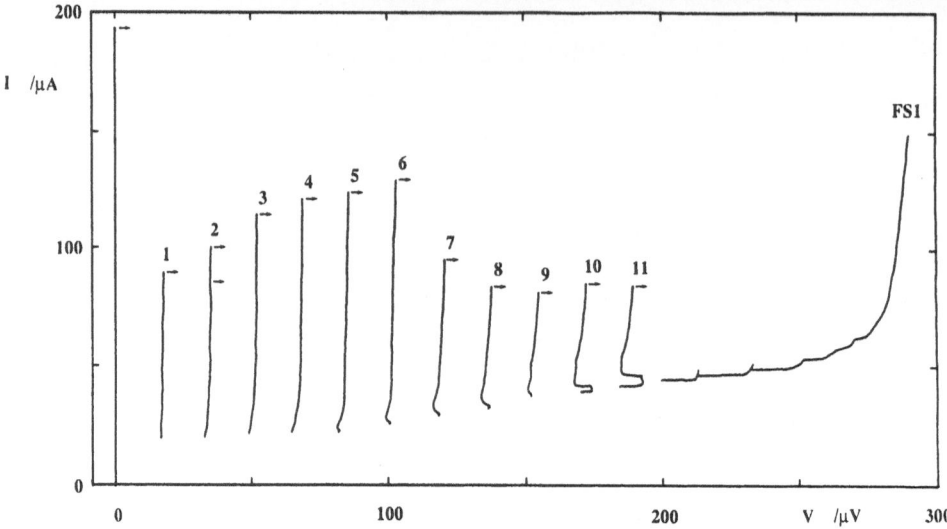

Figure 1. *Typical recorded dc IV-characteristic (sample H5-2A, $\ell = 5.5$ mm, $w = 200$ μm) showing the first mode step, MS1 and a series of parametrically generated mode steps, PMSk. The temperature was 4.2 K and an external magnetic field, $B = 272$ μT was applied. The arrows indicate the current at which the junction switches from the step. The structure to the right is the first Fiske step at 295 μV. The position of the foot-like structures visible on higher numbered PMSk agree with the voltages of the corresponding mode steps, MSn.*

In this limit we observe a series of large self-induced steps (PMSk, parametrically generated mode steps) excited by parametric pumping of the fundamental resonant mode by the kth subharmonic of the Josephson frequency. An example of a low-voltage IV-characteristic exhibiting such steps is shown in Fig. 1. The self-induced mode steps (MS), known from Josephson junctions interacting with resonant transmission line modes in the weak coupling limit,[2-5] are also observed. Our measurements show that in the multi-mode system with stronger coupling and large harmonic overlap between the mode resonances the MS seem to become unstable and to be dominated by the PMS. Parametrically generated steps in a single resonance system have been observed in a resistively shunted SQUID.[6] This type of steps are called harmonic steps in Ref. 1.

There are several possible applications of the strongly coupled junction-resonator system. Most obvious is to use the system as a narrow-band self-oscillating dc-biased microwave source. The available power is, as for all Josephson oscillators, approximately given by the product of the dc voltage and current at the bias point on the step. The generator impedance, however, is that of the resonator. What appears particularly attractive is that more such systems via next neighbour interaction can be coupled together in mutually phaselocked arrays. The usable power of a phaselocked array oscillator is proportional to the square of the number of systems, while the linewidth is inversely proportional to this number.

Previously, mutual phaselocking has been obtained in arrays of small tunnel junctions using an external impedance (eventually a common resonator) in parallel with the whole array[7,8]. Coupled arrays are advantageous also as microwave detectors, mixers and amplifiers due to their narrow linewidth and large saturation power. A straight forward application of the junction-resonator system is to determine *in situ* the high frequency propagation constants of thin-film transmission lines by the relatively simple dc measurement of the position and size of the MS and/or PMS in small test samples placed on the wafer.[9,10]

Finally the system is ideally suited for the basic study of nonlinear dynamics. Unlike a bare junction the system is of third or higher order, so bifurcations and chaos may occur.[11]

Stability analysis of single as well as arrays of strongly coupled systems is important especially with the future high current density submicron tunnel junctions where it will be very difficult to avoid interactions with resonances in the embedding network.

Parts of the work presented here has been submitted for publication elsewhere.[12]

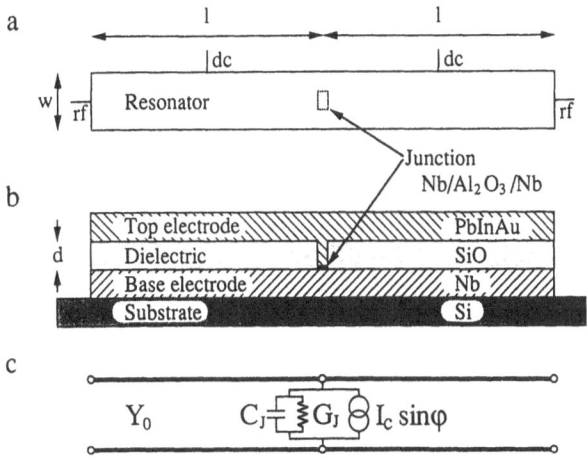

Figure 2. *(a) Top view, (b) side view, and (c) transmission line model of a microstrip resonator with a small Josephson tunnel junction. The width, half length and thickness of dielectric of the microstrip are w, ℓ, and d, respectively. The dimensions are out of scale.*

THEORY

In Fig. 2 is shown a sketch of a Josephson tunnel junction integrated in a microstrip resonator. The half length ℓ and the width w of the resonator are defined in the figure. The resonator consists of two superconducting thin-film electrodes separated by an insulating dielectric with thickness d. The rectangular junction is situated in the center of the resonator. The high impedance dc and rf connections are designed to interfere weakly with the resonator.

Also shown in the figure is a microwave model of the system. The resonator is an open-ended transmission line with characteristic admittance Y_0 and propagation constant $\gamma = \alpha + j\beta$. The junction is modelled by a parallel connection of the active element, the supercurrent channel $I_S = I_c \sin\varphi$, in the following called the bare Josephson element, and the passive admittance $Y_J(f) = G_J + j\,2\pi f C_J$, representing the junction capacitance C_J and the (in general nonlinear) quasi-particle conductance G_J. When dc-biased at a voltage V the corresponding Josephson oscillation has the fundamental frequency $f_J = (2e/h)V$ where $-e$ is the electron charge and h is Plancks constant. For the moment we neglect the admittance, Y_S, of the bare Josephson element as well as any admittance associated with the change in the current distribution in the films near the junction periphery. The former is valid for the uncoupled junction at frequencies well above its plasma frequency, f_p. We note that f_p in the measurements reported here is 3–5 times larger than the fundamental mode resonance frequency. However, as discussed below, Y_S seems to play a minor role.

The admittance $Y_e = 2Y_0 \tanh \gamma\ell$ of the two open-ended microstrip sections transformed to the position of the Josephson element can be calculated using ordinary transmission line theory.

The total admittance Y_Σ seen by the bare Josephson element then is

$$Y_\Sigma = Y_J + Y_e = Y_J + 2Y_0 \tanh \gamma\ell \tag{1}$$

and resonance occurs when the imaginary part equals zero; Im $Y_\Sigma = 0$. The corresponding current singularities, which show up in the dc IV-characteristic of the junction even if it is relatively weakly coupled to a resonator, we call mode steps (MS) in the following.

Assuming low microstrip losses, i.e. $\alpha\ell \ll 1$, Y_Σ can be approximated by

$$Y_\Sigma = G_J + G_e + j\left(2\pi f C_J + 2Y_0 \tan \pi \frac{f}{f_e}\right) \tag{2}$$

where $G_e = 2Y_0\,\alpha\ell(1 + \tan^2 \pi(f/f_e))$ is the real part of the admittance of the external resonator, and f_e is the resonance frequency of the unloaded resonator (without the junction), when one full wavelength of the oscillation equals the length of the resonator. f_e is given by

$$f_e = \frac{c_0}{2\ell \sqrt{\epsilon_r \chi_r}} \tag{3}$$

where c_0 is the light velocity in vacuum, ϵ_r the relative dielectric constant of the insulator, and $\chi_r = (d + \lambda_a + \lambda_b)/d$ accounts for the penetration of magnetic fields in the resonator electrodes. The penetration depths λ_a and λ_b should be corrected for the finite thickness the two electrodes.

As can be derived from Eq. (2), the microstrip sections will act as an inductance for frequencies between $(n - 1/2)\,f_e$ and $n\,f_e$, where n is an integer, so resonance only occurs in these intervals.

For the resonators examined here w is of the order some hundred microns, while d is about a factor of thousand smaller. Hence the effective dielectric constant ϵ_{eff} of the transmission line is equal to ϵ_r. Likewise, Y_0 is given by the asymptotic value

$$Y_0 = \frac{1}{120\pi} \sqrt{\frac{\epsilon_r}{\chi_r}} \frac{w}{d} \tag{4}$$

With the dimensions used here, Y_0 is typically a few Ω^{-1}. At least at the lowest resonance frequencies the admittance of the junction is much smaller than Y_0, and in

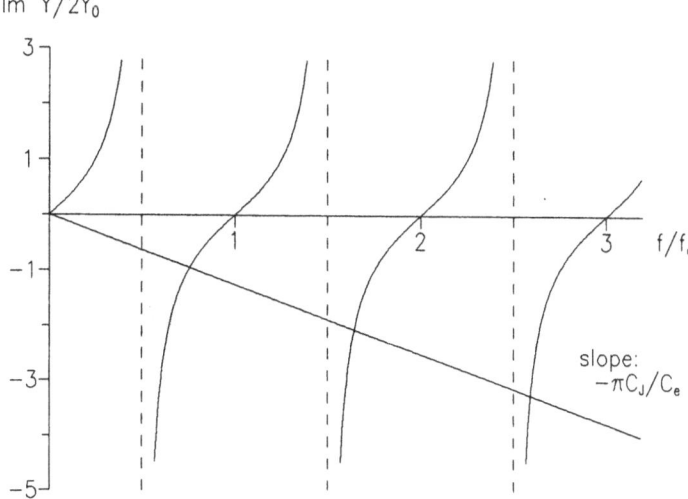

Figure 3. *Graphic solution of Eq. (5). The slope of the straight line is $-\pi C_J/C_e$. The intersection points determine the mode resonance frequencies f_n. From the figure it is obvious that $n\,f_1 < f_n$ for $n = 2, 3, \ldots$*

order to fulfill the resonance condition the zero admittances at the open ends of the microstrip therefore has to be transformed to a low admittance at the junction position. This configuration, with low admittance points at the ends and in the middle of the resonator, corresponds to a mode pattern with an integer number n of full wavelengths in the resonator. At higher frequencies $|\mathrm{Im}\,Y_e|$ has to be larger than Y_0 to meet the resonance condition, and the resonances are moved towards modes with $(n - 1/2)$ wavelengths.

The resonance condition (low-loss limit) can be written as

$$\tan \pi \frac{f}{f_e} = -\pi \frac{C_J}{C_e} \frac{f}{f_e} \tag{5}$$

where $C_e = \epsilon_r \epsilon_o (2\ell w)/d$ is the total capacitance of the microstrip resonator, and ϵ_o is the vacuum permittivity. The solutions to this equation are called f_n; the resonance frequency of the nth mode of the system. In the weak coupling limit the corresponding current singularities in the dc IV-characteristic will be the nth mode step (MSn).

In Fig. 3 the graphic solution of Eq. (5) is shown. Resonance occurs at frequencies where the tangent curve intersects the straight line. The slope of this line is $-\pi C_J/C_e$, thus, a larger C_J, or for the resonator (with given films and dielectric) larger d or smaller w, the mode resonance frequencies decrease. Here we have disregarded the dependence of d on χ_r. This is a good approximation for $d > 2(\lambda_a + \lambda_b)$. Generally, since ℓ enters in f_e longer resonators have lower resonance frequencies.

As described above, the first mode resonance frequencies are situated just below $n\,f_e$, while at higher frequencies the resonances lie closer to $(n - 1/2)f_e$. It is easily verified, solving Eq. (5) that $f_n > n f_1$ for $n = 2, 3, \ldots$ etc. The ratio $f_n/(n f_1)$ depends on the ratio C_J/C_e, i.e. the slope of the line in Fig. 3, and increases with C_J and d, and decreases with w and ℓ; the harmonics of the first (fundamental) mode frequency are closer to the frequencies of the mode resonances for long and wide resonators. We will return to the importance of the harmonic overlap at the end of this section.

Single Resonance Theory

Self-induced steps may be regarded as a sort of regenerative rf-induced steps. At resonance the field in the resonator is large, and the junction receives the resonator signal as driven from a microwave generator, which in turn is powered by the junction itself. This argument was used to explain the form and size of self-induced steps[1,13,14] in the much simpler system of a small Josephson junction coupled to a circuit with only a single resonance. The important assumption is that harmonics of the Josephson oscillations are negligible when operating near resonance. For use in the following we shall refer to these steps as single resonator-induced steps (RS, for brevity) and briefly review the main results of this *single resonance theory*.

Assuming a single frequency expression for the junction phase waveform

$$\varphi(t) = \varphi_0 + 2\pi f_J t + a \sin 2\pi f_J t \qquad (6)$$

where φ_0 is a constant, and f_J is the Josephson oscillation frequency. Furthermore, linearizing the loading admittance, Y_Σ, around the resonance frequency f_r determined by Im $Y_\Sigma(f_r) = 0$, gives

$$Y_\Sigma(f) \approx \text{Re } Y_\Sigma(f_r)(1 + j\,\xi), \quad f \sim f_r \qquad (7)$$

where $\xi = (f - f_r) \cdot (d \text{Im } Y_\Sigma(f_r)/df)/\text{Re } Y_\Sigma(f_r)$ is the normalized detuning.

The parameters a and φ_0 in Eq. (6) are now determined by a Harmonic Balance equation for the currents at dc and at frequency f_J in the Josephson element and the loading admittance.

Introducing the coupling parameter z defined as

$$z = \frac{I_c}{V_r \text{ Re } Y_\Sigma} \qquad (8)$$

where I_c is the critical current of the junction, and $V_r = (h/2e) f_r$ is the dc voltage corresponding to the resonance frequency f_r, it is possible to derive an expression for the step magnitude

$$I_s = I_c |J_n(a)| = \frac{a^2}{2z} \qquad (9)$$

Figures 12.5 and 12.6 in Ref. 1 ($m=1$) show the form and size of the RS as function of the relative detuning ξ/z and the coupling parameter z, respectively. A critical value of coupling exists for $z = 2.92$ where the normalized step size reaches its maximum, $I_s/I_c = 0.58$. When z exceeds certain threshold values (38, 130, ...) the RS exists in new, separate current intervals.

Also the form of the RS changes as z crosses 2.92. At weaker coupling the dc voltage across the junction increases continuously with increasing bias current, looking much like an ordinary resonance curve. With stronger coupling the RS become more cusp-like and may even show a backward bending.

For a given step z can be changed by tuning I_c; in our measurements we apply an external magnetic field. This leaves the denominator in Eq. (8) virtually unchanged.

In the undercritical region the height of the RS will decrease rapidly with decreasing I_c. If, on the other hand, z is larger than 2.92, I_s/I_c will increase with decreasing I_c, and the decrement in I_s will be slower.

The single resonance theory also predicts parametric subharmonic excitation of the single resonator system.[1] When the junction is (current) biased, say, on the quasi-particle curve at a dc voltage equal to nV_r, the resonator may be parametrically excited. This results in parametrically induced resonator steps (PRS) at these voltages, too. However, the analysis (Figures 12.6, for $m = 2$, and 12.10 in Ref. 1) shows that only the second PRS is soft excited, i.e. it will appear without external interference. The other PRS are hard excited, and will appear only, if an oscillation with the particular frequency already exists with some minimum amplitude. As mentioned such so-called harmonic steps have already been observed.[6]

Multi-mode resonator

After this review we will return to the microstrip junction-resonator system with its multiple resonances. In analogy with the single resonance theory the coupling parameter for the nth mode may be given as

$$z_n = \frac{I_c}{V_n(G_J + G_e)} \tag{10}$$

the low-loss limit. $V_n = (h/2e) f_n$ is the dc voltage corresponding to the mode resonance frequency f_n. The quality factor Q_n of the nth mode can be found from Eq. (2).

$$Q_n = \pi \frac{f_n}{f_e} \frac{f_e C_J + G_e/(\alpha\ell)}{G_J + G_e} \tag{11}$$

The microstrip losses in G_e are all contained in the real part α of the propagation constant, where α is the sum of three terms: surface losses α_c, dielectric losses α_d, and radiation losses α_r.

Harmonic overlap

In the experiments we have chosen a long and wide resonator, corresponding to a small value of C_J/C_e. Here the first harmonics $n f_1$ of the fundamental resonance are closer to the nth mode frequency f_n than the half-power bandwidth of the mode resonances. The resonator, loaded by the junction passive admittance, therefore comprises many high-Q resonant modes being spaced sufficiently constant in frequency to allow for a considerable overlap of harmonics. Consequently, when biased on a given mode resonance a large number of harmonics of the Josephson oscillation are likely to be excited in the system and the basic assumption of the single resonance theory is invalidated.

The existence of finite amplitude harmonics makes an analysis a complex problem due to the nonlinearity of the junction, and to our knowledge no such multi-resonator theory exists. Recent calculations and measurements[15] on Josephson junctions with rf drive at both the fundamental and the second harmonic have shown rf-induced steps being larger than the usual Bessel function prediction. So, again regarding the electromagnetic field in the resonator as an rf generator, one could expect larger steps than predicted above for the resonator induced steps, especially if the resonator is long and wide. The form and size of both the MS and the PMS in the harmonic overlap case presumably depend strongly on the amplitudes and mutual phase relations of a multitude of harmonics.

Whether the system choses to excite the mode resonance at which it is actually biased, or to generate a PMS by pumping the fundamental mode resonance—or in principle any of the modes available—depend in a complex way on several parameters in the multi-mode system. However, as discussed below large harmonic overlap and strong coupling seem to favourize the latter option.

SAMPLE GEOMETRY AND FABRICATION

The samples used in the measurements are based on an all-niobium ($Nb/Al_2O_3/Nb$) sandwich Josephson junction integrated in the center of a superconducting microstrip resonator. As seen in Fig. 2, the base Nb-layer of the junction is extended to form the bottom electrode of the resonator. The top electrode is formed by a third lead-indium-gold (PbInAu) alloy film, which is connected to the upper Nb-layer of the junction through a window in the SiO-layer. This window is made smaller than the junction to prevent shorts between the base layer and the PbInAu film. The SiO-layer is the separating dielectric in the microstrip resonator. The bottom electrode of the resonator is 100 μm longer and wider than the top electrode.

The dc connections to the junction are formed in the bottom and top electrodes perpendicular to the resonator at the points marked dc in Fig. 2. These points lie halfway between the junction and the ends of the resonator. When the resonator contains an integer number of wavelengths, the admittance transformed from the open end will be very large at the position of the dc connections. Thus these will not disturb the resonator, as verified by microwave calculations. The rf connection are 10 μm wide lines extending from the ends of the resonator. The lines are 10–20 times narrower than the resonator and outside the bottom electrode they are themselves microstrip lines with a thick copper foil positioned 0.5 mm above the sample as groundplane. Due to the large impedance mismatch the load on the resonator from the rf connections can be neglected. A tapered microstrip transformer reduce the line impedance to 50 Ω to match the microstrip-coaxial launcher at the edge of the chip. The rf circuit includes a dc block made on the chip.

Eight 25.4 mm × 12.7 mm samples are made in each batch on a 3-inch diameter high resistivity Si wafer. In a given batch the junctions show less than 5% variation in I_c. Only the resonator dimensions differ between the samples on a wafer, using $\ell = 2.75, 4.0,$ and 5.5 mm, and $w = 100$ and 200 μm. The size of the rectangular junction is 20 μm × 30 μm.

MEASURING SYSTEM

The sample is mounted in a small sampleholder with a launcher connecting to the coaxial waveguide leading to the room temperature microwave setup. A thick copper foil provides the groundplane in the microstrip circuit leading from the rf connections on the superconducting resonator (see Fig. 2) to the launcher. The sampleholder is placed in the center of a magnetic coil with a separate battery powered current supply. The magnetic field is applied in the plane of the junction parallel to the long side of the resonator. In the measurements reported here the sampleholder is emersed directly in liquid helium contained in a 10 l superinsulated cryostat equipped with a full length double cryoperm magnetic shield. The temperature can be stabilized in the range 4.2–1.5 K by regulating the bath pressure.

The dc current bias can be supplied either from an electronic unit with sweep and various offset facilities, or from a manually operated purely resistive circuit. The latter is used for the linewidth measurements (see below) where the bias noise is a critical parameter. A specially designed low-noise and minimum drift instrumentation amplifier detects the dc voltage across the junction. For accurate measurements of the step voltage or in situations where the voltage amplifier is disconnected we record the frequency of the emitted Josephson oscillation. Two junctions can be biased independently and measured simultaneously. The absolute calibration of recorded dc IV-characteristics relies on precision resistors, and the actual position of rf steps induced by microwave power applied to the junction from a synthesized sweep generator. The above-mentioned sweep and offset facilities are used also when we record the magnetic field dependence of the critical current and the size of the various rf- and self-induced steps.

All sensitive electronic parts are powered from batteries and large efforts have been made to reduce interference from external noise sources. The wires leading to the junction are carefully filtered, twisted pairs. The importance of the stability of the three dc bias parameters (temperature T, applied magnetic field B, and bias current I) is illustrated by the measured (typical) frequency tuning rates of the Josephson oscillation when biased on one of the PMS: $\Delta f/\Delta T = -0.15$ MHz/mK, $\Delta f/\Delta B = 0.5$ MHz/μT, and $\Delta f/\Delta I = 2$ MHz/μA (with a dynamic resistance of about 10 mΩ), respectively.

In order to characterize the samples and as a diagnostic tool for eliminating noise sources we measure the escape rate from the zero-voltage state. Usually we collect 50.000 escape events in an automated data processing system. Typically we find an effective temperature a few tenths of a Kelvin above the bath temperature. The escape rate measurements allows us to correct for noise suppression of the critical current.

EXPERIMENTAL RESULTS

dc properties

Fig. 4 show the low voltage part of in the dc IV-characteristic of two junctions (H5-2A and H5-5A) from the same wafer. Similar characteristics have been obtained for all the other junctions on the wafer. The results reported in this paper are typical also for junctions from other wafers, however, to avoid confusion and in order to emphasize the influence from the resonator we have chosen to discuss only results measured on these two samples with nearly identical junctions.

All the junctions on the wafer were measured to have the same parameters, e.g. area 30×20 μm^2 and $I_c = 775$ μA at 4.2 K. Sample H5-2A has the longest and widest resonator on the wafer, $\ell = 5.5$ mm, $w = 200$ μm, while H5-5A has the shortest and narrowest one, $\ell = 2.75$ mm, $w = 100$ μm. The critical current is corrected ($\sim 3\%$) for noise suppression found from escape rate measurements as described above.

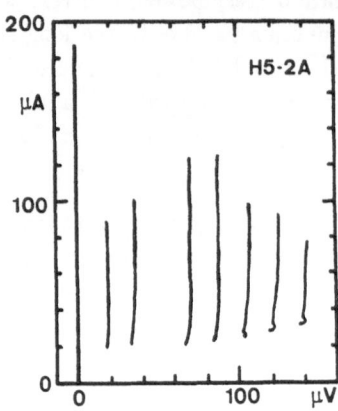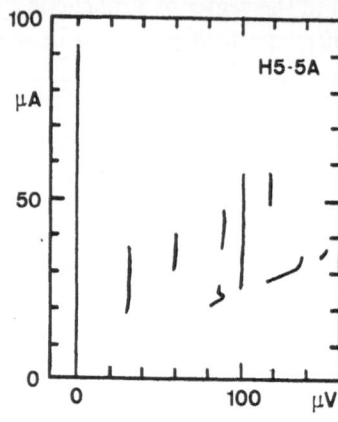

Figure 4. *Self-induced steps measured at 4.2 K on two samples with identical Josephson tunnel junctions: sample H5-2A, $\ell = 5.5$ mm, $w = 200$ μm, $B = 266$ μT, and sample H5-5A, $\ell = 2.75$ mm, $w = 100$ μm, $B = 295$ μT, respectively. Each picture is a reproduction of multiple exposure oscillographs taken with swept bias current.*

From the $I_c(B)$ minima and from the position of the Fiske step, at 295 μV, we derive the magnetic thickness, $d_J = 225$ nm, the capacitance $C_J = 28$ pF, and the Josephson penetration depth, $\lambda_J = 30$ μm at 4.2 K. The normalized length $L/\lambda_J = 1$ and the maximum plasma frequency $f_p = 46$ GHz. Corrected for the finite thickness (110nm) of the Nb-films and the Al-layer (20 nm) the value found for d_J is in accordance with the bulk value of the London penetration depth in niobium (86 nm).

In Fig. 4 the magnetic field was chosen give large steps, and clearly the steps are smaller for the H5-5A junction, both absolute and relative to I_c. Sample H5-2A has the longest and widest resonator of the two, so both the maximum coupling parameter and the contents of harmonics of the resonance oscillation frequency should be larger for this sample. On the dc IV-characteristic of junction H5-2A depicted in Fig. 4 the third step is missing. However, at other magnetic fields this step also appears.

The step structure extends to higher voltages as shown in Fig. 1. In sample H5-2A up to the first Fiske step at 295 μV. Actually we have in this sample occasionally observed steps also in the region between the first and second Fiske step, as well as a satellite structure with the same periodicity around the first Fiske step.

Biasing the junction on a step we may observe the radiation emitted. From the steps shown in Fig. 1 we observe radiation at a frequency corresponding to the fundamental mode of the loaded resonator system, 8.3 GHz. Hence, we conclude that all the steps are parametrically excited replica of the fundamental step.

Large signals are also radiated at the second harmonic of the fundamental mode frequency, i.e. at 16.8 GHz, on all steps. This further supports the presence of many large amplitude higher harmonics, and the invalidity of the single resonance theory. The foot-like structure visible at the bottom of each of the higher numbered PMSk agrees with the position of the corresponding ($n = k$) mode steps (MSn).

The frequency of the radiated signal from a step with fixed bias current can be tuned by varying the temperature and the applied magnetic field. At 8.4 GHz we find the same thermal tuning rate $\Delta f/\Delta T \approx -150$ kHz/μK on all steps, supporting the interpretation of a subharmonic pumping of the fundamental mode. The magnetic tuning rate at fixed bias current for PMSk increases linearly with k, as $\Delta f/\Delta B = (0.52 + k \times 0.13)$ MHz/ μT. We have no explanation for this dependence, but suspect that it is related to the change of bias current relative to the step size.

Since the single resonance theory does not apply to our experimental system with large amplitude harmonics of the Josephson oscillation a direct determination of z_n is not possible from the measured step form or size of the steps. In the following we device an alternative method to estimate the coupling parameter from measurements of the system losses using the junction as detector.

Mode coupling parameters and system losses

A small Josephson junction can be used as a very sensitive broadband microwave detector when biased at a fixed dc current on the quasi-particle curve in the vicinity of the superconducting gap. In the small signal limit the variation of the dc voltage across the junction is proportional to the power of the applied microwave signal. Fig. 5 shows voltage variation as function of frequency near the fundamental mode resonance for junction H5-2A. A constant amplitude microwave signal was supplied to the sample via the rf connection shown in Fig. 2. The quadratic power dependence has been carefully checked and the measurements have been carried out at different bias currents. Only the sensitivity changes with the bias point.

In the bias point (about 2.7 mV) the frequency of the Josephson oscillations is much larger than any system resonance frequency as well as the frequency of the applied microwave signal. Hence the admittance of the bare Josephson element can be neglected and the junction *in situ* measures the admittance of the resonator loaded by the passive admittance of the junction, Y_Σ in Eq. (2). This is strongly supported by the experimental fact that the curve shown in Fig. 5 is independent of applied magnetic field. The major source of error is the assumption of a constant power supplied by the high impedance rf connections at the end of the sample. In fact the detector response is folded with the frequency dependent transmission coefficient of the cables, isolators, etc., and a weak interference pattern superimposed on the resonance dip is visible in Fig. 5. The frequency at the maximum drop in voltage is $f_1 = 8.236$ GHz.

Biased at the top of MS1 with an applied magnetic field, $B = 272$ μT we observe Josephson radiation at 8.364 GHz. The discrepancy with the 8.236 GHz cannot be explained by magnetic tuning, and we propose two explanations:

(i) The highest frequency detected may come from a bias point being slightly below the top of the step. As the step is backward bending, similar to what we find in the single resonance theory,[1] external noise or thermal activation may prevent us from reaching the top of the step where the emitted frequency is 8.236 GHz.

(ii) Another possibility is the influence of the otherwise excluded admittance Y_S of the bare Josephson element. We can estimate this admittance at f_1 by including Y_S in Eq. (1). From the frequency difference at the top of the step we find that the pair current channel here acts as an inductance with $L_S = 190$ pH, which is ~ 450 times the minimum Josephson inductance, $L_J = \hbar/2eI_c = 0.425$ pH. This is in good agreement with our assumption of $Y_S \approx 0$.

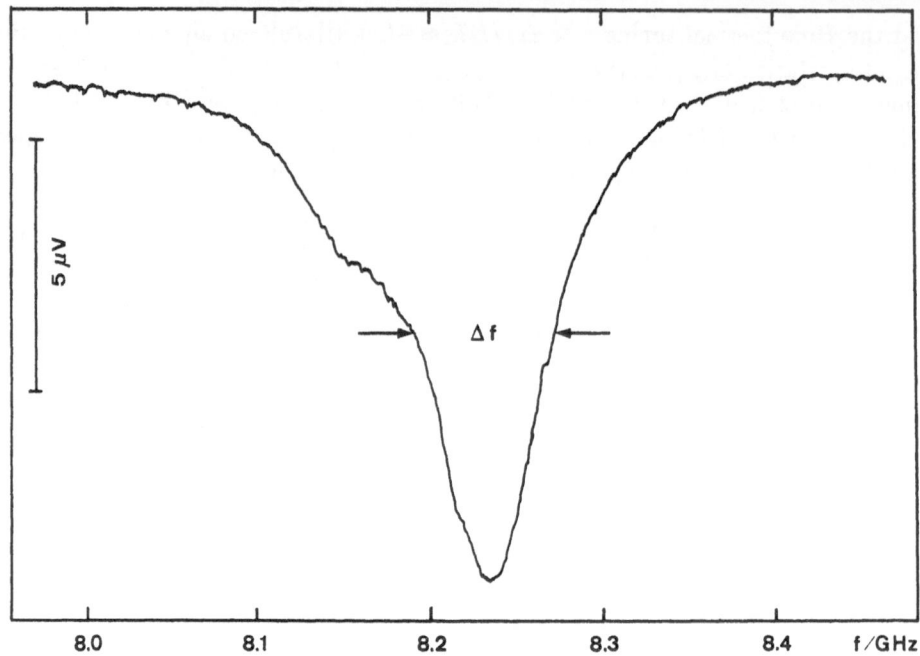

Figure 5. *The dc voltage response ΔV of the junction versus frequency of an applied microwave signal near the fundamental mode resonance frequency, f_1. The junction was used as a square-law detector biased just below the gap voltage, and a small swept microwave signal of constant power was supplied via the rf connections to the resonator. The quality factor Q_1 is evaluated from the frequency width Δf at the half power points.*

In other samples the top electrode of the resonator was designed to have a restriction at the junction, where the electrode is only as wide as the junction itself. Apart from the narrowing the resonator dimensions are $(\ell, w) = (4.0\ \text{mm}, 250\ \mu\text{m})$. In measurements with the junction as detector voltage drops at frequencies corresponding to half wavelength resonances were also measured. We believe the restriction decouples these modes from the junction. However, no steps were observed at the corresponding voltages.

SPECTRAL LINEWIDTH

The full half-power linewidth $\Delta \nu$ of the Josephson oscillations in a bias point $(I_{\text{dc}}, V_{\text{dc}})$ may be determined from the assumption that only thermal noise due to the quasi-particle conductance is present. Converting from current noise to frequency noise one readily obtains

$$\Delta \nu = 4\pi\, k_B T \left(\frac{2e}{h}\right)^2 \frac{R_d^2}{R_{\text{dc}}} \tag{12}$$

where R_d is the differential resistance and $R_{\text{dc}} = V_{\text{dc}}/I_{\text{dc}}$ is the dc resistance at the bias point, k_B is Boltzmann's constant and T the effective noise temperature.

The frequency and the linewidth $\Delta \nu$ of the emitted radiation with the junction biased on a step can be recorded on the spectrum analyzer. In Fig. 6 is shown $\Delta \nu$ in kHz versus R_d^2/R_{dc} in mΩ for sample H5-2A biased on different points along MS1. At

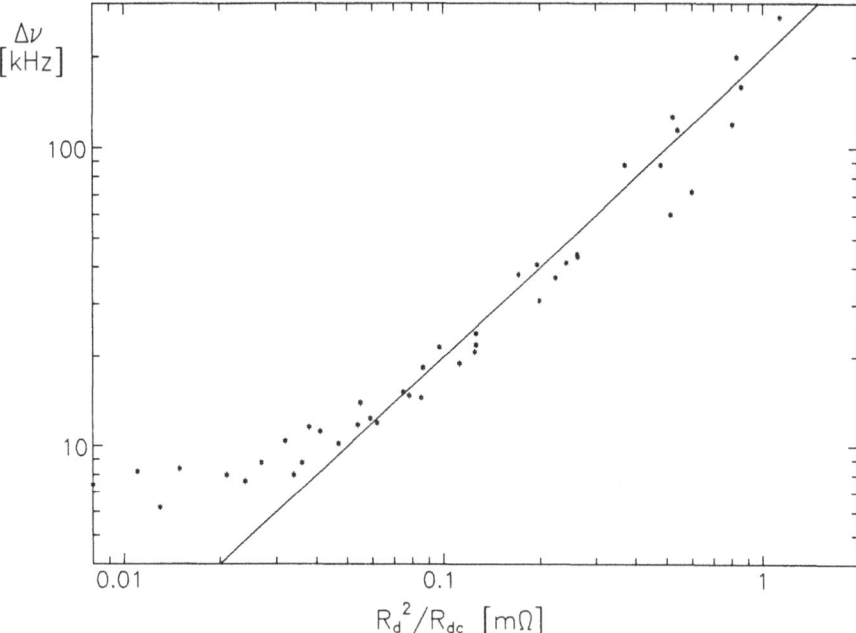

Figure 6. *Half-power linewidth $\Delta\nu$ versus the differential resistance R_d squared divided by the dc resistance R_{dc} (see text) of the radiation emitted from sample H5-2A with the junction biased on MS1. The inserted line has unity slope.*

large values of $\Delta\nu$ the points group around a straight line with unity slope as predicted by Eq. (12).

A resolution bandwidth of 10 kHz on the spectrum analyzer was used. This explains the saturation seen in the figure. From the measured $\Delta\nu$, R_d, and R_{dc} the noise temperature T in Eq. 12 can be found to 4.5 ± 1.5 K in agreement with the bath temperature (4.2 K) indicating a low level of external noise.

MUTUAL LOCKING

Here we use samples with two resonator/junction systems, both exhibiting the step structures described above. One will be denoted A, the second B in the following. Biasing junction A on MS1, and tracing out MS1 for junction B, frequency pulling and phaselocking is observed.

Fig. 7 shows a current-voltage trace of MS1 for junction B when junction A is biased at different points also on MS1. The leftmost trace shows MS1 with junction A biased at zero current/zero voltage. In the next trace junction A is biased at the point on MS1 where it radiates a signal with frequency 8.200 GHz with junction B biased at zero current/zero voltage. The circle on the trace denotes the point at which the phaselocked signal has the frequency 8.200 GHz. For the successive traces junction A is biased at the point where it radiates at 8.220, 8.240, ..., 8.380 GHz, respectively.

The maximum range of locking ($\Delta I = 40$ μA) is limited by the height of the step, and are centered around the point at which the locked signal frequency equals that of junction A unlocked (circles in Fig. 7). Locking occurs when the frequencies of the

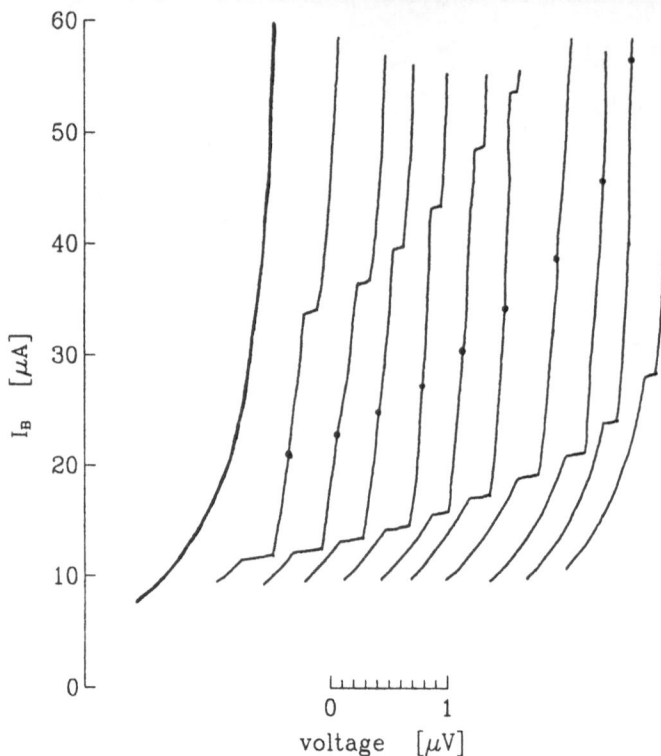

Figure 7. *Current-voltage trace of the first mode step MS1 for junction H5-2B, with junction H5-2A biased at different points along its MS1. The leftmost trace is for junction A biased at zero current/zero voltage, in the successive traces junction A is biased at the point where it radiated a signal with frequency 8.200, 8.220, ..., 8.380, when junction B is biased at zero current/zero voltage. Circles denote the point where the locked signal frequency equals that of junction A unlocked.*

signals from the two junctions differ less than 120 MHz at low bias to 70 MHz at high bias, and frequency pulling is clearly observed on the spectrum analyzer display already at a frequency difference of about 200 MHz.

Fig. 8 shows spectra of the locking process, when the bias current in junction B is increased. As seen from the figure, the power of the phaselocked signal is larger than the sum of the two signals before locking. Also linewidth narrowing has been observed, however, it has been difficult to characterize due to excessive bias current noise from one of the dc current sources used. The maximum locking range observed corresponds to about 20 % of I_c, however, with the supercurrent suppressed by a magnetic field. Locking is also observed when biasing the two junctions on other combinations of steps. First of all, interchanging the two junctions, keeping junction B at fixed bias and tracing MS1 for junction A, gives virtually no difference in locking ranges. The same is the case for the combinations of junction A and B on the steps MS1 and MS(-1). Also when biasing junction A on a higher order step, frequency pulling and locking is observed when tracing out a step for junction B. Generally, whenever the frequency difference of the signal radiated from the two states differ less than the order 100 MHz, locking occurs.

Frequency pulling and phaselocking was also observed when biasing junction A on the step believed to be MS2, where no signal at 8 GHz is radiated, and tracing out

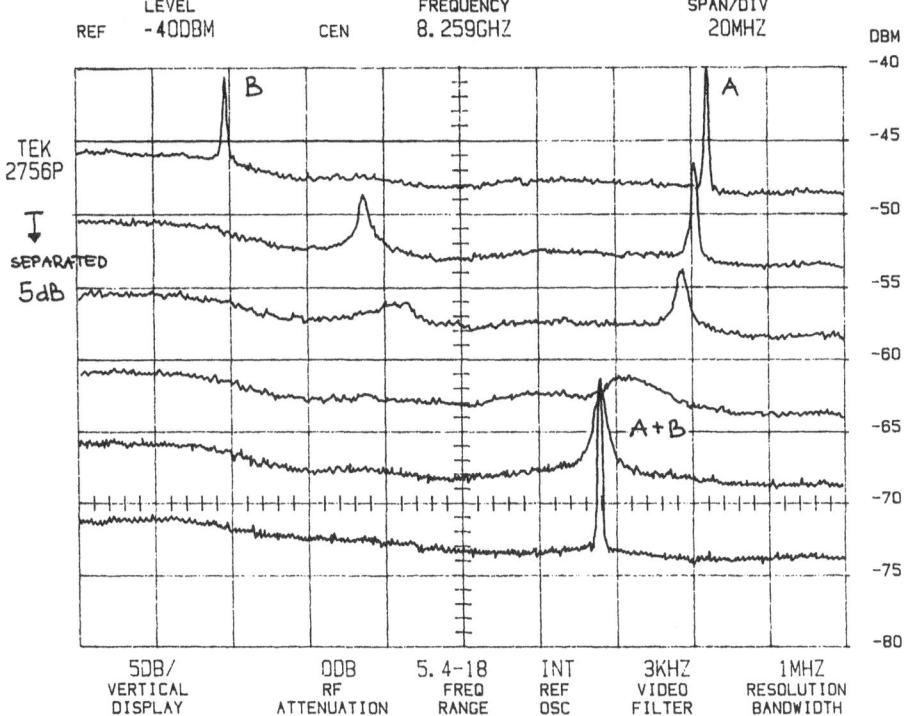

Figure 8. *Spectra of the locking process. The bias current is changed for junction B, pulling the signal frequency of junction A. After locking the power of the locked signal is larger than the sum of the two unlocked signals.*

MS1 for junction B. This is in agreement with the measurements of a large harmonic content of the Josephson oscillation on the steps. With both junctions biased on the supposed MS2, a maximum locking range of the same order as for MS1 is found.

CONCLUSION

Very close coupling between a small Josephson tunnel junction and a multi-mode microstrip resonator has been utilized to study self-induced mode steps, MSn, and parametrically generated mode steps, PMSk. The coupling parameter can be tuned with an applied magnetic field.

Compared with earlier experiments we have used longer resonators, i.e. lower resonance frequencies, which in itself favourizes a stronger coupling. Secondly, the junction capacitance loads the microstrip line and progressively shifts the resonant mode frequencies of the system. The shift, which depends on the ratio of the junction capacitance to the total capacitance of the external resonator, is kept relatively small in our experimental system with a small junction centered in a long low-impedance (wide and thin) stripline resonator.

We believe that the nearly equidistant resonant mode frequencies in combination with a large value of the coupling parameter ease the observed subharmonic parametric excitation of the fundamental mode resonance. The harmonic overlap also explains that both the PMSk and MSn are considerably larger than predicted in the single resonance theory. A large number of harmonics of the Josephson oscillations are nearly at resonance in the system, and hence their contributions to the nonlinear self-interactions can no longer be neglected.

The mode resonance frequencies and the quality factors, Q_n, are determined *in situ* using the junction as detector for externally applied microwave signals. The coupling parameter, which is deduced from the losses, is much larger than the critical value 2.92 found in the single resonance theory. From the difference between the fundamental mode resonance frequency and the frequency of the Josephson radiation detected on MS1 it is likely that we may infer the effective inductance of the pair-current channel in a dynamic state. This is of interest in the study of nonlinear system dynamics.

The linewidth of the microwave signal radiated from a state on a step was found to follow closely that expected from thermal current noise of the quasi-particle conductance converted to frequency noise. The noise temperature agrees well with the bath temperature.

Two closely spaced systems allow for mutual locking of the Josephson oscillations, and locking ranges up to the order of 20 % of the maximum supercurrent have been observed. In the locked state the junctions radiate more power than when oscillating seperately, and the linewidth of the signal is narrowed, indicating mutual phaselocking. We believe that such systems of resonator coupled junctions and phaselocking through next-neighbour interaction are good candidates for self-oscillating arrays.

ACKNOWLEDGEMENTS

We thank R. Fromknecht, J. Niemeyer, R. Pöpel, and M. R. Samuelsen for stimulating discussions, and A. Grunnet-Jepsen and A. Morgenstjerne for assistance with some of the measurements. This work was partly funded by the Commission of the European Communities, the Danish Research Academy, and the Danish National Council of Metrology.

REFERENCES

* Now at: Telecommunication Research Laboratory, DK-2970 Hørsholm, Denmark

1 K. K. Likharev, *Dynamics of Josephson Junctions and Circuits*, (Gordon and Breach, New York, 1986), Chapter 12.

2 G. Paterno, A. M. Cucolo, and G. Modestino, in *LT-17*, edited by U. Eckern, A. Schmid, W. Weber, and H. Wühl (North-Holland, Amsterdam, 1984) pp. 215-216.

3 V. Yu. Kistenev, L. S. Kuzmin, A. G. Odintsov, E. A. Polunin, and L. Yu. Syromyatnikov, in *SQUID-85*, edited by H. D. Hahlbohm and H. Lübbig (W. de Gruyer, Berlin, 1985) pp. 113-118.

4 A. D. Smith, B. J. Dalrymple, A. H. Silver, R. W. Simon, and J. F. Burch, IEEE Trans. Magn. **MAG-23**, 796 (1987).

5 A. Raisanen, W. McGrath, P. Richards, and F. Lloyd, IEEE Trans. Microwave Theory Techn. **MTT-33**, 1495 (1985).

6   L. Kuzmin, H. K. Olsson, and T. Claeson, in *SQUID-85* , edited by H. D. Hahlbohm and H. Lübbig (W. de Gruyer, Berlin, 1985) pp. 1017-1022.

7   J. E. Lukens, A. K. Jain, and K-L. Wan, in *Superconducting Electronics* NATO ASI Series, Vol. F59, edited by H. Weinstock and M. Nisenoff, (Springer Verlag, 1990), pp. 235-258.

8   P. Hadley, M. R. Beasley, and K. Wiesenfeld, Phys. Rev. B **38**, 8712 (1988).

9   A. Larsen, H. D. Jensen, and J. Mygind, (submitted to IEEE Trans. Microwave Theory Tech.)

10 A. D. Smith, J. A. Carpenter, D. J. Durand, L. P. S. Lee, IEEE Trans. Magn. **27** (1991), (to be published).

11 H. D. Jensen, A. Larsen, and J. Mygind, (submitted to J. Appl. Phys.)

12 A. Larsen, H. D. Jensen, and J. Mygind, (submitted to Phys. Rev. B.)

13 N. R. Werthamer and S. Shapiro, Phys. Rev. **164**, 523 (1967).

14 T. I. Smith, J. Appl. Phys. **45**, 1875 (1974).

15 R. Monaco, J. Appl. Phys. **68**, 679 (1990).

JOSEPHSON TRANSMISSION LINES COUPLING

M. Cirillo

Dipartimento di Fisica, Universita' di Roma *Tor Vergata*
00173 Roma, Italy

INTRODUCTION

The problem of the matching of a Josephson transmission line (JTL) to external loads presents noticeable difficulties because of the complexity of the nonlinear wave-propagation that takes place in this spatially distributed structure. In a continuum-limit approximation, the equation for a JTL is a sine-Gordon equation containing loss and forcing terms which is the equation describing the dynamics of a long Josephson junction (LJJ). Early studies of the impedance matching of JTLs were reported for a lossless line by Christiansen and Olsen[1] and for a line containing loss and forcing terms by Erne' and Parmentier[2].These authors considered the effect of a resistive load at the ends of JTLs. Successively, Newman and Davis[3] studied the propagation of fluxons along JTLs (containing loss and bias terms) coupled by resistive networks. We analyze in two cases a particular loading problem for Josephson transmission lines : we consider two JTLs coupled at one end by a lumped element that can be either an inductor or a capacitor. The analysis is performed on the lossless JTL and therefore on the unperturbed sine-Gordon equation because this approach allows isolating the phenomena that are generated by the boundary condition. We find good agreement between a model based on energy calculations and numerical simulations. The results of the analysis are compared with experimental data obtained from devices designed on the basis of our calculations. We also discuss our theoretical data and experiments in terms of realization of coherent arrays of long Josephson junctions, a subject which has recently attracted much attention.

MODEL EQUATIONS

The physical system that we consider is shown in Fig. 1. In the figure we have sketched two Josephson transmission lines (JTLs) having one boundary connected to the lumped element having impedance Z_0. The lossless JTL for the infinite spatial interval, in a continuum-limit approximation, is described by the sine-Gordon (SG) equation [4,5]

$$\phi_{xx} - \phi_{tt} = \sin \phi \qquad (1)$$

where the variable ϕ represents the phase difference between the two superconducting electrodes forming the Josephson junction. In eq. 1 space is normalized to the Josephson penetration depth $\lambda_j = (\Phi_0 / 2\pi l \, j_0)^{1/2}$ and time is normalized to the inverse of the Josephson

Nonlinear Superconductive Electronics and Josephson Devices
Edited by G. Costabile *et al.*, Plenum Press, New York, 1991

297

plasma frequency $\tau_j = (1/\omega_j) = (\Phi_0 c / 2\pi j_0)^{1/2}$. In the JTL model of a long Josephson junction the parameter l (resp. c) represents the inductance (resp. capacitance) per unit length of the superconducting striplines forming the junction. The parameter $\Phi_0 = 2.07 \times 10^{-15}$ Wb represents the quantum of magnetic flux (fluxon) and j_0 the maximum pair current per unit length. For the LJJ the time evolution of 2π - kinks solutions of eq. (1) is physically interpreted in terms of fluxon dynamics along the extended spatial dimension of the junction.

Line A

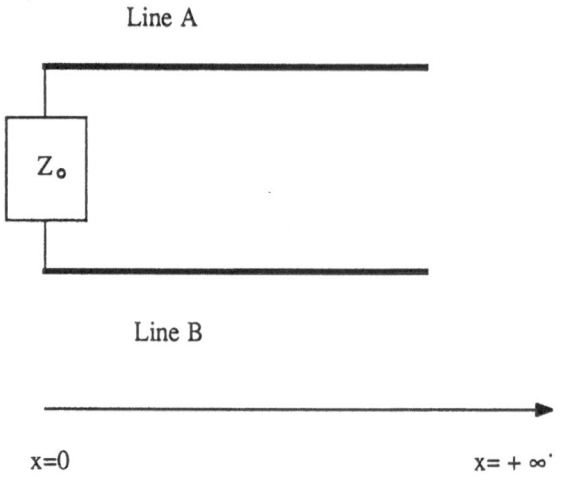

Line B

x=0   x= + ∞˙

Fig. 1. Transmission lines coupling scheme.

The coupling equation for the two JTLs of Fig. 1 can be obtained imposing that the current flowing through the coupling section having impedance Z_0 must be equal to the Josephson surface current at the end of the transmission lines

$$i (Z_0) = (\Phi_0/ 2\pi l) \ \phi_x \tag{2}$$

thus, we get, after substitution of the current terms $i (Z_0)$ and Josephson voltage equations (all normalized with respect to λ_j and τ_j) the equations [6,7]

$$\phi_x (0,t) = \xi (\phi^A (0,t) - \phi^B(0,t)) \tag{3}$$

with $\xi = (l \ \lambda_j)/L_0$ when the coupling element is an inductor having inductance L_0 and

$$\phi_x (0,t) = \zeta (\phi_{tt}^A (0,t) - \phi_{tt}^B (0,t)) \tag{4}$$

with $\zeta = C_0 / (c \ \lambda_j)$ when the coupling element is a capacitor having capacitance C_0.
Energy calculations based on integration of the sine-Gordon Hamiltonian [6,7] allow us to evaluate, on the basis of eq. (3) (resp. (4)) the following terms for the coupling energy of the two JTLs

$$\Delta (H(t) - H (-\infty))= (1/2) \ \xi (\phi^A (0,t) - \phi^B(0,t))^2 \tag{5}$$

$$\Delta (H(t) - H (-\infty))= (1/2) \ \zeta (\phi_t^A (0,t) - \phi_t^B(0,t))^2 \tag{6}$$

Starting from eq. (5) and (6) conditions for the reflections and transmission of the fluxons at the common boundary of the transmission lines can be formulated and we shall see in the next two sections that a reasonable agreement exists between numerical data, analytical predictions and experiments.

INDUCTIVE COUPLING

The conditions for reflection and transmission of fluxons along two JTLs coupled by eq. (3) were derived in ref. 6. We rewrite here the conditions for convenience. The reflections of fluxons take place when

$$\xi < \frac{1}{\pi^2}\left(\frac{1}{\sqrt{1-u^2}}-1\right)$$
(7)

and transmission when

$$\xi > \frac{4}{\pi^2}\left(2-\frac{1}{\sqrt{1-u^2}}\right)$$
(8)

However, transmission can also be expected when

$$\xi > \frac{4}{\pi^2}$$
(9)

i. e., when the coupling energy is higher than the soliton rest energy $H_0 = 8$. In eqs. (7-8) the parameter u represents the normalized velocity of the fluxons along the transmission lines. The three limit curves for the conditions (7), (8) and (9) are indicated in Fig. 2 (on a logarithmic ξ- scale) respectively by R (reflection), T (transmission), SFT (static fluxon transmission).

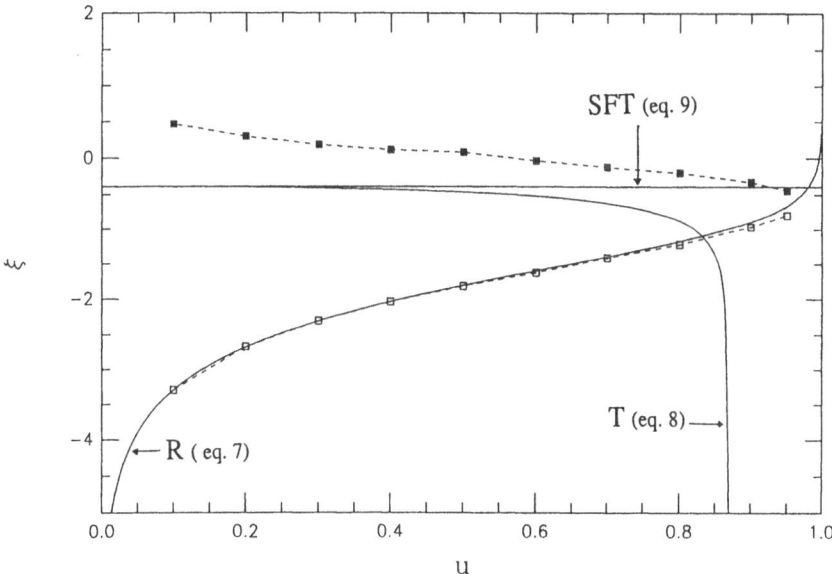

Fig. 2. Border curves for the observation of transmission and reflection phenomena for the inductive coupling. Empty (resp. full) squares represent the numerical threshold below (resp. above) which we observe reflections (resp. transmission) of fluxons.

In ref. 6 a qualitative comparison between the conditions (7-9) and the results of a numerical integration of a perturbed sine-Gordon equation (PSG) including loss and forcing terms was reported. In this paper we show the results of a direct comparison of the theoretical results (7-9) with a numerical integration of the sine-Gordon equation. We integrated a discretized version of eq. 1 over two finite spatial intervals by a fourth-order Runge-Kutta method[6] ; we coupled the two JTLs at one end the spatial interval by eq. 3, while the other end was left free[4] (i.e.$\phi_x = 0$) . This approximation for the semi-infinite problem set by eq. 1 and eq. 3 is reasonable if we consider sufficiently long (normalized) spatial intervals. For the simulations presented in this paper the length of the spatial intervals was never less than 10 and the time and space discretization steps were chosen depending on fluxon initial velocity (typically $\Delta x = \Delta t = 0.025$) .

Fig. 3a shows a result of a reflection experiment when the two JTLs are coupled by a parameter $\xi = 0.01$. We note that in Fig. 3 the spatial configuration of the two JTLs is topologically equivalent to the one of Fig. 1: the coupling point in Fig. 3 is in the center of the spatial interval. In Fig.3 the results are displayed in terms of the time evolution of ϕ_t.

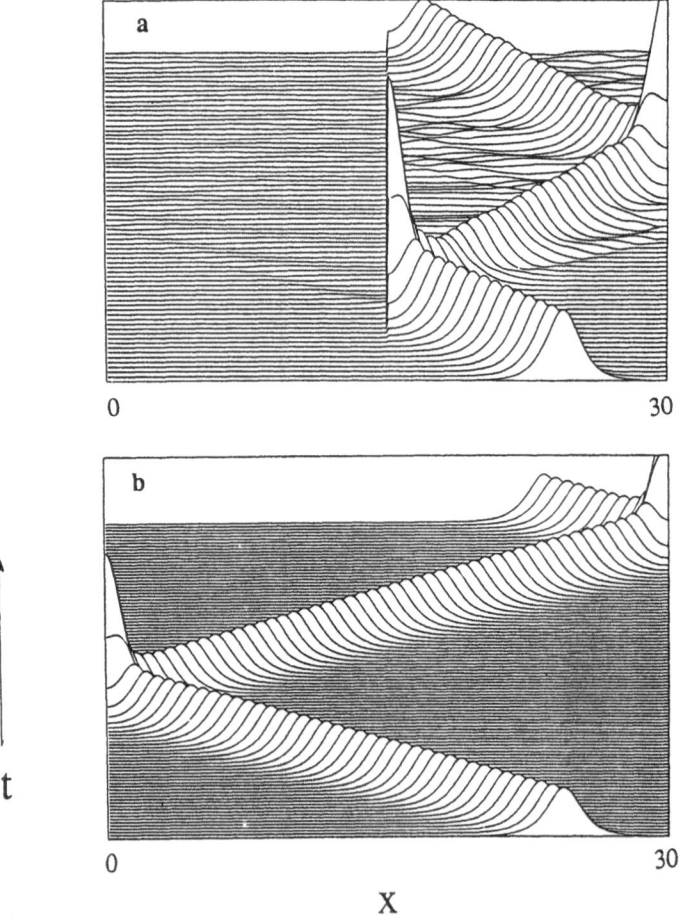

Fig. 3. Inductive coupling scheme : (a) result of a reflection experiment ; (b) result of a transmission experiment .

One fluxon is started with a given initial velocity u= 0.6 around the center of the line on the right side and moves toward the coupling point situated in the middle of the spatial interval. We see that the fluxon is reflected at the coupling point of the transmission lines (situated in the middle of the spatial interval) generating some radiation transmission in the left side transmission line while at the "free" end the fluxon is reflected back as expected[4,5]. The empty squares in Fig. 2 represent the numerical results obtained for different values of the initial velocities of the fluxons: below the dashed line, which represents only a guide to the eye, we observe numerically reflection of fluxons. We see that in this case the agreement between numerical results and theoretical predictions is excellent.

Similar experiments can be performed in order to observe the transmission of fluxons (see Fig. 3b) at the inductive discontinuity between the transmission lines. For the experiment shown in Fig. 3b we set u=0.6 and $\xi = 10$. In Fig. 4 we have reported also (full squares) the numerically observed threshold for transmission of fluxons : transmission is observed above the line connecting the full squares. The agreement with the theory in this case is not as good as in the previous one, however, we can see that the theoretical condition (8) is never violated. For high initial velocities (typically $u \gtrsim 0.9$) of the fluxons very complicated phenomena occur, such as the reflection of a soliton as a soliton, breather formations etc ; these effects are due perhaps to the fact that reflection and transmission curves cross each other ; a detailed analysis of these phenomena goes beyond the purposes of the present paper.

Before discussing the experimental results we point out now that in one-dimensional long Josephson junctions (LJJs), the continuous model for the JTLs, oscillations like the ones of Fig. 3 will give rise to current singularities in the current-voltage characteristics of the junctions appearing at asymptotic voltages , normalized to $(\Phi_0 / 2\pi)\omega_j$, $v_n = (n 2\pi u)/l$, where n is the step order number and $l = L/\lambda_j$ is a normalized parameter defined as the ratio between the physical length of the junctions L and λ_j. These singularities are known as zero-field steps (ZFS)[4,5] and, since their voltages are inversely proportional to the length where the oscillation takes place, an oscillation like the one of Fig. 3a will give rise to ZFS appearing at twice the voltage of those of Fig. 3b. In terms of electromagnetic waves propagation velocities and unnormalized quantities the asymptotic voltages of the ZFS can be written as $V_n = (n \Phi_0 \bar{c}) / L$ where $\bar{c}$ is the electromagnetic wave propagation velocity in the oxide barrier.

Experiments on long Josephson junctions have been performed in order to verify the analytical and numerical predictions. The results of these experiments have been mostly illustrated in ref. 6 and in ref. 8. These experiments were performed on Nb-NbOx-Pb Josephson tunnel junctions for which we estimated $\bar{c}= 0.05$ c , where c is the electromagnetic wave propagation velocity. The experiments were performed on parallel-biased long Josephson junctions coupled by a superconducting loop whose inductance could be calculated. In Fig. 4a we show a magnetic field diffraction pattern of a device consisting of the parallel connection of two long overlap-geometry Josephson tunnel junctions whose length was about 300 μm. The pattern of Fig. 4b is relative to the current height of a step appearing at the voltage of about 50 μV in the current-voltage characteristic of the device. This step had an amplitude of 25 μA and, due to the voltage position, could be interpreted as a fluxon being transmitted through the inductance discontinuity and therefore oscillating over a 600 μm length and generating the first (n=1) ZFS. This little step had a tiny modulation under the application of the magnetic field visible in Fig. 4b around zero magnetic field. The modulations that follow for increasing current are relative to a current singularity which was shifted in voltage of about 1 μV with respect to the little step. We believe indeed that the the loop modulations clearly visible in Fig. 4b (having the same periodicity of the modulations of the Josephson supercurrent height) are due to the phase locking of the Fiske steps[9] of the two 300 μm- long junctions more than the magnetic field dependence of the ZFS of a single 600 μm- long junction.

In any case, the patterns of Fig. 4 demonstrate that two Josephson junctions are locked in a voltage state generated by spatial modulations of the phase in each single junction. This phenomenon can be observed only when the value of the coupling parameter ξ is not much smaller than the unity ; this parameter, evaluated from stripline characteristics, is equal, for the experiments of Fig. 4, to 0.38 and therefore we are operating the device in a region where the oscillators are strongly coupled ,but the inductive load is still such that the reflections of flux-quanta can occur . More experimental data and details can be found in ref. 6, where another coupling configuration is also discussed. This latter coupling configuration is essentially a long Josephson junction with an inductance discontinuity which, in a first approximation can be analyzed in terms of eq. 1 and eq. 3. This kind of device can be considered in the framework of the general problem of inhomogeneities in long Josephson junctions (and related phenomenology), a subject which has been considered by other authors [10].

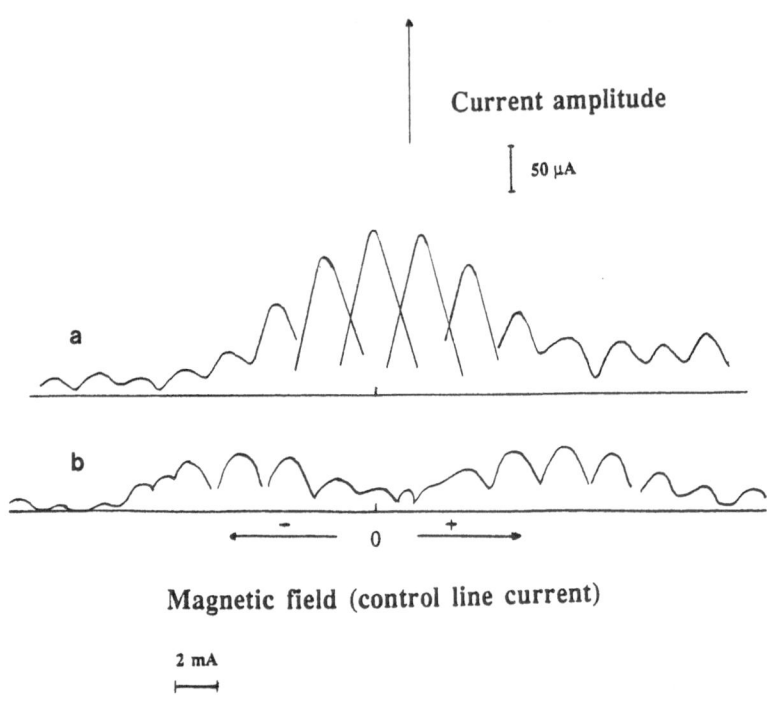

Fig. 4. (a) Josephson current diffraction pattern of a device consisting of the parallel connection of two long Josephson junctions; (b) magnetic field diffraction pattern relative to a current singularity appearing in the current-voltage characteristic of the two LJJ biased in parallel. The zero-current level for the vertical axis is indicated both in a) and b) by the horizontal straight lines.

CAPACITIVE COUPLING

In the case that the connection between the transmission lines is provided by a capacitor, a condition for the reflection of fluxons can be derived in the form [7]

$$\zeta < 1 - \sqrt{1 - u^2} \tag{10}$$

whereas a minimal condition for transmission can be put in the form

$$\zeta > \frac{2}{u^2}\sqrt{1-u^2} \tag{11}$$

The last condition can be obtained by using the same kind of arguments used in ref. 7 for the reflection of fluxons. In this case we evaluate $[\phi_t]^2$ from the waveform of a single moving fluxon and impose that the coupling energy (6) of the transmission lines is higher than the fluxon static energy ($H_0 = 8$). It is clear that, in deriving eq. (11) we assume that $\phi_t^B(0,t) = 0$ during the time interval in which the fluxon waveform passes through the end of line A. This is a rough approximation for evaluating the coupling energy (6), however, it is very unlikely that, if the condition (11) is not satisfied, the fluxons will be transmitted because we are surely underestimating the energy that can be absorbed by the capacitor.

In Fig. 5 we show the conditions (10) and (11) on a logarithmic ζ– scale, indicated by the arrows respectively with an R and a T, and compare them with numerical results obtained from the numerical integration of eq. 1 and eq. 4. The empty squares represent the points below which we observe reflection and the full squares represent the points above which we observe transmission. We can see that even in this case our approximate conditions are never violated and that now the agreement with the theoretical condition tends to be better for higher velocities of the scattering kink, as one would expect following the capacitor "charging" considerations discussed above and in ref. 7. We notice, however ,that we have an argument to believe that for $u \lesssim 0.1$ transmission of fluxons cannot take place, which indicates the existence of a more complex condition for transmission.

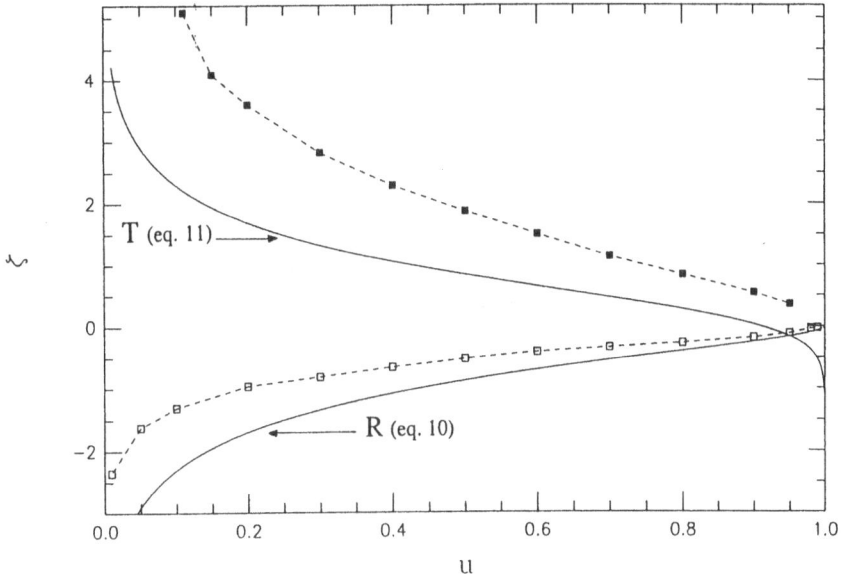

Fig. 5. Conditions for reflection and transmission for two capacitively coupled Josephson transmission lines. Below (resp. above) the line connecting the empty (resp. full) squares we observe reflection (resp. transmission).

We observed that transmission , for $u = 0.11$, takes place when $\zeta \gtrsim 12.5 \times 10^4$, but, for $u = 0.105$ and ζ–values up to 12.5×10^9 the fluxons were always absorbed by the discontinuity. We note that when the initial velocity is decreased from 0.115 to 0.11 the value of the parameter ζ allowing transmission varies only from 12.5×10^3 to 12.5×10^4. We performed halving tests in order to test that this was a real phenomenon and we concluded that, within the limits of the numerical uncertainty errors, there exist a discontinuity in the transmission experiments around $u=0.1$ that cannot be predicted only on the basis of eq. 11.

Although the comparison between numerical results and theoretical modeling is surely more complicated for the capacitive coupling, this latter is somewhat more interesting from the Josephson junctions point of view . One of the most relevant problems in using LJJ oscillators for high frequency applications is the limited power output of these devices operated in the fluxon-oscillator mode [4,5]. Thus, one of the possibilities for solving this problem is offered by the fabrication of coherent arrays of LJJs[11]. It is somewhat more convenient to fabricate series arrays than parallel-biased ones because the first offer no dc-bias current limitations. Therefore, if the LJJs of an array can be coupled by ac-techniques (for example our simple scheme employing the capacitor) it would be much better from the point of view of the applications. We are going to show now that our capacitive-coupling scheme may generate phase-locking between LJJs when the background equation is a perturbed sine-Gordon equation including bias and forcing terms.

For these phase locking experiments we consider the following perturbed sine-Gordon (PSG) equation

$$\phi_{tt} - \phi_{xx} + \sin \phi = \rho - \alpha\,\phi_t \tag{12}$$

with the boundary condition given by eq. 4.

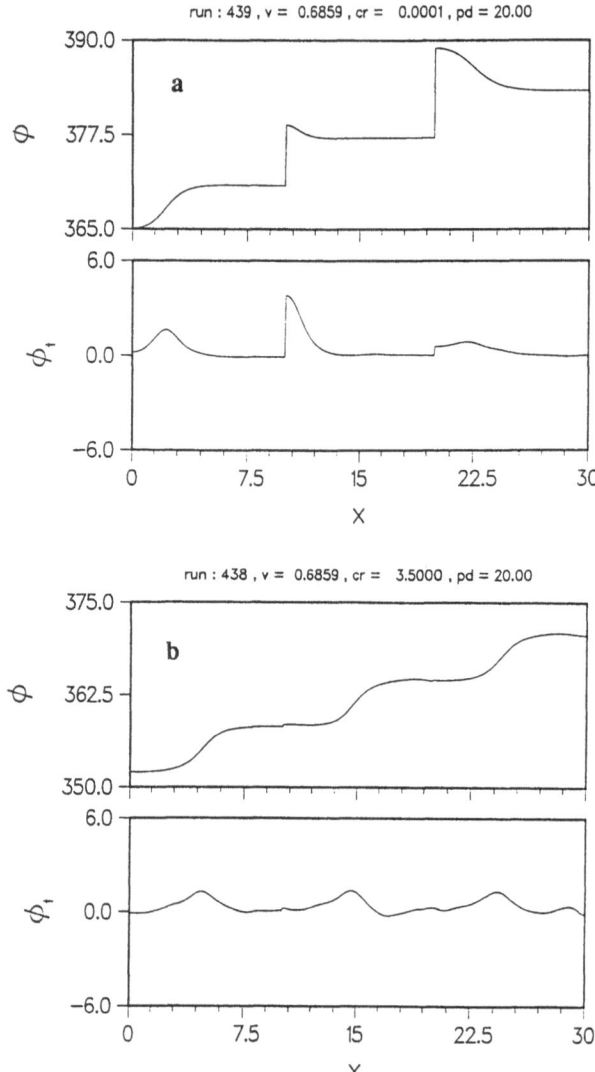

Fig. 6. (a) asyncronous motion of fluxons along three uncoupled transmission lines; (b) syncronous motion observed for an adequate choice of the coupling parameter.

The main idea in getting several LJJ locked is that a tiny amount of the radiation generated by the fluxons during the process of reflection couples into the other junctions generating a syncronous motion of the fluxons. This motion, in turn, will generate a higher voltage output due to syncronous reflections and the increased power could then be used to pump other devices. The practical way to solve this problem represents a complicated issue of millimeter wave engineering ; without going in the details of this problem we will show here how three capacitively coupled junctions can be phase-locked in the above mentioned terms. We will show that three junctions having slightly different characteristic frequencies can oscillate in syncronysm.

Two typical results of numerical-integration processes (third-order Runge-Kutta) for the eq. 12 and eq. 4 with $\Delta x = 0.05$ and $\Delta x = 0.025$ for three coupled junctions are shown in Fig. 6. For these simulations we set $\rho = 0.12$, $\alpha = 0.1$. We started the three fluxons, for the integration of both Fig. 6a and Fig. 6b, not far from the center of the three spatial intervals with the same polarity. However, their position is arbitrary and the fact that we choose to position them initially around the center is due basically to numerical stability and convergence criteria. At the leftmost and rightmost "uncoupled" ends of the transmission lines we impose free-end boundary conditions. The normalized lengths of the junctions are (from left to right) 10.05, 9.95, and 9.9. The coupling points of the junctions are situated around x=10 and x=20.

In Fig. 6a, a picture obtained after 1000 time units, the fluxons have not locked and mantain their characteristic frequencies of oscillation, as evident from the positions that they occupy along the lines. Instead in Fig. 6b we see that after 1000 time units the motion of the fluxons is well syncronized and they will undergo simultaneous reflections. The difference between the two cases is the fact that in Fig. 6a we set $\zeta = 5 \times 10^{-6}$ while for the experiments of Fig. 6b this same parameter was set to 0.175 in order to allow more radiation to be transmitted between the junctions. Thus, we conclude that it is possible to phase-lock several LJJs with an adequate choice of the capacitive coupling parameter ζ .

As described in ref. 7, devices fabricated on the basis of our calculations have shown very promising results. At present we are still involved in the fabrication of LJJ-based devices which employ a capacitive scheme for coupling the radiation between the junctions and we are trying to get a complete characterization on this scheme on the basis of experimental results.

ACKNOWLEDGEMENTS

The financial support of the Progetto Finalizzato *Tecnologie Superconduttive e Criogeniche (SUCRYTEC)* of the CNR (Italy) is gratefully acknowledged.

I wish to thank G. Paterno', S. Pace, S. Pagano, for giving me the permission to employ unpublished data (Fig. 4) and, for the same reason, I thank A.R. Bishop and P. S. Lomdahl (Fig. 6) . The suggestions and the comments of R. L. Kautz were important at the early stages of the work presented in this paper.

REFERENCES

1. P. L. Christiansen and O. H. Olsen, Reflection of fluxons in a Josephson line cavity , Physica D 1:412 (1980).
2. S. N. Erne' and R. D. Parmentier, Loading effects on Josephson junctions fluxon oscillators , J. Appl. Phys. 52:1608 (1981).
3. H. S. Newman and K. L. Davis, Fluxon propagation in Josephson transmission lines coupled by resistive networks , J. Appl. Phys. 53:7026 (1982).
4. R. D. Parmentier, Fluxon in long Josephson junctions, in "Solitons in Action," K. Lonngren and A. C. Scott, Eds. New York : Academic, New York (1978).
5. N. F. Pedersen, Solitons in Josephson transmission lines, in " Solitons", S. E. Trullinger, V. E. Zakharov, and V. L. Pokrovsky, Eds. : Elsevier Science , Amsterdam (1986).

6. M. Cirillo, Inductively coupled fluxon oscillators, <u>J. Appl. Phys.</u>, 58:3217, 1985.
7. M. Cirillo, A. R. Bishop, P. S. Lomdahl, and S. Pace, Analysis of the capacitive coupling of Josephson transmission lines, <u>J. Appl. Phys.</u>, 66:1772, 1989.
8. M. Cirillo, S. Pace, S. Pagano, and G. Paterno', Quantum interference phenomena between long Josephson junctions, <u>Phys. Lett.</u>, 109A:117, 1985.
9. A. Barone and G. Paterno', " Physics and applications of the Josephson effect ", (see chap. 9), J. Wiley , New York (1982).
10. B. A. Malomed and A. V. Ustinov , Pinning of a fluxon chain in a long Josephson junction with a lattice of inhomogeneities : Theory and experiment, <u>J. Appl. Phys.</u> ,67:3791, 1990.
11. R. Monaco, S. Pagano, and G. Costabile, " Superradiant emission from an array of long Josephson junctions ", <u>Phys. Lett .</u>, 131A: 122, 1988.

PHASE-LOCKING IN SERIES ARRAYS OF JOSEPHSON JUNCTIONS

Wolfram Krech and Hans-Georg Meyer

Physikalisch-Astronomische Fakultät
der Friedrich-Schiller-Universität Jena
6900 Jena, Germany

INTRODUCTION

To excite coherent oscillations in a dc biased series connection of Josephson junctions a long-range coupling is most favoured. Such phase-locked arrays are of practical interest as local oscillators in the very high-frequency electronics. Compared with the relatively low-ohmic single junction oscillator the array has the particular advantage that its impedance can be adapted to practical required values to get a maximum power output.

In the past years cooperative effects in series with two and also much more junctions were a subject of extensive experimental[1-11] as well as theoretical[11-23] investigations. The last mentioned papers refer to synchronization effects in arrays with microbridges characterized by a zero McCumber parameter $\beta_c = 0$. However, microbridges restrict the available oscillator frequencies to about 20GHz. With the application of tunnel junctions this limit can be increased more than one order of magnitude. Phase-locking in two junction arrays with a finite McCumber parameter $\beta_c > 0$ was treated numerically[24] and also analytically[25]. Large series arrays of tunnel junctions with randomly scattering junction parameters were considered in Ref. 26, and the stability of the coherent oscillations was investigated by a numerical solution of the corresponding Mathieu problem[27].

PHASE-LOCKED ARRAYS OF JOSEPHSON JUNCTIONS

To achieve phase-locking in the series connection a long-range coupling of the Josephson junctions is most useful. For this end the junctions must be shunted by a load as shown in Fig.1. Assuming identical parameters for all the junctions the current passing the nth component is given by

$$i_n = \beta_c \ddot{\phi}_n + \dot{\phi}_n + \sin\phi_n \qquad (n=1,\ldots,N) \tag{1}$$

within the frame of the well-known dimensionless version of the RCSJ model. E.g., the current is normalized to the critical current and the overdots denote the differentiation with respect to the dimensionless time $s = 2eRI_c t/\hbar$ etc. From the Kirchhoff laws we find for the equivalent circuit of Fig. 1 the equations

Nonlinear Superconductive Electronics and Josephson Devices
Edited by G. Costabile *et al.*, Plenum Press, New York, 1991

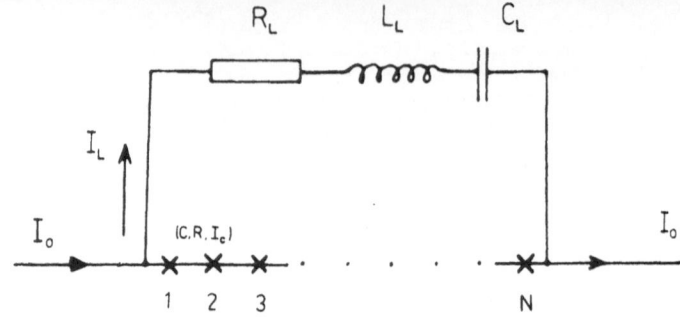

Fig. 1. Circuit diagram of a Josephson series array shunted by a
load. The load components are the resistance R_L, the
capacitance C_L and the inductance L_L. I_0 denotes the dc
bias and I_L the shunt current. The Josephson junctions
are marked by the crosses.

$$i_0 = i_L + i_n , \tag{2}$$

$$\sum_n \dot{\phi}_n = i_L/c_L + r_L \dot{i}_L + l_L \ddot{i}_L . \tag{3}$$

We suppose simple sinusoidal Josephson oscillations

$$\phi_n = \phi_n^o + \omega s - A_n \sin(\omega s - \phi_n^1) \tag{4}$$

where all the junctions are synchronized at the same angular frequency ω.
Finally, considering the consistency condition of the first harmonic
approximation with regard to the Fourier expansion of the $\sin\phi$ term,

$$A_n < 1/4 , \tag{5}$$

and balancing constant and harmonic components in Eqs. (2) and (3) one
gets

$$i_0 = \omega + \mathrm{Re} z_J/2\omega + A\sin(\phi_n^0+\vartheta)/2\omega . \tag{6}$$

This basic relation determines the array synchronization states. The
parameters $A \geq 0$ and ϑ are introduced in the following manner:

$$A\exp(-j\vartheta) = -jz\sum_n \exp(j\phi_n^0) \tag{7}$$

and the system impedance z can be expressed by the junction impedance
z_J and the load impedance z_L, respectively:

$$z = -z_J^2/(z_L + Nz_J) , \tag{8}$$

$$z_J = (1+j\beta_c\omega)^{-1}, \quad z_L = 1/jc_L\omega + r_L + jl_L\omega . \tag{9}$$

If in Eq. (6) A is greater than zero then each of the phases ϕ_n^0 takes one of the values ϕ or $\phi+\delta$

$$\phi_n^0 = \phi, \ \phi+\delta . \tag{10}$$

In the other case, if A = 0, it follows from the relation (6)

$$\sum_n \exp(j\phi_n^0) = 0 . \tag{11}$$

STABILITY OF THE PHASE-LOCKED STATE

For a more detailed study of the two solution types (10) and (11) their stability must be discussed. For this aim small perturbations $\varepsilon_n(s) = \hat{\varepsilon}_n\exp(\lambda s)$, $|\varepsilon_n(s)| \ll 1$ are added to the synchronized oscillations (4) which read now

$$\phi_n = \phi_n^0 + \omega s - A_n(\sin\omega s - \phi_n^1) + \varepsilon_n(s) . \tag{12}$$

We restrict our attention to the case where weakly time-dependent phase slip processes begin to destroy the phase-locked state, $|\dot{\varepsilon}_n| \ll \omega$. Then, proceeding with the modified ansatz (12) in the usual manner one obtains from Eqs. (2) and (3) the characteristic equation which yields the possible Lyapunov exponents λ. An inspection of this exponents show that among the solutions (10) only the one with $\phi_n^0 = \phi$ (uniform mode) can be stable on condition that[20,17]

$$\text{Im} z > 0 \tag{13}$$

(inductive regime). On the other hand, the mode (11) is stable only in the capacitive regime ($\text{Im} z < 0$).

POWER OUTPUT OF THE PHASE-LOCKED ARRAY

The rf components of the load voltage and the load current read

$$v_L^{\sim} = -\omega\sum_n A_n\exp(j\omega s - j\phi_n^1) , \ i_L^{\sim} = v_L^{\sim}/z_L \tag{14}$$

and the corresponding oscillation power is given by

$$P = \text{Re}(i_L^{\sim} v_L^{\sim *})/2 . \tag{15}$$

It can be shown that the solution type (11) is characterized by

$$\sum_n A_n\exp(-j\phi_n^1) = 0 \tag{16}$$

and, therefore, the incoherent mode can emit no power. In practical arrays this configuration can be avoided by violating its stability condition. For the stable solution (10) one finds

$$A_n = -j(z_J + Nz)\exp(j\phi + j\phi_n^1)/\omega \ , \tag{17}$$

in contrast to the incoherent mode all rf phases agree, cf. Eq. (4). The consistency condition derived from Eq. (5) is

$$|z_J + Nz|/\omega < 1/4 \tag{18}$$

and the emitted power can be written as

$$P = N^2/2 \cdot |z_J|^2 \mathrm{Re} z_L / |z_L + Nz_J|^2 \ . \tag{19}$$

Our next goal is to match the load to obtain a maximum power output for the uniform mode. With elementary methods we get

$$z_L = Nz_J^* \ . \tag{20}$$

Then, the maximum emitted power

$$P_{max} \simeq N/8 \tag{21}$$

is proportional to the number of junctions in the array. This result was already obtained in earlier papers[17,28], however, for the more special case of junctions with $\beta_c = 0$. For the matched load several expressions for the uniform mode take a very simple form. Eq. (20) yields

$$z_J + Nz = 1/2 \ , \tag{22}$$

then, the consistency condition is

$$\omega > 2 \tag{23}$$

and the stability condition reads $\beta_c \omega/(1+\beta_c^2\omega^2) > 0$, which is fulfilled in any case.

For a more detailed study of the stability we use not the simple condition (13) but the most relevant Lyapunov exponent

$$\lambda_{rel} = -1/2\beta_c + [1 - 2\beta_c^2/(1+\beta_c^2\omega^2)]^{1/2}/2\beta_c \ . \tag{24}$$

The uniform mode is most stable if the real component of this exponent is most negatively. We note that because of the consistency condition $\omega > 2$ after an expansion λ_{rel} can be approximated quite-well by

$$\lambda_{rel} \simeq -\beta_c/2(1+\beta_c^2\omega^2) \ . \tag{25}$$

Then, the maximum stability is reached at $\beta_c\omega \simeq 1$ and ω as small as possible, that is $\omega = 2$. Both the consistency condition (23) which means at least that higher harmonics in the junction oscillations can be dropped and the minimum of the Lyapunov exponent (25) indicating the stability maximum in the array give rise to an optimum operation region $\beta_c \simeq 0.5$ and $\omega \simeq 2$ (or $i_0 \simeq 2.13$). These values are not very different from those ones derived in[27] numerically solving the Mathieu problem for an array with a purely ohmic shunt.

By the way, a simple resistive load gives not the maximum power output. At the point of maximum stability ($\beta_c\omega \simeq 0.77$) the power is only 87% of the maximum possible one.

ORDER PARAMETER CONCEPTION

From the stability behaviour of the different operation modes (10,11) it can be supposed that there is a thermodynamic phase transition in a large series array of Josephson junctions. Therefore, Eq. (1) must be completed by a fluctuation current ξ_n producing weakly time-dependent variations of the phase ϕ_n^0: $|\dot{\phi}_n^0| \ll \omega$. We obtain for the noise-induced phase motion the so-called reduced equation

$$\beta_c\ddot{\phi}_n^0 + \dot{\phi}_n^0 - A\sin(\phi_n^0+\vartheta)/2\omega + i_0 - \omega - \mathrm{Re}z_J/2\omega + \xi_n = 0 \qquad (26)$$

with the correlation properties

$$\langle\xi_n(s)\rangle = 0 \ , \ \langle\xi_n(s)\xi_m(\bar{s})\rangle = K\delta_{nm}\delta(s-\bar{s}) \ . \qquad (27)$$

The correlation strength

$$K = 4ek_BT(1+\mathrm{Re}z_J/2\omega)/ \ I_c \qquad (28)$$

contains a term due to the down conversation of noise. Now, turning back to Eq. (7) and supposing small fluctuations of the quantities ϕ_n^0 we obtain for large series arrays $N \gg 1$ in a consistent manner that the fluctuations of A and ϑ can be neglected. In this way, we see that the quantity $A\exp(j\vartheta)$ or the load current, respectively, can be considered as an order parameter. Starting from Eqs. (7) and (26) we find the following order parameter conception ($A \ll 1$)[23]:

$$\dot{\phi}^0 + i_0 - \omega - \mathrm{Re}z_J - A\sin\phi^0/2\omega + \xi = 0 \ , \qquad (29a)$$

$$A\mathrm{Re}z - N|z|^2\langle\sin\phi^0\rangle = 0 \ , \qquad (29b)$$

$$A\mathrm{Im}z - N|z|^2\langle\cos\phi^0\rangle = 0 \ . \qquad (29c)$$

For simplicity, we have dropped in the above equations the subscript n and the transformation $\phi^0+\vartheta \to \phi^0$ was used. A power series expansion of the mean-values $\langle\sin\phi^0\rangle$, $\langle\cos\phi^0\rangle$ within the Fokker-Planck apparatus yields up to the third order in A

$$A\mathrm{Re}z - N|z|^2\{4\omega I[(2\omega I)^2+(2\omega K)^2]A-\omega I A^3\}/$$
$$4[(2\omega I)^2+(\omega K)^2][(2\omega I)^2+(2\omega K)^2] = 0 \ , \qquad (30a)$$

$$A\mathrm{Im}z - N|z|^2\{2\omega K[(2\omega I)^2+(2\omega K)^2]A+\omega K A^3\}/$$
$$4[(2\omega I)^2+(\omega K)^2][(2\omega I)^2+(2\omega K)^2] = 0 \ , \qquad (30b)$$

with the abbreviation $I = i_0 - \omega - \mathrm{Re}z_J/2\omega$. Matching together these equations we obtain in the limit $A \to 0$ the relation

$$2\omega K(T_0,\omega) = N\mathrm{Im}z(\omega) \qquad (31)$$

311

which determines the temperature T_0 of the phase transition point. Finally, from Eqs. (30) we also obtain the standard formulation

$$\alpha(T)A + \beta(T)A^3 = 0 \tag{32}$$

with the coefficients

$$\alpha(T) = dK/dT\big|_{T_0} (T-T_0) \ , \tag{33a}$$

$$\beta(T) \sim \beta(T_0) = 3\omega K^2(T_0)/$$

$$8Imz[(2\omega I)^2+(\omega K(T_0))^2][(2\omega I)^2+(2\omega K(T_0))^2]. \tag{32b}$$

Therefore, the conception offers the well-known features of the Landau theory for a second-order phase transition provided that $Imz > 0$. Consequently, the coherent mode with a load current different from zero is favoured for $T < T_0$ if the inductive regime is realized.

In the case of a matched load one obtains for the transition temperature

$$T_0 = \beta_c\omega/[1+2\omega(1+\beta_c^2\omega^2)]\cdot I_c/4ek_B \ . \tag{34}$$

The maximum critical temperature is given by $T_0 = 0.1\ I_c/4ek_B$ where in agreement with condition (23) $\omega = 2$ ($\beta_c = 1.12$) was chosen. For critical currents $I_c > 3.2\mu A$ the transition temperature exceeds 4.2K.

REFERENCES

1. D. W. Jillie, J. E. Lukens, and Y. H. Kao,
   IEEE Trans. Magn. MAG-13:578 (1977).

2. D. W. Jillie, M. A. H. Nerenberg, and J. A. Blackburn,
   Phys. Rev. B21:125 (1980).

3. C. Varmazis, R. D. Sandell, A. K. Jain, and
   J. E. Lukens, Appl. Phys. Lett. 33:357 (1978).

4. P. E. Lindelof, J. Bindslev Hansen, J. Mygind,
   N. F. Pedersen, and O. H. Soerensen,
   Phys. Lett. 60A:45 (1977).

5. P. E. Lindelof and J. Bindslev Hansen,
   J. Low Temp. Phys. 29:369 (1977).

6. L. G. Neumann, D. Y. Dai, and Y. H. Kao,
   Appl. Phys. Lett. 39:648 (1981).

7. D. W. Palmer and J. E. Mercereau,
   Phys. Lett. 61A:135 (1977).

8. A. K. Jain, P. M. Mankiewich, and J. E. Lukens,
   Appl. Phys. Lett. 36:774 (1980).

9. A. K. Jain, P. M. Mankiewich, A. M. Kadin, R. H. Ono,
   and J. E. Lukens, IEEE Trans. Magn. MAG-17:99 (1981).

10. J. Bindslev Hansen, P. E. Lindelof, and T. F. Finnegan,
    IEEE Trans. Magn. MAG-17:95 (1981).

11. A. K. Jain, K. K. Likharev, J. E. Lukens, and
    J. E. Savageau, Phys. Rep. 109:310 (1984).

12. B. Giovannini, L. Weiss-Parmeggiani, and B. T. Ulrich, Helv. Phys. Acta 51:69 (1978).

13. M. A. H. Nerenberg, J. A. Blackburn, and D. W. Jillie, Phys. Rev. B21:118 (1980).

14. L. S. Kuzmin, K. K. Likharev, and G. A. Ovsyannikov, Radio Eng. Electron. Phys.(USSR) 26:1067 (1981).

15. W. Krech, Ann. Phys.(Leipzig) 39:50 (1982).

16. D. Y. Dai and Y. H. Kao, J. Appl. Phys. 52:4135 (1981).

17. K. K. Likharev, L. S. Kuzmin, and G. A. Ovsyannikov, IEEE Trans. Magn. MAG-17:111 (1981).

18. M. A. H. Nerenberg and J. A. Blackburn, Phys. Rev. B23:1149 (1981).

19. G. A. Ovsyannikov, L. S. Kuzmin, and K. K. Likharev, Radio Eng. Electron Phys.(USSR) 27:1613 (1982).

20. W. Krech, Ann. Phys.(Leipzig) 39:117 (1982).

21. W. Krech, Ann. Phys.(Leipzig) 39:349 (1982).

22. M. A. H. Nerenberg, J. A. Blackburn, and S. Vik, Phys Rev. B30:5084 (1984).

23. W. Krech, K. K. Likharev, and K.-H. Berthel, in:

    "LT-17" (Conference Proceedings), U. Eckern, A. Schmid, W. Weber, H. Wühl, eds., Elsevier Science Publishers B. V., Amsterdam (1984), p. 910.

24. G. S. Lee and S. E. Schwarz, J. Appl. Phys. 55:1035 (1984).

25. W. Krech, M. Riedel, Ann. Phys.(Leipzig) 44:329 (1987).

26. V. K. Kornev, K. K. Likharev, E. S. Soldatov, and Yu.P. Tsargorodtsev, in: "SQUID '85" (Conference Proceedings),

    H. D. Hahlbohm, H. Lübbig, eds., de Gruyter, Berlin, New York (1985), p. 1059.

27. P. Hadley, M. R. Beasley, and K. Wiesenfeld, Appl. Phys. Lett. 52:1619 (1989).

28. K. K. Likharev and B. T. Ulrich, "Systems with Josephson Contacts"(Russian Edition), Moscow State University Publishers, Moscow (1978), chap.6.

SOLITONS IN LONG JOSEPHSON JUNCTIONS WITH INHOMOGENEITIES

A. V. Ustinov

Institute of Solid State Physics, USSR Academy of Sciences
Chernogolovka, Moscow district, 142432 USSR

1. INTRODUCTION: WHY INHOMOGENEITIES ?

The motion of magnetic flux quanta in long Josephson junctions has been studied intensively in the last decade and became a very interesting area of both theoretical and experimental soliton physics. In a long Josephson junction (LJJ) a magnetic flux quantum (fluxon) corresponds to a soliton solution of the perturbed sine–Gordon equation:

$$\varphi_{xx} - \varphi_{tt} = \sin\varphi + \alpha\varphi_t - \beta\varphi_{xxt} + \gamma\,, \qquad (1)$$

Here $\varphi(x,t)$ is a spatially– and time–dependent superconducting phase difference between electrodes of the LJJ, x is normalized to the Josephson penetration depth λ_j and t is normalized to the inverse plasma frequency ω_0^{-1}. Dynamics of solitons in *homogeneous* LJJ have been studied in great detail so far [1].

Spatial *inhomogeneities* in LJJ may effect essentially the dynamics of solitons. In the theoretical model the inhomogeneities may be taken into account through the spatial dependence of particular term(s) in Eq. (1) . For example, the LJJ with a spatial variation of the critical current density may be described by a modified equation:

$$\varphi_{xx} - \varphi_{tt} = f(x)\sin\varphi + \alpha\varphi_t + \gamma\,. \qquad (2)$$

Note, that here and further in this paper we do not take β–dissipative term into account. A local inhomogeneity at $x = x0$ of the dimension much less than λ_j may be described by the following function:

$$f(x) = 1 + \varepsilon\,\delta(x - x_0)\,.$$

According to the perturbation theory [2] such an inhomogeneity acts upon the soliton as an effective potential

$$U_1(\xi - x_0) = 2\,\varepsilon\,cosh^{-2}(\xi - x_0)\,, \qquad (3)$$

Nonlinear Superconductive Electronics and Josephson Devices
Edited by G. Costabile *et al.*, Plenum Press, New York, 1991

315

ξ being soliton coordinate. Thereby a microshort ($\varepsilon > 0$) repels , and a microresistor ($\varepsilon < 0$) attracts the soliton irrespective of its polarity.

While passing through the local inhomogeneity, the moving soliton emits electromagnetic waves with plasma dispersion law [2] and it may be also trapped by the inhomogeneity [3,4]. The both effects have been intensively studied theoretically (for a review see [5]) . The trapping of a soliton by the microresistor–type inhomogeneity was recently observed experimentally [6]. An inductive–type inhomogeneity was also studied both in experiment and theoretically [7].

In the present contribution I overview a group of new effects we have recently investigated in the LJJs with artificially prepared inhomogeneities, concentrating mainly on the experiments and the numerical simulations. We have studied a special case of LJJs with *spatially periodic inhomogeneities.*

The interest in LJJ with a regular lattice inhomogeneities was induced by theory [2, 8], where the effect of radiation of the plasma waves by moving soliton was discussed. In Eq. (2) periodic inhomogeneities are corresponding to

$$f(x) = 1 + \varepsilon \sum_{j=1}^{N} \delta(x - ja) . \tag{4}$$

As it was first pointed out by Pedersen and Welner [9] and what later we showed by numerical simulations [10] in the presence of the plasma wave radiation one may expect the resonances between soliton and plasma modes which yield the fine structure in the current-voltage characteristics (CVC). Because this fine structure may be observed in experiment, such a prediction was our original motivation to perform the experimental study of LJJs with periodic artificial inhomogeneities. Later we found [11,12] that multi-soliton dynamics in such a LJJ is even more interesting, and it is actually the main group of the phenomena I will review here. At the end of the chapter I discuss the connection of our studies with the thin-film experiments by Martinoli [13] and the recent investigation of "giant Shapiro steps" and the motion of superlattices in two-dimensional Josephson junction arrays by Benz *et al.* [14] (see also in this book).

2. EXPERIMENTAL

We have studied the overlap $Nb - NbO_x - Pb$ junctions (Fig. 1) of the dimensions $L \times W = 500 \times 20 \ \mu m^2$. The samples were made employing the conventional thin-film technology including magnetron sputtering of Nb, photolithography, plasma oxidation of Nb and thermal evaporation of Pb. The periodic modulation of the critical current density was performed by SiO strips which were fabricated on the Nb film using lift-off lithography before the evaporation of the Pb layer [6,15]. The strips were of the width $w_{in} = 10 \ \mu m$ and of the thickness d_{in} between 800 Å and 1500 Å for different samples. These strips were configured perpendicularly to the larger dimension of the LJJ and served as inhomogeneities. The insulating layer between the superconducting electrodes in the area of inhomogeneities was thick enough to prevent any tunneling, therefore the critical current density was equal to zero.

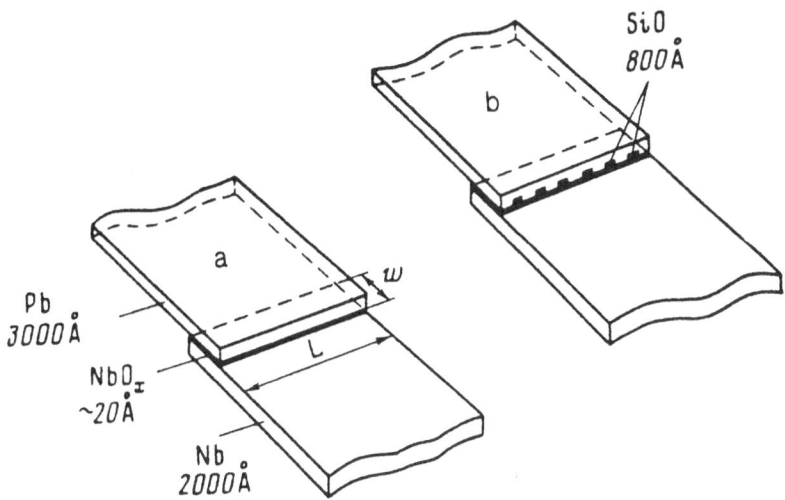

Figure 1. Schematic view of experimentally studied (a) *homogeneous* and (b) *inhomogeneous* LJJs of the overlap geometry.

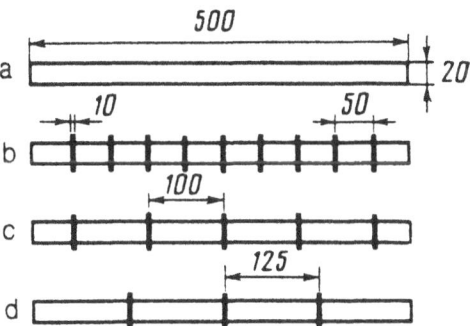

Figure 2. Top view of LJJs area with different geometry of inhomogeneities: the homogeneous junction (a), and the junctions with $N = 9$ (b), $N = 5$ (c), and $N = 3$ (d) inhomogeneities (all the dimensions are in μm).

Each substrate served for making several homogeneous LJJs (Fig. 1a) and eight junctions with the lattice of inhomogeneities (Fig. 1b). The number of inhomogeneities N was equal to 3, 5, or 9 inhomogeneities for different junctions (Fig. 2) with the corresponding spacing between them $a_{in} = a\lambda_j = 125, 100$, or $50\ \mu m$. The details of the measurement technique were described earlier in Ref. [16]. For multi–soliton measurements the external magnetic field was applied in the plane of LJJs. The measurements were performed at the temperature $T = 4.2K$, the Josephson penetration depth λ_j for different samples was in the range of $30 \div 45\ \mu m$.

3. NUMERICAL SIMULATIONS

In order to model the behaviour of a LJJ with inhomogeneities we were solving Eq. (2) numerically. In order to eliminate the influence of boundaries we have examined the problem with periodic boundary conditions [17], and only few simulations have been performed for open–end overlap geometry for more a detailed comparison with [18] experiment. Both single– and multi–soliton cases were investigated with the following boundary conditions:

$$\varphi(\ell, t) = \varphi(0, t) + 2\pi n\,, \quad \varphi_x(\ell, t) = \varphi_x(0, t)\,,$$

where $\ell = L/\lambda_j$ and n is the number of solitons. The simulations have been performed with the normalized junction length $\ell = 8 \div 12$ and with the dissipation coefficient $\alpha = 0.01 \div 0.1$. The lattice (4) consisting of typically $N = 5$ inhomogeneities was approximated by the function [10]:

$$f_{num}(x) = 1 + sign\,(\varepsilon) \sum_{j=1}^{N} \left(1 - \tanh^2 \frac{2(x - ja - \tilde{x})}{\varepsilon} \right)\,. \tag{5}$$

In order to compare our results with experiment we were interested mainly in simulating the lattice of microresistors ($\varepsilon < 0$) and have used ε in the range between -0.1 and -1.0. The details about the initial conditions, numerical algorithm and CVC calculations may be found elsewhere [12]. In the numerical data presented below the averaged voltage u is given in normalized units, so that the motion of a single soliton with the normalized Swihart velocity $v = 1$ is corresponding to $u = 1$ in LJJ.

4. PINNING OF SOLITONS

Because each inhomogeneity may be considered as a pinning center for solitons, our inhomogeneous LJJ represents an example of one-dimensional system with flux pinning on regularly spaced pinning centers. For the first time such a system was realized in two dimensions for Abrikosov vortices using thin superconducting film with modulated thickness [13].

In sufficiently large external magnetic field H solitons enter the LJJ and the average spacing between them d is dependent on the field H^{-1}. In our case it was easily possible to measure the value of the pinning force in order to compare the experimental results with the perturbation theory calculations [19].

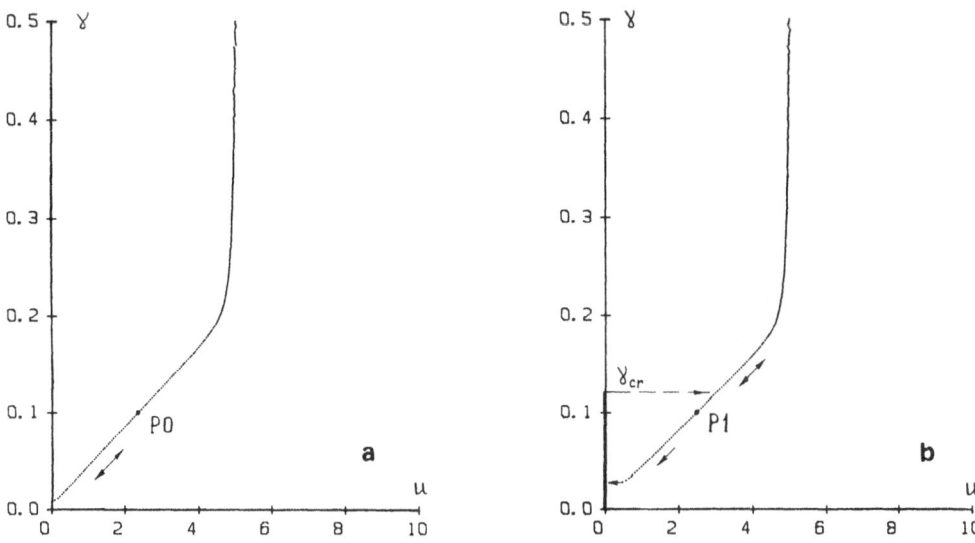

Figure 3. Numerically calculated CVCs of annular LJJ of the length $\ell = 8$ with $n = 5$ solitons: (a) homogeneous junction with $f(x) \equiv 1$; (b) inhomogeneous junction with $N = 5$ inhomogeneities.

The maximum value of the critical current I_c (i.e. the strongest pinning of solitons) in LJJ is achieved when the spacing between solitons

$$d \approx \frac{\Phi_0}{H\left(\lambda_{Nb} + \lambda_{Pb}\right)} \qquad (6)$$

and the period of the lattice of inhomogeneities a are commensurate:

$$p\, d \; = \; q\, a, \qquad (7)$$

where p and q are integer, Φ_0 is the magnetic flux quantum, λ_{Nb} and λ_{Pb} are the magnetic field penetration depths in superconducting electrodes.

Fig.3a shows the numerically calculated CVC of a homogeneous junction with five "frozen" solitons. The value of the critical current for homogeneous annular LJJ is zero. The average voltage $u \sim \langle \varphi_t \rangle$ first increases with external current approximately linearly, and then approaches the asymptotic value $u \approx n$ where the solitons move with the highest velocity $v \approx 1$.

Fig.3b presents the same CVC, but when the lattice of inhomogeneities is present in LJJ. The case of $n = N = 5$ corresponds to the commensurate configuration with equal periods of the soliton chain d and of the lattice of inhomogeneities a. The lowest energy state of such a system is realized when every soliton is trapped at an inhomogeneity. One can see that the CVC has hysteresis and a considerable critical current $\gamma_{cr} \sim 0.12$ is observed. The value γ_{cr} is proportional to the pinning force of the soliton chain on the lattice of inhomogeneities.

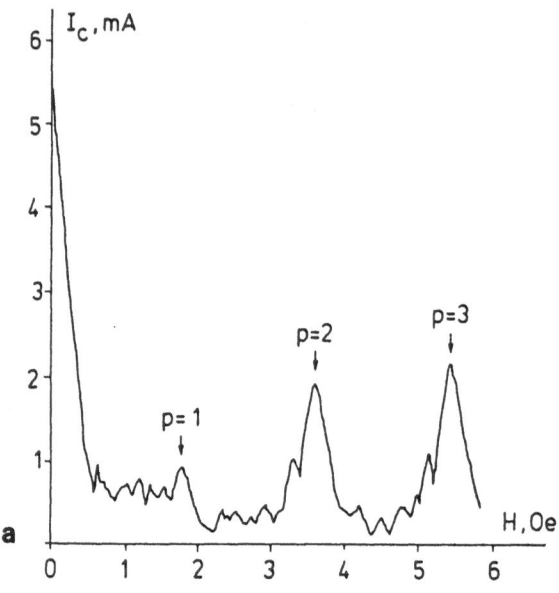

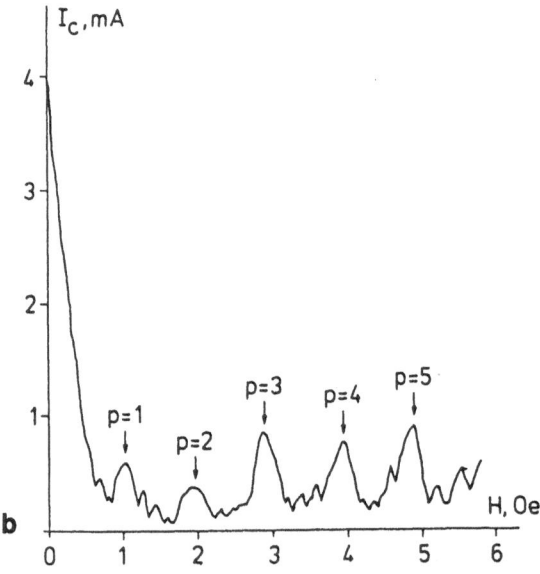

Figure 4. Dependence of the critical current I_c on magnetic field H for the LJJ with $N = 9$ (a) and $N = 5$ (b) inhomogeneities. The p–values are determined by Eq.(7) with $q = 1$.

We have also studied experimentally the dependence of the critical current I_c of our junctions on the external magnetic field H [20] and compared the results with analytical theory [19]. Two typical experimental curves are presented in Fig.4. The critical field for solitons to enter the LJJs was about $H_{c1} \approx 0.7$ Oe. With the increasing of the field H the spacing between solitons decreases according to (6). At certain values of H Eq.(7) holds and a commensurate configuration takes place. These values of field correspond to pronounced maxima of I_c observed in the curves. According to Eq.(7) the field spacing between the maxima is constant and depends on the lattice period a. The positions of the maxima with p-values indicated in Fig.4 are fitting well to Eq.(7) with $q = 1$. We have not seen any significant pinning for $q > 1$. By analogy with two–dimensional arrays [14,21] it means, that we see clearly omly the pinned states for integer values of the frustration parameter f. We believe that the reason is due to that our junctions are not long enough and contain only few inhomogeneities.

Using the perturbation theory it is possible to calculate the value of the pinning force starting from Eq.(2) and (4). For the commensurate case (7) with $q = 1$ one gets a complicated expression for the critical value of the distributed bias current [19] :

$$\gamma_{cr} = \frac{4\varepsilon^2}{27\pi^2 p^2 k^6} \left[(1 + k^2)(2k^2 - 1)(2 - k^2) + 2(1 + k^4 - k^2)^{3/2} \right] . \tag{8}$$

The parameter k is determined by the equation :

$$d = 2\, k\, \mathcal{K}(k) , \tag{9}$$

where $\mathcal{K}(k)$ is the complete elliptic integral of the first kind and k is it's modulus. The limit of $k \to 0$ corresponds to strongly overlapped solitons in the chain, where expression (8) takes a more simple form:

$$\gamma_{cr} \simeq \frac{\varepsilon}{pd} \left[1 + \frac{5}{4} \left(\frac{d}{2\pi} \right)^4 \right] . \tag{10}$$

The strength of soliton chain pinning on inhomogeneities may be estimated from the height of the peaks observed in the $I_c(H)$ curves. Taking the different junctions and varying the scale of λ_j with temperature, we were able to measure the dependence of the pinning force as a function of the normalized lattice spacing a. Fig.5 shows the comparison between formula (10) and our experimental data obtained at several temperatures $4.2K < T < 6.8K$ for the LJJs with the different number of inhomogeneities. Experimentally we have measured the relative height of peaks $I_c(H_p)/I_c(0)$. The theory is in a reasonable agreement with our experimental results . With the decreasing of the spacing between solitons the chain becomes more rigid, and the pinning force in commensurate state increases. Note, however, that the comparison of theory and our experiment is relevant only in the particular range of magnetic field:

$$\frac{\Phi_0}{\lambda_j(\lambda_{Nb} + \lambda_{Pb})} < H < \frac{\Phi_0}{w_{in}(\lambda_{Nb} + \lambda_{Pb})} .$$

In small field H the pinning is determined by LJJ boundaries, whereas for the large field range soliton spacing is becoming comparable with the dimension of inhomogeneities and the approximation (4) is not valid any more. A detailed discussion may be found elsewhere [19].

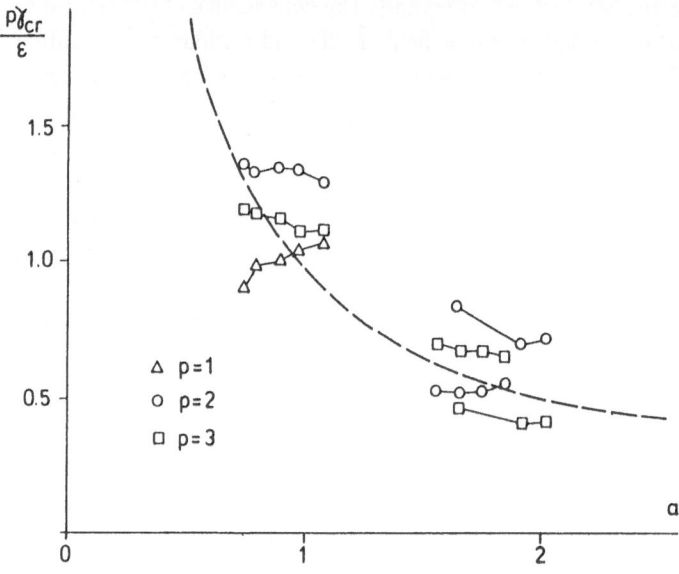

Figure 5. The comparison of theory and experiment: points indicate the experimental results for different p–commensurate states, the dashed line corresponds to the theoretical formula (10).

5. RADIATION AND RESONANCES

Single soliton dynamics

In zero magnetic field a shuttle-like motion of a single soliton manifests itself by a zero-field step (ZFS) on CVC of a LJJ. We have studied the first ZFSs for inhomogeneous LJJs both by numerical simulations and experimentally.

Fig.6 shows the calculated part of the ZFS for the LJJ with the lattice of five inhomogeneities [10] and the same ZFS for a homogeneous LJJ. One can see that the two CVCs are essentially different. The ZFS of the inhomogeneous junction displays a set of pronounced additional steps at certain voltages. In order to understand the origin of these steps, in our numerical simulations we inspected the time dependence of voltage $\varphi_t(x_v, t)$ at the different points of CVC. The pictures corresponding to the points A, B, C, D in Fig.6 are shown in Fig.7. In addition to the soliton oscillations the presence of high frequency oscillation plasma mode is obvious. As seen from the curves A and B, the plasma mode amplitude becomes larger with increasing γ along the same step. The temporal period of plasma oscillations is the integer number of times smaller than the soliton circulation period:

$$m\,\omega_{sol} = i\,\omega_{pl}, \qquad (11)$$

where ω_{pl} is the plasma wave frequency, and the soliton frequency is

322

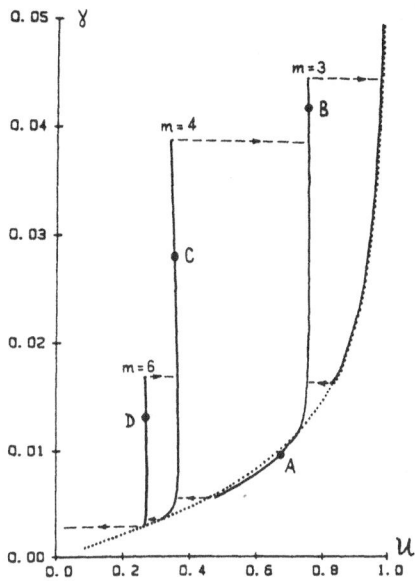

Figure 6. Solid lines – numerically calculated part of the first ZFS for the annular LJJ with $N = 5$ inhomogeneities. Dots show the ZFS for homogeneous LJJ with the same parameters.

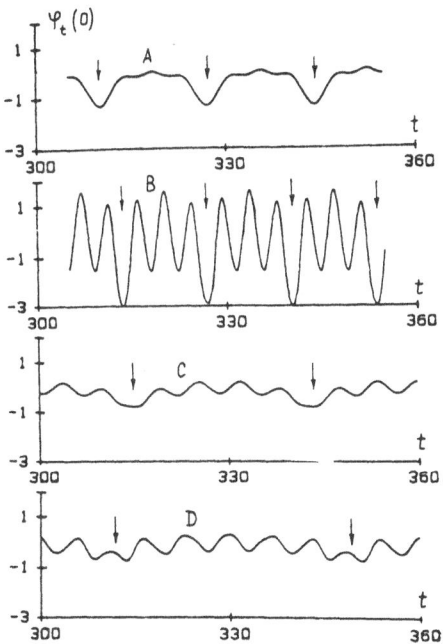

Figure 7. Voltage evolution at the point $x_v = 0$ for inhomogeneous LJJ biased at the different regions of the ZFS shown in the previous figure. The arrows show the moments when the soliton is passing through the point x_v.

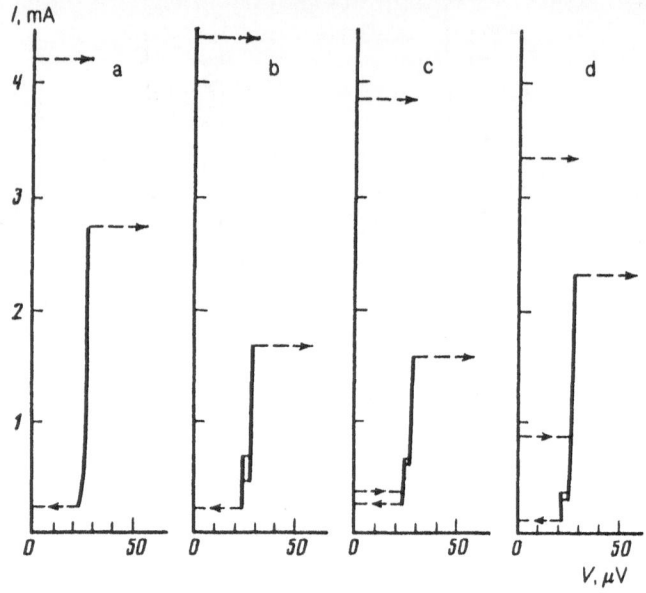

Figure 8. Experimentally measured first ZFSs for different LJJs: (a) the homogeneous junction, and the junctions with $N = 9$ (b), $N = 5$ (c), and $N = 3$ (d) inhomogeneities.

$$\omega_{sol} = \frac{2\pi v}{\ell}.\tag{12}$$

The integer m is changing for different steps in the CVC shown in Fig.7. Another integer i in this case may be considered as equal to 1 because ω_{sol} is several times smaller than $\omega_{pl} \sim 1$ and only the lowest harmonics are important.

Hence, the physical origin of the fine–structure steps in ZFS of the junctions with inhomogeneities is the resonance between the soliton and the plasma modes. According to the perturbation theory [2,8], the frequency of plasma waves ω_{pl} radiated by moving soliton is dependent on the inhomogeneities spacing a and the soliton velocity v. It is possible to calculate the voltages where resonances occur [10,16]:

$$u_m = \sqrt{\left(1 - \frac{2\ell}{am}\right)^2 + \left(\frac{\ell}{\pi m}\right)^2}.\tag{13}$$

Formula (13) is in a good agreement with our numerical simulations.

In Fig.8 experimentally measured ZFSs for different LJJs are shown. We have observed all first ZFSs approximately at the same voltage for different LJJs we prepared. This means that solitons were always able to penetrate through the inhomogeneities and to propagate in the whole junction. In LJJs without inhomogeneities we have not observed any visible fine structure of ZFS (Fig.8a). On the other hand, in the agreement with our numerical simulations in LJJs with inhomogeneities we clearly have seen pronounced splitting of the first ZFS. The shape of additional fine–structure steps was dependent on the geometry of inhomogeneities in the LJJ. The discussion of the correlation between theory and experiment may be found in [16,18].

Multi–soliton dynamics

Let us come back to the case of sufficiently large external magnetic field H. The statics of multi–soliton state in a LJJ has been already discussed in Sec.4. Unipolar solitons now occupy the whole junction and at a sufficiently large bias current $I > I_c(H)$ soliton chain start to move along the LJJ. Again, during the motion of solitons because of their interaction with inhomogeneities the plasma waves are generated in the LJJ. The formula (13) may be used again, but now the soliton frequency (12) is different and is determined by the spacing d between solitons in the chain:

$$\omega_{sol} = \frac{2\pi v}{d}. \tag{14}$$

Hence, in multi–soliton case the frequency ω_{sol} is not fixed by the geometrical length ℓ of the junction and the soliton velocity v, but it is possible to change ω_{sol} by varying d with the external magnetic field according to (6).

We have studied the CVCs in the multi–soliton case experimentally. In Fig.9a the CVC of a homogeneous LJJ are shown. A single broad step well-known as a flux-flow step [22] is observed. This step is corresponding to the motion of a soliton chain with the velocity close to the Swihart velocity ($v \sim 1$). The CVCs of the LJJ with five inhomogeneities for the same experimental conditions are presented in Fig.9b. As compared with the previous three curves, a set of additional sharp steps is observed in this case. The shapes of the steps and their positions along the voltage axis depend on the magnetic field H. However, that an average voltage interval between the steps at least at high voltages is approximately constant and is about $100\mu V$. This value coincides very well with Fiske step voltage spacing [23,24] one can expect for a junction with the length equal to the distance between neighboring inhomogeneities.

In order to explain the step structure at high voltages, a more detailed theoretical approach taking into account real geometry of inhomogeneities was developed [25]. In the area of an artificial inhomogeneity in our experimental system not only the *tunneling* is suppressed, but also there is a drastic change of the local value of *capacitance* in the junction. We have considered further modification of the perturbed sine-Gordon equation (2):

$$\varphi_{tt} - \varphi_{xx} + \sin\varphi = -\alpha\varphi_t - \gamma + \underbrace{(\epsilon_1 \sin\varphi + \epsilon_2\varphi_{tt})\sum_{j=1}^{N}\delta(x - ja)}_{inhomogeneities}, \tag{15}$$

where $\epsilon_1 = (1 - \varepsilon) > 0$ and $\epsilon_2 > 0$, and according to the reasons discussed in Ref.[25] we still neglect here the spatial dependence of the terms φ_{xx}, $\alpha\varphi_t$, and γ .

As follows from Eq.(15), each inhomogeneity "is seen" by moving soliton as an effective potential different from the potential (3):

$$U_2(\xi) = -2\left(\epsilon_1 + \frac{\epsilon_2 v^2}{1 - v^2}\right)\cosh^{-2}\frac{\xi - x_0}{\sqrt{1 - v^2}}, \tag{16}$$

325

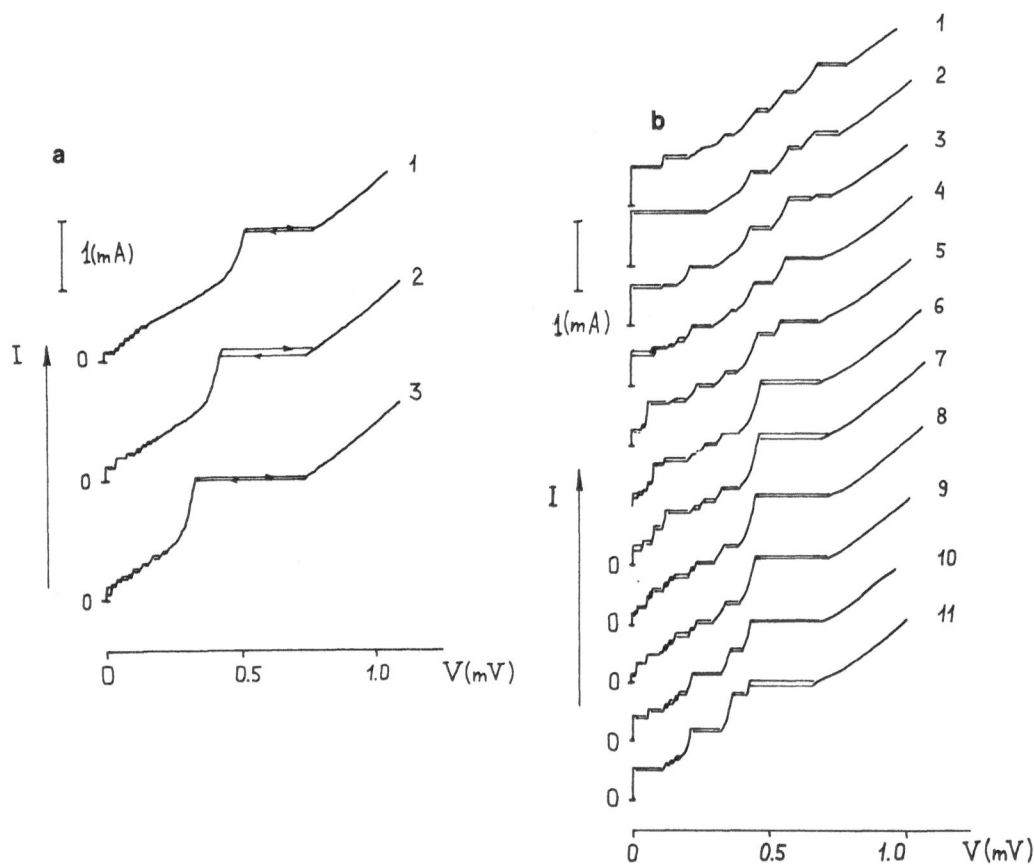

Figure 9. Experimental CVCs of two LJJs in the different magnetic fields H: (a) the homogeneous junction at $H = 2.70\,Oe$ (1), $2.25\,Oe$ (2), $1.80\,Oe$ (3); and (b) the junction with $N = 5$ inhomogeneities at the field decreasing from (1) to (11) in the range of $2.84\,Oe > H > 1.78\,Oe$.

For $\epsilon_{1,2} > 0$ the potential (16) cannot reflect the solitons and they move freely through the LJJ. In contrast, the plasma waves with the dispersion law $\omega_{pl}^2 = 1 + \mathbf{k}^2$ have a reflection coefficient

$$\beta = \frac{\epsilon_{eff}^2}{4\mathbf{k}^2 + \epsilon_{eff}^2} \; ; \; \epsilon_{eff} = \epsilon_1 - (1 + \mathbf{k}^2)\epsilon_2 \, .$$

With our particular experimental parameters of inhomogeneities in a good approximation $\epsilon_1 \approx \epsilon_2 = \epsilon$ and we obtain finally:

$$\beta = \frac{\mathbf{k}^2 \epsilon^2}{4 + \mathbf{k}^2 \epsilon^2} \, . \tag{17}$$

At high voltages for relativistic solitons $(1 - v^2) \ll 1$ the characteristic wave numbers of radiated plasma waves are $\mathbf{k} \sim (1 - v^2)^{-1}$. According to (17) in this case inhomogeneities become practically impenetrable for the plasma waves. Hence, the plasma radiation is reflecting from inhomogeneities and the wave number $\mathbf{k}$ must be quantized by the spatial distance between inhomogeneities. Such a dynamic regime is corresponding to the Fiske modes exited in every short part of the LJJ between adjacent inhomogeneities. A detailed discussion was presented elsewhere [25].

We conclude, that in our multi–soliton experiment solitons move across the hole inhomogeneous junction and generate a coherent radiation, which is localized between inhomogeneities . In other words, at high voltages in LJJ we have the regime of synchronized Fiske steps, which are exited in the short intervals of the LJJ between inhomogeneities.

It is important to note, that the phase shift of oscillations between neighboring short junctions may be tuned easily in our experiment by changing the spacing between solitons in the running chain with the external magnetic field. When the phase shift is equal to the integer number of 2π, we obtain a perfect *super–radiative* state in the system. The advantages of synchronized Josephson junctions for signal generation are well-known [26,27]. I believe that a possibility of using LJJ with inhomogeneities as a local oscillator for microwave detectors is very promising [28].

6. SUPERSOLITONS

In the discussion of the static properties of the multi–soliton chain in Sec.4 we considered only the case of commensurate spatial periods of soliton chain and lattice of inhomogeneities. On the other hand, while discussing the dynamic properties in the previous section we assumed that the spacing between solitons in the chain is not changing along the chain. This limit is corresponding to undeformed "rigid" chain is relevant for a reasonably high magnetic field. Now let us discuss the case, when the soliton chain is sufficiently "soft" and is able to be deformed.

As it was first found by numerical simulations [12] and than confirmed experimentally [29], when the spatial periods of soliton chain d and lattice of inhomogeneities a are incommensurate, a motion of chain defect-like excitations of the magnetic flux

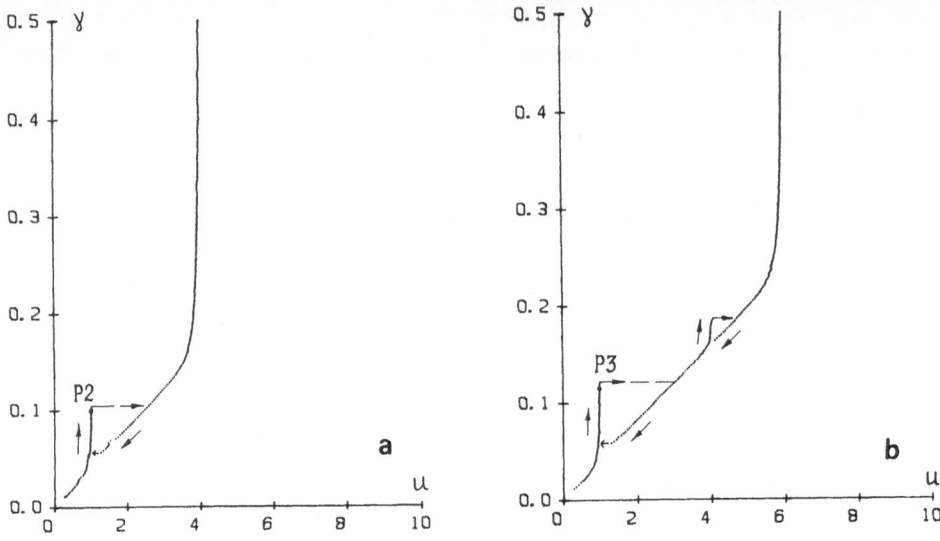

Figure 10. Numerically calculated CVCs for two different incommensurate configurations of a chain of n solitons in the annular LJJ with the lattice of $N = 5$ inhomogeneities: $n = 4$ (a) and $n = 6$ (b).

("supersolitons") may occur because of the deformations of the soliton chain in the incommensurate potential.

Fig.10 shows numerically calculated CVCs for two cases of the soliton chain incommensurable with the lattice of inhomogeneities ($N = 5$). One can see in Fig.10a, that for the number of solitons $n = 4$ besides the Swihart velocity step at the voltage $u \approx 4$ the CVC displays also a vertical step at $u \approx 1$. For n = 6 (Fig.10b) the similar clearly defined low-voltage step appears in the CVC. This step is resembling the previous one and positioned also at $u \approx 1$. There exists also a small step at $u \sim 4$.

The results presented in Fig.10 may be well understood in the ansatz of "supersolitons" [12]. In Fig.11 the instantaneous spatial distributions of the normalized magnetic field $\varphi_x(x)$ for different dynamic regimes in LJJ are shown. The commensurate state corresponding to the point P1 in Fig.3b is shown in Fig.10a. Five equidistant solitons form a non-deformed chain in this case. In the incommensurate case the picture is drastically different and one finds the soliton chain strongly deformed. In Fig.11b for $(n - N) = -1$ a distinct singularity of a *decreased* value of φ_x (a supersoliton of the negative polarity) is seen, which propagates in the LJJ with approximate velocity $v \sim -1$. Thereon in Fig.11c for $(n - N) = 1$ the singularity of the *increased* value of φ_x (a supersoliton of the positive polarity) is obvious, and it propagates in the opposite direction with approximate maximum velocity of $v \sim 1$. The small step at $u \sim 4$ in Fig.10b may be also understood as motion of four supersolitons of the negative polarity, which are the excitations from the commensurate state with $p = 2$ and for them $(n - 2N) = -4$. Fig.12 shows the development of normalized magnetic field $\varphi_x(x, t)$ with time for commensurate (a) and incommensurate (b) states in the LJJ.

The supersolitons may be described as a kind of defects, or dislocations, in the

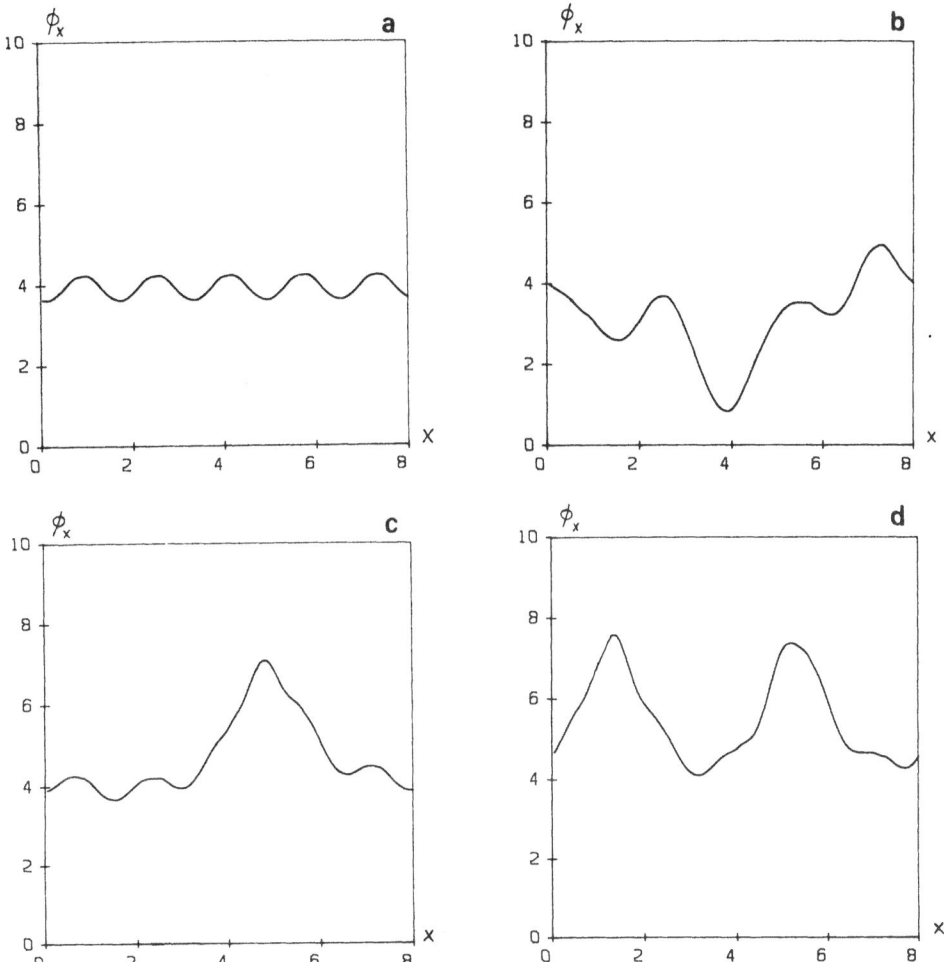

Figure 11. Normalized magnetic field distribution $\varphi_x(x)$ in the annular LJJ for different points of CVCs: (a) point P1 (Fig.3b); (b) point P2 (Fig.10a); (c) point P3 (Fig.10b); (d) two supersolitons for the resonance at $u \approx 2$ with $n = 7$.

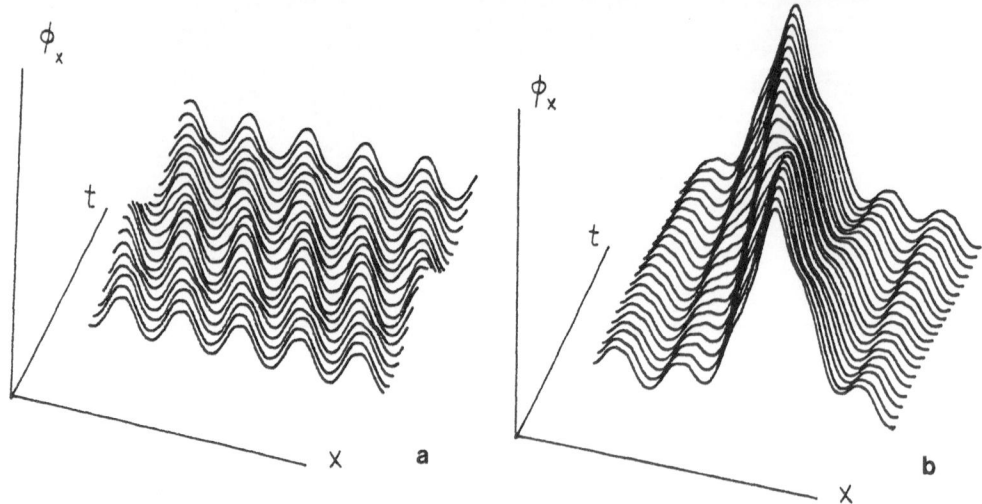

Figure 12. Evolution of normalized magnetic field distribution $\varphi_x(x,t)$ in the annular inhomogeneous LJJ corresponding to different points of CVCs: (a) point P1 (Fig.3b); (b) point P3 (Fig. 10b).

chain of solitons which is placed in spatially incommensurate potential. The reason we call the supersolitons is that they are essentially soliton–like excitations in the chain of solitons. The sine–Gordon–type equation describing supersolitons was recently derived by Malomed [30]. Following him, a deformable soliton chain may be described by a solution of unperturbed equation (1):

$$\varphi(x,t) = \pi - 2\,am\left(\frac{x - \xi(x,t)}{k}, k\right) \; ; \; 0 < k < 1 \,, \tag{18}$$

where am is the Jacobi elliptic amplitude with the same modulus k as the parameter in (9), and the chain coordinate ξ is dependent both on spatial coordinate x and time t. Putting the solution (18) in Eq.(1) and using (4) the following equation for the variable ξ has been derived:

$$\xi_{tt} - \xi_{xx} + \frac{4\varepsilon}{a\rho k}\,sn\frac{\xi}{k}\,cn\frac{\xi}{k}\,dn\frac{\xi}{k} = \underbrace{\frac{G}{\rho} - \alpha\xi_t}_{perturbations} \,, \tag{19}$$

where sn, cn, and dn are standard elliptic functions and

$$\rho = \frac{4k^2\mathcal{E}(k)}{\mathcal{K}(k)}; \; G = -\frac{\pi\gamma k}{\mathcal{K}(k)} \,,$$

$\mathcal{E}(k)$ being the complete elliptic integral of the second kind.

This new equation (19) for ξ is different from the classical sine–Gordon equation (1) and may be naturally called as the *elliptic sine-Gordon* equation. The equation describes the deformations of the sine–Gordon soliton chain. Solitary wave solutions of Eq.(19)

has been also obtained in [30]. In fact, there are two different solutions for $\varepsilon > 0$ and for $\varepsilon < 0$. In Ref.[31] we have investigated numerically some soliton properties for both unperturbed and perturbed Eq.(19).

Finally, here I present the experimental results showing a convincing evidence for supersolitons in real LJJs. The CVCs of the LJJ with the lattice of $N = 9$ inhomogeneities were studied in the intermediate magnetic field range near the commensurability peaks of the critical current I_c [29]. Near the commensurate state at $p = 1$ in the magnetic field $H > H_{p=1}$ a low-voltage step appears in the CVC (Fig.13). A new feature about this particular step as compared with other steps was that voltage of this step was drastically dependent on the value of the magnetic field H. This fact is indicated in Fig.13a for four values of H, the step is marked by asterisk (*). Similar step (**) was also observed for the magnetic field $H > H_{p=2}$ (Fig.13b). Note, that super–radiant Fiske steps discussed in Sec.5 at the voltage values multiple to $200\,\mu V$ were also observed, but they were not changing their position with magnetic field.

The results of measurements of the steps are summarized in Fig.14a, and Fig.14b shows the dependence $I_c(H)$ for this particular LJJ. Fig.14a represents the magnetic field dependence of the maximum voltage for each step. The left group of the data points corresponds to the ordinary multi–soliton flux-flow Swihart step shown previously in Fig.9. The maximum voltage of this step is approximately proportional to the number of solitons, i.e. to the external magnetic field H. Other two groups of points display the behaviour of the steps (*) and (**) from Fig.14a. The maximum voltages of these steps are also approximately proportional to H, but they have different offset along the H axis. Comparing Fig.14a and Fig.14b one can easily see, that the offset field for each step is very close to the commensurate fields $H_{p=1} \approx 1.5\,Oe$ and $H_{p=2} \approx 3.0\,Oe$, where the maxima of $I_c(H)$ are observed. In accordance with the theory, the slopes dV/dH of all these three steps are approximately equal.

In the experiment supersolitons start their motion from one edge of the junction, pass through the whole junction and escape at the other edge of it. This process may be considered as a supersoliton flux-flow regime. The voltage of the supersoliton flux-flow step is proportional to the number of supersolitons.

7. CONCLUSIONS

We have studied statics and dynamics of solitons in long quasi–one–dimensional Josephson junctions *with periodical artificial inhomogeneities* by experiment, numerical simulation and theory. Here I summarize the most essential results we obtained in our study of multi–fluxon dynamics in inhomogeneous LJJs.

Superradiation (rigid soliton chain)

- Experimentally we have observed a new type of wide stable steps on the CVC of LJJ in magnetic field,

- voltage spacing of these steps corresponds to Fiske steps of short Josephson junctions between inhomogeneities,

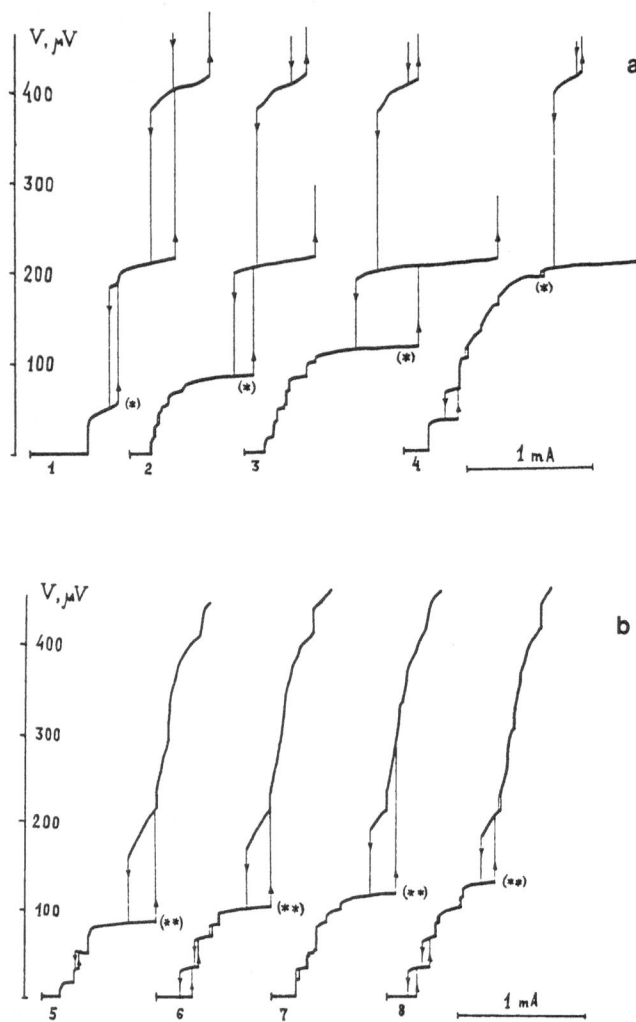

Figure 13. Low-voltage part of the CVC of LJJ with $N = 9$ inhomogeneities in the varying external magnetic field H: (a) in the vicinity of the first $I_c(H)$ peak: $1.83\,Oe$ (1), $1.98\,Oe$ (2), $2.14\,Oe$ (3), and $2.36\,Oe$ (4); (b) in the vicinity of the second $I_c(H)$ peak: $3.41\,Oe$ (5), $3.47\,Oe$ (6), $3.53\,Oe$ (7), and $3.63\,Oe$ (8).

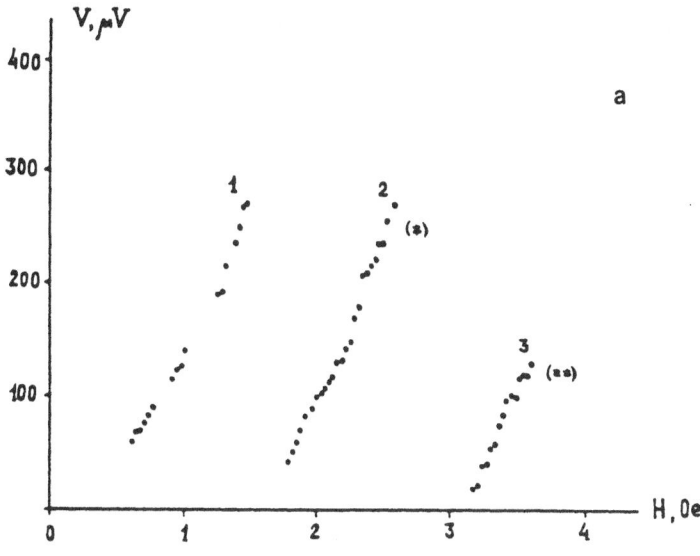

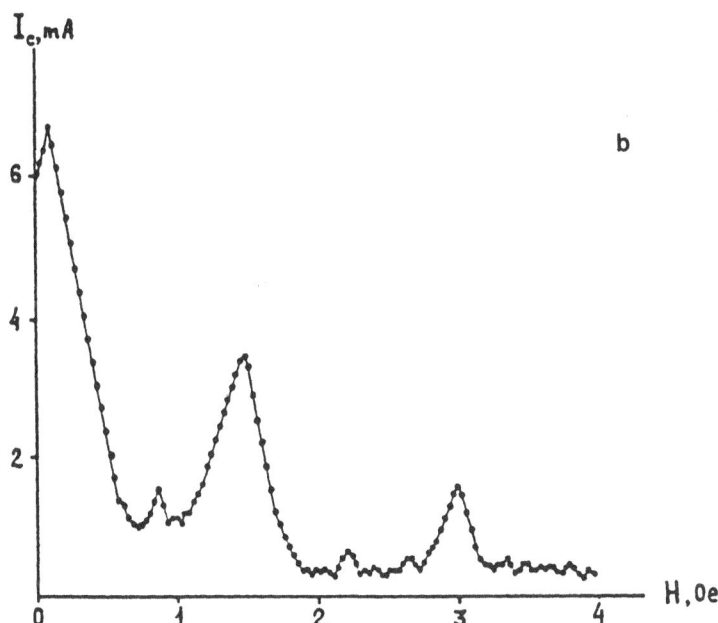

Figure 14. (a) Magnetic field dependence of the maximum voltage of the steps in the CVC: (1) ordinary soliton flux-flow step, (2) step (*) in Fig.a, (3) step (**) in Fig.a. (b) Critical current I_c of the LJJ as a function of the external magnetic field H (the line was drawn for an eye).

- using the perturbation theory we explain the steps as due to the *coherent radiation of plasma waves* by a sequentially moving solitons in LJJ with inhomogeneities,

- the Fiske modes are *synchronized* by soliton chain moving through the whole LJJ.

Supersolitons (soft soliton chain)

- Numerical simulations of annular junction revealed the motion of particle–like and hole–like *defects in the soliton chain*,

- new *elliptic sine-Gordon equation* describes supersolitons analytically,

- straight evidence of the flux-flow of supersolitons was observed experimentally.

Here I want also to point out the connection of the supersoliton phenomena I have reviewed here with the other commensurability–incommensurability studies of magnetic flux in superconducting structures. The pinning of the Abrikosov vortex lattice in a thin superconducting film with modulated thickness was studied by Martinoli [13]. In his experiments the two–dimensional vortex lattice was pinned by one–dimensional structure of channels on the surface. In our case both the pinning potential and the vortex chain are quasi–one–dimensional, but the physics is actually the same. On the other hand, recent studies of the frustrated magnetic flux in two–dimensional Josephson SNS junction arrays and especially the investigation of "giant Shapiro steps" in this systems by Benz *et al.* [14] represent another example of the same sort. In the experiment they observed the rf–current stimulated motion of the two-dimensional flux superlattice, which was formed in the array when the amount of flux per unit cell (frustration parameter f) was not integer. This effect is resembling very much the flux–flow of supersolitons in our LJJs with inhomogeneities. The *main difference* of our experimental system from the others is *much lower dissipation* accompanying the motion of flux quanta in LJJs. We were able to observe the motion of supersolitons just applying the dc bias current and looking at the resonances on the CVC. I believe, that experimental study of two–dimensional arrays of tunnel Josephson junctions in magnetic field could show many interesting resonance phenomena of flux dynamics.

8. ACKNOWLEDGEMENTS

I would like to thank all my colleges and co–authors of the quoted papers, and especially Boris Malomed, continuous collaboration with whom was very stimulating for this work. I am also grateful to J. Bindslev Hansen, A. Davidson, M. Cirillo, G. Costabile, J. C. Fernandez, C. Lobb, R. Monaco, J. Mygind, J. Niemeyer, R. D. Parmentier, N. F. Pedersen, G. Reinisch, M. R. Samuelsen, and other participants of the workshop in Capri for taking interest in the work and fruitful discussions. The support of the Alexander von Humboldt Stiftung at the final stage of this work is gratefully acknowledged.

References

[1] N. F. Pedersen. In A. Barone, editor, *Josephson Effect - Achivemens and Trends*, World Scientific, Singapore, 1986.

[2] D. W. McLaughlin and A. C. Scott. *Phys. Rev.*, B 18:1652, 1978.

[3] Yu. S. Galpern and A. T. Filippov. *Sov. Phys. - JETP*, 86:1527, 1984.

[4] J. C. Fernandez, M. J. Goupil, O. Legrand, and G. Reinisch. *Phys. Rev. B*, 34:6207, 1986.

[5] Yu. S. Kivshar and B. A. Malomed. *Rev. Mod. Phys.*, 61:763, 1989.

[6] Yu. F. Drachevsky, V. P. Koshelets, I. L. Serpuchenko, and A. N. Vystavkin. *Fiz. Nizk. Temp.*, 14:646, 1988.

[7] M. Cirillo. *J. Appl. Phys.*, 58:3217, 1985.

[8] G. S. Mkrtchyan and V. V. Schmidt. *Solid St. Commun.*, 30:791, 1979.

[9] N. F. Pedersen and D. Welner. *Phys. Rev.*, B 29:2551, 1984.

[10] A. A. Golubov and A. V. Ustinov. *IEEE Trans. Magn.*, 23:781, 1987.

[11] B. A. Malomed, I. L. Serpuchenko, M. I. Tribelsky, and A. V. Ustinov. *Sov. Phys. - JETP Lett.*, 47:591, 1988.

[12] A. V. Ustinov. *Phys. Lett.*, A 136:155, 1989.

[13] P. Martinoli. *Phys. Rev.*, B 17:1175, 1978.

[14] S. P. Benz, M. S. Rzchowski, M. Tinkham, and C. J. Lobb. *Phys. Rev. Lett.*, 64:693, 1990.

[15] I. L. Serpuchenko and A. V. Ustinov. *Sov. Phys. JETP Lett.*, 46:549, 1987.

[16] A. A. Golubov, I. L. Serpuchenko, and A. V. Ustinov. *Sov. Phys. - JETP*, 67:1256, 1988.

[17] A. Davidson, B. Dueholm, and N. F. Pedersen. *J. Appl. Phys.*, 60:1447, 1986.

[18] A. A. Golubov, A. V. Ustinov, and I. L. Serpuchenko. *Phys. Lett.*, A 130:107, 1988.

[19] B. A. Malomed and A. V. Ustinov. *J. Appl. Phys.*, 67:3791, 1990.

[20] I. L. Serpuchenko and A. V. Ustinov. *Solid St. Commun.*, 68:693, 1988.

[21] M. S. Rzchowski, S. P. Benz, M. Tinkham, and C. J. Lobb. *Phys. Rev.*, B 42:2041, 1990.

[22] J. Quin, K. Enpuku, and K. Yoshida. *J. Appl. Phys.*, 63:1130, 1988. See also references therein.

[23] M. D. Fiske. *Rev. Mod. Phys.*, 36:221, 1964.

[24] B. Dueholm, E. Joergensen, O. A. Levring, J. Mygind, N. F. Pedersen, M. R. Samuelsen, O. H. Olsen, and M. Cirillo. *Physica*, B 108:1303, 1981.

[25] B. A. Malomed and A. V. Ustinov. *Phys. Rev.*, B 41:254, 1990.

[26] J. Bindslev Hansen and P. L. Lindelof. *Rev. Mod. Phys.*, 56:431, 1984.

[27] R. Monaco, S. Pagano, and G. Costabile. *Phys. Lett.*, A 131:122, 1988.

[28] A. V. Ustinov. To be published.

[29] V. A. Oboznov and A. V. Ustinov. *Phys. Lett.*, A 139:481, 1989.

[30] B. A. Malomed. *Phys. Rev.*, B 41:2616, 1990.

[31] B. A. Malomed, V. A. Oboznov, and A. V. Ustinov. *Zh. Exp. Teor. Fiz.*, 97:924, 1990. English translation in Sov. Phys. - JETP.

FLUXON DYNAMICS IN SINGLE AND COUPLED
LONG JOSEPHSON JUNCTIONS WITH INHOMOGENEITIES

T. Skiniotis and T. Bountis   St. Pnevmatikos

Department of Mathematics   Research Center of Crete
University of Patras   P. O. Box 1527
Patras 26110 Greece   71110 Heraclion,Crete,Creece

ABSTRACT

     Fluxon dynamics in a variety of Long Josephson Junction
(LJJ) systems is studied using the collective coordinate
analysis of McLaughlin and Scott, and the results of the ODE
predictions are compared with the numerical solution of the
full PDE for each problem. The LJJ systems we have studied
include: Single LJJ's with several microresistor (or
microshort) inhomogeneities, an LJJ model proposed by Reinisch
and Fernandez with added inhomogeneities and a system of two
coupled LJJ's introduced by Kivshar and Malomed. In all cases,
the effects of forcing and damping are taken into account and
interesting phenomena of fluxon trapping and bifluxon
formation are observed and discussed.

INTRODUCTION

     As the papers presented at this Workshop and the
extensive literature cited therein clearly demonstrate, there
has been, over the last few years a rapidly increasing
activity in the field of Josephson Junction effects [1-5] and
their implementation in present day technology [6-8].
     In the mathematical analysis of Josephson Junctions a
common point of departure has been the perturbed Sine-Gordon
equation [4-5]

$$\varphi_{tt} - \varphi_{xx} + \sin\varphi = -\alpha\varphi_t + \beta\varphi_{xxt} - \gamma, \qquad (1.1)$$

which, in its dimensionless form (1.1), describes the time
(and space) variation of the difference $\varphi(x,t) = \varphi_1 - \varphi_2$ of the
phases of the wave functions of the two superconductors. The
first two terms on the right hand side of (1.1) account for
normal electron tunneling and surface resistance losses
respectively and γ denotes a uniformly distributed bias
current. We shall be dealing here with Long Josephson
Junctions (LJJ),whose length $L \gg \lambda_j$,the later being, the
Josephson penetration length [5]. From a mathematical point of
view, the solution of eq. (1.1) can be studied using the
perturbation theory of McLaughlin and Scott [9], when α,β,γ
are small.
     Such studies, however, need to be complemented by

Nonlinear Superconductive Electronics and Josephson Devices
Edited by G. Costabile *et al.*, Plenum Press, New York, 1991

337

numerical simulations solving the full p.d.e. (1.1), in order to establish the range of validity of the perturbative approach and determine the precise effect of the boundary conditions at x=0, L.

In two recent papers [10,11] we investigated these questions on LJJ's containing a number of point-like inhomogeneities [9]

$$\varphi_{tt} - \varphi_{xx} + \sin\varphi = -\alpha\varphi_t + \beta\varphi_{xxt} - \gamma + \sum_{i=1}^{N} \mu_i \, \delta(x-a_i), \qquad (1.2)$$

of the "microshort" ($\mu_i > 0$) or "microresistor" ($\mu_i < 0$) type, in the case $\beta = 0$. Our results clearly show that the motion of "kink-solitons" or "fluxons" of the form

$$\varphi_{\pm}(x,t) = 4\tan^{-1}\left[\exp\left(\pm \frac{x - X}{(1-U^2)^{1/2}}\right)\right] , \qquad (1.3)$$

on such LJJ's is accurately described by the reduced o.d.e's for the "collective coordinates" X(t), U(t), obtained by the McLaughlin and Scott procedure [9], for a considerable range of values of the inhomogeneity "strengths" μ_i.

In this paper, we begin in section 2 by extending our earlier results to the case $\beta \neq 0$ and presenting our computation of the first Zero Field Step (ZFS1) on the I-V characteristic of an LJJ with 3 inhomogeneities. We then go on to apply the collective coordinate approach to another LJJ model with inhomogeneities,

$$\varphi_{tt} - \varphi_{xx} + [1+w(x)]\sin\varphi = -\alpha\varphi_t - \gamma + \sum_{i=1}^{N} \mu_i \delta(x-a_i) , \qquad (1.4)$$

due to Reinisch and Fernandez [12], involving a potential $w(x) = \epsilon x^2/2$ imposed on the junction.

As we demonstrate in section 3, the phase plane analysis of the o.d.e's for X(t) and U(t) provides considerable information on the dynamics of fluxons, which agrees very well with simulation studies of the full p.d.e.(1.4)

Finally, in section 4, we address the problem of 2 inductively coupled LJJ's [13,14]

$$\varphi_{tt} - \varphi_{xx} + \sin\varphi = -\alpha_1\varphi_t - \gamma_1 + \epsilon\psi_{xx}$$
$$\psi_{tt} - \psi_{xx} + \sin\psi = -\alpha_2\psi_t - \gamma_2 + \epsilon\varphi_{xx} \qquad (1.5)$$

for which we investigate, using collective coordinates, the phenomenon of bifluxon formation in the general case, i.e. not only in the nonrelativistic limit discussed in [13]. Moreover, allowing for the presence of an ac-bias current in the first junction, we find evidence of phase-locking and homoclinic chaos [15] in the bifluxon oscillations of the system.

2. FLUXON INTERACTIONS WITH INHOMOGENEITIES

In recent studies [10,11] we have observed that fluxon

interaction with inhomogeneities is well described by the phase plane analysis of the system of o.d.e's.

$$\dot{U} = \frac{\pm \pi \gamma}{4}(1-U)^{3/2} - \alpha\, U\,(1-U^2) - \frac{\beta}{3}\,U + (1-U^2)\sum_{i=1}^{N}\mu_i \operatorname{sech}^2\vartheta_i \tanh\vartheta_i$$

$$\dot{X} = U - \frac{1}{2}\,U \sum_{i=1}^{N}\mu_i (X-a_i)\operatorname{sech}^2\vartheta_i \tanh\vartheta_i \qquad (2.1)$$

where $X(t)$, $U(t)$ are the fluxon "center of mass" position and velocity coordinates, cf. (1.3),

$$\vartheta_i \equiv (X-a_i)\Big/ \sqrt{1-U^2}, \quad i=1,2,\ldots N \qquad (2.2)$$

and μ_i, a_i are the "strengths" and locations of the inhomogeneities along the junction respectively.

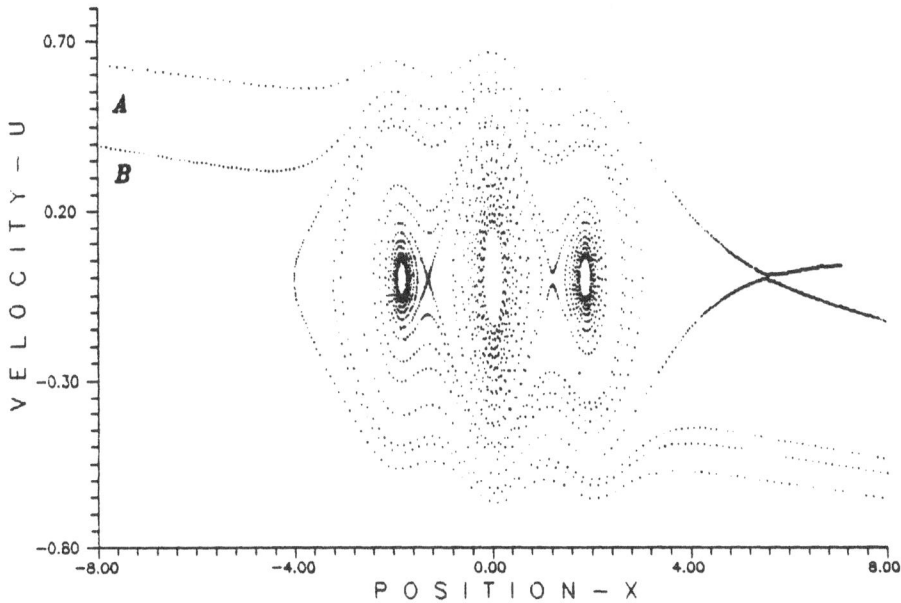

Figure 1a.   Phase plane with $\mu_i = -0.9, -1.5, -1.0$ microresistors, $\beta = 0.0$
$\alpha := 0.033, \gamma := 0.002$, position of impurities $X_i := -2.0, 0.0, 2.0$

We have found that, for experimentally relevant values of the parameters α, γ, μ_i and $\beta = 0$, there generally exist, in the X,U phase plane, near every inhomogeneity, two fixed points: one stable, $(X_i^s, 0)$, and one unstable, $(X_i^u, 0)$ $i=1,2,\ldots,N$.

In the microshort case , $\mu_i > 0$, the unstable fixed point is very close to the inhomogeneity and repels fluxons passing by, while in the microresistor case , $\mu_i < 0$, $(X_i^u, 0)$ and $(X_i^s, 0)$ "exchange places" and the inhomogeneity becomes attracting.

In Figure 1α we show a phase plane picture depicting the separatrices (or invariant manifolds) of the 3 unstable

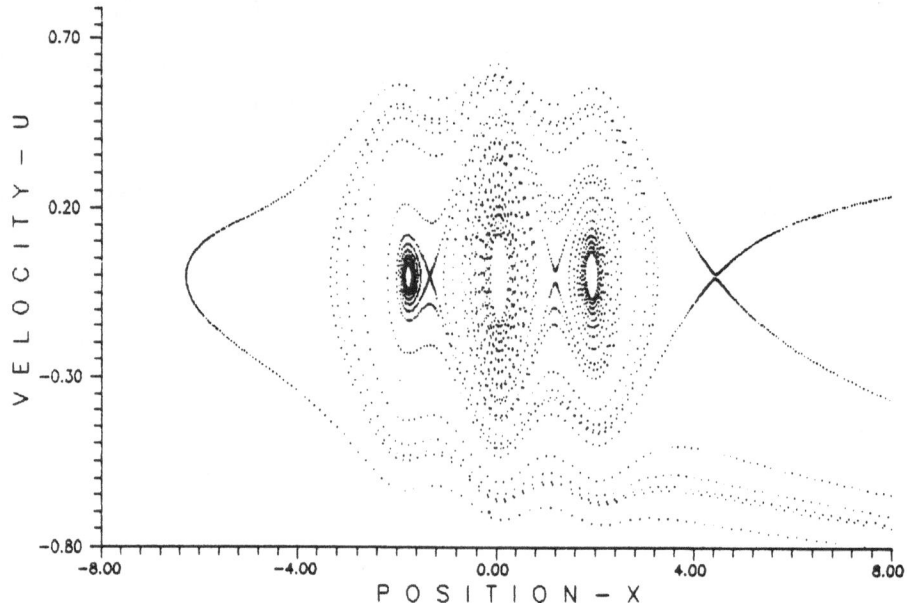

Figure 1b. Same as Fig.1a with $\gamma := 0.02, \beta := 0.002$.

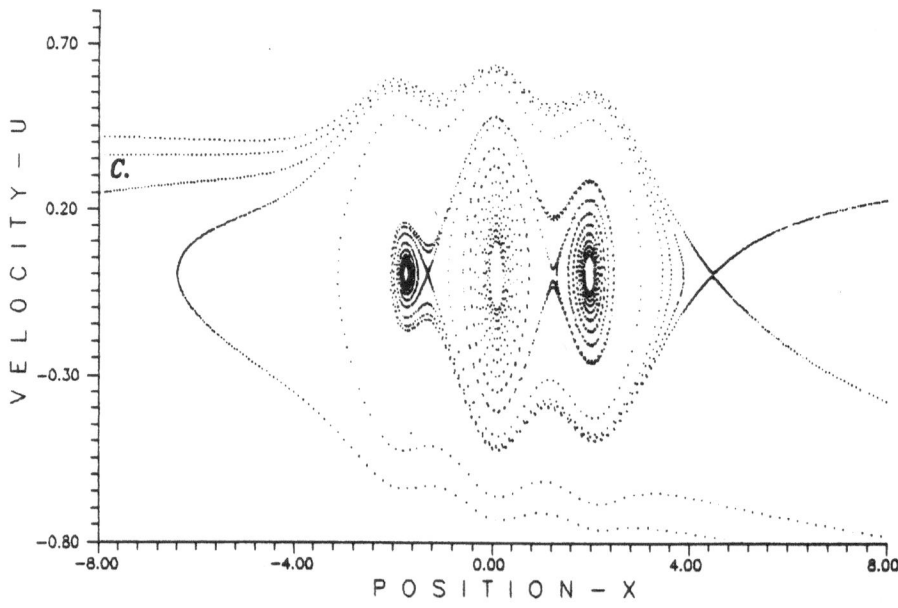

Figure 1c. Same as Fig.1a with $\gamma := 0.02, \beta := 0.02$.

fixed points $(X_i^u, 0)$ associated with 3 microresistors, in the case $\beta = 0$. Solving the full p.d.e. (1.2) for this problem, with boundary conditions

$$\varphi_x(0, t) = \varphi_x(L, t) = 0 \qquad (2.3)$$

we have indeed observed that a fluxon (1.3), when started with X(0), U(0) at the points A,B, shown in Figure 1a, follows the corresponding trajectory predicted by the figure.

Allowing now for surface losses on our junction by letting $\beta \neq 0$ in (2.1), we plot again the separatrices of the same problem of Figure 1a, in Figures 1b,c. The main observation here is that including such dissipation effect does not significantly alter the fluxon dynamics in the junction. As the value of β is increased, of course, one notices a more pronounced "upward" slope of the separatrices on the upper left part of Figures 1b,c, as fluxons (coming from the left) experience enhanced damping and tend to a limiting velocity U_∞ satisfying

$$\gamma = \frac{4U_\infty}{\pi(1-U_\infty^2)^{1/2}} \left[\alpha + \frac{\beta}{3(1-U_\infty)^2} \right] \qquad (2.4)$$

evidently _lower_ than the $\beta = 0$ case.

However starting with X(0), U(0) at point C shown in Figure 1c, the actual fluxon was observed to be attracted by the stable fixed point of the second inhomogeneity, instead of that of the third one near the predicted by the o.d.e.'s. Such discrepancies do occur when the "corridors" between the invariant manifolds are particularly narrow, and are attributed to the fact that across separatrices into the basins of attraction of more than one inhomogeneity.

We have also computed I-V characteristics of this LJJ, in the presence of 3 inhomogeneities. In Figure 2 we present the results of one such computation showing the occurrence of the first Zero Field Step (ZFS1) appearing at the value $V_{ZFS1} \cong 5.66 \times 10^{-6}$ of the (time and space) averaged voltage

$$V = \frac{1}{N(t_\kappa - t_o)} \sum_{i=1}^{N} \left[\varphi(x_i, t_k) - \varphi(x_i, t_o) \right] \qquad (2.5)$$

where N is the number of steps in space, used in our finite difference integration scheme, and $T_k = t_k - t_o$ is the time elapsed, until a "steady state" on the junction has been observed, for every value of γ (changed by an adaptive small step size).

The exceptionally low value of V at which the ZFS1 appears in Fig.2 is due to the fact that our fluxon eventually gets trapped by the second inhomogeneity ! It is interesting that, even in that case, ZFS1 has the usual shape. Currently, we are studying the more physically relevant case of I-V characteristics of one or more fluxons starting with γ values for which they eventually escape from the inhomogeneities and start moving freely back and forth along the LJJ [17].

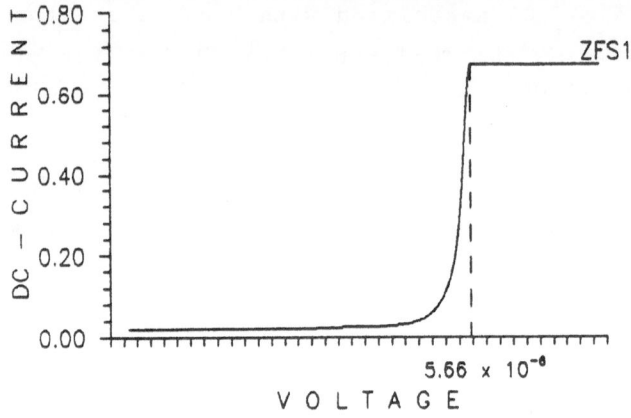

Figure 2. ZFS—1 for the case of three impurities with parameters:
$\alpha := 0.033, \mu_i := -0.9, -1.5, -1.0, X_i := 25.0, 27.0, 29.0, \gamma_{init} := 0.001$

3.   FLUXON DYNAMICS IN THE REINISCH-FERNANDEZ MODEL

We now present one more application of the collective coordinate approach to a model introduced by Reinisch and Fermandez [12], in which we have also included 3 inhomogeneities of the microresistor type ($\mu_i < 0$). The o.d.e's for the $X(t)$, $U(t)$ coordinates of a fluxon, derived from eq.(1.4) of this model in the usual way [9], are :

$$\dot{U} = \frac{\pi \gamma}{4} (1-U^2)^{3/2} - \alpha U(1-U^2) + \frac{\epsilon}{2} \; X \; (1-U^2)^2 + \frac{1-U^2}{2} \sum_{i=1}^{3} \mu_i \, sech^2 \vartheta_i \, tanh \vartheta_i$$

$$\tag{3.1}$$

$$\dot{X} = U - \frac{U}{2} \sum_{i=1}^{3} \mu_i (X-a_i) sech^2 \vartheta_i \, tanh \vartheta_i - \frac{\epsilon}{8} (1-U^2)U \left[X^2 + \frac{\pi^2}{4} (1-U^2) \right]$$

where $\beta = 0$, ϵ is the coefficient of the potential $w(x)$ imposed on the junction,

$$w(x) = \epsilon x^2 / 2 \qquad , \qquad \epsilon > 0 \qquad\qquad (3.2)$$

and the ϑ_i's are as defined in (2.2).

As Reinisch and Fernandez have observed in their numerical simulations [12], the presence of this modulation in the maximum Josephson current density can cause fluxons to oscillate about the potential minimum at $X=0$. What we wish to investigate here is how such a region of stable oscillations changes in the presence of a number of point like inhomogeneities in the junction.

Clearly, the fixed points for this model, $X(t) \equiv X = const$, $U=0$, are given by the roots of the transcendental equation

$$0 = \frac{\pi \gamma}{2} - \epsilon X + \sum_{i=1}^{3} \mu_i \, sech^2 (X-a_i) \, tanh(X-a_i) \qquad (3.3)$$

where we have taken

$$\mu_1 = -0.9, \quad \mu_2 = -1.5, \quad \mu_3 = -1.0 \qquad (3.4)$$

and the $a_i's$ spaced 2 units apart, e.g. $a_1=10$, $a_2=12$, $a_3=14$. Note that, besides the usual fixed point pairs $(X_i^{u,s},0)$ associated with the ith inhomogeneity (see section 2), there is now on additional one at $(X_o,0)$, with

$$X(t) \equiv X_o \simeq -\pi\gamma/2\epsilon \qquad , \qquad (3.5)$$

at which the sum terms in (3.3) are very small.

As $\epsilon \rightarrow 0$, of course, this fixed point moves off to $X_o=-\infty$ and the phase plane portrait of the dynamics is similar to what has been observed in the past (see Figure 3a). As $\epsilon>0$ grows, however, $(X_o,0)$, which is a <u>stable-spiral</u> fixed point, moves closer to the origin and the large basin of attraction surrounding it becomes the dominant feature of the phase plane pictures (see Figure 3b, for $\epsilon=0.025$).

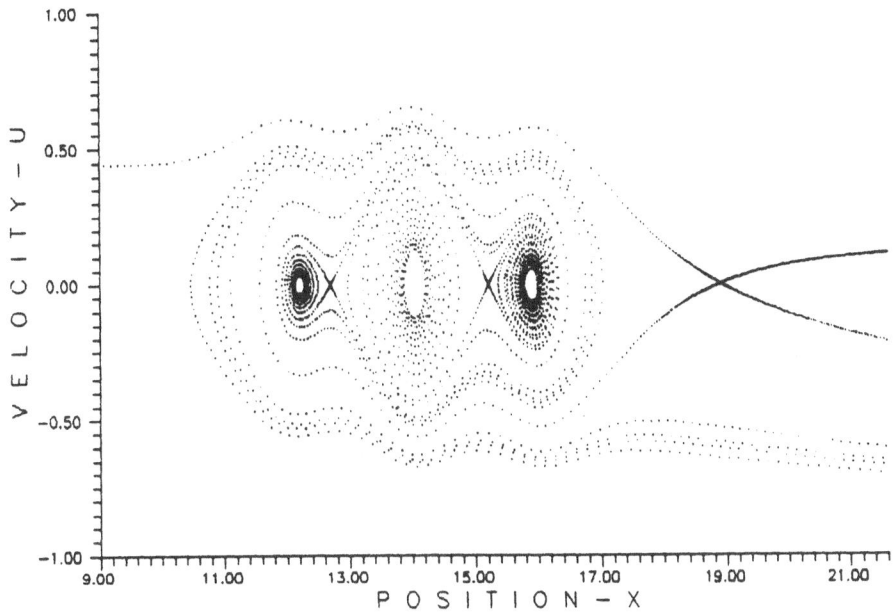

Figure 3a. (X,U) plane for Reinisch and Fernandez model for
$\mu_i=-0.9,-1.5,-1.0,X_i:=12.0,14.0,16.0,\epsilon:=0.001$.

On the other hand, with increasing ϵ, the basins of attraction of the stable points $(X_i^s,0)$ near the inhomogeneities decrease rapidly in size, and may even disappear altogether via saddle - node bifurcations (see Figure 3c). In Figure 3c for example, where $\epsilon=0.03$, the last fixed point pair $(X_1^{u,s},0)$ is about to disappear, allowing thus the basin of attraction around $(X_o,0)$ to occupy practically the full extent of the plane.

In order to investigate the usefulness of this analysis, we now solve the p.d.e. of the problem (1.4), looking for fluxon solutions (1.3), with different initial conditions X(0), U(0). The results are quite interesting and provide

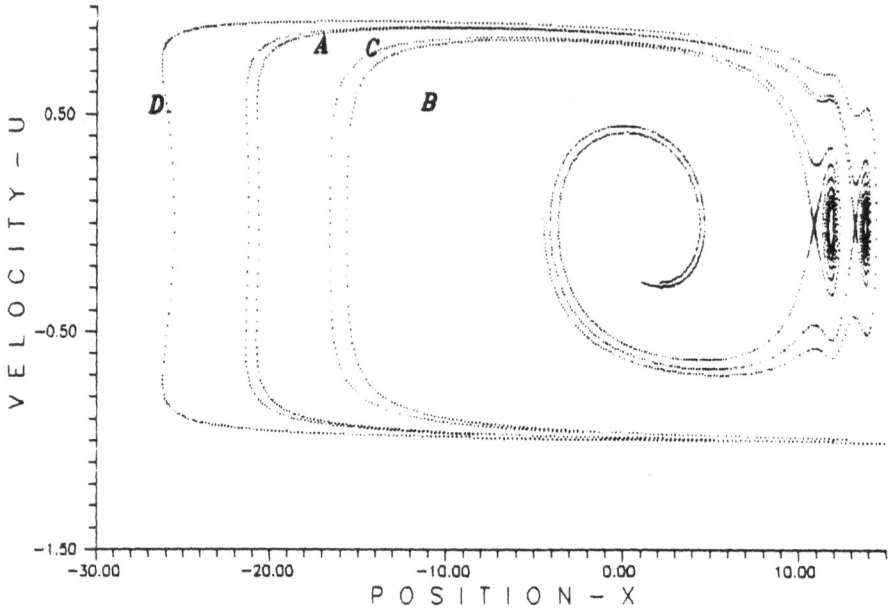

Figure 3b.   Same as Fig.3a with $\varepsilon := 0.025$.

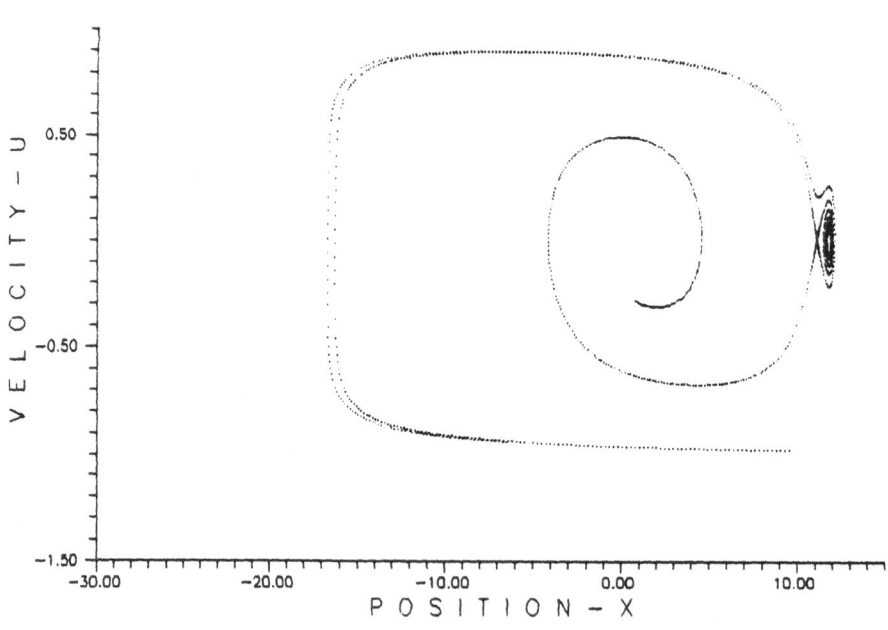

Figure 3c.   Same as Fig.3a with $\varepsilon := 0.03$

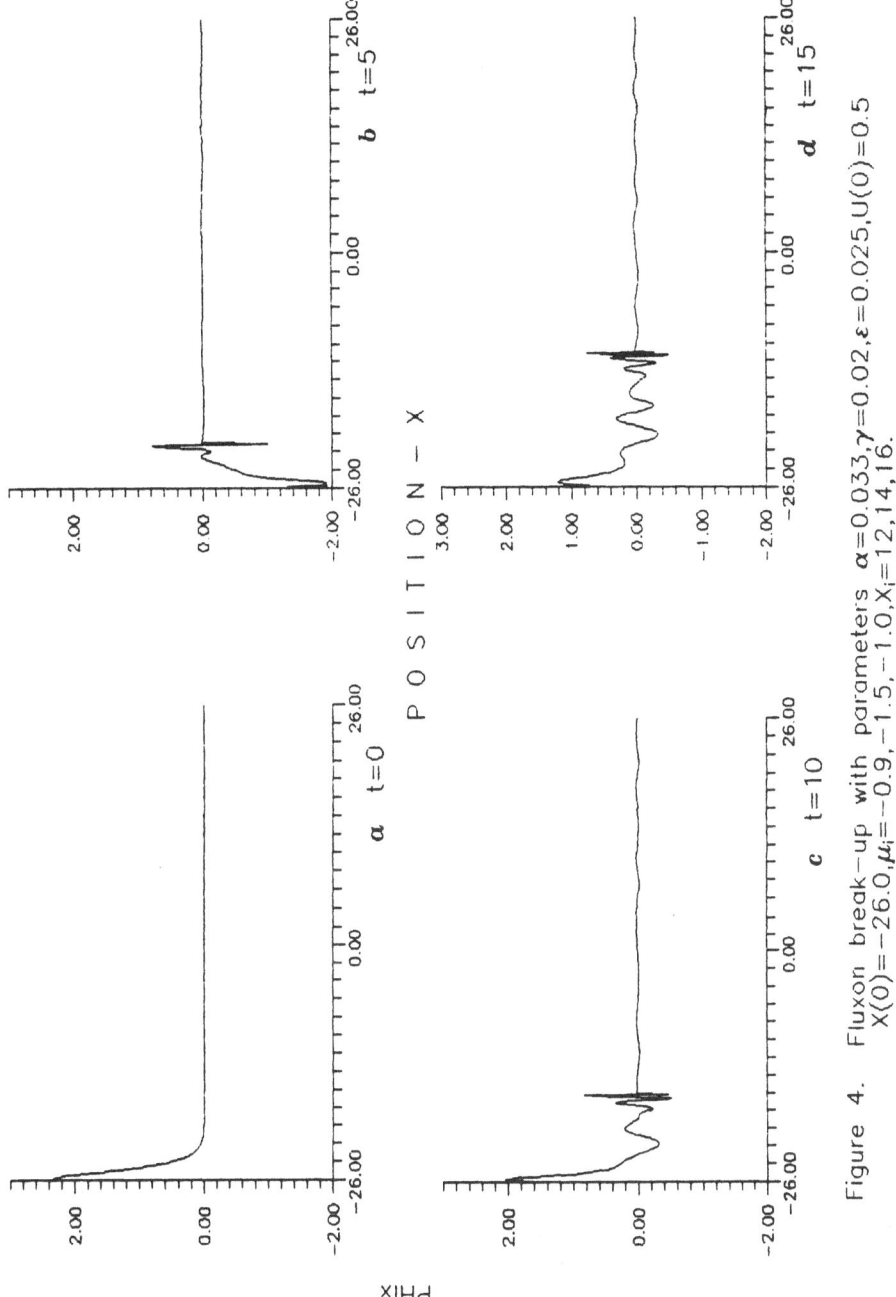

Figure 4. Fluxon break-up with parameters $\alpha=0.033, \gamma=0.02, \varepsilon=0.025, U(0)=0.5$ $X(0)=-26.0, \mu_i=-0.9, -1.5, -1.0, X_i=12,14,16$.

further evidence for the validity of the o.d.e. (collective variable) approach in these problems :

Starting e.g. with a fluxon having initial conditions $(X(0),U(0)) = (-17,0.80)$, i.e. at point A of Fig. 3b, we find that it goes around the basin of attraction of the first inhomogeneity and is eventually attracted by the stable fixed point near the origin. Similarly, starting at point B the fluxon was attracted by $(X_0,0)$. Due to the narrowness of the corresponding corridor (see remark below (2.4)), beginning at C, the fluxon ended up again at $(X_0,0)$.

On the other hand, when we placed our fluxon at $(-26,0.5)$, i.e point D of Fig. 3b, it behaved as if there was a "wall" in front of it on which it kept bouncing and then reflecting off the left end of the junction again and again until it finally broke up! This interesting phenomenon, anticipated by the o.d.e's, is graphically depicted in the solution of the p.d.e's shown in Figure 4.

4.   BIFLUXON OSCILLATIONS IN COUPLED LJJ's

We have also undertaken a study of fluxon dynamics in 2 coupled LJJ's whose individual phase differences $\varphi(x,t)$ and $\psi(x,t)$ satisfy the p.d.e's [13,14]

$$\varphi_{tt} - \varphi_{xx} + \sin\varphi = -\alpha_1 \varphi_t - \gamma_1 + \epsilon_1 \psi_{tt} + \epsilon_2 \psi_{xx} \qquad (4.1a)$$

$$\psi_{tt} - \psi_{xx} + \sin\psi = -\alpha_2 \psi_t - \gamma_2 + \epsilon_1 \varphi_{tt} + \epsilon_2 \varphi_{xx} \qquad (4.1b)$$

where the ϵ_1 and ϵ_2 terms represent a possible capacitative and inductive coupling between the junctions respectively. As phase locking phenomena between coupled LJJ's are of particular importance for practical applications (see other papers in this volume), it would be interesting to study fluxon dynamics in model systems like (4.1) and compare the results with actual experimental observations [16].

As a first step in such a study, we have chosen to investigate by collective coordinates the behavior of fluxons in a model with only inductive coupling ,i.e $\epsilon_1 = 0$ in (4.1) , considered already by Kivshar and Malomed in the nonrelativistic limit of very slow fluxons [13]. We begin, as usual, by substituting in the equations of motion (4.1) the two fluxon forms

$$\left.\begin{array}{c}\varphi\\\\\psi\end{array}\right\} = 4\tan^{-1}\left\{\exp\left(\sigma_i \frac{x - X_i}{\sqrt{1 - U_i^2}}\right)\right\}, \quad \sigma_i = \pm 1 , \qquad (4.2)$$

where i=1,2 refers to a fluxon in the first (φ) or second (ψ) junction respectively. Following the perturbation approach of McLaughlin and Scott [9], the fluxon coordinates $X_i(t)$ and $U_i(t)$ are found to satisfy the o.d.e's

$$\dot{X}_1 = U_1 \ , \ \dot{X}_2 = U_2 \qquad\qquad (4.3\alpha)$$

$$\dot{U}_1 = -\alpha_1(1-U_1^2)U_1 + \frac{\sigma_1\gamma_1\pi}{4}(1-U_1^2)^{3/2} + \frac{\sigma_1\sigma_2}{2}\frac{1-U_1^2}{(1-U_2^2)^{1/2}}\ \varepsilon_2 I \quad (4.3b)$$

$$\dot{U}_2 = -\alpha_2(1-U_2^2)U_2 + \frac{\sigma_2\gamma_2\pi}{4}(1-U_2^2)^{3/2} - \frac{\sigma_1\sigma_2}{2}\frac{1-U_2^2}{(1-U_1^2)^{1/2}}\ \varepsilon_2 I \quad (4.3c)$$

where

$$I = \int_{-\infty}^{\infty} \text{sech}\xi \ \tanh\xi \ \text{sech}[\rho(\xi+\vartheta)]\ d\xi, \qquad (4.4)$$

with

$$\vartheta \equiv \frac{X_2-X_1}{(1-U_2^2)^{1/2}}\ , \qquad \rho \equiv \left(\frac{1-U_2^2}{1-U_1^2}\right)^{1/2} \qquad (4.4\alpha)$$

We note parenthetically, at this point, that eqns (4.3α) should contain, on their rhs, terms $O(\varepsilon_2)$ which have been omitted at this approximation. We have verified in our computations, however, that taking these terms into account does not significantly affect the results which we describe below.

In the nonrelativistic limit $|U_i|\ll 1$, $i=1,2$, considered in the literature [13], U_i^2 terms are neglected, $\rho\simeq 1$ and the integral (4.4) can be analytically evaluated to yield.

$$I = I_{nr} = \frac{2}{\sinh X}\left(1-\frac{X}{\tanh X}\right), \quad X \equiv X_2-X_1\ , \qquad (4.5)$$

In that case, and assuming identical junctions i.e,

$$\alpha_1 = \alpha_2 = \alpha \qquad\qquad , \qquad (4.6a)$$

with

$$f \equiv \frac{\pi}{4}(\sigma_2\gamma_2-\sigma_1\gamma_1), \quad g \equiv \frac{\pi}{4}(\sigma_1\gamma_1+\sigma_2\gamma_2)\ , \qquad (4.6)$$

eqns (4.3b,c) can be subtracted to obtain, using (4.3α)

$$\ddot{X} = -\alpha\dot{X} + f - \frac{2\varepsilon_2\sigma_1\sigma_2}{\sinh X}\left(1-\frac{X}{\tanh X}\right) \qquad , \qquad (4.7)$$

Thus, the distance between the two fluxons, X(t) may be viewed, in this approximation, as a nonlinear damped oscillator moving in the potential

$$V(X) = -fX + AX/\sinh X, \quad A\equiv 2\varepsilon_2\sigma_1\sigma_2 \qquad , \qquad (4.8)$$

which we plot in Fig.5 for f>0, A>0 and A<0.

More specifically, let us consider the case of an antifluxon in junction 1 and a fluxon in junction 2, i.e. $\sigma_1=-1$ and $\sigma_2=1$, with $\varepsilon_2<0$, so as to be in the A>0 situation,

shown in Fig.5α. It is clear that if the fluxons are started
with initial conditions such that

$$U_1(0) = U_2(0) = 0, \quad X_L < X(0) < X_R \quad , \quad (4.9)$$

X_L =-20.075, X_R =-0.143, cf. Fig.5α, they will approach each
other until they are at a distance $X_{min} \cong -5.1$ and then they
will separate, forming a bifluxon which will continue to
execute damped oscillations (for α>0), within the potential
well shown in the figure. This is as far as the fluxons
relative motion is concerned.

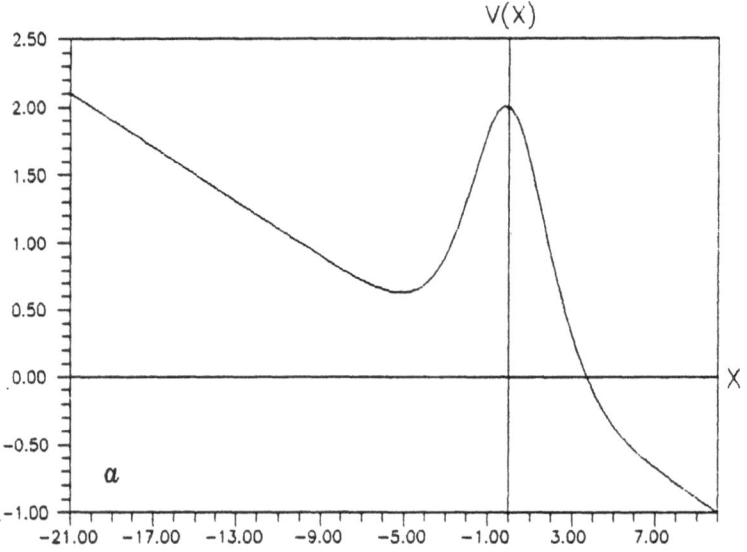

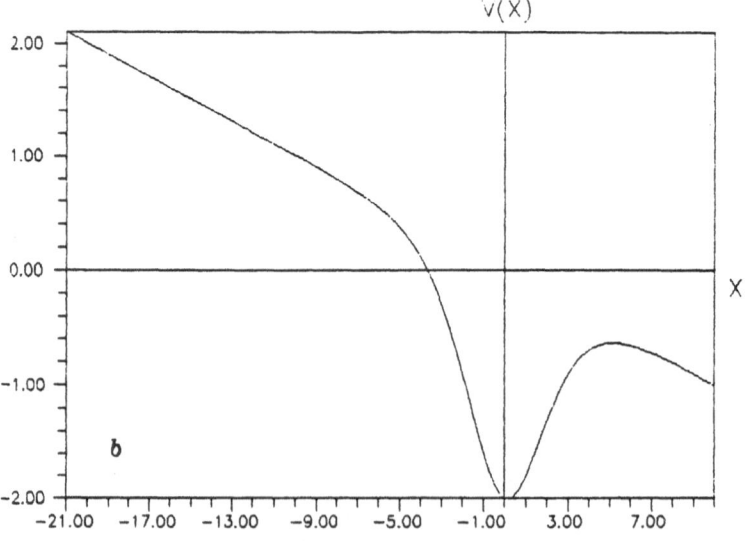

Figure 5 . A plot of the potential V(X) vs. X of (4.8)
for f=0.1 and *(a)* A=2, *(b)* A=-2.

On the other hand, their "center of mass" motion can be deduced by adding eqns (4.3b,c), whence one finds in the limit $|U_i| \to 0$:

$$\ddot{X}_1 + \ddot{X}_2 = -\alpha(\dot{X}_1 + \dot{X}_2) + g \quad , \qquad (4.10)$$

cf. (4.6), which can be integrated to give

$$\dot{X}_1 + \dot{X}_2 = c\, e^{-\alpha t} + g/\alpha \quad , c = \text{arb.const} \qquad (4.11)$$

This demonstrates how the bifluxon will move as $t \to \infty$. For convenience, we may choose $\gamma_1 = \gamma_2$ which makes g=0 in (4.6b) and places us in the rest frame of the "center of mass" of the system.

We have carried out a similar analysis in the general case, for all U_i, and have found entirely analogous results with the nonrelativistic case $|U_i| \ll 1$, as follows : Let us first define the new variables

$$V_i \equiv U_i / (1 - U_i^2)^{1/2}, \quad i = 1,2 \qquad (4.12)$$

in terms of which the equations of motion (4.3 b,c) take the symmetric form

$$\dot{V}_1 = -\alpha_1 V_1 + \sigma_1 \gamma_1 \frac{\pi}{4} + \frac{A}{4}(1 + V_1^2)^{1/2}(1 + V_2^2)^{1/2} I \qquad (4.13a)$$

$$\dot{V}_2 = -\alpha_2 V_2 + \sigma_2 \gamma_2 \frac{\pi}{4} - \frac{A}{4}(1 + V_1^2)^{1/2}(1 + V_2^2)^{1/2} I \qquad (4.13b)$$

with $A = 2\epsilon_2 \sigma_1 \sigma_2$, and I given by (4.4), for all U_i. We can then solve equations (4.13) by a standard 4th order Runge-Kutta scheme, evaluating numerically, at every step, the integral I=I(ρ,ϑ). Fortunately, this is not as time-consuming a task as it might at first appear, because the integrand in (4.4) takes appreciable values only over very short ranges of ξ, outside of which it goes to zero exponentially.

We thus obtained results which were qualitatively identical and quantitatively very similar to the nonrelativistic case. For example, adding eqns (4.13α,b), in the symmetric case (4.6a) , an integral of the motion analogous to (4.11) is immediately found

$$V_1 + V_2 = c\, e^{-\alpha t} + g/\alpha, \qquad c = \text{arb const}, \qquad (4.14)$$

For an antifluxon in LJJ1 and a fluxon in LJJ2, with $\gamma_1 = \gamma_2$(g=0), and initial conditions within the potential wells of Figure 5, a bifluxon is formed as described above , which oscillates forever, if $\alpha = 0$, or comes to rest as $t \to \infty$, if $\alpha > 0$.

Choosing, for example,

$$-f_1 = f_2 = 0.1, \quad \text{with} \quad f_i \equiv \sigma_i \pi \gamma_i / 4, \quad i = 1,2 \qquad (4.15)$$

and $\epsilon_2 = -0.5$, we found solving (4.13) in the undamped case, that a bifluxon state, with $U_1(0)=U_2(0)=X_1(0)=0$, is formed if $-5.2<X_2(0)<0$, whereas the nonrelativistic approximation gave $-4.97<X_2(0)<0$. On the other hand, no bifluxon state is found if $\epsilon_2 \geq -0.3225$.

Finally, we have observed evidence of chaotic oscillations associated with the presence of homoclinic orbits, when an ac bias current is applied on the junctions. For example, with

$$f_1 = -0.1 + \epsilon\cos\omega t, \quad f_2 = 0.1 \qquad (4.16)$$

and $\omega=0.2$, starting with initial conditions

$$U_1(0)=U_2(0)=X_1(0) = 0, \quad X_2(0)= -5.1 \qquad (4.17)$$

and $\epsilon=0.02$ in (4.16), the fluxon-antifluxon pair oscillated many times, reaching maximum and minimum distances which were completely aperiodic, before it eventually broke-up with the fluxon and antifluxon passing through each other and going to $\pm\infty$ respectively. These distances, as well as the number of oscillations occurring before the bifluxon breakup, were seen to depend very sensitively on the initial conditions and parameter values listed above.

Introducing a slight amount of dissipation in the problem, by setting $\alpha_1 = \alpha_2 = 0.01$ in (4.13), we observed the presence of a large basin of attraction about a stable fixed point of the Poincare map $X(t_k)$, $U(t_k)$, $t_k = 2\pi k/\omega$, $k=0,1.2,...$ This fixed point is none other than a simple periodic orbit to which the bifluxon phase-locks and oscillates with frequency $\omega=0.2$ as $t \rightarrow \infty$.

An unstable fixed point of this Poincare map also exists and is associated with the maximum in Fig.5α at which the fluxon and antifluxon nearly lie on top of each other . With $\alpha_1 = \alpha_2 = 0.01$, $\omega=0.2$ and (4.16), we observed a homoclinic tangency of the invariant manifolds, of this unstable fixed point at $\epsilon \cong 0.0575$, signalling the onset of transient chaotic oscillation in that region [15].

5. CONCLUSIONS

We have studied in this paper, using the collective coordinate approach, a number of single LJJ systems with inhomogeneities, and one system of 2 inductively coupled LJJ's. Interesting phenomena of fluxon trapping by inhomogeneities were observed and the formation of bifluxons was predicted in the coupled case. Moreover, these bifluxons, when driven by an ac bias current, were seen to execute phase-locked or chaotic oscillations depending on the parameter values and initial conditions of the system.

We have also compared the predictions of the odes of this approach with the results obtained by numerically solving the corresponding pdes, and have found, in all cases we tried, very satisfactory agreement. In the one notable exception of the 2 coupled LJJs, we have just managed to overcome certain

stability problems of our algorithms and will therefore
present our simulations of bifluxon oscillations in a future
publication [17].

Much remains to be done: We are currently working on
single and coupled LJJ systems with constant bias current, in
the presence of an rf microwave field, affecting the boundary
conditions in the usual way [5].

$$\varphi(0,t)+\beta\varphi_{xt}(0,t)=\varphi(L,t)+\beta\varphi_{xt}(L,t)=\eta_0\cos(\omega t+\delta)$$

Of primary concern in our studies will be the understanding of
such important phenomena as phase-locking and multifluxon
oscillations in these cases, especially as they relate to the
I-V characteristics of each system [17].

6. ACKNOWLEDGEMENTS

We have benefited from many useful discussions with
P.Christiansen, M.Soerensen, S.Pagano. R.Parmentier, G.Rotoli,
A.Ustinov and many other participants of this very interesting
Workshop. We also wish to acknowledge here the financial
support provided to our research programme by the Greek
Ministry of Industry, Energy and Technology .

7. REFERENCES

[1] A.Barone and G.Paterno, "Physics and Applications of the
     Josephson Effect", (Wiley, Interscience, New York, 1982).

[2] K. K. Likharev, "Dynamics of Josephson Junctions and
     Circuits", (Gordon and Breach, New York, 1986).

[3] N. F. Pedersen, Phys. Scr. T13 (1986) 129.

[4] P. L. Christiansen and N. F. Pedersen, in "Singular
     Behavior and Nonlinear Dynamics" ed. St. Pnevmatikos, T.
     Bountis and Sp. Pnevmatikos (World Scientific, Singapore,
     1989).

[5] S.Pagano,"Nonlinear Dynamics in Long Josephson Junctions"
     Ph.D thesis, Danish Center for Appl. Math. and Mechanics,
     Techn. Univ. of Denmark (1987).

[6] I. L. Serpuchenko and A. V. Ustinov, JETP Lett.46 (11),
     (1987) 549.

[7] A. A. Golubov, I. L. Serpuchenko and A. V. Ustinov, JETP
     94 (1988) 297.

[8] See many articles and their references in this volume.

[9] D.W.McLaughlin and A.C.Scott, Phys. Rev. A18 (1978) 1652.

[10] T.Bountis and St.Pnevmatikos, Phys.Lett. A143 (4,5)
      (1990) 221.

[11] T.Bountis, St.Pnevmatikos, St.Protogerakis, G.Sohos
      "Fluxon Trapping By Inhomogeneities in Long Josephson
      Junctions", to appear in Lect.Notes in Phys. 353 (1990)
      203.

[12] G. Reinisch and J. C. Fernandez in same volume as Ref. [4].

[13] Y. S. Kivshar, B. A. Malomed, Phys. Rev. B37 (16) (1988) 9325.

[14] N. Grønbech - Jensen, M. R. Samuelsen, P. S. Lomdahl, J. A. Blackburn, Phys. Rev. B42 (7) (1990).

[15] J. Guckenheimer and P. J. Holmes, "Nonlinear Dynamical Systems, Bifurcations and Vector Fields", Springer, New York (1983).

[16] T. Holst, J. Bindslev Hansen, N. Grønbech-Jensen, and J. A. Blackburn Phys. Rev. B42 (1) (1990) 127.

[17] T. Skiniotis, T. Bountis and St. Pnevmatikos, in preparation.

IMAGING OF SPATIAL STRUCTURES OF FISKE- AND SHAPIRO-STEP STATES

AND PHOTON-ASSISTED TUNNELING STATES IN JOSEPHSON JUNCTIONS

T. Doderer, D. Quenter, B. Mayer, C.A. Krulle, A.V. Ustinov[#], and
R.P. Huebener

Physikalisches Institut, Lehrstuhl Experimentalphysik II,
Universität Tübingen, D-7400 Tübingen, Fed. Rep. of Germany
[#]Permanent address: Institute of Solid State Physics, USSR
 Academy of Sciences, Chernogolovka

J. Niemeyer, R. Fromknecht, and R. Poepel

Physikalisch-Technische Bundesanstalt, D-3300 Braunschweig, FRG

U. Klein, P. Dammschneider, and J.H. Hinken[$]

Institut für Hochfrequenztechnik, Technische Universität
Braunschweig, D-3300 Braunschweig, FRG
[$]Present address: Hans Kolbe & Co., FUBA-Forschungszentrum,
 D-3202 Bad Salzdetfurth, FRG

We report the first experimental results obtained by Low Temperature Scanning
Electron Microscopy (LTSEM) on dynamic states of Josephson tunnel junctions.
The spatially resolved measurements dealing with self-resonant modes
(*Fiske-steps*) revealed information about the distribution of the rf-magnetic field
amplitudes inside the Josephson junction. We imaged one- and two-dimensional
cavity-modes inside a square junction. By coupling a Josephson element phase-
coherently to a microwave field we imaged the distribution of the maximum current density of the generated constant-voltage steps (*Shapiro-steps*). For junction
designs used as modern voltage standards we measured homogeneously distributed
step-current densities up to the fourth step. By irradiating superconducting tunnel
junctions with microwaves *photon-assisted tunneling* of quasiparticles takes place.
Using LTSEM we imaged the standing microwave pattern inside the junction. As a
result, the wavelength of the microwave inside a single tunnel junction and the
wavelength inside a superconducting stripline is displayed directly. The spatial
resolution achieved by LTSEM is about 1 - 2 μm.

1. INTRODUCTION

Josephson tunnel junctions show interesting dynamical behaviour in a
magnetic field or if they are coupled to a rf-source [1,2]. In both cases one
observes not only a change of the dc-Josephson current but also structures at
finite voltages in the current-voltage-characteristics (IVC). With a magnetic field
applied parallel to the junction barrier Fiske-steps are generated at discrete

Nonlinear Superconductive Electronics and Josephson Devices
Edited by G. Costabile *et al.*, Plenum Press, New York, 1991

353

voltages. As resonant dynamic modes of the junction, the corresponding voltages are only determined by the dimensions and the material of the superconducting tunnel junction. On the other hand, by applying microwave irradiation Josephson tunnel junctions show at least two distinct peculiarities. Firstly, due to phase-coherent coupling of the Josephson oscillator to the external oscillator rf-induced constant-voltage steps (Shapiro-steps) develop. These steps are the basis of voltage standards [3]. Secondly, near the energy gap region ($V = 2\Delta/e$) photon-assisted tunneling of the quasiparticles takes place [4,5].

Most of the experimental studies of the statics and dynamics of Josephson junctions are engaged in measuring the IVC under various experimental conditions. However, the IVC contains only spatially averaged information of the static or dynamic processes inside the junction. By applying spatially resolved measuring techniques for investigating those various states it is possible to get a deeper insight into the corresponding junction behaviour. Here, the achieved spatial resolution of the applied technique plays a crucial role. Among the competing methods Low Temperature Scanning Electron Microscopy (LTSEM) offers the best spatial resolution [6] (1 - 2 μm) and represents a very versatile tool for investigating low temperature sample properties [7,8].

Previously, LTSEM was applied to the investigation of static states of Josephson junctions. The spatial distribution of the maximum Josephson current density in one- [9] and two-dimensional [10] junctions was measured and images of vortex states and trapped flux quanta in the junctions were obtained. Besides the Josephson effects superconducting tunnel junctions show interesting behaviour concerning the tunneling of quasiparticles. By applying LTSEM to quasiparticle tunneling it is possible to image the distribution of the tunneling conductivity or the distribution of the energy gap [11,12]. Here, we report on our latest results of spatially resolved studies of Josephson tunnel junctions. Dynamical self-resonant modes with junctions in an external static magnetic field and rf-induced states with junctions coupled to a microwave source were investigated.

In section 2 the basic experimental procedures will be discussed. After that, we present experimental results combined with theoretical considerations dealing with Fiske-steps, Shapiro-steps, and photon-assisted tunneling states (section 3, 4, and 5, respectively). In section 6 we summarize our conclusions.

2. EXPERIMENTAL PROCEDURES

For scanning a sample at low temperatures with an electron beam it is necessary to have a special cold sample stage. To obtain stable mechanical and thermal conditions a specific cryostage was constructed [13]. With this cryostat it is possible to scan the sample with the electron beam of a conventional scanning electron microscope (SEM) while the sample is kept at about 4 K. The thin film sample is mounted on a substrate with high thermal conductivity which separates the liquid helium from the vacuum of the SEM. The sample film is either deposited directly on this high thermal conductivity substrate or the sample substrate is attached to a single-cristalline sapphire disk using high thermal conductivity glue. By minimizing the incoming thermal radiation by radiation shields and minimizing the dissipated power in the sample, sufficiently low temperatures are achieved. Temperatures down to about 1.6 K can be reached by pumping the helium gas above the liquid. For studying sample properties at higher temperatures a small bifilarily wound heater can be attached to the substrate.

For our studies of the rf-coupled Josephson junctions a new cryostat was constructed [14]. The main modifications are the installed microwave waveguide (60 - 90 GHz), a larger liquid helium tank inside the sample chamber of the SEM, and a magnetic shielding at room temperature around this tank. With this larger tank it is possible to attach larger sample substrates to the sample holder. Most of our measurements dealing with rf-coupled Josephson junctions used modern sample designs which need much more space than a single junction. The samples consisted of a dipole antenna, a superconducting microstripline which is terminated by a lossy stripline, dc-electrodes which are connected to the tunnel junctions by thin film rf-rejection filters, and of course the Josephson junctions. The basic rf-design of our samples is taken from the modern Josephson voltage standard. A detailed description of the design and the fabrication of the samples for Josephson voltage standard maintenance can be found in [15]. Additionally, this larger liquid helium tank gives the possibility to install two pairs of superconducting coils inside the tank. With these coils a sufficiently homogeneous magnetic field could be generated within the plane of the tunnel junction barrier in any direction.

Following [16], the effect of electron-beam irradiation on superconducting niobium or lead thin film samples can be considered as local heating at the e-beam focus. The lateral extension of the beam-induced thermal perturbation of the specimen film is given by the thermal healing length

$$\eta = \left(\frac{\varkappa\,d}{\alpha}\right)^{1/2}. \tag{2.1}$$

Here $\varkappa$ and d are the heat conductivity and the thickness of the specimen film, respectively. The quantity α is the heat transfer coefficient describing the heat flux from the film into the substrate. The presented experimental results were obtained by the response signals generated by means of the beam-induced thermal perturbation of the Josephson junction. Hence, the spatial resolution of our imaging technique is limited by the healing length η. For the samples used in our experiments the spatial resolution is about 3 μm. As a typical e-beam power we used about 2.5 μW (25 kV, 100 pA). For further details on the thermal model for treating the beam-induced sample perturbation we refer to previous summaries [7,8].

As outlined above, the focused electron beam irradiation results in *local* heating the Josephson tunnel junction. The corresponding response signal is the voltage- or current-change of the *whole* junction. These changes are extracted from the current-voltage-characteristics of the junction by different techniques which will be described in the following sections. We obtained images of different junction properties by scanning the junction with the electron beam and by recording the e-beam induced voltage- or current-changes as a function of the beam focus position. To understand these response signals it is essential to have a specific model of signal generation for each of the experimental conditions. These models will be discussed in the following sections together with the experimental results.

3. FISKE-STEPS

Using rectangular small Josephson junctions (length ℓ and width w are small compared to the Josephson penetration depth λ_J) and applying a magnetic field parallel to the junction barrier Fiske-steps are generated in the IVC at voltages [17,18]

$$V_{m;n} = \frac{h\,\overline{c}}{4\,e}\left[\left(\frac{m}{w}\right)^2 + \left(\frac{n}{\ell}\right)^2\right]^{1/2}. \tag{3.1}$$

Here h, $\overline{c}$, and e are Planck's constant, the Swihart velocity, and the elementary charge, respectively. m and n are integers. The corresponding modes are one-dimensional if m or n vanishes. This is expected if the magnetic field is parallel to an edge of the junction. Otherwise we expect two-dimensional modes.

The detailed features of the current-voltage-characteristics of the Fiske-steps are determined by the quality factor Q of the cavity. For our samples only the surface losses of the superconducting electrodes (Q_s) and the losses due to the tunneling of quasiparticles (Q_q) play an important role [19,20]

$$\frac{1}{Q} \approx \frac{1}{Q_s} + \frac{1}{Q_q} \quad . \tag{3.2}$$

The *local* irradiation of the Josephson junction with the focused electron beam results in a decrease of the quality factor Q of the *whole* junction. Scanning the homogeneous junction with the e-beam a spatially independent decrease of Q_q but a spatially dependent decrease of Q_s is expected. The losses inside the superconducting electrodes are determined by the real part of the surface impedance Re[Z(x,y)] and the rf-surface currents. These rf-currents are correlated with the distribution of the rf-magnetic field amplitude parallel to the surface of the electrodes inside the superconducting cavity $H_\parallel(x,y)$. As a result of the local temperature increase due to the e-beam irradiation Re[Z(x,y)] increases. With homogeneous junction electrodes this increase of Re[Z(x,y)] is independent of the irradiation position (x,y). Hence, irradiating the junction with the electron beam at (x,y) the decrease $-\Delta Q_s(x,y)$ of Q_s is proportional to the square of $H_\parallel(x,y)$.

Due to the decrease of Q the step voltage $V_{m;n}$ and the maximum step current $I^{max}_{m;n}$ change. These changes offer the possibilty to obtain two distinct images by LTSEM: The *voltage-image* $\Delta V_{m;n}(x,y)$ and the *current-image* $\Delta I^{max}_{m;n}(x,y)$. Assuming only a weak perturbation due to e-beam irradiation at (x,y) the decrease $-\Delta V_{m;n}(x,y)$ of the step voltage is proportional to the decrease of the quality factor Q(x,y). Hence, for two-dimensional cavity-modes in rectangular small Josephson junctions we expect the voltage-image

$$-\Delta V_{m;n}(x,y) = c_1 + c_2\left[\left(\frac{n\pi}{\ell}\right)^2 \sin^2\left(\frac{n\pi}{\ell}x\right)\cos^2\left(\frac{m\pi}{w}y\right) + \right.$$
$$\left. \left(\frac{m\pi}{w}\right)^2 \cos^2\left(\frac{n\pi}{\ell}x\right)\sin^2\left(\frac{m\pi}{w}y\right)\right] \quad . \tag{3.3}$$

Here m and n denote the mode for the component of the external static magnetic field in x- and y-direction, respectively. c_1 is the spatially independent voltage-signal due to the decrease of Q_q and the shift of the bias point in the IVC. This shift results from the increase of the tunneling conductivity due to the electron beam irradiation of the current biased junction [11]. The latter contribution will be explained in section 5. c_2 is the amplitude of the spatially dependent voltage-signal and is proportional to the increase of Re[Z]. For the current-image $\Delta I^{max}_{m;n}(x,y)$ we need a more sophisticated model to understand the measurements. This results from the sensitive dependence of the maximum Fiske-step current on various experimental parameters, e.g. the quality factor, the maximum Josephson current, or the external magnetic field [1,18]. A more detailed description of the signal generation and the experimental results will be given in [21].

The sample used in our experiments was a 50x50 μm^2 Nb/Al$_2$O$_3$/Nb-tunnel junction. The relative length and width of the junction was $\ell/\lambda_J = w/\lambda_J = 2.0$. Imaging the spatial distribution of the tunneling conductivity and of the energy gaps of the superconducting electrodes [11] reveals no inhomogeneity. Imaging the spatial distribution of the maximum Josephson current I_c by scanning the junction with the electron beam and by recording the beam induced change of I_c as a function of the beam focus, we observed only a slight deviation from a homogeneous current distribution due to the self field of the Josephson current [10]. Hence, the junction is of intermediate size.

Fig. 1 shows the voltage-images $-\Delta V_{0;n}(x,y)$ of six one-dimensional modes (n = 1 ... 6). As indicated in the figure, the external static magnetic field is directed parallel to the y-axis and thus parallel to one edge of the junction. The value of the magnetic field was chosen to obtain almost maximum step currents in every case. For imaging the Fiske-step states we scanned the current-biased junction with the electron beam and recorded the beam induced change of the corresponding step voltage. The current value was chosen to be about half of the maximum step current. The voltage-images were obtained by chopping the electron beam (20 kHz, 50 % duty cycle) and measuring the voltage phase-coherently by a lock-in technique. Typical beam induced voltage signals $-\Delta V_{0;n}(x,y)$ were 10 - 1500 nV, whereas the voltage of the first Fiske-step $V_{0;1}$ is 210 μV. Hence, the e-beam induced voltage signals are clearly less than 1 % of the absolute values of the step voltages. Comparing these voltage signals with our calculation (3.3), we see reasonable agreement for the first three steps. For the fourth and higher steps there is an increasing deviation near the left and right hand edges of the junction. Here, the absolute value of the voltage signal is smaller than the expected one [21]. Probably this comes from the weaker influence of the electron beam on the junction due to the finite diameter of the beam.

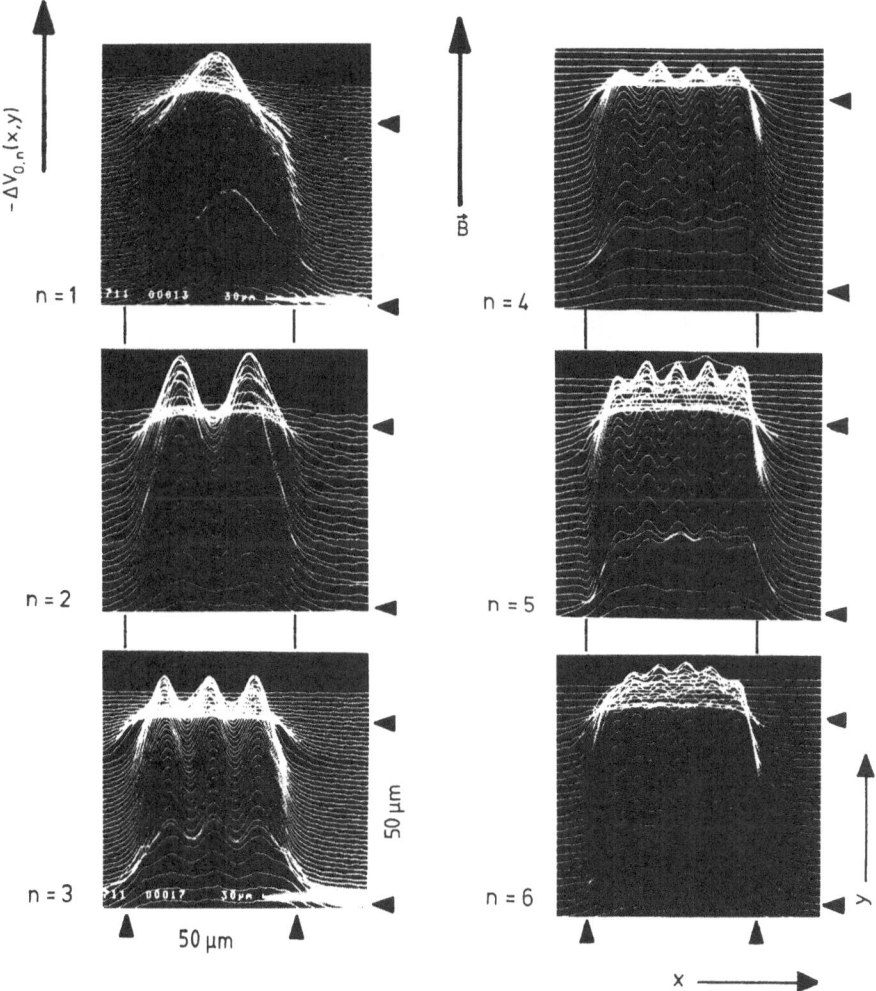

Fig. 1. Voltage-images $-\Delta V_{0;n}(x,y)$ of the first six one-dimensional Fiske-modes. As indicated, the external magnetic field is parallel to one edge of the Josephson junction. The arrows indicate the borders of the junction.

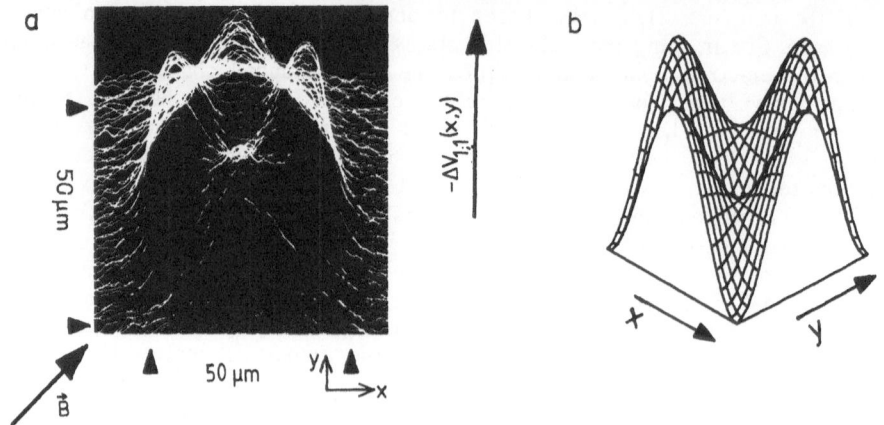

Fig.2. (a) Voltage-image $-\Delta V_{1;1}(x,y)$ of the first two-dimensional Fiske-mode. The external magnetic field is oriented nearly diagonal across the square of the junction. (b) Calculated voltage signal $-\Delta V_{1;1}(x,y)$ from (3.3). Only the modulated part is shown ($c_1 = 0$). Here the junction is rotated by 45° in comparison to (a) to show this figure more clearly.

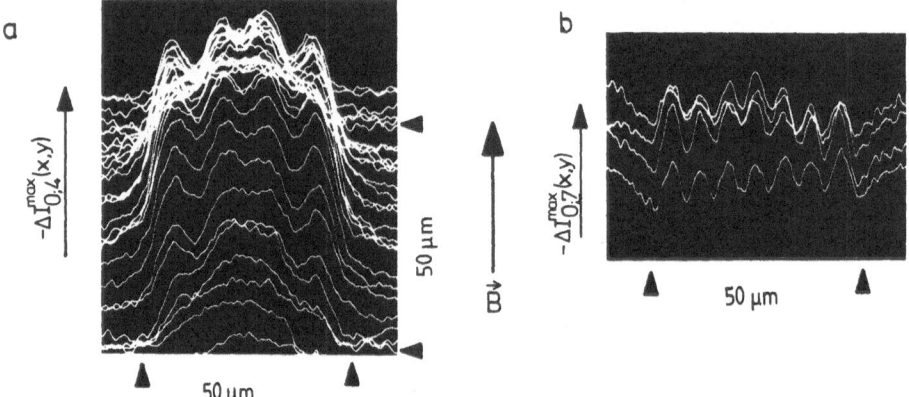

Fig. 3. Current-images $-\Delta I_{0;n}^{max}(x,y)$ of (a) the fourth ($n = 4$) and (b) the seventh ($n = 7$) one-dimensional Fiske-mode. The direction of the external magnetic field is parallel to one edge of the junction.

Fig. 2a shows the voltage-image $-\Delta V_{1;1}(x,y)$ of the first mixed-mode Fiske-step. Here, the external magnetic field is oriented nearly diagonal across the square of the junction. The step voltage $V_{1;1}$ was 300 μV. The calculated voltage signal $-\Delta V_{1;1}(x,y)$ is plottet in Fig. 2b using (3.3). Here we used $c_1 = 0$ to show the modulated part only and c_2 value is arbitrary. Comparing Figs. 2a and 2b, we see reasonable agreement of the measurement with our calculation.

Two examples of the current-images $-\Delta I_{0;n}^{max}(x,y)$ are shown in Fig. 3 for the fourth (Fig. 3a) and seventh (Fig. 3b) one-dimensional mode. The value of the external magnetic field was chosen to obtain nearly maximum step currents. For obtaining a current-image of a Fiske-step, we sweep the corresponding part of the IVC with a frequency of a few kHz. During each cycle the maximum step current is measured. The spatial distribution of the electron beam induced change of the maximum step current $-\Delta I_{m;n}^{max}(x,y)$ is obtained by scanning the junction with the beam and recording the maximum step current as a function of the beam focus position. Previously, the same imaging procedure was applied to the Josephson current and for details of the measuring technique we refer to [9]. As for the voltage-images the typical time needed for one line-scan is 1 s. Therefore, the maximum step current is measured a few thousand times during each line-scan. For the current-images we have not applied lock-in detection of the signal. Therefore, the noise level is increased in comparison with the voltage-images. Irradiation the Josephson junction with the electron beam does not only change the maximum step currents but also increases the quasiparticle current [11,22]. In the sub-gap region of the IVC this increase is more pronounced for higher voltages. If the maximum Fiske-step current is decreased by the e-beam irradiation we see a spatially independent shift of the modulated signal $-\Delta I_{m;n}^{max}(x,y)$ to lower values. This is clearly visible in Fig. 3b. The change of the maximum step current due to the e-beam irradiation is about 5 % of the absolute value of the step current.

Fig. 4 shows the current-image $-\Delta I_{1;2}^{max}(x,y)$ of a two-dimensional mode. As in Fig. 2a the magnetic field is oriented nearly diagonal. The corresponding step voltage $V_{1;2}$ was 470 µV. This image is obtained by modulating the brightness of the SEM-screen. Bright regions indicate large values of $-\Delta I_{1;2}^{max}(x,y)$ and vice versa. The typical time needed for one line-scan in brightness modulation technique is 100 ms. Increasing the number of line-scans in comparison to the y-modulation technique by a factor of ten, again one image takes about one minute. Taking ideal square junctions the IVC contains no information whether the (1;2)- or the (2;1)-mode is excited. Using our imaging technique it is easy to distinguish between the two equivalent modes. On the other hand, most of the real experimental conditions favor one mode above the other due to asymmetrical boundary conditions.

If the magnetic field is parallel to one edge of the junction, the spatial variation of the e-beam induced current and voltage signals appears only perpendicular to the direction of the field. On the other hand, we observed two-dimensional modes for nearly diagonal direction of the magnetic field. The reasonable agreement of the voltage-images with our calculations clearly supports our interpretation in terms of the cavity-mode excitations in the Josephson junction of intermediate size used in our experiments. These results are in agreement with spatially resolved studies of zero-field current steps of one-dimensional Josephson junctions [23]. For further details on our studies of the Fiske-steps we refer to [21].

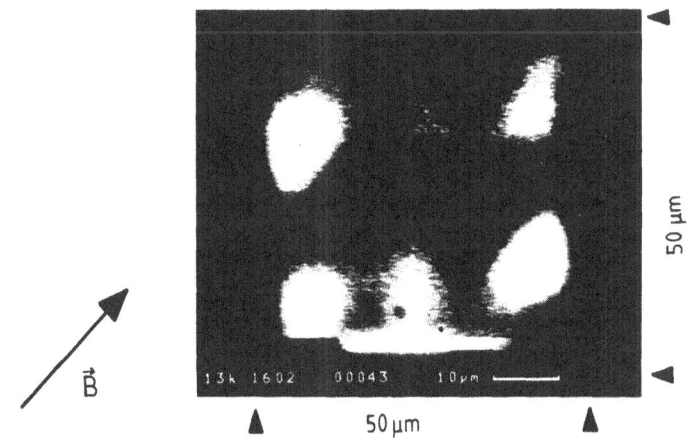

Fig. 4. Current-image $-\Delta I_{1;2}^{max}(x,y)$ of the (1;2) two-dimensional Fiske-mode shown in brightness modulation technique. The external magnetic field is oriented nearly diagonal across the square of the junction.

4. SHAPIRO-STEPS

Next we report on the first spatially resolved investigations of rf-coupled Josephson junctions. In contrast to Fiske-steps, the Shapiro-steps offer only the possibility to use the current-imaging technique due to the constant voltage of the steps. If the junction is phase-locked to an external microwave source, irradiating the junction with the electron beam does not result in a measureable change of the step-voltage (larger than 1 nV for a typical e-beam power). For obtaining current-images of the Shapiro-step states we use the same procedure as for the Fiske-steps described above. The main influence of local e-beam irradiation at a position (x,y) is to decrease the critical Josephson current density j_c and to increase the London penetration depth λ_L at that position. For homogeneous tunnel junctions the change of j_c and of λ_L are independent of the e-beam focus position (x,y). Assuming a spatially dependent rf-voltage $V_{rf}(x,y)$ inside the junction the usual analysis of the step-current dependence on V_{rf} [1] leads to a spatial dependent Shapiro-step current density $j_n(x,y)$ of the nth step

$$j_n(x,y) = (-1)^n j_c(x,y) J_n\left(\frac{2e V_{rf}(x,y)}{\hbar \omega_{rf}}\right) \sin\left(\varphi_0(x,y)\right) . \tag{4.1}$$

J_n, ω_{rf}, and φ_0 are the nth-order Bessel function, the frequency of the applied rf, and the phase difference between the Josephson oscillation and the $(n-1)^{th}$ harmonic of the applied rf, respectively. Here we used the voltage-bias model which is correct if the drive frequency ω_{rf} is much larger than the plasma frequency ω_p of the junction [24]

$$\omega_{rf} \gg \omega_p . \tag{4.2}$$

Hence, irradiating the junction with the electron beam at (x,y) the beam induced change of the Shapiro-step current $-\Delta I_n(x,y)$ is given by

$$-\Delta I_n(x,y) = \int (-1)^n \left(-\Delta j_c(\tilde{x},\tilde{y})\right) J_n\left(\frac{2e V_{rf}(\tilde{x},\tilde{y})}{\hbar \omega_{rf}}\right) \sin\left(\varphi_0(\tilde{x},\tilde{y})\right) d\tilde{x}\, d\tilde{y} . \tag{4.3}$$

Because of the local nature of $-\Delta j_c$, $-\Delta I_n$ has the same spatial dependence as the unperturbed current density $j_n(x,y)$. Therefore, by measuring the change of the maximum Shapiro-step current as a function of the beam focus (x,y) we measure the distribution of the maximum Shapiro-step current density

$$-\Delta I_n^{max}(x,y) \sim j_n^{max}(x,y) . \tag{4.4}$$

As for measuring the spatial distribution of the maximum Josephson current [25,26], in the dynamic state we must also consider possible nonlocal effects. Nonlocal effects could be an e-beam induced change of V_{rf} or of φ_0. These contributions may result from the increase of the London penetration depth λ_L and the decrease of the critical current density j_c due to the local heating by the electron beam. For example, the local increase of λ_L results in a local decrease of the Swihart velocity and, therefore, the local wavelength of the external rf-drive λ_{rf} will become shorter inside the tunnel junction. If the dimensions of the junction are not small compared to λ_{rf} this would result in a change of $V_{rf}(x,y)$, as discussed in section 5.

We used PbInAu/PbInoxide/PbAu-tunnel junctions embedded in a superconducting stripline well coupled to a receiving antenna and terminated by a lossy stripline. The microwave frequency was 70 GHz which results in a step voltage of 145 μV. The junction design with respect to the critical current density, the width (40 μm) and the length (20 μm) - measured in the direction of rf-current flow - of the junction is chosen to have the same values as for Josephson voltage standards based on the zero-current-bias approach [3,15,24]. Following these requirements [24] we expect a homogeneously distributed Shapiro-step current density. Neglecting the nonlocal effects, this will result in a spatially independent e-beam induced signal $-\Delta I_n^{max}(x,y)$.

Fig. 5a shows the spatial distribution of the critical Josephson current density $j_c(x,y)$ measured without external rf-power. With this junction ($j_c = 21$ A/cm^2, $\omega_p/2\pi = 23$ GHz, $\lambda_J = 80$ μm) the dimensions are smaller than λ_J. Hence, by recording the e-beam induced change of the maximum Josephson current, $-\Delta I_c$, as a function of the beam focus position, we obtained $j_c^{max}(x,y)$ [9]. In fig. 5b and c the current-images $-\Delta I_n^{max}(x,y)$ of the first and second Shapiro-step are shown. Here the microwave enters the junction from the lower edge across the whole width. For obtaining the images, the smallest microwave power was used to maximize the direct current of each step. Within the resolution of our imaging technique the maximum Shapiro-step current densities of the first and second step are homogeneously distributed. Using another junction with the same critical current density ($j_c = 21$ A/cm^2) the maximum step current densities showed the same spatial distribution as the critical Josephson current density up to the fourth Shapiro-step.

Imaging the distribution of the maximum Shapiro-step current density using a junction which is inhomogeneously coupled to the stripline (only near one corner) showed an inhomogeneously distributed step current density of the second step (Fig.6) whereas the current of the first step and the critical Josephson current were distributed quite homogeneously.

A Josephson junction with a trapped magnetic flux quantum shows a decreased critical Josephson current. Applying microwaves of proper power to the junction, the Josephson current is also decreased in the same way. Both times the IVC of that junction is nearly the same. By imaging the distribution of the maximum Josephson current, we obtained images as in reference [10] for trapped flux quanta. Here the Josephson current density is strongly modulated at the position of the flux quantum. In contrast, the distribution of the Josephson current, reduced to the same value due to rf-input, was quite homogeneous.

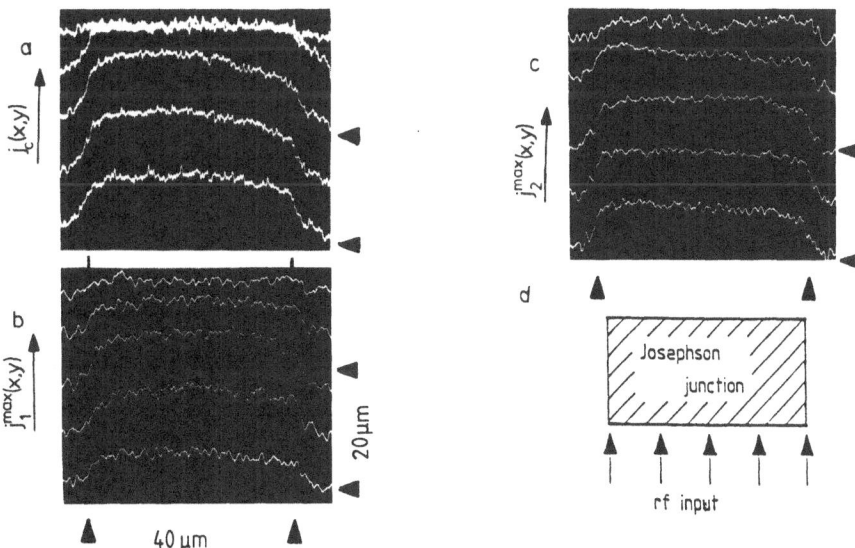

Fig. 5. Spatial distribution of (a) the critical Josephson current density j_c without microwave power and of the maximum Shapiro-step current densities j_n^{max} of (b) the first and (c) the second step. In Fig. (b) and (c) the rf enters the junction across the whole width from the lower edge (d).

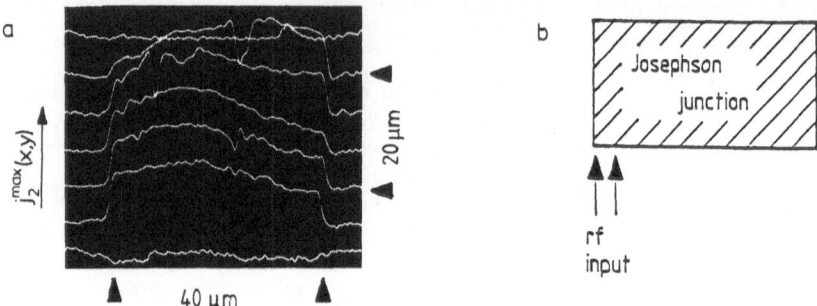

Fig. 6. Spatial distribution of (a) the maximum Shapiro-step current density j_2^{max} of the second step. The sharp drops in the signal are due to surface irregularities. The rf enters only near the lower left corner (b).

5. PHOTON-ASSISTED TUNNELING - IMAGING OF STANDING MICROWAVES

For many rf-applications of Josephson tunnel junctions like SIS-mixers, direct detectors (SIS-videodetectors), or voltage standards [2] it is highly desirable to know their interaction with the external microwave system and the propagation of the microwaves inside the devices. Recent investigations of rf-driven long Josephson junctions [27 - 30 and references therein] indicate the inverse ac Josephson effect due to synchronization of the fluxon oscillations and the externally applied rf-oscillation. In most of the theoretical studies the influence of the external microwaves is introduced by the boundary conditions or is assumed to be uniform across the whole junction. But in general, the distribution of the rf field inside the junction is not known [30]. If the frequency of the externally applied rf is larger than the plasma frequency of the junction, the electromagnetic field of the microwaves will penetrate into the junction [31]. The large impedance mismatch between the junction and the surrounding structure results in a large reflection of the microwaves at the boundaries. Hence, coupling of a junction to a microwave field of frequency larger than the plasma frequency results in standing electromagnetic waves inside the superconducting tunnel junction [32] and therefore the theoretical studies need a spatially inhomogeneous, temporally harmonic ansatz for describing the influence of the applied rf unless the dimensions of the junction are very small compared to the rf-wavelength inside the junction.

Here we report spatially resolved studies on the distribution of the rf-fields inside Josephson tunnel junctions. Obviously, these experimental investigations include all the losses of the microwaves inside the junction and can easily be extended to two-dimensional systems which is very complicated for the theoretical studies. Our imaging technique is based on photon-assisted tunneling (PAT) of quasiparticles which takes place in rf-coupled superconducting tunnel junctions [4,5]. If there is a standing microwave inside the junction, we expect an inhomogeneously distributed PAT-current density. Hence, the formerly developed technique for imaging of the current density distribution in superconducting tunnel junctions [11,22] can be extended to rf-coupled junctions.

By scanning the electron beam across the current-biased junction and recording the beam induced voltage signal we imaged the standing microwave pattern inside the junction. Fig. 7 shows the e-beam induced electronic processes within the current-biased junction. Here a schematic plot of the IVC of the e-beam irradiated subjunction is given. Irradiating the junction with the electron beam results in a change of the quasiparticle current $\Delta I(x,y)$ in the subjunction at the point of irradiation (x,y) due to local heating. If the temperature rise is large enough, the

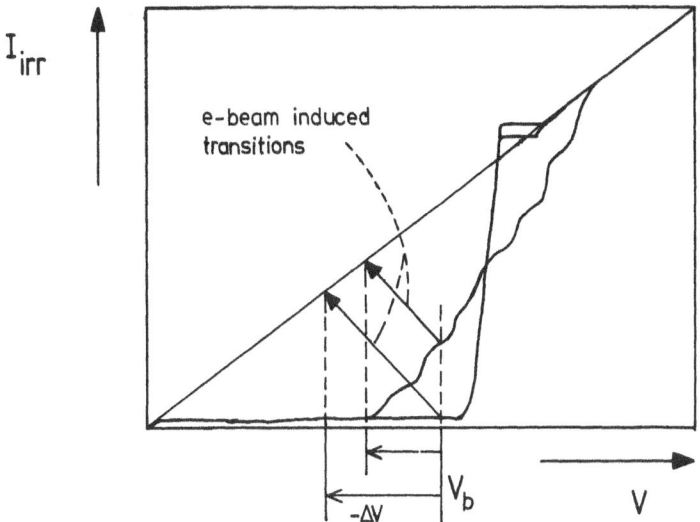

Fig. 7. Schematic presentation of the electronic processes within the e-beam irradiated part of the current-biased junction. The IVC is shown with and without rf-input. A more detailed description is given in [11].

irradiated subjunction will switch to the normal-state asyptote of the IVC as shown in Fig. 7. Because the junction is current-biased, the nonirradiated part of the junction acts as a load resistor and the differential resistance at the bias point determines the slope of the transition in Fig. 7 and, hence, the amplitude of the e-beam induced voltage signal $\Delta V(x,y)$. Under the assumption that the irradiated area of the junction is small compared to the nonirradiated area and that the resulting voltage signal is small, we have

$$\Delta V(x,y) \;=\; -\left(\frac{\partial V}{\partial I}\right)_{V_b} \Delta I(x,y) \;. \tag{5.1}$$

Here V_b denotes the bias point of the junction defined by the IVC and the value of the current bias. Hence, the magnitude of the voltage signal is proportional to the e-beam induced current change at the irradiated point. If the frequency of the applied microwaves is large enough, a step-like increase of the current near $V = 2\Delta/e$ is generated (PAT-steps) [4,5]. Therefore, the differential resistance at the bias point V_b depends sensitively on V_b. Hence, according to (5.1) the amplitude of the e-beam induced voltage signal ΔV varies strongly if we change V_b.

From Fig. 7 we expect larger voltage signals $-\Delta V(x,y)$ from areas with a smaller amount of absorbed microwave power and vice versa if $V_b < 2\Delta/e$. On the other hand, if $V_b > 2\Delta/e$ we expect the opposite. However, with Nb/Al-Al$_2$O$_3$/Nb-tunnel junctions which show a pronounced proximity effect in the IVC, the dependence of the voltage signal ΔV on the local rf-power is more complicated if a bias point $V_b \gtrsim 2\Delta/e$ is used because the normal-state asymptote can cross the IVC of the junction. Therefore the e-beam induced voltage signals ΔV will show a more complicated behaviour. For all images presented in this paper we used $V_b < 2\Delta/e$.

Concerning the origin of the modulation of the PAT-current density there is a close analogy to the imaging of Fiske-steps. Following the description given in section 3, we expect a large PAT-current density at positions inside the junction where the rf-magnetic field is large and vice versa in agreement with our experimental results partly presented in the following. For this we used the standing wave model for the distribution of the rf-magnetic field with the usual boundary conditions.

The samples used in our experiments were Nb/Al-Al$_2$O$_3$/Nb-tunnel junctions of different geometries concerning the dimensions of the junctions and the rf-coupling. The frequency of the applied microwave was 70 GHz, whereas the junction plasmafrequency was about 35 GHz. Therefore the penetration of the rf into the junction is expected [31]. For recording the images presented in this paper, we used a bias point $V_b \approx 2$ - 2.5 mV and the e-beam induced voltage signals were about 1 μV. Fig. 8a shows the voltage image without microwave input [11,22]. Obviously, the junction (20x400 μm^2) is very homogeneous. Fig. 8b shows the image of the standing microwave pattern inside this narrow long junction. The rf enters the junction from the right-hand side. Therefore, the left-hand side acts as the passive end, whereas the right-hand side acts as the active end of the rf-driven junction and, hence, the amplitude of the rf-magnetic field at the left edge is vanishing leading to a local maximum of $-\Delta V(x,y)$. From this figure we obtain directly the rf-wavelength λ_{rf} = 123 μm and the normalized Swihart velocity $\bar{c}/c$ = 0.029 inside the junction. The values are close to the expected ones [33]. Note, the spatial dependence of $-\Delta V(x,y)$ does not represent the rf-magnetic field in a direct way, but is dependent on the *square* of the rf-magnetic field amplitude (section 3). Therefore the spatial modulation of $-\Delta V(x,y)$ shows $\lambda_{rf}/2$. Whereas the imaging of Fig. 8b was done at about 4 K, Fig. 8c shows the voltage-image at a higher temperature T^+. From this image we obtain $\lambda_{rf}(T^+)$ = 105 μm and $\bar{c}/c$ = 0.025. At a higher temperature the London penetration depths of the two superconducting junction electrodes will be larger and, hence, the Swihart velocity will be smaller [33]. From the measured value of $\lambda_{rf}(T^+)$ we calculate $T^+ \approx 7.6$ K. A detailed analysis of our measurements will give information on the damping of the rf inside the junction, clearly indicated in Fig. 8b and 8c, and on the rf-coupling of the junction to the microwave feeding at the right-hand side. This analysis is not done up to now but we will extend our investigations to those problems in the near future.

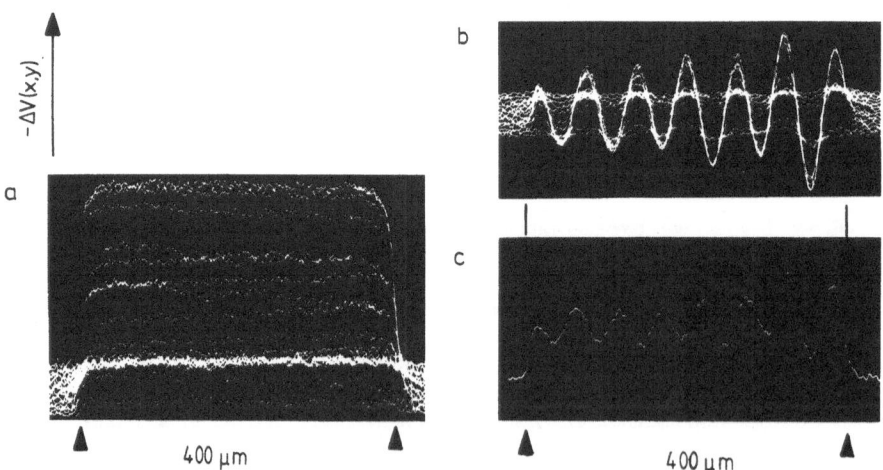

Fig. 8. (a) Voltage image without rf-power [11,22]. As shown, the junction (20x400 μm^2) is very homogeneous. (b) Voltage-image of the standing microwave pattern inside the same junction at about 4 K. The rf enters from the right-hand side. (c) One linear scan along the length of the junction at an increased temperature. The rf-wavelength inside the junction becomes shorter due to the increase of the London penetration depths of the superconducting electrodes.

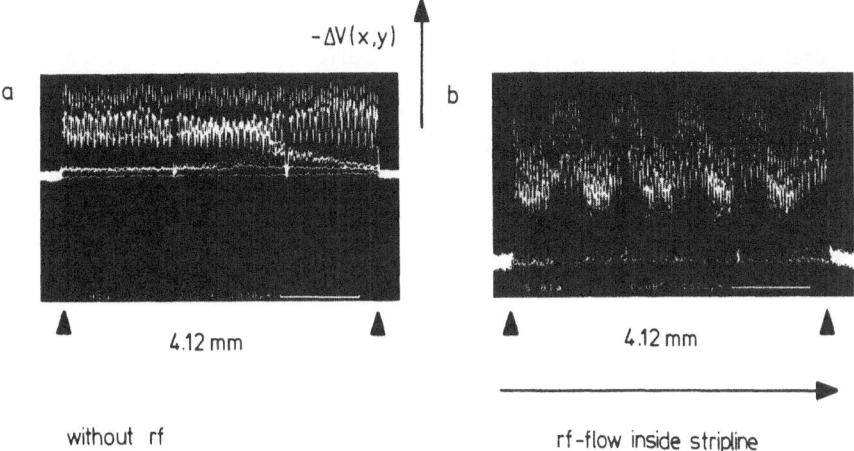

Fig. 9. (a) Voltage-image of a series-array of 100 Josephson-junctions without rf-input: Distribution of the tunneling conductivity. Each signal-peak represents one junction. Obviously, two of the junctions are shorted. (b) Voltage-image of the same array but with rf-input. The modulation of the signal amplitudes represents the standing microwave inside the underlying microstripline.

Next our investigations of a series-array of 100 Nb/Al-Al$_2$O$_3$/Nb-Josephson junctions will be described. The design of the individual junction and the rf-coupling to the microstripline is taken from the present voltage standard of the Physikalisch-Technische Bundesanstalt in Braunschweig, FRG. Each of the junctions is 22 μm in length and 46 μm in width and individually coupled to the underlying stripline. The total length of the array is 4.12 mm. Fig. 9a shows the distribution of the tunneling conductivity along the array [34]. Each of the signal peaks represents one tunnel junction. Obviously, two of the junctions are shorted. If rf-power is turned on, we obtain Fig. 9b. The microwave propagates inside the underlying stripline from the left to the right. The description of the signal generation given above for a single junction holds also for the series-array. Therefore, there is a standing microwave inside the stripline. The standing wave probably comes from an inhomogeneity of the lossy stripline. This inhomogeneity of the resistive load is located close to the superconducting stripline. For obtaining the image, each of the 98 junctions acts as a detector for the rf-power within the microstripline just at the position of the junction. From Fig. 9b we obtain directly the wavelength inside the Nb/Nb$_2$O$_5$-SiO-MgO/Nb-stripline λ_{rf}^{MSL} = 1.68 mm and the normalized Swihart-velocity $\bar{c}/c$ = 0.39. Again, this wavelength is close to the expected value [33]. As expected [15,35], Fig. 9b does not indicate any damping of the microwave along the array.

Note, for imaging the standing microwave inside a single junction we use just the same junction as the cavity and as the detector of the rf. On the other hand, for imaging the standing microwave inside the stripline we use the stripline as the cavity and each junction of the series-array as a detector of the rf.

6. CONCLUSIONS

Using Low Temperature Scanning Electron Microscopy (LTSEM), we investigated spatially resolved dynamic states of Josephson tunnel junctions. By imaging self-resonant modes (Fiske-steps) in a square junction of intermediate size, we obtained evidence for the excitation of one- and two-dimensional cavity-modes. Our imaging technique reveals the distribution of the square of the rf-

365

magnetic field amplitude inside the junction. Studying Shapiro-steps by LTSEM, the spatial distribution of the maximum step current density was measured. Using samples of the Josephson voltage standard design, the maximum direct currents of the constant-voltage steps are homogeneously distributed across the junction area up to the fourth step. LTSEM offers the possibility to image the standing microwave pattern inside Josephson junctions and inside superconducting microstriplines. For this, we imaged the spatial distribution of the photon-assisted tunneling current of the quasiparticles. The spatial resolution achieved by this low temperature imaging technique is about 1 - 2 µm.

ACKNOWLEDGEMENTS

We would like to thank K.K. Likharev for valuable discussions and M. Koyanagi for manufacturing and sending the square junctions of high quality. During the preparation of our own samples two of the authers (D. Q. and T. D.) enjoyed assistance from A. Larsen, H. Dalsgaard Jensen, S. Maggi, and W. Meier. One of us (T. D.) is pleased to acknowledge the help from L. Grimm concerning the microwave devices and the stimulating discussions with J. Bosch and R. Gross. One of the authors (A. V. U.) wants to thank for the support from the Alexander von Humboldt-Stiftung. Financial support of this work from the Deutsche Forschungsgemeinschaft is gratefully acknowledged.

REFERENCES

[1]   A. Barone, G. Paternò, Physics and Applications of the Josephson Effect, John Wiley & Sons, New York (1982)

[2]   K.K. Likharev, Dynamics of Josephson Junctions and Circuits, Gordon and Breach Science Publishers, New York (1986)

[3]   J. Niemeyer, L. Grimm, Metrologia **25**, 135 (1988)

[4]   A.H. Dayem, R.J. Martin, Phys. Rev. Lett. **8**, 246 (1962)

[5]   P.K. Tien, J.P. Gordon, Phys. Rev. **129**, 647 (1963)

[6]   J. Bosch, R. Gross, R.P. Huebener in: Josephson Effect - Achievements and Trends, A. Barone, ed., World Scientific Publishing, Singapore, 1986, p. 174

[7]   R.P. Huebener, Rep. Prog. Phys. **47**, 175 (1984)

[8]   R.P. Huebener in: Advances in Electronics and Electron Physics, Vol. **70**, P.W. Hawkes, ed., Academic Press, New York (1988), p. 1

[9]   J. Bosch, R. Gross, M. Koyanagi, R.P. Huebener, J. Low Temp. Phys. **68**, 245 (1987)

[10]   J. Mannhart, J. Bosch, R. Gross, R.P. Huebener, J. Low Temp. Phys. **70**, 459 (1988)

[11]   H. Seifert, R.P. Huebener, P.W. Epperlein, Phys. Lett. **95A**, 326 (1983)

[12]   R. Gross, D.B. Schmid, R.P. Huebener, J. Low Temp. Phys. **62**, 245 (1986)

[13]   H. Seifert, Cryogenics **22**, 657 (1982)

[14] T. Doderer, H.-G. Wener, R. Moeck, C. Becker, R.P. Huebener, Cryogenics **30**, 65 (1990)

[15] J. Niemeyer in: Superconducting Quantum Electronics, V. Kose, ed., Springer-Verlag, Berlin (1989) p. 228

[16] R. Gross, M. Koyanagi, J. Low Temp. Phys. **60**, 277 (1985)

[17] M.A.H. Nerenberg, P.A. Forsyth, Jr., J.A. Blackburn, J. Appl. Phys. **47**, 4148 (1976)

[18] H. Svensmark in: SQUID '85, Superconducting Quantum Interference Devices and their Applications, H.D. Hahlbohm, H. Lübbig, eds., Walter de Gruyter, Berlin (1985) p. 471

[19] T.C. Wang, R.I. Gayley, Phys. Rev. B **18**, 293 (1978)

[20] R.F. Broom, P. Wolf, Phys. Rev. B **16**, 3100 (1977)

[21] B. Mayer, T. Doderer, R.P. Huebener, to be published

[22] P.W. Epperlein, H. Seifert, R.P. Huebener, Phys. Lett. **92A**, 146 (1982)

[23] S. Meepagala, J.T. Chen, J.-J. Chang, Phys. Rev. B **36**, 809 (1987)

[24] R.L. Kautz, C.A. Hamilton, F.L. Lloyd, IEEE Trans. Magn., **MAG-23**, 883 (1987)

[25] J.-J. Chang, C.H. Ho, D.J. Scalapino, Phys. Rev. B **31**, 5826 (1985)

[26] J. Bosch, R. Gross, M. Koyanagi, R.P. Huebener, Phys. Rev. Lett. **54**, 1448 (1985)

[27] N.F. Pedersen, A. Davidson, Phys. Rev. B **41**, 178 (1990)

[28] G. Costabile, R. Monaco, S. Pagano, G. Rotoli, Phys. Rev. B **42**, 2651 (1990)

[29] M. Cirillo, F.L. Lloyd, J. Appl. Phys. **61**, 2581 (1987)

[30] G. Costabile, R. Monaco, S. Pagano, J. Appl. Phys. **63**, 5406 (1988)

[31] J.-J. Chang, Phys. Rev. B **34**, 6137 (1986)

[32] C.A. Hamilton, S. Shapiro, Phys. Rev. B **2**, 4494 (1970)

[33] J.C. Swihart, J. Appl. Phys. **32**, 461 (1961)

[34] J. Bosch, R. Gross, R.P. Huebener, J. Niemeyer, Appl. Phys. Lett. **47**, 1004 (1985)

[35] J. Niemeyer, J.H. Hinken, R.L. Kautz, Appl. Phys. Lett. **45**, 478 (1984)

CRITICAL CURRENT IN LONG JOSEPHSON JUNCTIONS :

A PHASE SPACE APPROACH

S. Pagano, B. Ruggiero, M. Russo, and E. Sarnelli

Istituto di Cibernetica del Consiglio Nazionale delle Ricerche
I-80072, Arco Felice, Naples, Italy

INTRODUCTION

Long Josephson junctions,[1] i.e. junctions with at least one geometrical dimension (L,W) larger than the Josephson penetration depth λ_j have been matter of a great deal of theoretical[2,3] and experimental[4,5] work.

Potential applications are in radioastronomy and in space communications,[6] where it has been proposed to use long junctions as local oscillators for millimeter wave integrated superconducting receiver.[7]

The electrodynamics of a long Josephson tunnel junction is described by a Perturbed Sine Gordon Equation (PSGE).[8] Assuming the junction lying in the x,y-plane, the PSGE can be written in terms of the superconducting phase difference ϕ and in normalized units as:

$$\phi_{xx} - \phi_{yy} - \phi_{tt} - \sin\phi = \alpha\phi_t - \beta\left(\phi_{xxt} + \phi_{yyt}\right) \tag{1}$$

where the subscripts indicate the spatial (x,y) and the time (t) derivatives. The variables x,y and t are normalized to λ_j and to the inverse of the plasma frequency $\omega_j = 2\pi c/\lambda_j$ respectively; α and β are dimensionless parameters representing the junction shunt conductance and the electrode surface conductance.

Eq. (1) cannot be solved analytically, as a consequence numerical and perturbative methods[9,10] have been proposed.

In the following it will be considered the time independent case, so that Eq. (1) reduces to:

$$\phi_{xx} - \phi_{yy} - \sin\phi = 0 \tag{2}$$

In this framework the present paper reports a theoretical analysis and experimental results for overlap-type tunnel junctions, (see Fig. 1a). Since in any real junction there is a current flowing through all the borders, a more general geometrical configuration has to be considered, (see Fig. 1b).

Investigating Eq.(1) many approaches[11,12] have been used with different assumptions on the bias current distribution. In most of the cases a homogeneous current distribution along each border of the junction is assumed.

A simple way to take into account of a non-uniform bias current distribution was developed by Barone et al,[13] using a linearization (sin $\phi \sim \phi$) of the Sine-Gordon Equation (SGE) for small variation of the phase.

Nonlinear Superconductive Electronics and Josephson Devices
Edited by G. Costabile *et al.*, Plenum Press, New York, 1991

369

In a recent paper,[14] a bias current distribution with singularities at the edges of the junction electrodes has been considered, making the situation very similar to a mixture of an overlap and an in-line configuration. More recently,[15,16] a new semi-analytical approach to solve the SGE has been developed. The main idea is to perform a phase-space analysis of the time-independent PSGE.

In this framework the first section deals with a general approach to the problem. The application of this technique to the semi-infinite overlap, finite length overlap, and mixed overlap/in-line junction configurations is matter of the following sections. A comparison between theoretical predictions and experimental data is also reported in the semi-infinite and mixed overlap/in-line case which well represents the real junction configurations.

THEORETICAL APPROACH

The static phase configurations of a long Josephson junction are described by Eq. (2) with the boundary conditions:

$$\phi_x(0,y) = \underline{h} \cdot \underline{y}\Big|_{x=0} \tag{3a}$$

$$\phi_x(l,y) = \underline{h} \cdot \underline{y}\Big|_{x=l} \tag{3b}$$

$$\phi_y(x,0) = -\underline{h} \cdot \underline{x}\Big|_{y=0} \tag{3c}$$

$$\phi_y(x,w) = -\underline{h} \cdot \underline{x}\Big|_{y=w} \tag{3d}$$

where $l = L/\lambda_j$, $w = W/\lambda_j$, $\underline{h} = H/\lambda_j J_c \mu_0$ is the normalized magnetic field ($\underline{H}_e$) in the junction plane; J_c is the Josephson current density, $\underline{x}$ and $\underline{y}$ are unitary vectors. For the geometry considered in Fig. 1b, assuming an uniform current distribution, the contribution to the magnetic field due to the bias current can be separated from the external magnetic field, and the boundary conditions become:

$$\phi_y(0,y) = \eta - \mu\gamma l \tag{4a}$$

$$\phi_x(l,y) = \eta + \nu\gamma l \tag{4b}$$

$$\phi_y(x,0) = -\delta\gamma w \tag{4c}$$

$$\phi_y(x,w) = \sigma\gamma w \tag{4d}$$

where $\gamma = I_c/LWJ_c$ is the normalized bias current and η is the normalized external magnetic field assuming $H_e = (0, H_e, 0)$.

The parameters μ, ν, σ, δ indicate the fraction of the bias current flowing through the junction edges verifying the Ampere's law : $\mu + \nu + \sigma + \delta = 1$.

The present description allows to take into account both "pure" overlap ($\mu = \nu = 0$) and in-line ($\sigma = \delta = 0$) configurations.

In the one dimensional case ($l >> 1 >> w$), the y dependence of the phase ϕ can be factorized and assumed parabolic, therefore Eq.(2) is reduced to a 1+1 differential equation system. For a symmetrical bias current configuration ($\mu = \nu$), Eqs. (2) and (4) become respectively:

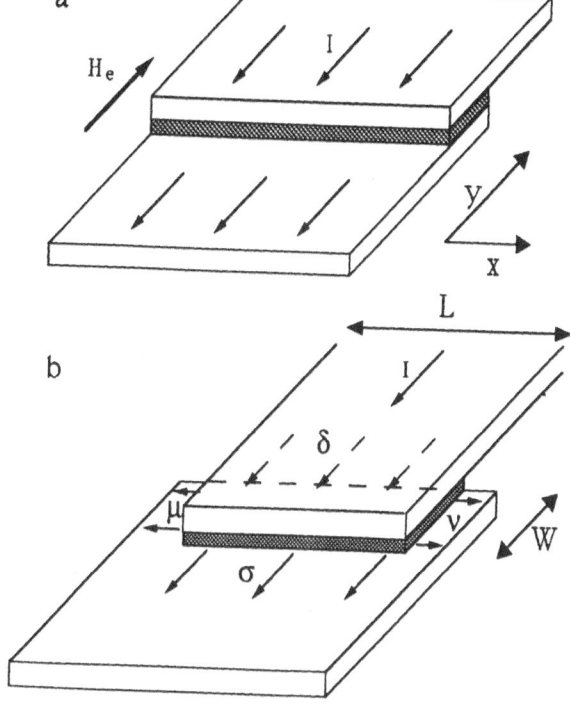

Fig. 1 Sketch of the considered junction geometries. a) "pure"
overlap; b) mixed overlap/in-line configuration. $\mu, \upsilon,$
σ, δ indicate the fraction of the bias current flowing
through the junction borders.

$$\phi_{xx} - \sin\phi = -(1 - 2\mu)\gamma \qquad (5)$$

$$\phi_x(0) = \eta - \mu\gamma l \qquad (6a)$$

$$\phi_x(l) = \eta + \mu\gamma l \qquad (6b)$$

By varying the parameter μ the boundary conditions of a "pure" overlap ($\mu = 0$) and of a mixed overlap/in-line ($0 < \mu < 0.5$) configuration can be obtained. Eq.(5) is a pendulum-type equation, the only difference consisting in the substitution of the time variable t with the space variable x and in a shift of the phase by π. By integrating Eq.(5) it is possible to obtain the analytical expression of the phase-space $\phi_x(\phi)$ as:

$$\phi_x(\phi) = \pm \sqrt{2(k - (1 - 2\mu)\gamma\phi - \cos\phi)} \qquad (7)$$

where k is an integration constant. All the possible phase configurations represent solutions of Eq. (5) if they verify the boundary conditions, Eqs.(6). A picture of the phase space is shown in Fig. 2a. The phase plane is symmetric with respect to the $\phi_x = 0$ axis and replicates itself under a phase shift of $2\pi m$

(with m integer), provided that the integration constant k is redefined as k' = k - 2mπγ.

In Fig. 2a $A_m = (\sin^{-1} \gamma \pm 2m\pi, 0)$ and $B_m = (\pi - \sin^{-1} \gamma \pm 2m\pi, 0)$ are two infinite sequences of fixed points of Eq. (5). Two families, constituted by open and closed curves, are also shown. They are separated by curves passing through the fixed points A_m (the separatrices). In Fig. 2b separatrices corresponding to different values of the bias current γ at a given value of μ, are shown.

The study of the phase plane allows predictions on the critical current dependence on the magnetic field and junction length.

In the following sections, some relevant cases, characterized by different values of the parameter μ, are discussed.

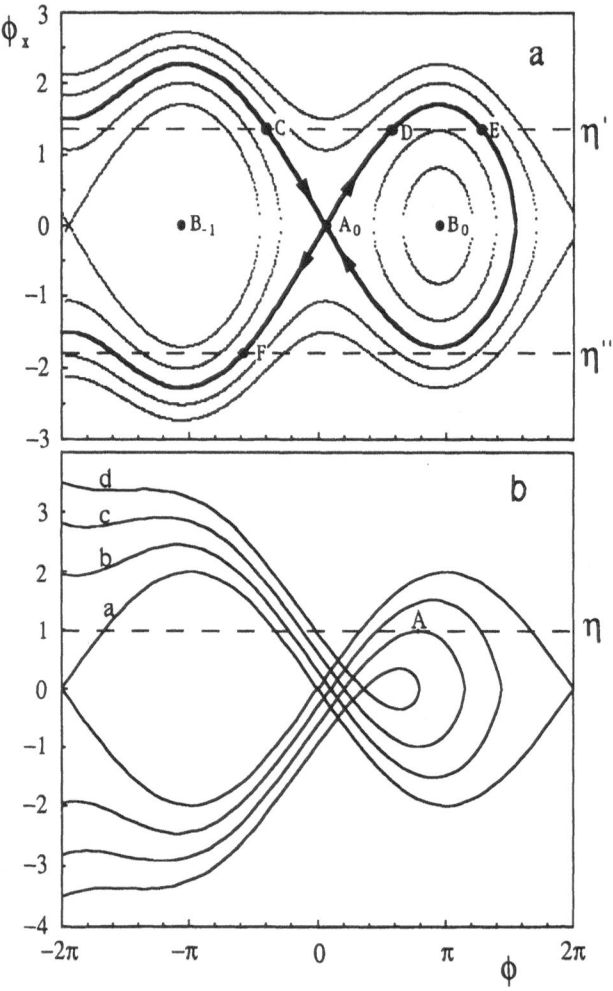

Fig. 2 a) Phase space configurations for $\gamma = 0.5$, $\mu = 0.3$, and for different values of the integration constant k. The solid line is the separatrix and $A_m = (\sin^{-1} \gamma \pm 2m\pi, 0)$ and $B_m = (\pi - \sin^{-1} \gamma \pm 2m\pi, 0)$, with m integer, are fixed points of Eq. (7)
b) Separatrices for $\mu = 0.2$ and different values of the normalized bias current γ: a) $\gamma = 0$, b) $\gamma = 0.3$, c) $\gamma = 0.6$, d) $\gamma = 0.9$.

In the semi-infinite length limit of an overlap junction ($\mu = 0$) the external magnetic field is assumed to influence only one edge ($x = 0$) of the junction, while on the other one ($x \to \infty$) there is no field ($\eta = 0$). At this border the phase relaxes to the equilibrium value $\phi = \sin^{-1}\gamma$. The boundary conditions become:

$$\phi_x(0) = \eta \tag{8a}$$

$$\phi_x(x \to \infty) = 0 \tag{8b}$$

$$\phi(x \to \infty) = \sin^{-1}\gamma \pm 2m\pi \tag{8c}$$

Referring to Fig. 2a, all the solution curves are parts of the separatrix. Inserting Eqs. (8) in Eq. (7) the analytical expression for the separatrix is obtained:

$$\phi_x(\phi) = \pm \sqrt{2\left(\sqrt{1-\gamma^2} + \gamma \sin^{-1}\gamma - \cos\phi - \gamma\phi\right)} \tag{9}$$

Referring to Fig. 2a, branch solutions in the phase space are the curve DA_0 for a positive applied magnetic field η' and FA_0 for a negative magnetic field η''.

By an analysis of the existence range of the branches DA_0 and FA_0 it is possible to evaluate the dependence of the critical current on the external magnetic field, $\gamma_c(\eta_c)$. The critical value η_c is obtained when the line $\phi_x = \eta'$ becomes tangent to the separatrix, (see Fig. 2b). The tangent point co-ordinates ($\pi - \sin^{-1}\gamma$, η_c) inserted in Eq.(9) give the corresponding value of γ_c. It is worth noting that this critical situation can occur for the branch DA_0 and not for the branch FA_0, since the latter has always an intersection with the line $\phi_x = \eta''$. This circumstance leads to the occurrence of a "plateau" effect for the critical current at negative values of the magnetic field. The γ_c vs. η_c theoretical dependence is shown in Fig. 3 (solid line). Such an anomalous dependence is due to the possibility of injecting an arbitrary number of fluxons into the junction with a negative magnetic field keeping the balance by the bias current.

Practically a semi-infinite configuration can be simulated by a very long junction in which the magnetic field influences only one edge. From the experimental point of view such a situation can be achieved by means of a control current generating the magnetic field near one edge of the junction; in this way the boundary conditions (Eqs.(8)) approximately hold. The experimental magnetic field dependence of the Josephson critical current $I_c(H_e)$ is reported in Fig. 3. The data refer to a very long Nb/NbO$_x$/Pb overlap-type junction ($l > 10$). Two different magnetic field configurations are considered. When the external magnetic field is uniformly distributed over the junction it is evident that the $I_c(H_e)$ dependence (crosses) is symmetric with respect to $H_e = 0$. In the case of an asymmetrically distributed field a highly asymmetrical $I_c(He)$ dependence (squares) is observed. The discrepancy between the experimental data and the theoretical behavior is obviously due to an insufficient length of the junction.

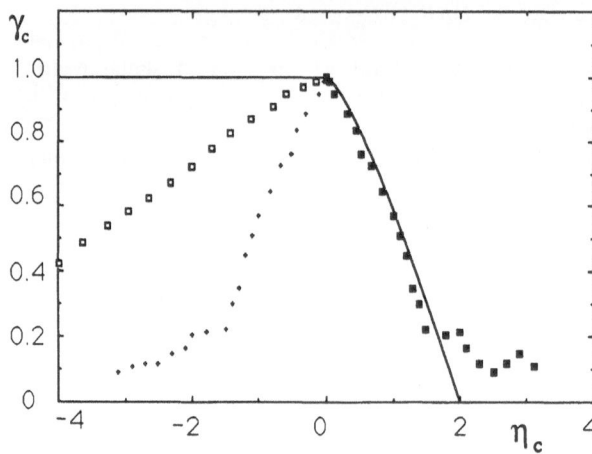

Fig. 3 Magnetic field dependence of the maximum dc Josephson current for a semi-infinite overlap junction. The solid line is the theoretical behavior. The experimental results are reported in normalized units. Two different configurations of the external magnetic field are considered: in the former case, the field is uniformly distributed over the junction (crosses), and in the latter it is localized mainly near only one junction edge (squares).

FINITE LENGTH OVERLAP JUNCTION

In the case of a finite length junction with an overlap geometry, the phase configuration is described by Eq. (5), under the boundary conditions reported in Eqs. (6) with $\mu = 0$ (overlap limit).

The normalized length l of the junction is implicitly given by the integral:

$$l = \int_{\phi(0)}^{\phi(l)} \phi_x^{-1} d\phi \tag{10}$$

computed along the phase space branch satisfying the boundary conditions Eqs.(6).

For each curve in the phase space, $\phi(0)$ and $\phi(l)$ are the values of the phase at the intersections between the curve and the straight line $\phi_x = \eta$ (see Fig. 4).

It is worth nothing that solutions related to m magnetic flux quanta (fluxons) have to satisfy the condition:

$$2\pi m \leq |\phi(l) - \phi(0)| \leq 2\pi(m+1) \tag{11}$$

In this picture[15] there is a minimum (maximum) value of the integration constant k in Eq. (6) k_{min} ($k^m{}_{max}$), for which it is possible to find an intersection between $\phi_x = \eta$ and a curve $\phi_x(\phi)$ on the right (left) of the point A_0. It is also possible to see[12] that only the maximum value of k it is different for different values of m. Hence all the possible solutions lay on curves with: $k_{min} < k < k^m{}_{max}$ (see Fig. 4).

The maximum and the minimum length are:

$$I_{max}(\gamma,\eta) = \int_{\phi(0)}^{\phi(l)} \phi_x^{-1}(k_{min},\gamma,\eta)d\phi \qquad (12a)$$

$$I_{min}^m(\gamma,\eta) = \int_{\phi(0)}^{\phi(l)} \phi_x^{-1}(k_{max}^m,\gamma,\eta)d\phi \qquad (12b)$$

By solving numerically the equations $I_{max}(\gamma,\eta) = 1$ and $I_{min}^m(\gamma,\eta) = 1$ the complete dependence of the critical current on the magnetic field is computed. In Fig. 5 it is shown the theoretical dependence $\gamma_c(\eta_c)$ for $l = 6$.

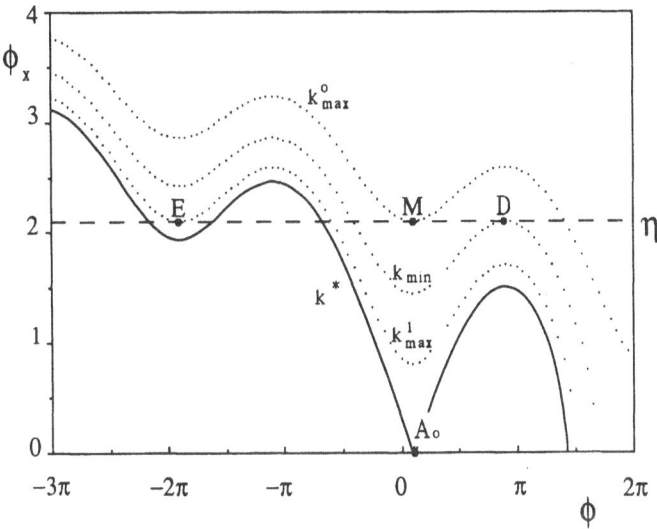

Fig. 4 Phase space diagram for $\gamma = 0.5$ and $\mu = 0.2$. k_{min}, k^0_{max}, k^1_{max}, and k^* identify the curves tangent to $\phi_x = \eta$ at the first maximum at the right of A_0 (point D), at the point M, at the first minimum on the left of the A_0 (point E), and the separatrix, respectively.

MIXED OVERLAP/IN LINE CONFIGURATIONS

A mixed overlap/in-line configuration is characterized by a fraction of the bias current flowing through the junction borders, perpendicularly to the junction long direction. In the present picture, it implies $\mu \neq 0$ in the boundary conditions, Eqs.(6).

The phase space plot is quite similar to the case $\mu = 0$, nevertheless it requires a rather different analysis. In spite of a difficult general treatment, relevant results can be obtained in some significant limits. Assuming $\eta = 0$ the boundary conditions become:

$$\phi_x(0) = -\mu\gamma l \qquad (13a)$$

$$\phi_x(l) = +\mu\gamma l \qquad (13b)$$

Referring to Fig. 6, a solution branch must connect a positive and a negative value of ϕ_x. Since a finite length junction is considered, a solution curve cannot contain the point A_0, so that all the stable solution branches must belong to the closed curve family, between the curve a and the separatrix c. For a given choice of the parameters μ, γ, and l, the solution is the branch satisfying Eq. (10). It is worth nothing that in the general case ($\eta \neq 0$) these considerations are valid until the condition

$$\eta < \mu\gamma l \qquad (14)$$

is verified.

In order to obtain the dependence of the maximum dc Josephson current on the junction length $\gamma_c(\eta=0, l)$ for different values of the parameter μ, the value of k such that γ reaches its maximum value γ_c has to be evaluated. Increasing the current, the separatrix becomes closer and closer to the fixed point H_e (see Fig. 2b), and the straight lines given by Eq. 13 separate each other. Referring to Fig. 6, a critical state, corresponding to the maximum γ_c value allowed, is obtained when Eq. (10) is satisfied, ($k = k_a$). The corresponding curve is the tangent to the straight lines $\phi_x = \pm \mu \gamma l$ (boundary conditions). Straightforward calculations show that, for a long enough junction ($l > 3$), the curve a is very close to the separatrix so that $k_a = k_c$. This circumstance justifies the assumption $k_a = k_c$.

The expressions for the two integration constants are:

$$k_c = \sqrt{1-\xi^2\gamma^2} + \xi\gamma\sin^{-1}\xi\gamma \qquad (15a)$$

$$k_a = \xi\gamma\pi + \frac{(\mu\gamma l)^2}{2} - \xi\gamma\sin^{-1}\xi\gamma - \sqrt{1-\xi^2\gamma^2} \qquad (15b)$$

where $\xi = 1-2\mu$. So that:

$$\frac{(\mu\gamma_c l)^2}{4} = \sqrt{1-\xi^2\gamma_c^2} + \xi\gamma_c\sin^{-1}\xi\gamma_c - \frac{\xi\gamma_c\pi}{2} \qquad (16)$$

From Eq. (16) the $\gamma_c(l,\mu)$ in zero external magnetic field can be computed, (see Fig. 7).

376

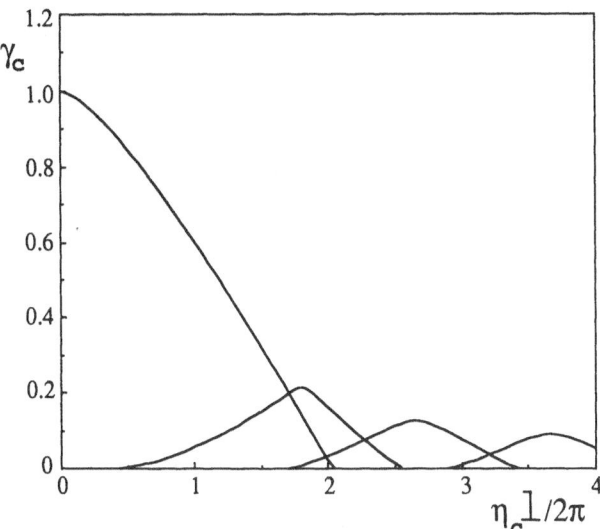

Fig. 5 Theoretical dependence of the maximum dc Josephson current on the external magnetic field for a long overlap junction with $1 = 6$.

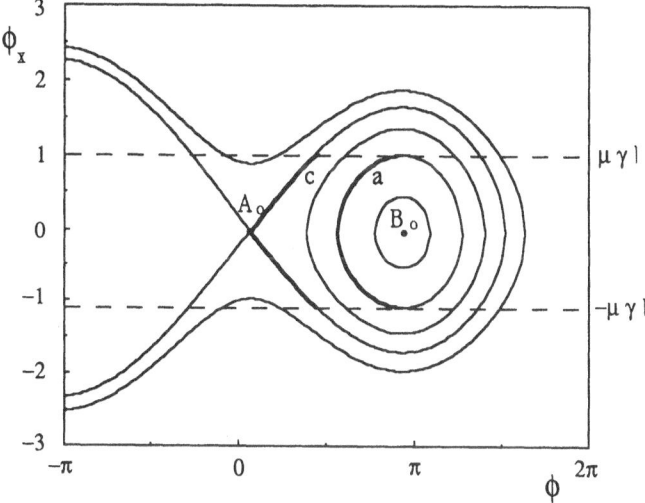

Fig. 6 Phase-space configuration at $\eta = 0$ for a junction with a mixed overlap/in-line geometry. The curves refer to different values of the integration constant k for $\gamma = 0.5$ and $\mu = 0.3$. Curve a is the tangent to the straight lines representing the boundary conditions; curve c is the separatrix.

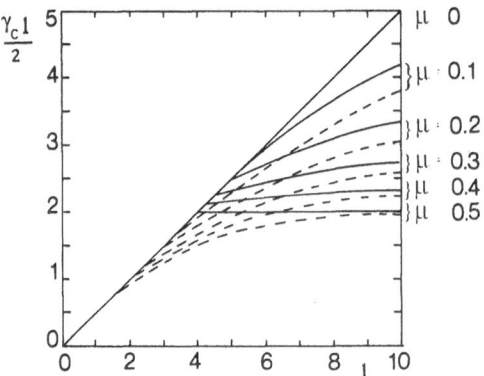

Fig.7 Dependence of the critical current on the junction length computed from Eq. (16) for different values of the parameter μ (solid line). The curves are compared with the calculations reported in Ref. (13) (dashed lines).

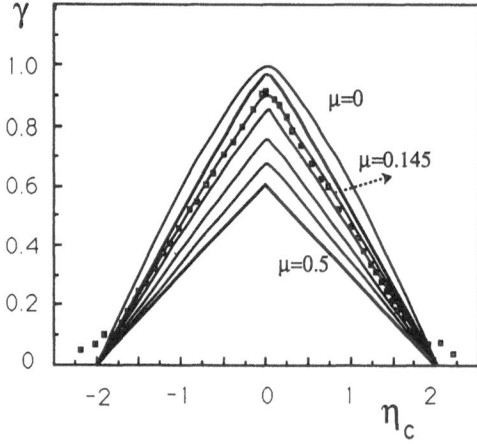

Fig.8 Theoretical magnetic field dependence of the Josephson critical current for different values of the in-line parameter μ (0, 0.1, 0.2, 0.3, 0.4, 0.5). In the figure it is also shown the comparison between the theoretical behavior at the measured value of μ ($\mu = 0.145$) and the experimental dependence (dots). Experimental data refer to a V-VO_x-Pb junction, (see text).

As far as concerns the dependence of the maximum Josephson current on the external magnetic field, the occurrence of a non zero value of the parameter μ results in elaborate calculations. However an analytical expression for the central lobe (m = 0) of $\gamma_c(\eta)$ can be considered by simply changing the first term of Eq. (16) into $(\eta + \mu\gamma_c l)^2/4$, obtaining:

$$\eta(\gamma_c) = 2\sqrt{\sqrt{1 - \xi^2 \gamma_c^2} + \xi\gamma_c \sin^{-1}\xi\gamma_c - \frac{\xi\gamma_c \pi}{2}} - \mu\gamma_c l \qquad (17)$$

The central lobe (m = 0) of the dependence $\gamma_c(\eta)$ for different values of the parameter μ is reported in Fig. 8.

From the experimental point of view the reported theoretical behaviors can be used to evaluate the degree of deviation from a pure overlap geometry of an actual junction. In this framework a V-VO$_x$-Pb overlap-type junction has been considered to take into account its non-canonical $I_c(H_e)$ dependence. By estimating J_c = 0.87 A cm^{-2} and the magnetic penetration depth d = 378 nm from a small junction lying on the same substrate, λ_j = 282 μm, l = 6.5, and w = 1.1 are obtained. These values and the measured value of the critical current $I_c(0)$ give $\gamma_c(0)$ = 0.87. So from Eq. (16) a value μ = 0.145 is evaluated.

The knowledge of the μ value allows to compute the $\gamma_c(\eta)$ dependence, (see Eq.(17)). The comparison between the theoretical behavior and the experimental data exhibits a quite good agreement, as shown in Fig. 8.

CONCLUSIONS

The static configurations of the phase in long Josephson junctions have been investigated by a phase space analysis. Assuming aN uniform bias current, the developed method allows to compute the magnetic field dependence of the Josephson critical current in overlap junctions for any length.

This analysis also predicts the occurrence of a "plateau" in the magnetic field dependence of the critical current, depending on the boundary conditions. Preliminary experimental results confirm the theoretical predictions.

An approach to mixed overlap/in-line configurations has been developed. Such an analysis is restricted to the case of zero magnetic flux quanta threading the junction. Differences with respect to the pure overlap case have been shown. In particular theoretical and experimental results have been reported concerning the dependence of the maximum Josephson current on the junction length and on the external magnetic field as a function of the in-line parameter μ. A good agreement between theory and experiments has been found. Work is in progress to extend the analysis to the case of an arbitrary number of fluxons.

ACKNOWLEDGEMENTS

This work has been partially supported by The National Research Council of Italy (C.N.R.) under the Progetto Finalizzato "Superconductive and Cryogenic Technologies".

REFERENCES

1   A. Barone, and G. Paternò, "Physics and Applications of the Josephson Effect," Wiley, New York (1982).
2   D.W. McLaughlin and A.C. Scott, Phys. Rev. A18:1652 (1978)
3   O.M. Olsen, N.F. Pedersen, M.R. Samuelsen, M. Svensmark and D. Welner, Phys. Rev. B33:168 (1986).

4   G. Costabile, A.M. Cucolo, S. Pace, R.D. Parmentier, B. Savo and R. Vaglio, in :
     "SQUID '80 Superconducting Quantum Interference Devices and Their
     Applications," H.D. Hahlbohm and H. Lubbig eds., de Gruyter, Berlin (1980),
     p. 147.

5   I.O. Kulik and I.K. Yanson, "Effekt Dzhozefsona Sverkhprovodyashchikh
     Tunnel'aykh Struktruakh (The Josephson Effect in Superconducting
     Tunneling Structures).," Nauka, Moskow (1970). (Engl. Transl., Holsted
     Press, New York, 1972)

6   T. Nagatsuma, K. Empuku, F. Irie and K. Yoshida, J. Appl. Phys. 58:441 (1985)

7   D.G. Creté, P. Fesutrier, M. Honus, R. Monaco and P.J. Encrenaz, in:
     "Stimulated Effects in Josephson Devices", M. Russo and G. Costabile eds. ,
     World Scientific Publishing, Singapore (1990), pp. 87-98.

8   R.D. Parmentier, in : "Solitons in Action", K. Lonngren and A. Scott eds. ,
     Academic Press, New York (1978), pp. 173-199.

9   C.S. Owen, D.J. Scalapino, Phys. Rev. 164:538 (1967).

10   O.M. Olsen and M.R. Samuelsen, J. Appl. Phys. 54:6522 (1983).

11   A. Barone, F. Esposito, K.H. Likharev, V.K. Semenov, B.N. Todorov and R.
     Vaglio, J. Appl. Phys. 53:5802 (1982).

12   T. Yamashita, M. Kunita, and Y. Onodera, Jpn. J. Appl. Phys. 7:288 (1967).

13   A. Barone, W.J. Johnson and R. Vaglio, J. Appl. Phys. 46:3628, (1975).

14   M. R. Samuelsen and S. A. Vasenko, J. Appl. Phys. 57:110 (1984).

15   S. Pagano, B. Ruggiero and E. Sarnelli, Phys. Rev. B43:5364 (1991).

16   E. Sarnelli, S. Pagano, B. Ruggiero and M. Russo, IEEE Trans. Mag. 27:2716
     (1991).

MAGNETIC TUNING OF PLASMA OSCILLATIONS

IN LONG JOSEPHSON JUNCTIONS

Thorsten Holst and Jørn Bindslev Hansen

Physics Laboratory I / MIDIT
The Technical University of Denmark
DK-2800 Lyngby, Denmark

INTRODUCTION

The main purpose of this work is to study the influence of an applied static magnetic field on the dynamics of Josephson tunnel junctions. In order to investigate the spatial dependence of the Cooper pair phase difference (the Josephson phase), we have carried out experiments on junctions of various normalized lengths. A Josephson junction is called *small* if both the length L and the width W of the tunneling area are small compared to the Josephson penetration depth, λ_J . We have *long*, quasi-onedimensional junctions if one of the physical dimensions is larger then λ_J , i.e. W $<< \lambda_J <$ L.

The plasma oscillations in the zero-voltage state were chosen for the study as a measure of the basic dynamics of the junction. Plasma oscillations in Josephson tunnel junctions, first reported by Dahm et al.[1], involve a resonant exchange of energy between the electric field and the kinetic energy of the Cooper pairs. With typical junction parameters the plasma frequency may range from 1 to 100 GHz. The plasma frequency plays an important role in the study of non-linear dynamics in Josephson junctions, where the ratio between the frequency of the driving signal and the plasma frequency is an important parameter.[2] Some noise properties, such as the life-time of the zero-voltage state, are also effected by the junctions plasma frequency.[3] To keep things simple we will only be concerned with small amplitude plasma oscillations. Measurements of large amplitude plasma oscillations in the non-linear regime are reported in Ref. 4.

The dynamical effect of a magnetic field was studied by comparing two kinds of measurements. The plasma oscillations were tuned by changing the critical current, either by applying an external magnetic field or by raising the temperature. In the experiments, the functional dependence of the plasma oscillations on the applied magnetic field exhibited a cross-over from small to long junction behavior.

Nonlinear Superconductive Electronics and Josephson Devices
Edited by G. Costabile *et al.*, Plenum Press, New York, 1991

This behavior was not observed when the plasma oscillations were tuned by an increase of the temperature. Numerical simulations of the governing partial differential sine-Gordon equation were also carried out. The theoretical results agree with the experiments and allow us to interpret the observed cross-over in terms of the spatial variation of the Josephson phase.

THEORY

The theoretical concept of plasma oscillations is based on the simple and widely used resistor-capacitor shunted junction (RCSJ) model. The current of Cooper pairs is modelled by the Josephson equations, while the loss, due to thermally excited pairs, is represented by a shunt conductance 1/R. A shunt capacitance C is formed by the geometric overlap of the two superconducting electrodes. Since the impedance of the junction is much smaller than the impedance of the external circuit both in the microwave regime and in the dc power circuit, we may assume that current sources, I(t), are used to bias the junction.

Small Junctions

The equation of motion for the Josephson phase $\phi(t)$ is obtained by calculating the various current contributions through the tunneling barrier and using the Josephson equations

$$I_J(t) = I_{co}\sin\phi(t), \quad V(t) = \frac{\hbar}{2e}\phi_t$$

where I_J is the Cooper pair current, I_{co} is the critical current, and V(t) is the voltage across the barrier. The subscript t denotes differentiation with respect to time. The equation for $\phi(t)$ may be recast in dimensionless form:

$$\phi_{tt} + \alpha\phi_t + \sin\phi = i_0 + i_1\cos\Omega t \tag{1}$$

where the externally applied current i(t) on the right hand side is taken to be in the specified form. The unit frequency is now the maximum plasma frequency $\omega_{po} = (2eI_{co}/\hbar C)^{1/2}$ and time is normalized to ω_{po}^{-1}. Currents are in units of I_{co} and the loss parameter α is equal to $(\omega_{po}RC)^{-1}$. The structure of Eq.(1) is similar to the equation for a damped pendulum in a gravitational field driven by an externally applied torque, i(t). In the zero-voltage state we can linearize Eq.(1) by assuming $\phi(t) = \phi_0 + \epsilon\phi_1(t)$, which is valid for a small-amplitude oscillating driving force, i.e. $i_1 \ll 1$. Formally Eq.(1) now has the same form as the equation of motion for a driven damped linear resonance circuit

$$\phi_{1tt} + \alpha\phi_{1t} + \cos\phi_0 \cdot \phi_1 = i_1\cos\Omega t$$

with the resonance frequency given by

$$\Omega_p = \sqrt{\cos \phi_0} = \sqrt[4]{1 - i_0^2} \qquad (2)$$

for $\alpha \ll 1$ (experimentally, α is typically of the order 10^{-2}). We will identify Ω_p with the (normalized) frequency of plasma oscillations in the linear, small amplitude limit.

If an external magnetic field H_{ex} is applied in the plane of the barrier, ϕ will vary spatially perpendicular to the field. We assume that the magnetic field is applied uniformly in the y-direction (see Fig. 1), so ϕ depends only on the x-direction.

Since the spatial dimension of the Josephson phase is neglected in the equation of motion for small junctions, the problem of calculating the plasma frequency reduces to a matter of proper spatial averaging. This is equivalent to calculating the *center-of-mass* motion for a *rigid* body in classical mechanics. The magnetic field will cause the critical current to vary in a Frauenhofer-like diffraction pattern, $I_c(H_{ex})$. It is easy to show [1,5,6], that I_{co} should be replaced by $I_c(H_{ex})$ in the expression for the plasma frequency.

Long Junctions

For long junctions the spatial dimension of the tunneling structure must be taken into account. The equation of motion can be set up by representing the junction as a lumped transmission-line equivalent of infinitesimally small RCSJ-circuits, see for example Ref. 7. The result is the *perturbed sine-Gordon equation*

$$\phi_{tt} - \phi_{xx} + \alpha \phi_t + \sin \phi = i_0 + \epsilon i_1 \cos \Omega t \qquad (3)$$

where the space coordinate x is normalized to the Josephson penetration depth, $\lambda_J = (\hbar/2e\mu_o d J_{co})^{1/2}$. Here $J_{co} = I_{co}/WL$ and the magnetic thickness $d = \lambda_1 + \lambda_2 + t_{ox}$, where the λ's are the London penetration depths for the two superconductors, and t_{ox} is the thickness of the oxide barrier. We have taken the applied current to be uniformly distributed along the length of the junction (overlap geometry). The external magnetic field is accounted for through the boundary conditions: $\phi_x(\pm l/2, t) = \eta$.[7] Here, the (dimensionless) quantity η is the external field in units of $H_{ex}/J_{co}\lambda_J$, and l is the normalized length of the junction.

In the previous section we treated the dynamics of plasma oscillations as a center-of-mass problem for a rigid chain of pendula. Since the mechanical analog to a long Josephson junction is a *flappy* chain of pendula, the approach is now somewhat different. The purpose is to find the amplitude of the pendulum oscillations along the x-direction as a function of the external field.

We present here in the section "NUMERICAL SIMULATIONS" the results of numerical calculations, which agree with perturbation theory.[6]

EXPERIMENTAL TECHNIQUES

The samples used were 70×70 or 800×20 μm^2 Nb-Nb_xO_y-Pb tunnel junctions incorporated into a 50-Ω microstrip structure on a Corning 7059 glass substrate. The thin film pattern is shown in Fig. 1. The microwave coupling to the sample was established through coaxial launchers at the ends of the microstrip. The microwave current used to stimulate the plasma resonance was provided from a microwave generator (a frequency synthesizer) connected to one end of the microstrip. In order to suppress microwave resonances, the other end of the microstrip was terminated by a 50-Ω load at helium temperature. For convenience, the plasma resonance frequency was tuned by varying the dc bias, I_0 (see Eq. (2)), while the frequency of the applied microwave current was fixed. Using this method the plasma resonance manifested itself as a dip in the reflected microwave signal as a function of I_0. A circulator at room temperature separated the incoming signal from the reflected signal which was amplified and detected by two low-noise FET amplifiers followed by a digital spectrum analyzer (SA in Fig. 1).

The sample was placed in a vacuum can immersed in a pressure-regulated liquid helium bath. A pair of Helmholtz coils was used to create a static magnetic field in the direction indicated in Fig. 1. The current and voltage leads connected to the sample were carefully shielded and filtered against external noise sources. The ambient magnetic field was reduced by a double mu-metal shield and the entire experimental set-up was placed in an rf-shielded room.

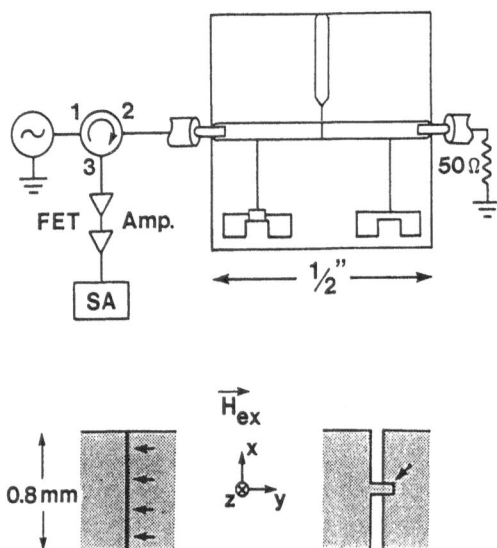

Fig. 1.   In the upper part a schematic of the experimental microwave system is shown ("SA" stands for spectrum analyzer), together with the thin film microstrip design. The enlargement in the bottom part shows the junction geometries (the overlap of the two superconducting films is indicated by arrows).

EXPERIMENTAL RESULTS

To ascertain that we only stimulated the plasma oscillations in the small-amplitude limit, we decreased the microwave current, i_0, until the position of the resonance dip as a function of the bias current did not change for further reduction of i_0. With our microwave coupling and detecting system the applied microwave power ranged from 10^{-15} W to 10^{-11} W.

From Eq.(2) it is clear that a plot of $f_p^4 = (\omega_p/2\pi)^4$ vs. I_0^2 should yield a straight line which intersects the axes in f_{po}^4 and I_{co}^2. Figure 2 shows a plot of this kind in the case of no applied magnetic field for a small junction. Measurements of the plasma resonance offer a precise way to determine the critical current and the capacitance. Our values of I_{co} are always a few percent larger then the value obtained by a slow sweep of the I-V characteristic, where premature switching from the zero-voltage state, induced by thermal noise, gives an upper bound on the critical current which is smaller than the actual I_{co}.

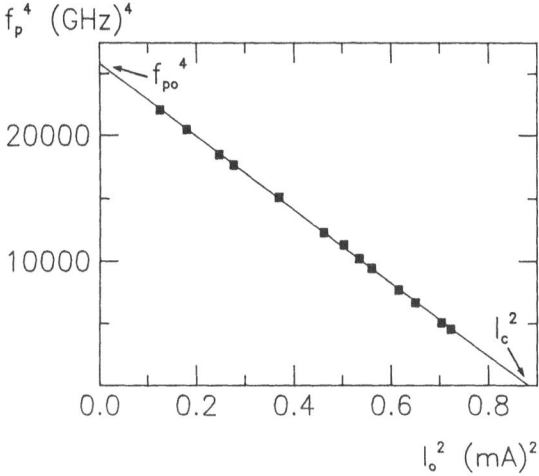

Fig. 2.   Experimental bias current tuning of the plasma resonance frequency in a small junction in the absence of a magnetic field. The intersections with the axes give I_{co}(4.2 K) = 938.9 μA, f_{po} = 12.67 GHz, and therefore C = 450 pF. The straight line is a least-squares fit.

In order to examine the effect of a static magnetic field on the dynamics of the junction we compared measurements for two ways of reducing I_{co} - either by applying an external magnetic field or by raising the temperature T. For every value of $I_c(H_{ex})$ or $I_c(T)$ a plot like Fig. 2 was made in order to find f_{po} and I_c (at least four well-spaced frequencies were used for the least-squares fit). The results are presented as a plot of f_{po}^2 vs. I_c, where the theory for small junctions predicts a straight line through origo. The top graph in Fig. 4 shows the results for a small junction.

For longer junctions the behavior deviates from that of the small junctions. In Fig. 3 we notice two important characteristics. First, we still have straight lines which intersect the x-axis in I_c^2, both when the temperature was increased or a magnetic field was applied. Second, and most important, the *slope* of the lines changes with the applied field. Since we still have straight lines in Fig. 3 we feel justified to make a least-squares fit and extract informations about f_{po} and I_c. The results are shown in the middle and bottom part of Fig. 4. A plot of f_{po}^2 vs. I_c should still yield a straight line through origo for long junctions with a *uniform* current distribution and in the absence of magnetic fields. The $I_c(T)$-series of measurements in Fig. 4 support our assumption of a uniform current distribution.

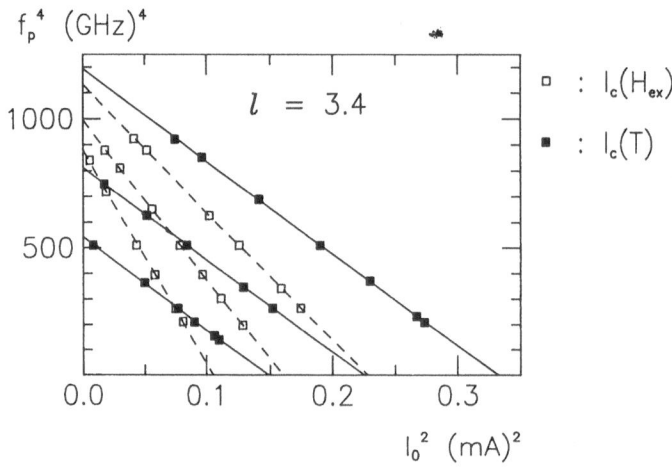

Fig. 3.   Experimental bias current tuning of the plasma resonance frequency in a long junction. For comparison the critical current was reduced in two ways: either by applying an external magnetic field (□) or by raising the temperature (■). The parameters estimated from the intersection with the axes are $I_{c0}(4.2 \text{ K}) = 576.5 \ \mu\text{A}$, $f_{po} = 5.88$ GHz, and $C = 1283$ pF. All lines are least-squares fits.

There can be up to a 22 % difference in plasma frequency between the cases where the critical current is tuned by the magnetic field or by the temperature (bottom part of Fig. 4). In this example the normalized junction length was 3.4 and the critical current was decreased to 57 % of I_{c0} at 4.2 K. The corresponding reduction of I_{c0} for a junction of normalized length 1.1 was 7.5 %.

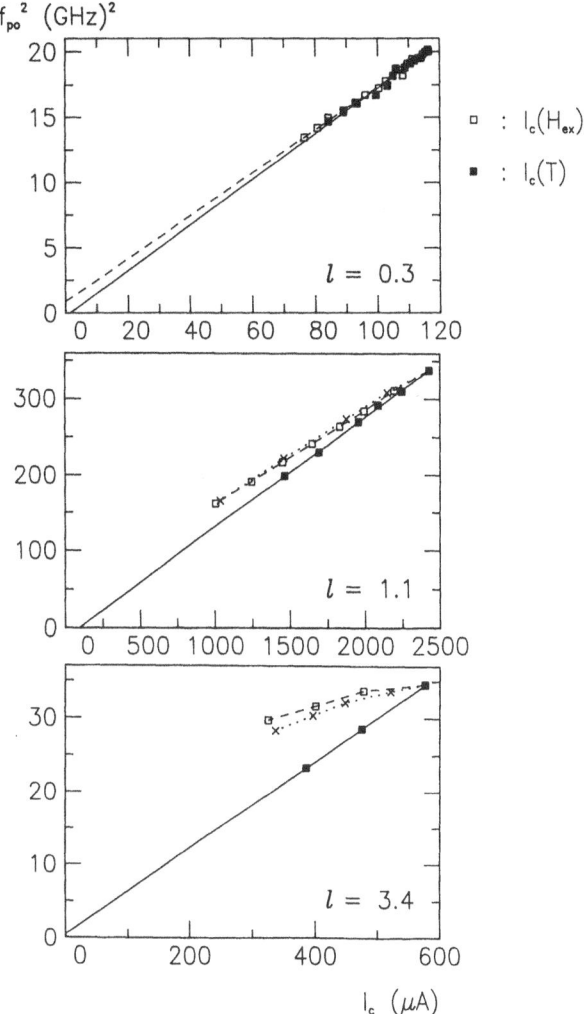

Fig. 4. Experimental results plotted as the squared maximum plasma frequency vs. the critical current for junctions of various normalized length, l. The intersections with the axes in a f_p^4 vs. I_0^2 plot were used to determine f_{po} and I_c (see Figs. 2 and 3). For comparison the critical current was reduced by either applying a magnetic field ($\square$) or by raising the temperature ($\blacksquare$). All solid lines and the dashed line in the top part of the figure are least-squares fits. The crosses ($\times$) show the results from a numerical simulation of the perturbed sine-Gordon equation, Eq. (3), together with the evaluation of the parameter A, Eq. (4) ($\alpha = 0.1$, $i_1 = 10^{-3}$, and the relevant normalized length).

NUMERICAL SIMULATIONS

Numerical simulations of Eq.(3) with the boundary condition $\phi_x(\pm l/2, t) = \eta$ were obtained using an explicit difference scheme. A time grid of 0.005 and a space grid of 0.02 were used in the integration.

To compare the numerical simulations with the experiments we have to establish a relation between the calculations and the reflected signal measured by the detection system. As a reasonable parameter we take the spatial average of the voltage.

$$A(t, \Omega, \eta) = \int_{-l/2}^{l/2} \phi_t(x, t) dx \qquad (4)$$

The amplitude of the parameter A was determined by recording the maxima and minima of A's oscillations in time over a period of approx. 40 time units. The point of resonance for A was then found by maximizing A with respect to Ω for a given value of the magnetic field η. Doing this for $i_0 = 0$ we found the maximum plasma frequency as a function of the magnetic field η. The resonance points for the parameter A are shown as crosses in Fig. 4. The numerical simulations were performed with the relevant normalized lengths. An example of the stationary and dynamical distribution of the Josephson phase is shown in Fig. 5.

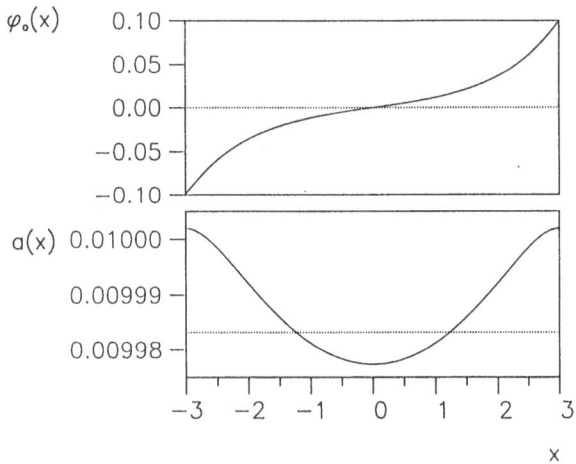

Fig. 5.   **Top part:** The magnetic field, which is modelled through the boundary conditions $\phi_x(\pm l/2, t) = \eta$, perturbs the static distribution of the Josephson phase, $\phi_0(x)$. This perturbation of the phase then effects the dynamics of the plasma oscillations.
Bottom part: Spatial distribution of the small-amplitude plasma oscillations, a(x). These oscillations are superimposed on the stationary background shown in the top part of this figure.
The solid lines are simulation with magnetic field and the dashed horizontal lines indicate the uniform variation of the Josephson phase in the absence of a magnetic field. For all curves the simulation was performed with the parameters: $l = 6$, $\eta = 0.1$ or 0, $\Omega = 0.9971$, $\alpha = 0.1$, $i_0 = 0$, and $i_1 = 10^{-3}$.

CONCLUSIONS

We have experimentally demonstrated that the functional dependence of plasma oscillations in small tunnel junctions ($L/\lambda_J = 0.3$) is very well accounted for by the resistor-capacitor shunted junction model. This is the case when the critical current is decreased either by raising the temperature or by applying a magnetic field. If the junction is approximately one Josephson penetration depth long, deviations from the theory for small junctions start to show up. The deviations become more pronounced for still longer junctions. We emphasize that the above-mentioned deviation only appears in the presence of a magnetic field. More specificly, when an external magnetic field is applied, the square of the plasma frequency does not decrease as fast as the critical current. This again means that the plasma frequency is larger in a long junction exposed to a magnetic field than for the same junction with the critical current decreased by raising the temperature. An effect like this should show up as corrections in expressions involving the plasma frequency, for example the life-time of the zero-voltage state.

Numerical simulations were performed on the governing sine-Gordon equation. They reproduced the experimental data quite well. This allows us to give the following qualitative interpretation of our results: the critical current is the bias current value, where the Josephson phase tilts over and starts to rotate. In other words, the critical current is a static property of the junction, whereas the plasma oscillation is obviously a dynamical property of the junction. A small junction can be treated as a rigid body, and it is not possible to distinguish between the two properties. For long junctions the static and dynamic behavior in the above-mentioned sense become uncoupled. The critical current is determined by the amount of magnetic field at the boundaries. Since the field only penetrates a distance of the order of one Josephson penetration depth into the junction, the interior of the junction is unaffected by the field. This leaves the plasma oscillations in the main part of the junction unperturbed by the magnetic field. Thus, as observed with increasing junction length the detuning of the plasma frequency with applied field becomes smaller.

ACKNOWLEDGEMENT

We thank G.F. Eriksen for technical assistance and I. Rasmussen and S. Hjort for sample fabrication. N. Grønbech-Jensen is also acknowledged for helpful advice on the simulations. One of us (T.H.) would like to thank the Carlsberg Foundation for a generous scholarship grant during the initial stages of this work. This work was partly supported by the Danish Natual Science Research Council.

REFERENCES

1. A.J. Dahm, A. Denenstein, T.F. Finnegan, D.N. Langenberg, and D.J. Scalapino, *Study of the Josephson Plasma Resonance*, **Phys. Rev. Lett.** 20:859 (1968).
2. R.L. Kautz and R. Monaco, *Chaos and Catastrophe near the Plasma Frequency in the rf-biased Josephson Junction*, **IEEE Trans. Mag.** MAG-25:1399 (1989).

3. M.H. Devoret, J.M. Martinis, and J. Clarke, *Resonant Activation from the Zero-Voltage State of a Current-Biased Josephson Junction*, **Phys. Rev. Lett.** 53:1260 (1984).

4. A.J. Dahm and D.N. Langenberg, *Nonlinear Effects in the Josephson Plasma Resonance*, **J. Low Temp. Phys.** 19:145 (1975).
O.P. Balkashin and I.K. Yanson, *Nonlinear Plasma Oscillations in Josephson Tunnel Junctions*, **Sov. J. Low Temp. Phys.** 2:143 (1976).
T. Holst and J. Bindslev Hansen, *The Plasma Resonance in Microwave-Driven Small Josephson Tunnel Junctions*, **Physica B** 165 & 166:1649 (1990).

5. N.F. Pedersen, T.F. Finnegan, and D.N. Langenberg, *Magnetic Field Dependence and Q of the Josephson Plasma Resonance*, **Phys. Rev. B** 6:4151 (1972).

6. T. Holst, to be published elsewhere.

7. P.S. Lomdahl, O.H. Soerensen, and P.L. Christiansen, *Soliton Excitations in Josephson Tunnel Junctions*, **Phys. Rev. B** 25:5737 (1982).
N.F. Pedersen, *Solitons in Josephson Transmission Lines*, in: "Solitons", eds. S.E. Trullinger, V.E. Zakharov, and V.L. Pokrovsky, North-Holland, Amsterdam (1986).

SOFT AND HARD TURBULENCE

IN LONG RF-BIASED JOSEPHSON JUNCTIONS

Luis E. Guerrero and Miguel Octavio

Centro de Física
Instituto Venezolano de Investigaciones Científicas
Apartado 21827, Caracas 1020A, Venezuela

1. INTRODUCTION

This chapter is concerned with the turbulent behavior, an interesting and fundamental dynamical state not well understood. Once it had been established that *chaos is not turbulence*, the turbulent phenomenon has received a great deal of attention in recent years[1-6].

Systems with many degrees of freedom can develop coherent structures (spatiotemporal patterns). In this way the system reduces the effective number of degrees of freedom and is able to exhibit periodic and low-dimensional chaotic *(in time)* behavior. The activation of new degrees of freedom can give rise to turbulent-like dynamics (incoherent in space and *disordered in space as well as in time*).

In Fig.1 we present the contrast between these different dynamics: Fig. 1(a) shows a typical low-dimensional Poincaré map for a chaotic dynamic, whereas Fig. 1(b) corresponds to fully developed spatiotemporal turbulence in which the attractor completely loses fractality and autosimilarity (intimately related with the underlying mechanism of the chaotic phenomena). The low-dimensional strange attractor is destroyed due an activation of an increased number of effective degrees of freedom. Actually the notion that an attractor underlies turbulent behavior is under suspect: according to Crutchfield and Kaneko[2], long transients preclude observation of the behavior ruled by the asymptotic invariant measure and the nature of the attractor is irrelevant to the observed behavior.

In 1987, Heslot, Castaing and Libchaber[1] reported a Rayleigh-Bénard experiment which followed the transition fron chaos to soft and hard turbulence. The soft turbulence regime can be regarded as a dynamical state globally disordered in space and in time. This state contrasts with the hard turbulence

Nonlinear Superconductive Electronics and Josephson Devices
Edited by G. Costabile *et al.*, Plenum Press, New York, 1991

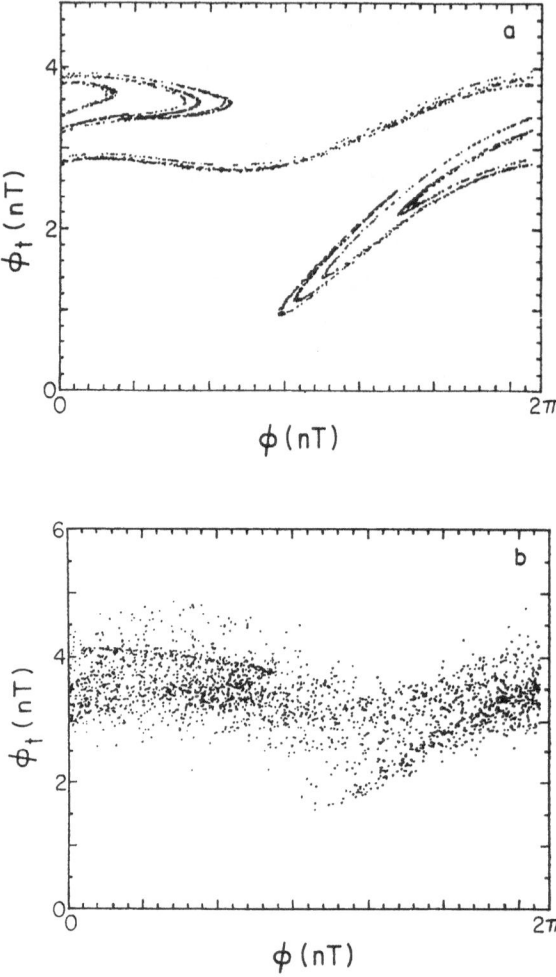

Fig. 1. From chaos to disorder. (a) Poincaré map showing a typical strange attractor for low-dimensional chaos. (b) The strange attractor is destroyed as spatial correlation decays.

regime in which the pattern formation and conversion induces a reemergence of coherence in a narrow frequency range.

The solitonlike character of the the long Josephson junction can play a fundamental role as an activating mechanism of chaos and introduces a rich variety of new and interesting spatiotemporal phenomena[7,8]. Even richer behavior can also be exhibited by the the long Josephson junction, like regimes of soft and hard turbulence[6]. The aim of this chapter is to present the study of turbulent regimes in long Josephson junctions as a field which provides an insight of what determines the ability of a system to become turbulent.

This chapter is organized as follows: in Sec. 2 we present a description of the forced long Josephson junction. In Sec. 3 we explore the transition to the soft turbulent state. In Sec. 4 we contrast the soft and the hard turbulence regimes. Finally, in Sec. 5 we summarize and present our conclusions.

2. THE LONG JOSEPHSON JUNCTION

The forced long Josephson junction considered by us has been discussed extensively[9]. We model this system with the usual sine-Gordon-like equation,

$$\phi_{xx} - \phi_{tt} - \sin\phi = \alpha\phi_t - \rho\sin(\Omega_d t) \qquad (1)$$

where $\phi = \phi(x,t)$ is the phase difference of the superconducting order parameter between each side of the barrier and its derivative in time is the voltage across the junction. The term $\alpha\phi_t$ represents quasiparticle loss. The distance is normalized to the Josephson penetration depth λ_J, time is normalized to the inverse of the Josephson plasma frequency, the rf amplitude ρ is normalized to the critical current and Ω_d is the normalized applied frequency. The external applied field is taken into account through,

$$\phi_x(0,t) = \phi_x(L,t) = \eta \qquad (2)$$

where L is the junction length and η is a measure of the external magnetic field.

The perturbed sine-Gordon equation is the simplest wave-equation for a periodic medium and occurs frequently in solid-state physics. The system described by eqs.(1,2) is in fact analogous to a chain of coupled pendula forced by an external torque. This analogy indicates that the turbulent behavior can be present not only in fluids but also in solid-state and mechanical systems described by relatively simple models.

3. ONSET OF SOFT TURBULENCE

The study of the route to turbulent-like behavior can establish what determines the ability of the system to become turbulent, providing in this fashion a better understanding of the turbulent state.

In this section we explore the onset of soft turbulence in long Josephson junction as the amplitude of a radio-frequency (rf) drive is increased. Parameter values are $L = 10\lambda_J$, $\alpha = 0.252$ and $\Omega_d = 0.65$. We employ open boundary conditions, $\eta = 0.0$, spatially uniform drive and flat initial conditions. Thus the pattern formation phenomena is fully spontaneous in contrast with Rayleigh-Bénard experiments in which symmetry breaking and consequent pattern formation is induced via boundary conditions. In addition, our results were obtained in the absence of thermal noise.

3.1. Spontaneous Pattern Formation and Symmetry Breaking

In most physical situations the development of turbulence is preceded by the formation of spatiotemporal structures. Hereafter we show that in the case of the long Josephson junction the pattern formation phenomenon can be autonomously excited.

For a long Josephson junction with $\rho = 2$ we find a transition from a transient with period 1 to a period 2 regime. This transition at first sight might appear to be a simple bifurcation in time as in low-dimensional systems. However, as shown in Fig. 2(a) (where we plot the phase difference $\Delta\phi$ between two points of the junction for each period of the rf drive as a function of time) the voltage is no longer homogeneous in space. The fact that $\Delta\phi$ repeats only every second period shows that the spatial extent of the junction has now become significant. Figure 2(b) reveals the spatiotemporal profile sustained by the system after the homogeneous transient: a "virtual" breather (a solitonlike state with an internal degree of freedom) centered at the edge of the junction is formed with only half of it present inside the junction. This "virtual" breather switches at each period between its two states, thus yielding a steady state in which the spatial symmetry is broken, much like the symmetry-breaking precursor (in phase space) of the usual transition to chaos. Fig. 2(b) can be interpreted to be the inability of the system to sustain a soliton commensurate with its size, jumping instead into a sort of bifurcation in space. The system thus succeeds in breaking the symmetry in real space and not in phase space as in temporal chaos.

3.2 The Two-Frequency Route to Turbulence

Below we present the development of turbulence as concerning the involvement of new degrees of freedom via the breakdown of coherence of the spatiotemporal profile.

As ρ is increased, once the breather forms from the initial flat condition there is long transient both in space and in time: there is an increasing difficulty of the system to attain a response which oscillates locked to the frequency of the driving force. The inner bands in Fig. 3 corresponds to a regime ($\rho = 2.0079$) that after such a long transient finally reaches the state with a breather oscillating with a frequency equal to half the driver (Fig. 2(b)).

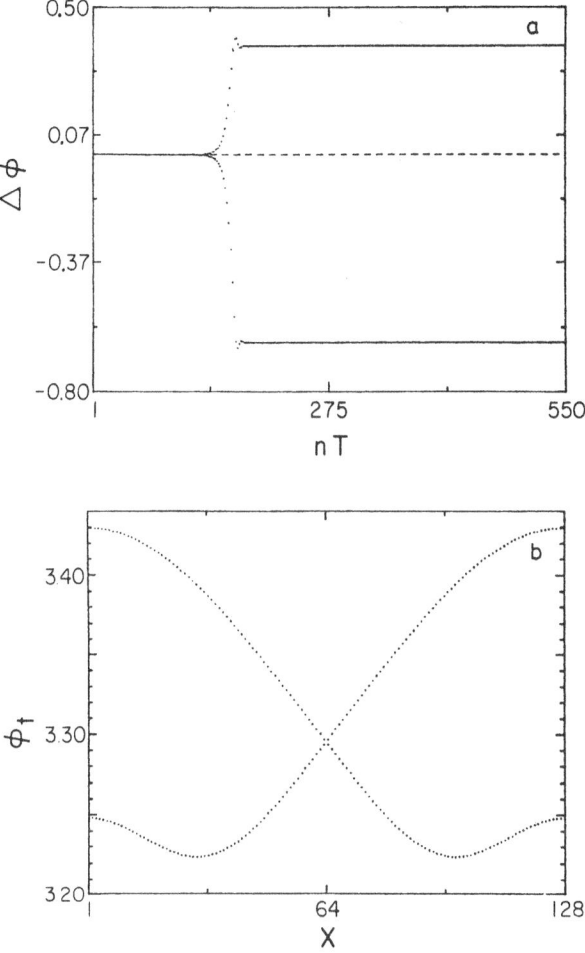

Fig. 2. Spontaneous Symmetry Breaking. (a) The difference of the phase between two points of the junction $\Delta\phi$ *vs* nT (T is the period of the harmonic drive). For $n > 180$, $\Delta\phi$ oscillates between the upper and lower curves revealing the development of a spatiotemporal profile. (b) Strobed profile $\phi_t(x, nT)$ *vs* x showing nonsymmetric breather oscillation.

As the system is driven harder the system ceases the long transient behavior in space and in time via generation of a quasiperiodic response. This quasiperiodic regime is possible because the spatiotemporal excitation oscillates at a frequency inconmensurate with the external drive. The outer bands in Fig. 3 presents for $\rho = 2.01$ the generation at the very onset of pattern formation of a breather-like excitation unlocked to the frequency of the driving force.

The coherence of the spatial profile decreases as the forcing is increased further as we present in Fig. 4(a). Thus the final state is disordered not only in time but in space and can be regarded as a soft-turbulent like regime, a regime in which there is a breakdown of the pattern formation ability of the system[1,6].

The underlying mechanism of the transition to the soft turbulent regime is the following[10]: in the quasiperiodic regime, different points of the junction differ in the way the different frequencies linearly combine. This has the effect of reducing the coherence of the spatiotemporal profile which decreases as the forcing is increased further.

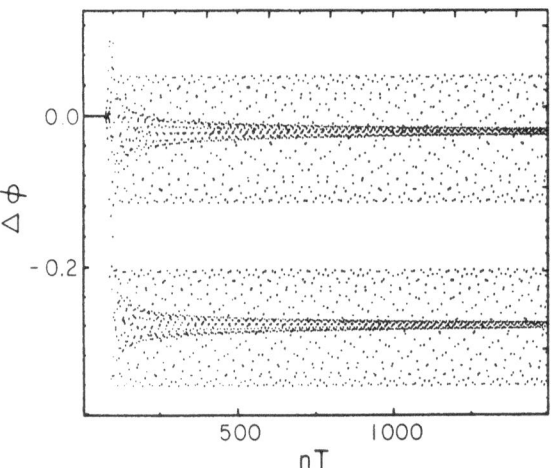

Fig. 3. Difference of the phase between two points of the junction $\Delta\phi(nT) \, vs \, nT$ for a regime with a long transient in space and time at $\rho = 2.0079$ (inner bands) and a quasiperiodic regime at $\rho = 2.01$ (outer bands).

4. SOFT AND HARD TURBULENCE

An applied magnetic field can excite pattern formation and destruction phenomena in the long Josephson junction. This system exhibits a hard turbulence regime when parameter values are $L = 5\lambda_J$, $\alpha = 0.252$, $\Omega_d = 0.65$, $\rho > 0.7375$ and $\eta = 1.25$. With these constraints the dynamical behavior corresponds to the regime of creation and destruction of fluxons[8].

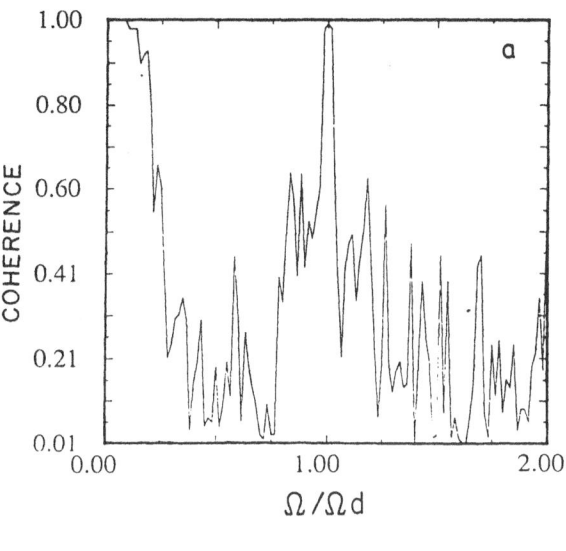

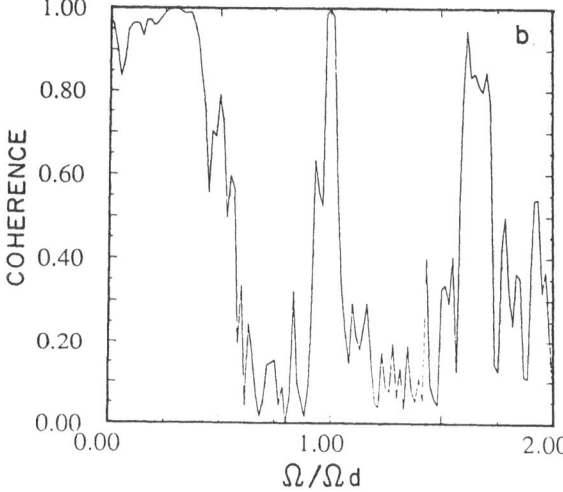

Fig. 4. *Coherence spectra.* (a) Soft turbulence: spatial coherence dissapears for all values but those of very low frequencies and the driving frequency. (b) Hard turbulence: spatial coherence dissapears for all but one frequency (other than the driving frequency).

In Fig. 4(b) we present the coherence spectrum for this fluxonic regime: coherence dissapears for all but one frequency (other than the drive frequency). This is in contrast with the soft turbulent regime (Fig. 4(a)) in which spatial coherence dissapears completely.

The statistics of the voltage recordings are very different in the two types of turbulent behaviors as is shown by the histograms of fluctuations in both regimes (Fig.5(a,b)). However, these histograms are also different from those reported for the Rayleigh-Bénard system[11] suggesting that this characterization might not provide a universal clear-cut distinction between the two regimes.

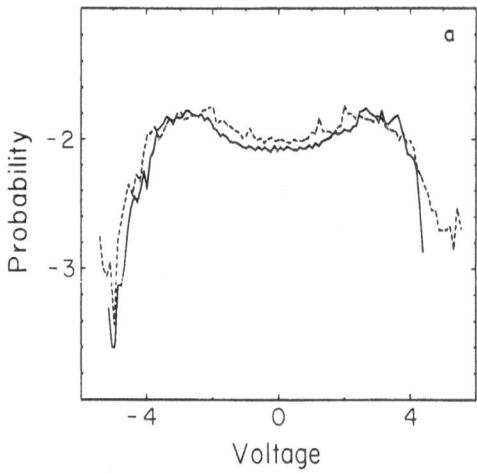

Fig. 5(a). Histogram of voltage fluctuations for soft turbulence. The continuous line represents the fluctuations for the signal at $x = L/2$ whereas the dashed lines corresponds to the signal at $x = L$ '.

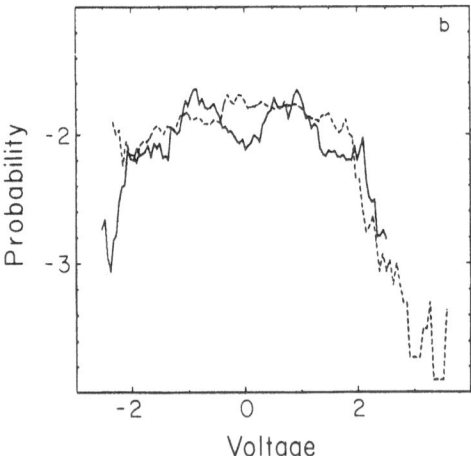

Fig. 5(b). Histogram of voltage fluctuations for hard turbulence. The continuous line represents the fluctuations for the signal at $x = L/2$ whereas the dashed lines corresponds to the signal at $x = L$ '.

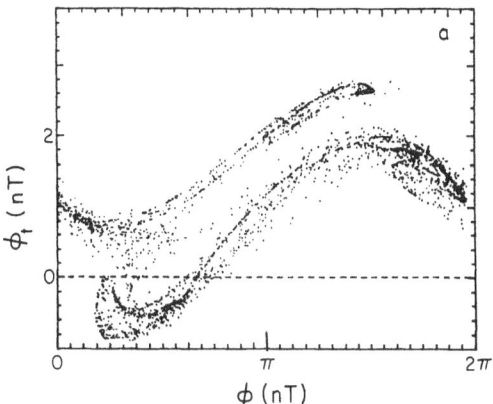

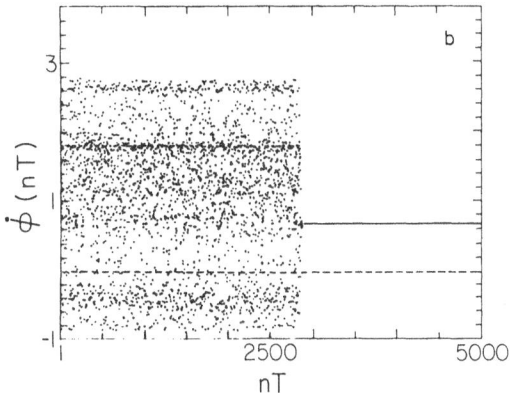

Fig. 6. *Hard turbulence and pattern conversion.* (a) Poincaré map
$\phi_t(L/2,T)\,vs\,\phi(L/2,T)\,(\mathrm{mod}\,2\pi)$ for the hard turbulent
regime $(\rho = 0.7375)$. (b) Strobed time series
$\phi_t(L/2,nT)\,vs\,nT$ corresponding to a sudden pattern
conversion after a long transient $(\rho = 0.8)$.

At $\rho = 0.8$ the pattern formation and conversion mechanism underlying the hard turbulence regime, manifests in a striking way: the fluxonic (traveling-wave) can give rise to a localized breather-like oscillation locked to the driving frequency[8]. Fig. 6(a) presents the dynamical attractor that is suddenly removed after a long transient as is shown in Fig. 6(b).

5. SUMMARY

Soft turbulent behavior can be autonomously excited in the long Josephson junction, with a single oscillating drive. The spatiotemporal symmetry breaking, the spontaneous pattern formation and the ability of the system to sustain an spatiotemporal profile unlocked to the driving force determines the ability of the system to achieve the soft-turbulent regime.

In turn, pattern formation and conversion appears as the pervasive feature of the hard turbulent regime.

The route to soft turbulence begins with a period doubling bifurcation accompanied with pattern formation. This route differs from Feigenbaum's period doubling cascade as the control parameter is varied: the period 2 regime exhibits a direct transition to a two-frequency quasiperiodic regime and further increase of the control parameter destroys coherence in space. The knowledge of the differences between the onset of chaos and the onset of turbulence can provide a way to distinguish these two regimes in experimental situations.

ACKNOWLEDGEMENTS

This work has been partially supported by CONICIT under project S1-1828 and by the EEC contract CI1*/ 0506-DK (AM). One of us (L.E.G.) gratefully acknowledges the hospitality of the Physics Laboratory I at The Technical University of Denmark.

REFERENCES

1. F. Heslot, B. Castaing and A. Libchaber, Transitions to turbulence in helium gas, Phys. Rev. A 36:5870 (1987).
2. J. P. Crutchfield and K. Kaneko, Are attractors relevant to turbulence?, Phys. Rev. Lett. 60:2715 (1988).
3. H. Chaté and P. Manneville, Spatiotemporal intermittency in coupled map lattices, Physica D 32:409 (1988).
4. M. H. Jensen, Fluctuations and scaling in a model for boundary-layer-induced turbulence, Phys. Rev. Lett. 62:1361 (1989).
5. T. Bohr and O. B. Christensen, Size dependence, coherence, and scaling in turbulent coupled-map lattices, Phys. Rev. Lett. 63:2161 (1989).

6. M. Octavio and L. E. Guerrero, Turbulence in Josephson junctions, to appear in Phys. Rev. A.

7. L. E. Guerrero and M. Octavio, Quasiperiodic and chaotic behavior due to competition between spatial and temporal modes in long Josephson junctions, Phys. Rev. A 37:3641 (1988).

8. L. E. Guerrero and M. Octavio, Spatiotemporal effects in long rf-biased Josephson junctions: chaotic transitions and intermittencies between dynamical attractors, Phys. Rev. A 40:3371 (1989).

9. A. C. Scott, F. Chu and S. Reible, Magnetic-flux propagation on a Josephson transmission line, J. Appl. Phys. 47:3272 (1976).

10. L. E. Guerrero and M. Octavio, Quasiperiodic route to soft turbulence in long Josephson junctions, Physica B 165&166:1657 (1990).

11. M. Sano, X. Z. Wu and A. Libchaber, Turbulence in helium-gas free convection, Phys. Rev. A 40:6421 (1989).

APPLICATION OF THE BÄCKLUND–

TRANSFORMATION IN THE SINE–GORDON SYSTEM

M.G. Forest and E.A. Overman II

Department of Mathematics
The Ohio State University
Columbus, Ohio 43210, U.S.A.

P.L. Christiansen and M.P. Soerensen

Laboratory of Applied Mathematical Physics
The Technical University of Denmark
DK–2800 Lyngby, Denmark

N. Grønbech–Jensen

Physics Laboratory I
The Technical University of Denmark
Dk–2800 Lyngby, Denmark

R. Flesch

Department of Mathematics
Heriot–Watt University
Riccarton–Edinburgh, Scotland

D.W. McLaughlin

Department of Mathematics and Program in
Applied and Computational Mathematics
Princeton University
Princeton, New Jersey, U.S.A.

R.D. Parmentier

Department of Physics
Salerno University
Salerno
84081 Baronissi, Italy

S. Pagano

Istituto di Cibernetica del Consiglio Nazionale delle Ricerche
80072 Arco Felice, Italy

Nonlinear Superconductive Electronics and Josephson Devices
Edited by G. Costabile *et al.*, Plenum Press, New York, 1991

403

INTRODUCTION

In recent years the spectral transform for the sine–Gordon equation has been carefully investigated [1–3]. At the same time the perturbed sine–Gordon system has been used as a model for long Josephson tunnel junctions [4], being able to reproduce experimental results concerning I–V characteristics, radiation spectra, line width, stability etc [5–8].

The present work was initiated when the authors were visiting at the Center for Modelling, Nonlinear Dynamics and Irreversible Thermodynamics (MIDIT) at the Technical University of Denmark. The motivation is to investigate to what degree some of the complicated exact solutions to the unperturbed sine–Gordon equation, obtained by application of the spectral transform, will serve as attractors in the dynamics of the perturbed sine–Gordon system. A particular goal is to investigate the role of multikink solutions on a running background in the explanation of the anomalously high current steps observed experimentally on long Josephson junctions (only) when there are subjected to large amplitude microwave drive [9].

Here we derive some of the exact solutions to the sine–Gordon equation with periodic boundary conditions. The Bäcklund transformation is used to obtain a 4π–kink solution on a running background. The decline of this solution into multikink solutions in the presence of loss and uniform dc–bias is illustrated. A full account of the results with our conclusions will be presented elsewhere.

The sine–Gordon equation and the Bäcklund transformation

Let u(x,t) be a solution to the sine–Gordon equation

$$u_{xx} - u_{tt} - \sin u = 0 \tag{1.a}$$

with spatial period L

$$e^{iu(x+L,t)} = e^{iu(x,t)} . \tag{1.b}$$

Then we can obtain a new solution, U(x,t), by the Bäcklund transformation

$$U(x,t) = u(x,t) + 2i \ln \left[i \frac{\psi_2(x,t;\lambda)}{\psi_1(x,t;\lambda)} \right] , \tag{2}$$

where $\psi_1(x,tz,\lambda)$ and $\psi_2(x,t,\lambda)$ are solutions to the scattering equations [3].

$$\left[\begin{bmatrix} 0 & -1 \\ 1 & 0 \end{bmatrix} \frac{\partial}{\partial x} + \frac{i}{4}(u_x + u_t) \begin{bmatrix} 0 & 1 \\ 1 & 0 \end{bmatrix} + \frac{1}{16\lambda} \begin{bmatrix} e^{iu} & 0 \\ 0 & e^{-iu} \end{bmatrix} - \lambda \right] \begin{bmatrix} \psi_1 \\ \psi_2 \end{bmatrix} = 0 \tag{3a}$$

and

$$\left[\begin{bmatrix} 0 & -1 \\ 1 & 0 \end{bmatrix} \frac{\partial}{\partial t} + \frac{i}{4}(u_x + u_t) \begin{bmatrix} 0 & 1 \\ 1 & 0 \end{bmatrix} - \frac{1}{16\lambda} \begin{bmatrix} e^{iu} & 0 \\ 0 & e^{-iu} \end{bmatrix} - \lambda \right] \begin{bmatrix} \psi_1 \\ \psi_2 \end{bmatrix} = 0 , \tag{3b}$$

where λ is a critical point in the spectrum of the L–operator in the spectral transform. Note that U will not be real unless

$$| \psi_2(x,t,\lambda)/\psi_1(x,t;\lambda) | = 1 . \tag{4}$$

Furthermore, it is easily verified that the product $\psi_1 \psi_2$ satisfies the linear partial differential equation

$$(\psi_1 \psi_2)_{xx} - (\psi_1 \psi_2)_{tt} - (\cos u) \psi_1 \psi_2 = 0 . \tag{5}$$

From now on we restrict ourselves to consider solutions of Eq. (1) <u>without</u> spatial structure. That is, we let $u(x,t) = f(t)$. Here

$$f(t) = 2\arcsin(\text{ksn}(t;k)) \equiv 2\arcsin(\text{sn}(kt;1/k)) \ , \tag{6}$$

where the modulus in the Jacobi elliptic function, k, is arbitrary $k > 0$. The energy of the solution (6) is found by insertion into the Hamiltonian for Eq. (1)

$$H = \int_0^L \left[\tfrac{1}{2}u_x^2 + \tfrac{1}{2}u_t^2 + 1 - \cos u \right] dx \ , \tag{7}$$

yielding $H = 2Lk^2$. For $0 < k < 1$, $f(t)$ oscillates periodically, while for $k > 1$, $f(t)$ becomes a running solution (the McCumber solution). In both cases the period of f (mod 2π) becomes $T = 2K(1/k)/k$ $(k < 1)$ where K is the complete elliptic integral of first kind. It is useful to note that the derivative of $f(t)$ with respect to time becomes

$$\dot{f}(t) = \sqrt{2} \sqrt{H/L - 1 + \cos f(t)} \ . \tag{8}$$

Eqs. (3a) and (6) now yield

$$\psi_{1,xx} + \nu^2 \psi_1 = 0 \ \text{and} \ \psi_{2,xx} + \nu^2 \psi_1 = 0 \tag{9a,b}$$

where the constant ν is given by

$$\nu = \sqrt{\Lambda^2 + h/8} \ . \tag{10}$$

Here we have introduced

$$\Lambda = \lambda - \frac{1}{16\lambda} \ \text{and} \ h = H/L \tag{11a,b}$$

for convenience. (Note that $\lambda^2 \geq 0$ when λ is real, $-\frac{1}{4} < \Lambda^2 < 0$ for $|\lambda| = \frac{1}{4}$, and $\Lambda^2 \leq -\frac{1}{4}$ when λ is imaginary). H is the energy of the waveform per unit length.

The Floquet determinant [3]

$$\Delta(\lambda) = 2\cos L\nu \tag{12}$$

equals ± 2, and $\Delta'(\lambda) = 0$ since λ is a critical point under the Bäcklund transformation.

Trivial solutions to Eq. (9a,b) are

$$\psi_1(x,t;\lambda) = e^{\pm i\nu x}\varphi_1^{\pm}(t;\lambda) \ \text{and} \ \psi_2(x,t;\lambda) = e^{\pm i\nu x}\varphi_2^{\pm}(t;\lambda) \tag{13a,b}$$

where the time dependencies are denoted $\varphi_1^{\pm}(t)$ and $\varphi_2^{\pm}(t)$. (The exponential factors in Eq. (13) have $\mp$ rather than $\pm$ in them so that the solution, $U(x,t)$, in Eq. (2) increases by 4π). Using Eq. (13) in Eq. (3a) we get a relationship between $\varphi_1^{\pm}$ and $\varphi_2^{\pm}$

$$\frac{\psi_2^{\pm}(x,t;\lambda)}{\psi_1^{\pm}(x,\iota,\lambda)} = \frac{\varphi_2^{\pm}(t;\lambda)}{\varphi_1^{\pm}(\iota,\lambda)} = \frac{\pm \ i\nu + \frac{i}{4}\dot{f}(t)}{-\frac{1}{16}\lambda e^{-if(t)} + \lambda} = \frac{\frac{1}{16\lambda}e^{if(t)} - \lambda}{\pm \ i\nu - \frac{i}{4}\dot{f}(t)} \ . \tag{14}$$

405

Thus we may choose $\varphi_1^\pm(t;\lambda)$ and $\varphi_2^\pm(t;\lambda)$ so that

$$\psi_1^\pm(x,t;\lambda) = e^{\mp i\nu x} \sqrt{\frac{\pm i\nu + \frac{i}{4}\dot{f}(t)}{-\frac{1}{16\lambda}e^{-if(t)} + \lambda}} \sqrt{\varphi^\pm(t;\lambda)} \tag{15a}$$

and

$$\psi_2^\pm(x,t;\lambda) = e^{\mp i\nu x} \sqrt{\frac{-\frac{1}{16\lambda}e^{-if(t)} + \lambda}{\pm i\nu + \frac{i}{4}\dot{f}(t)}} \sqrt{\varphi^\pm(t;\lambda)} \ . \tag{15b}$$

With these choices of $\psi_1^\pm$ and $\psi_2^\pm$ we get

$$\psi_1^\pm(x,t;\lambda)\,\psi_2^\pm(x,t;\lambda) = e^{\mp 2i\nu x}\,\varphi^\pm(t;\lambda) \tag{16}$$

which inserted into Eq. (5), yields the Lamé equation for $\varphi^\pm$

$$\ddot{\varphi}^\pm(t;\lambda) + [4\nu^2 + \cos f(t)]\,\varphi^\pm(t;\lambda) = 0 \ . \tag{17}$$

A first order equation for $\varphi^\pm(t;\lambda)$ may also be obtained from Eq. (3b)

$$\frac{\dot{\varphi}^\pm(t;\lambda)}{\varphi^\pm(t;\lambda)} = F^\pm(f;\lambda) + F^\pm(-f;\lambda) \tag{18a}$$

with

$$F^\pm(f;\lambda) = (\mp i\nu + \tfrac{i}{4}\dot{f})\frac{\frac{1}{16\lambda}e^{if} + \lambda}{\frac{1}{16\lambda}e^{if} - \lambda} \ . \tag{18b}$$

We now introduce the complex constant α by

$$\text{sn}(\alpha,k) = 2i\Lambda/k, \quad \text{cn}(\alpha,k) = 2\nu/k, \quad \text{dn}(\alpha,k) = 2\,\Lambda \ . \tag{19a,b,c}$$

Then, as shown in Appendix A, we obtain

$$\frac{\dot{\varphi}^\pm}{\varphi^\pm} = \pm\frac{\text{sn}(t,k)}{\text{sn}(\alpha,k)\text{sn}(t\mp\alpha,k)} \mp \frac{\text{cn}(\alpha,k)\text{dn}(\alpha,k)}{\text{sn}(\alpha,k)} \tag{20}$$

whose solution is

$$\varphi^\pm(t) = C\,\frac{H(t\mp\alpha,k)}{\theta(t,k)}\,e^{\pm t Z(\alpha,k)} \ , \tag{21}$$

where Z, θ, and H denote Jacobi's zeta, theta. and eta functions, respectively, and C is an integration constant.

So far we have not used the fact that $U(x,t)$ is periodic in x with period L. From Eq. (13) we get

$$\nu = \pi/L \quad n = 1,2,\dots \ , \tag{22}$$

and from Eq. (12) $\Delta(\lambda) = \pm 2$ $(\Delta'(\lambda) = 0)$ in accordance with the fact that λ is a critical point. From Eqs. (4) and (14) follows that in order that $U(x,t)$ be real λ must be purely imaginary. Therefore the energy density $h = 2k^2$ must be so large that

$$\Lambda^2 \equiv (\pi/L)^2 - h/8 \le -\tfrac{1}{4} \ . \tag{23}$$

or $\quad k^2 \ge 4\pi^2/L^2 + 1$.

So far, $\psi_2(x,t)/\psi_1(x,t)$ is a function of t only (Eq. (14)). However, spatial dependence can be obtained by letting

$$\psi_1 = e^{i\gamma}\psi_1^+ + \psi_1^- \text{ and } \psi_2 = e^{i\gamma}\psi_2^+ + \psi_2^- , \tag{24a,b}$$

where γ is a constant. Here our new solution

$$U(x,t) = f(t) + 2i \ln \left[i \, \frac{e^{i\gamma}\psi_2^+ + \psi_2^-}{\psi_1^+ + e^{i\gamma}\psi_1^-} \right] \tag{25}$$

will be real for γ real and λ imaginary. Fig. 1 shows a plot of the solution at time $t = 0$ for $L = 12$ and $H = 31.2$ ($h = 2.6$), making condition (23) fulfilled by 3 per cent. The 4π–kink is stable and travels through spatial interval with a velocity, $v_{4\pi} > 1$.

Other Solutions of the Sine–Gordon Equation

The energy of the 4π–kink solution (25) is rather high. We therefore consider simpler solutions of Eq. (1a,b) with lower energies. For $L = \infty$ the simple 2π–kink, travelling with velocity v $(-1 < v < 1)$ is given by

$$U_{kink}(x,t) = 4 \arctan \exp \frac{x - vt}{\sqrt{1 - v^2}} \tag{26}$$

with energy $H = 8/\sqrt{1 - v^2}$. The corresponding travelling wave solution in the periodic case is a kink–train given by

$$U_{kink-train}(x,t) =$$

$$2 \arcsin \left\{ sn \left[\sqrt{\frac{r}{2(\kappa^2 - \omega^2)}} \, (\kappa x - \omega t), \, i\sqrt{\frac{2}{E}} \right] \right\} , \tag{27}$$

where $\kappa = 2\pi/L$, and the periodicity condition $U_{kink-train}(x + L,t) = U_{kink-train}(x,t) + 2\pi$ determines ω such that

$$\kappa^2 - \omega^2 = \frac{\pi^2 r}{2K^2(i\sqrt{2/r})} , \tag{28}$$

r being a parameter smaller than approximately $32 \, e^{-L}$ in order that ω be real. $E = E(i\sqrt{2/r})$ is the complete integral of second kind. The energy of the kink–train (27) is

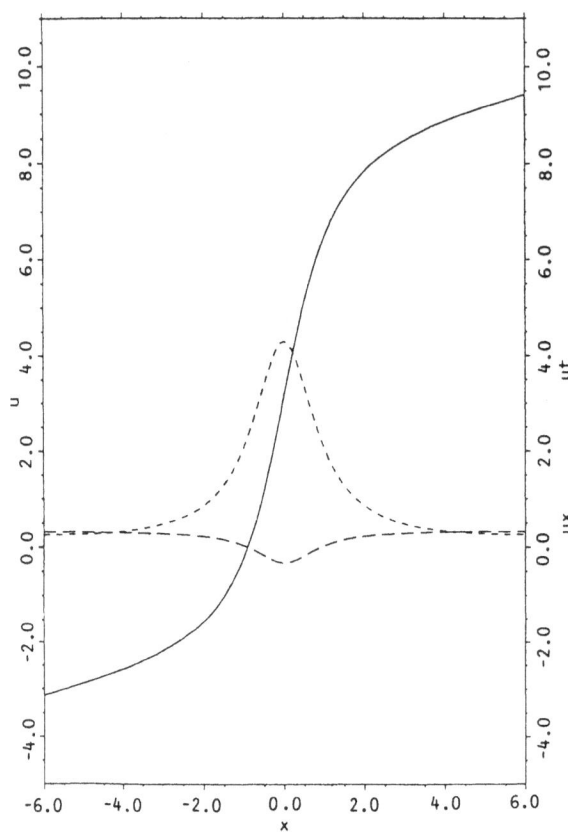

Fig 1. 4π–kink solution (25) for L = 12, h = 2.6. $U(x,0)$ full curve, $U_x(x,0)$ short dashed curve, $U_t(x,0)$ long dashed curve.

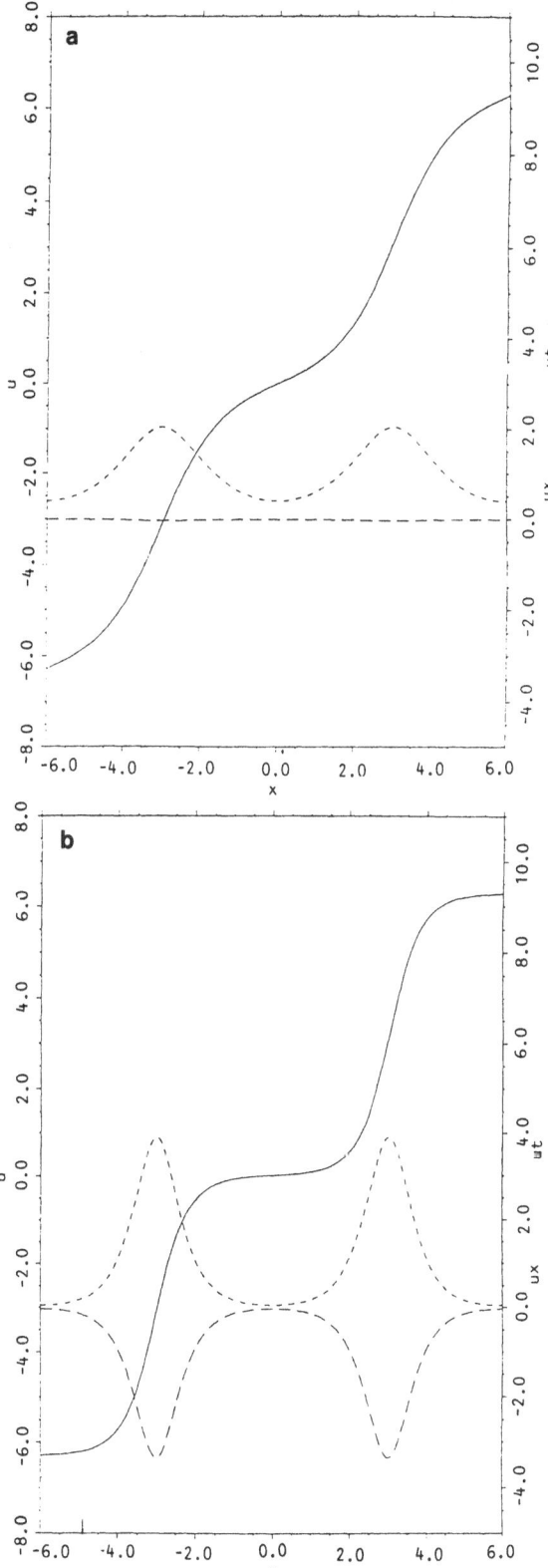

Fig. 2.   Two kink–trains (27) for L/2 = 6, (a) h = 1.35 (minimal energy), (b) h = 2.6
          (same energy as in Fig. 1) u(x,0) full curve, u_x(x,0) short dashed curve, u_t(x,0)
          long dashed curve.

409

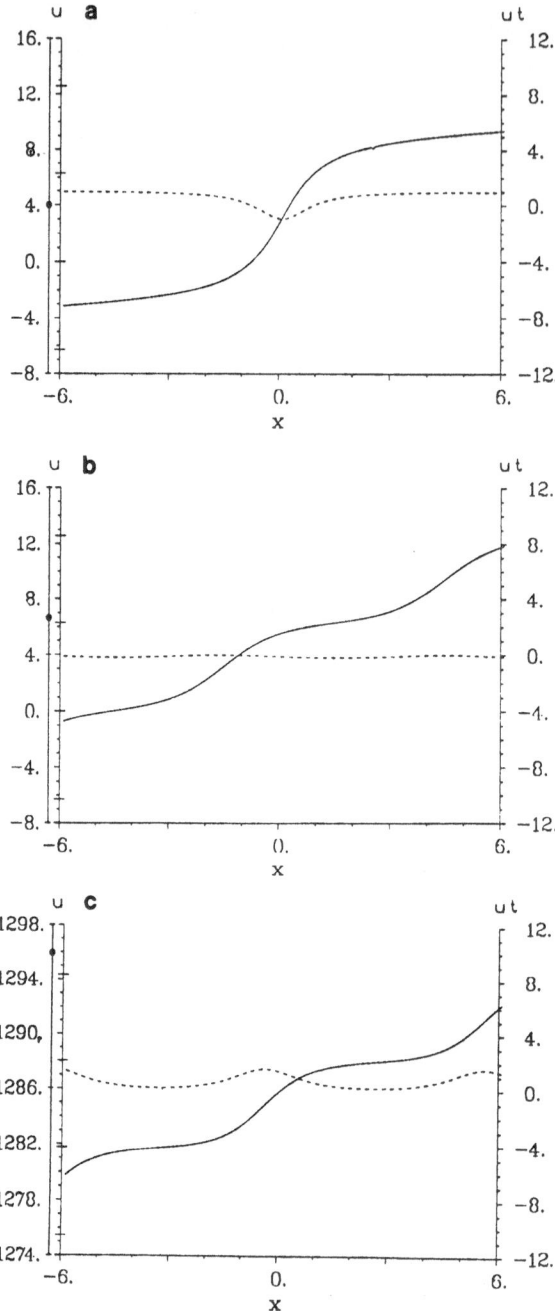

Fig. 3.   Numerical solutions of (30) for L = 12. (a) u(x,0) = 4π–kink waveform (25), (b) u(x, 2000) for $\alpha = 0.02$, $\gamma = 0$, (c) u(x, 2000) for $\alpha = 0.02$, $\gamma = 0.05$. u full curve, u_t dashed curve.

$$H_{kink} = -rL + 8\sqrt{r/2}\ E(i\sqrt{2/r})\ .\tag{29}$$

The minimum energy of the kink–train min $H_{kink} \to 8+$ for $L \to \infty$.

To compare this kink–train with our 4π–kink solution we must put two kink–trains together, i.e. we must compare a kink–train with period $L/2$ to a 4π–waveform with period L so that they both vary by 4π. Fig. 2 shows plots of kink–trains for two periods at time t = 0 for $L/2 =$ 6 and in Fig. 2a for the minimal energy H = 16.16 (h = 1.35) and the same energy as in Fig. 1 H = 31.2 (h = 2.6).

Numerical Solutions of the Perturbed Sine–Gordon Equation

We now use the 4π–kink solution (25) as initial condition in numerical solutions of the perturbed sine–Gordon equation

$$u_{xx} - u_{tt} - \sin u = \alpha u_t - \gamma\tag{30a}$$

with spatial period L

$$e^{iu(x+L,t)} = e^{iu(x,t)}\ .\tag{30b}$$

The initial condition is shown in Fig. 3a.

The case of loss and bias ($\alpha > 0$, $\gamma = 0$) is simplest because here energy is being removed from the system, and the solution approaches the minimum energy configuration. Since the stationary kink–train has the minimum energy configuration for the exact sine–Gordon equation it is to be expected that the 4π waveform breaks up into two humps and approaches a two kink–train wave form. Fig. 3b illustrates this development at t = 2,000 in the case $\alpha = 0.02$. For the same loss, but with some dc–bias ($\alpha = 0.02$, $\gamma = 0.05$) we get the steeper two kink–train shown in Fig. 3c.

Our studies of the effect of other perturbations like loss of type βu_{xxt} and rf–bias $\Gamma \cos \omega t$ in Eq. (30a), time varying boundary conditions and open and boundary conditions instead of Eq. (30b) will be published in a fuller account elsewhere.

ACKNOWLEDGMENTS

The authors (MGF, EAO, DWMcL) and (PLC) acknowledge the hospitality of the Technical University of Denmark and the Center for Nonlinear Studies, Los Alamos National Laboratory, respectively. Financial support from the Danish Technical Research Council and Thomas B. Thriges Fond is acknowledged by the authors (RF, RDP) and (SP), respectively.

APPENDIX A

Substituting Eq. (6) into (18) we obtain

$$\frac{\dot{\varphi}^{\pm}}{\varphi^{\pm}} = \frac{\pm\ 2i\Lambda^2 + \dfrac{k^2}{4}\ sn(t,k)cn(t,k)dn(t,k)}{\dfrac{k^2}{4}sn^2(t,k) + \Lambda^2}\tag{A1}$$

or

$$\frac{\dot{\varphi}^{\pm}}{\varphi^{\pm}} = \frac{\pm\ \text{sn}(\alpha,k)\text{cn}(\alpha,k)\text{dn}(\alpha,k)\ +\ \text{sn}(t,k)\text{cn}(t,k)\text{dn}(t,k)}{\text{sn}^2(t,k) - \text{sn}^2(\alpha,k)}. \qquad (A2)$$

Then application of the identity

$$\text{sn}(u \pm v) = (\text{sn}^2 u - \text{sn}^2 v)/(\text{snu cnv dnv} \mp \text{snv cnu dnu}) \qquad (A3)$$

yields Eq. (20). (Eq. (A3) is found in Ref. 10, page 64, but <u>not</u> in Ref. 11).

Using Ref. 10 (and including its equation numbers for easy reference) we obtain, following [12],

$$\frac{\dot{\varphi}^{\pm}}{\varphi^{\pm}} = k^2 \text{sn}(t,k)\text{sn}(\pm\ \alpha + iK',k)\text{sn}(t \mp\ a + iK',k)$$

$$\mp \frac{\text{cn}(\alpha,k)\,\text{dn}(\alpha,k)}{\text{sn}(\alpha,k)}\ ,\quad (122.07)$$

where $K'(k) = K(\sqrt{1 - k^2})$,

$$\frac{\dot{\varphi}^{\pm}}{\varphi^{\pm}} = Z(t \mp\ \alpha + iK',k) - Z(t,k) - Z(\mp\ \alpha + iK',k)$$

$$\mp \frac{\text{cn}(\alpha,k)\text{dn}(\alpha,k)}{\text{sn}(\alpha,k)}\ ,\ (142.02)$$

and

$$\frac{\dot{\varphi}^{\pm}}{\varphi^{\pm}} = Z(t \mp\ \alpha + iK',k) - Z(t,k) - Z(\mp\ \alpha,k) + \frac{\pi i}{2K}\ .\ (141.01)$$

Now

$$Z(u,k) = \frac{\theta'(u,k)}{\theta(u,k)}\ , \qquad\qquad (1053.01)$$

$$Z(u + iK',k) = H'(u,k)/H(u,k) - \frac{i\pi}{2K}\ ,$$

where H' and θ' mean the partial derivative with respect to the first argument. Thus

$$\frac{\dot{\varphi}^{\pm}}{\varphi^{\pm}} = \frac{H'(t \mp\ \alpha,k)}{H(t \mp\ \alpha,k)} - \frac{\theta'(t,k)}{\theta(t,k)} \pm Z(\alpha,k) \qquad (A.4)$$

for which we obtain the solution (21).

REFERENCES

1. M.G. Forest and D.W. McLaughlin, <u>J. Math. Phys</u>. 23:1248 (1982).
2. E.A. Overman II, D.W. McLaughlin, and A.R. Bishop, <u>Physica D</u> 19:1 (1986).
3. N. Ercolani, M.G. Forest, and D.W. McLaughlin, <u>Physica D</u> 43:349 (1990)
4. D.W. McLaughlin and A.C. Scott, <u>Phys. Rev. A</u> 18:1652 (1978).

5. P.S. Lomdahl, O.H. Soerensen, and P.L. Christiansen, <u>Phys. Rev. B</u> 25:5737 (1982)

6. M.P. Soerensen, R.D. Parmentier, P.L. Christiansen, O. Skovgaard, B. Dueholm, E. Joergensen, V.P. Koshelets, O.A. Levring, R. Monaco, J. Mygind, N.F. Pedersen, and M.R. Samuelsen, <u>Phys. Rev. B</u> 30:2640 (1984).

7. F. If, P.L. Christiansen, R.D. Parmentier, O. Skovgaard, and M.P. Soerensen, <u>Phys. Rev. B</u> 32:1512 (1985).

8. S. Pagano, M.P. Soerensen, P.L. Christiansen, and R.D. Parmentier, <u>Phys. Rev. B</u> 38:4677 (1988).

9. G. Rotolì, G. Costabile, and R.D. Parmentier, <u>Phys. Rev. B</u> 41:1958 (1990).

10. A. Cayley, "An Elementary Treatise on Elliptic Functions" 2nd ed., Dover Publications, Inc. (New York 1961).

11. P.F. Byrd and M.D. Friedman, "Handbook of Elliptic Integrals for Engineers amd Physicists", Springer (New York 1954).

12 E.T. Whittaker and G.N. Watson, "A Course of Modern Analysis", 4th ed. (Cambridge University Press, Cambridge, 1927).

DYNAMICS OF VORTICES MOTION IN GRANULAR HIGH T_C THIN FILM BRIDGES

G.A.Ovsyannikov[#], K.Y. Constantinian[$] and L.E. Amatuni[$]

[#]Institute of Radio Engineering and Electronics
USSR Academy of Sciences,Moscow 103907, USSR
[$]Institute of Radio Physics and Electronics,
Armenian Academy of Sciences, Ashtarack-2, 378410 Armenia,
USSR

INTRODUCTION

High T_c superconductors (HT_cS) could be considered as a Josephson media (JM) which consists of superconducting grains with mean size a_0 connected by Josephson junctions with critical current I_c. The theoretical study of JM were curried out by Giovanini and Weiss[1], Rosenblatt and co workers[2] and recently essentially contributed by Clem[3] and Sonin E.B. with co workers (S-theory)[4].

HT_cS bridge structures (BS) with constriction dimensions considerably larger than coherence length ξ exhibit phenomena inherent to Josephson effects, such as Shapiro steps on I-V curve under microwave radiation due to the coherent motion of magnetic vortices driven by bias current[5-7]. These phenomena in BS made from homogeneous superconducting thin films is well known[8-10]. The distinctive feature of granular HT_C BS are vortices formation with large magnetic radius, called hypervortices[4] and their coherent motion as well in absence of external microwave field[5,6].

So far, there are no analysis the dynamics of such vortices motion, particularly the processes of microwave self-radiation and nonlinear behavior of BS but Jung and Konopka[11] observed radiation from HT_cS film.

THEORY

Let us consider the transport properties of HT_cS BS made from thin film by means of two cuts in direction perpendicular to bias current. Within London electrodynamics of granular thin films at the distance much larger than a_0 S-theory give the relationship between supercurrent j_s and mean phase of order parameter ϕ and vector-potential $\mathbf{A}$:

$$j_s = (2e/\hbar) \cdot g \cdot (\nabla\phi - 2\pi\mathbf{A}/\phi_0), \tag{1}$$

Nonlinear Superconductive Electronics and Josephson Devices
Edited by G. Costabile *et al.*, Plenum Press, New York, 1991

415

where $g^{-1}=16\pi^3\lambda_L^2/\phi_0^2+(E_Ja_0)^{-1}$, λ_L - London penetration depth, $E_J=\hbar j_c^g/2e$ energy of a Josephson link per unit area, j_c^g - critical current density of the link, $\phi_0=hc/2e$ - magnetic flux quanta. Here S-theory suppose that ϕ - changes only at grains boundaries.

For the case of homogeneous film with thickness $d\leq\lambda_L$, from (1) we get well-known equation for j_s by the changing $(2e/\hbar)g=I_0/\pi d$, $(I_0=c\phi_0d/8\pi\lambda_L^2$ - critical current). At $I=I_0$ Meissner state becomes unstable and Abrikosov vortices already begin to bear with electromagnetic interaction radius:

$$\lambda=\lambda_\perp=\lambda_L cth(d/2\lambda_L). \qquad (2)$$

In the case of weak coupling $(I_0>j_c^ga_0d)$ $\quad\lambda$ is[4]:

$$\lambda = \begin{cases} \lambda_J & a_0>>\lambda_J>>\lambda_L \\ \lambda_J & \lambda_J>>a_0>>\lambda_L \\ \lambda_J(\lambda_L/a_0)^{1/2} & \lambda_L>>a_0, \quad \lambda_J>>(a_0\lambda_L)^{1/2} \end{cases} \qquad (3)$$

Two last cases of (3) are belong to hypervortices which are very similar to several Josephson vortices with overlapping magnetic field. In these cases the critical field $H_{c1}=4\pi^2g/\phi_0\ln(\lambda/a_0)$ is very weak ~ 0.1 Oe[4]. The expression for j_s is invariant for replacing $(2e/\hbar)g$ by $I_0/\pi d$ so for analysis BS of granular HT_c one can use the results of Alamazov and Larkin (AL) theory[9] developed for BS made of homogeneous superconducting thin film in case $\xi<W<\lambda$, $L\to0$ (W - width of BS and L - it's length). Thus, using results of papers[4,12-14] also for the main features of HT_cS BS we get follow results:

1) Supercurrent distribution in BS is not homogeneous even in case $W<\lambda$. For $I>I_0'=(2e/\hbar)dg$ a potential barrier exists for vortices creation up to critical current of BS:

$$I_c\approx dg(2e/\hbar)(W/a_{ef})^{1/2}, \qquad (4)$$

a_{ef} is the effective core area which equal to ξ for Abricosov vortices and $a_{ef}\approx a_0$ for JM .

2) If a pinning or some inhomogeneities of structure are absent a vortex creates at the edge of BS and moves viscously across BS under Lorenz force. A vortex creation time is proportional to ΔI $(I-I_c)$ and independent of BS dimensions, viscous motion time $t_\eta=\pi\eta eW^2/4\hbar I_c \sim W^{3/2}$, η - friction coefficient.

3) For the currents slightly exceed I_c the I-V curve of BS is hyperbolic $V\sim\Delta I^{1/2}$. A plato exists at $V=V_0=\hbar\pi/et_\eta$ when a single flux is moving across BS. The current enhancement accompanied by vortex number (N) increa-

sing and BS differential resistance R_d make a jump and peaks observed on I-V curve at $V=NV_0$. For larger voltages (N>>1) the I-V curve is almost linear and then it is parabolic $V \sim V_0 I^2/I_c I_0'$.

4) Taking into account spatial fluctuations of critical current in JM $G^2 = <\delta I_c^{*2}>/I_c^2$ the conditions for Shapiro steps existence in long BS have been developed[12]:

$$L < \lambda_J (I/I_c)^2 (\lambda_J/d)^2 (a_0^2 g^2/\lambda_J^2)^{-1} \qquad (5)$$

5) Accoding to AL-theory vortex local excitation take place due to local current rise. In the case of planar Josephson junction lattice with a inhomogeneity the first vortex row begins to move in a area just close to a local inhomogeneity, the next rows appear in neighboring areas on the distance comparable with λ[13], so strong vortex-vortex interaction observed in spite of large BS length $L \gg \lambda_J$.

6) Arbitrary located pinning centers depress the coherent motion of vortex lattice. But regularity of vortex motion still exists and reveals in sharp impedance singularities at microwaves[14].

SAMPLES AND EXPERIMENTAL PROCEDURE

Two types of $YBa_2Cu_3O_{7-x}$ (YBCO) BS with different L and W were investigated. In one case, BS (#G and #F series) were patterned in YBCO films using photolithography and Ar ion milling. The films of thickness d=300 nm were deposited on hot MgO substrates by in-situ laser ablation. They were highly c-axis oriented and had grains size of about a_0=200 nm[15]. In the second type of BS the film of thickness d=1μm with $a_0 \simeq 1$μm was deposited on sapphire substrates with ZrO_2 interlayer by magnetron sputtering[16] were used.

I-V and dV/dI curves were measured at temperatures between 4.2 and 100K in autonomous case and in present of small magnetic fields (H<30 Oe) and low level of external microwave power (at frequency range 7-10 GHz and 18-22 GHz). The magnetic field was applied perpendicular to the film plane. The sample was shielded from external electromagnetic field by μ-metal – screen which reduced the earth's field at least an order of magnitude.

The BS was either placed in (i) a microstrip line which in one end was coupled to a coax and in other to a quarter-wave open circuit (8-10 GHz), or (ii) a waveguide terminated by a plunger that was adjusted to give a maximum output (18-22 GHz). Self radiation from the BS was registered in former case by means of X-band amplifiers with a total noise temperature of 800K and a bandwidth of 3 GHz followed by a spectrum analyzer with bandwidth Δf=1MHz or a power meter with a YIG filter (Δf=25 MHz). In the latter case a receiver with bandwidth Δf=300 MHz and fluctuation sensitivity δT≈0.1 K and integrating time τ≈1 sec was used in the frequency range f≈18÷22GHz. The coaxial output was provided for radiometer receiver at f≈1.7 GHz with ΔF≈36 MHz and δT≈0.02 K.

The parameters at 4.2K of some of our samples are presented in Table 1.

Table 1. BS parameters. j_c and λ_J at T=4.2K

Sample	W	L	T_c	j_c	λ_J
	μm	μm	K	A/cm²	μm
G7.2	5	10	81	$6 \cdot 10^6$	0.1
G3.1	2	10	71	$8 \cdot 10^4$	0.9
G3.2	3	14	71	$8 \cdot 10^4$	0.9
G3.3	7	12	71	$7 \cdot 10^4$	1.0
G4.4	6	20	92	$7 \cdot 10^6$	0.1
F3.1	9	10	60	$4 \cdot 10^4$	1.2
A1.1	50	120	70	$2 \cdot 10^3$	5.6
J1.1	45	110	35	60	17
J1.2	50	120	40	$2 \cdot 10^2$	32

The sizes of BS and the grains were measured using a scanning electron microscope (SEM). Critical current densities j_c were obtained by dividing the
maximum zero-voltage current (DC voltage of BS <5μV) by the cross section of
the bridge (W·d). Critical temperature T_c determined by extrapolating the
$j_c(T)$ dependence to zero. The Josephson penetration depth λ_J, which is a an
important JM [4] parameter, was estimated from the formula for a sandwich type
Josephon junction:

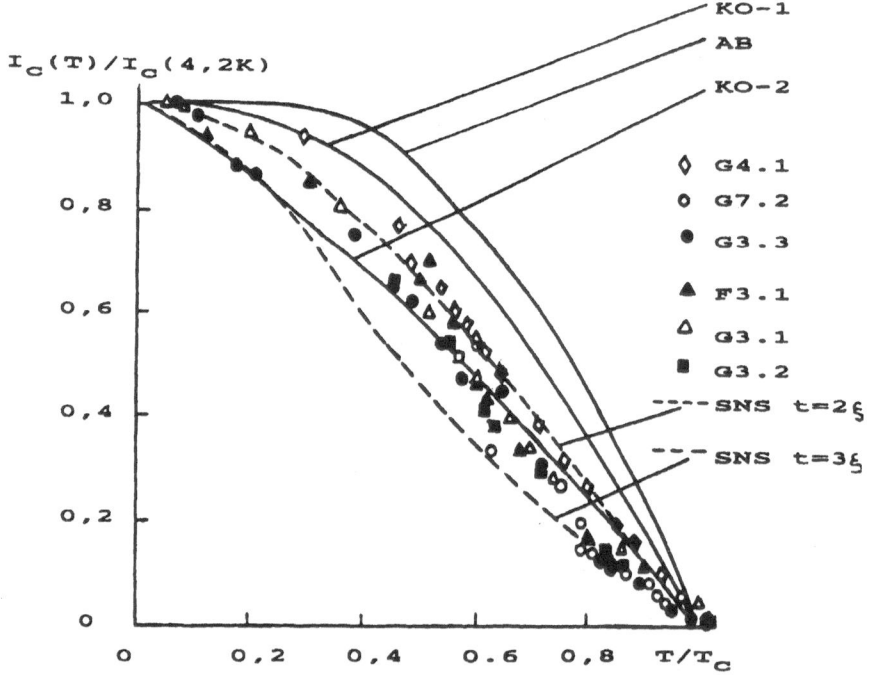

Fig. 1. The temperature dependencies of normalized critical current   for
        several BS. The solid lines represents the Ambegaokar-Baratoff
        (AB) theory for tunnel junctionsand for weak links with
        immediate conductivity by Kulik and Omenlyanchuk for dirty
        (KO-1) and pure (KO-2) cases. The dashed line - theory for SNS
        junctions in dirty limit for different ratios t/ξ.

$$\lambda_J = (c\phi_0/8\pi^2 J_c t_{ef})^{1/2}, \qquad (6)$$

where $t_{ef} = t + 2\lambda_L$ is the effective magnetic field penetration depth between two grains. For YBCO we suppose $\lambda_L = 0.2$ μm and it is larger then the thickness of interlayer between grains $t \sim \xi = 0.5-2$ μm.

RESULTS AND DISCUSSION

Critical current and I-V curve

The normalized critical current $I_c(T)/I_c(T=4.2K)$ as a function of normalized temperature T/T_c for several BS are shown in Fig.1. The theoretical curves are also presented in Fig.1, solid lines: $I_c(T)$ of the Ambegaokar-Baratoff (AB) theory for tunnel junctions and of the Kulik-Omelyanchuk - theory for weak links with direct conductivity in the dirty (KO-1) and - clean (KO-2) limit respectively[17]. We observe reasonable agreement with the theory KO-2 theory which is valid for short and clean weak links ($t/\xi < 1$, $l \gg t$, l - the mean free path) as well as with the theory of SNS junctions in dirty limit ($l \ll t$) with ratio $t/\xi \approx 2 \div 3$ (dashed lines). For small enough critical current densities $j_c < 2 \cdot 10^6$ A/cm^2 for $\lambda_L \approx 0.2$ μm, $a_0 \approx 300$ nm, in expression (4) only parametere g is temperature depended value, determined by critical current between two neighboring grains. Hence, the experimental curves in Fig.1 indicate inter granular critical current temperature dependencies. Because of small $\xi \sim t < 1$ we suppose that inter granular coupling is a weak link of clean type, but not dirty one. Nonlinear behavior $I_c \sim (T_{co} - T_c)^n$ (n=1.5-2) near T_c can be explained by thermal fluctuations (kT$\sim E_J$)[1,3].

Fig. 2 shows I-V curves of BS G3.2 at various temperatures. The curves do not show hysteresis over a broad temperature range (4.2K<T<60K). An -

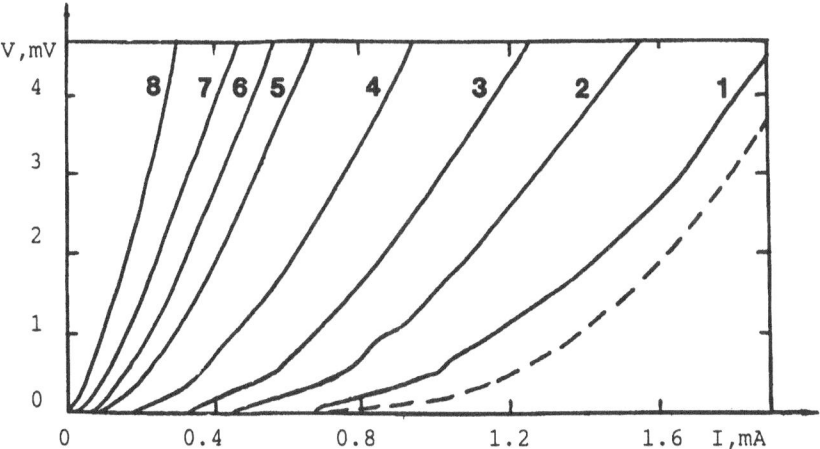

Fig. 2. The I-V curve of BS G3.2 at temperatures: 1 - T= 4.2 K, 2 - 31, 3 - 44 , 5 - 49, 6 - 53, 7 - 57,, 8 - 59. The dashed line is the I-V curve for G7.2 at T=16.4 K with scales I/3, V/·5.

excess current I_{ex} is exists at high voltages (V>1mV, for T=50K) when I-V curves are shifted relatively $V=IR_N$ by the voltage $V_{ex}=I_{ex}R_N$. This behavior is typical for weak link[17] with direct conductivity and supports our assumption of JM with this type of inter granular conductivity weak links in the BS. Note that the I-V curves at low T, consist of several segments with practically constant deferential resistance R_d, separated by unstable and hysterical parts. The relation $\lambda_J > a_0 \approx \lambda_L$ is fulfilled in wide temperature range $T \simeq 4 \div 50$ K for this BS and according to S-theory $\lambda \approx \lambda_J$. The irregularities of the structure of the BS material (order of a_0) have a small effect on the vortex motion process (weak pinning) hence the flow of vortices, and induced voltage is proportional to the force (current) and a constant dynamic resistance results. A constant R_0 corresponds to a flux flow regime of one vortex row in BS. $R_0 \sim n_v/\eta$, where n_v is the density of vortices in BS. We observed $R_0 \sim I_c^{-0.5}$ as the temperature is varied. This agrees with with AL theory for flux flow in absence of external magnetic field (H=0, n_v= const, $I \gg I_c$) if we assume $\eta \sim \lambda_J^{-1}$ as for Josephson vortices. At higher current the number of vortices and velocity v increase (N~I, v~I) and the I-V curve has a parabolic form $V \sim I^2$.

A flux flow regime can hardly be realized in BS with $\lambda_J < a_0$ or for very inhomogeneous film (on scale of λ) due to strong pinning. BS G7.2 has high j_c, $\lambda_J \simeq 0.1$ μm, $a_0 \simeq 0.3$ μm and I-V curve is smooth (see the dashed line in Fig.2). It is mainly determined by by thermally activated flux creep[6].

Electromagnetic radiation

Fig.3 shows I-V curve of BS and microwave radiation power vs bias voltage dependence P(V) obtained in autonomous regime. Self radiation peaks extended by V axis up to 0.5 mV. The power distribution P(V) differs consi-

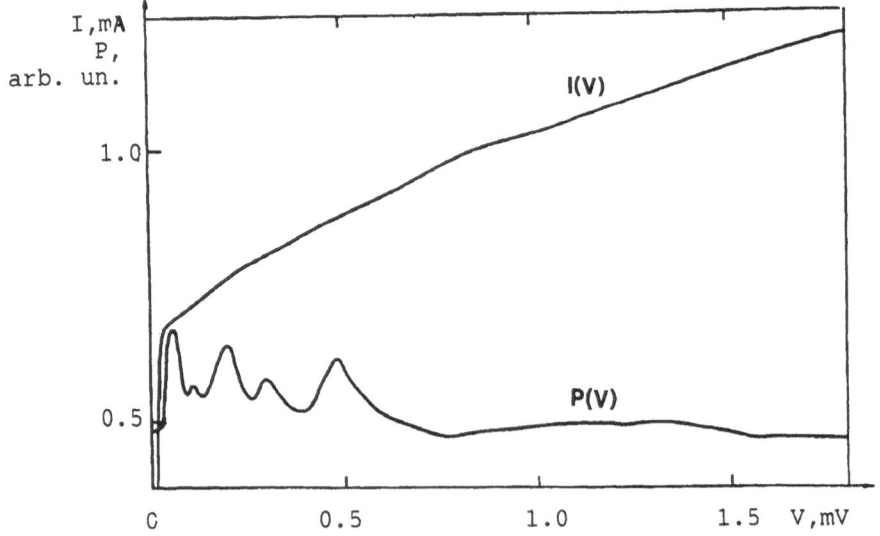

Fig. 3.   I-V curve and emitted microwave power at f=7.63 GHz, Δf=25 MHz for BS G3.2 at T=4.2 K.

derably from the one typical for Josephson microbridges[18]. The first peak of radiation occur at a voltage given by the Josephson relation with frequency of receiver f and m/n=1:

$$V_{m,n} = (m/n)(h/2e)f. \qquad (7)$$

This fact and the linear part of I-V curve with $R_d = R_0 = $const for $V \le$ 200 μm indicates a single row existence in the BS. For fixed bias current the vortices motion is periodical and during a vortex passing across BS the quantum phase difference of the BS electrodes changes by 2π. As a result every vortex induces ac voltage with frequency, determined by vortices motion velocity v:

$$f = Nv/b, \qquad (8)$$

where b is a characteristic dimension of vortex lattice, $b \approx w$[9,10] The first peak in P(V) corresponds to the motion of one vortex across the BS at weak part of the BS for example near some defect in structure. Successive peaks in P(V) are due to changes in N(I) and v(I) and also to the appearance of new rows in case L>λ as the bias current is increased. At larger V, when a

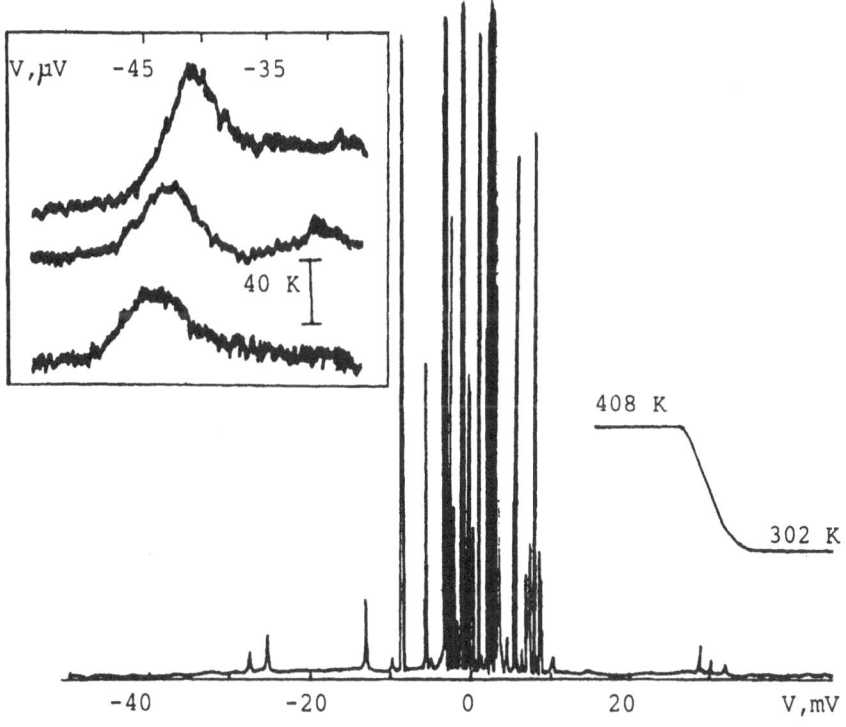

Fig. 4.   Radiation power vs bias voltage at T=4.2K, f=21.15 GHz for BS A1.1. The calibration temperatures of the receiver sensitivity are given at the right - the temperatures of the matched loads are denoted. The insert shows P(V) for an edge-type Josephson junction at frequencies f≈17.9; 19.3; 21.8 GHz.

considerable number of vortices exist, P(V) dependence exhibit nonequidistant peaks. In the long (L>>λ) BS as follows from digital result[13] all successive rows appeare in vicinity of the first one and strong interact with it. But as we see from Fig.3 for G3.4 there is no increase in power for increasing V as should be due to increasing N in the BS[10]. This may be due to a variation in phase between the moving row; in fact an antiphase radiation of in neighboring rows.

A similar results was obtained for the other type of BS A1.1 with a relatively larger BS areas and smaller j_c ($\lambda_J \approx 5.6$ μm) as displayed in Fig.4. Microwave emission (at 21 GHz) was observed up to $V_{max} \approx 8$ mV. The relation $V_{max}/V_{1,1} \geq 180$ is consistent with the filling of the BS by moving vortices $WL/\lambda_J^2 \approx 190$, if we suggest that every next peak of radiation is due to the increasing of vortices number in the BS. Note, that radiation intensity increases with magnetic field $H \approx 10$ Oe up to 5 times.

The bandwidth of radiation $\Delta F=300$ MHz, was measured for G3.2 by a spectrum analyzer in presence of small magnetic field. It is only two times higher than the ΔF estimated from a resistively shunted model of Josephson junction with shunt resistance $R_N=R_0 \approx 1$ Ohm corresponding to the flux flow resistance. The observed increase of ΔF may be due to external noise[*1]. The maximum output power of BS A1.1 was estimated to $3 \cdot 10^{-11}$ W, after taking into account the impedance mismatch between BS and the wave guide section. This value is considerably larger than those for Josephson microbridge[18]. We also compare (the insert of Fig.4) with the smaller maximum power emitted from an edge-type Josephson junction[19] in the same experimental set-up.

External affection by microwave

The microwave induced structure on the I-V curve of a BS $\lambda_J > a_0$ (see Fig.5) is well pronounced and strongly asymmetric (for positive and negative bias). The flow of vortices is synchronized by the external microwave field. Harmonic and sub-harmonic current steps appear at voltages related to the external frequency by the Josephson relation(7). From Fig.5 it is seen that the peak at P(I) dependence, corresponded to V=43 μV ($m=n=1$ in eq.(7)) is divided into two peaks, located symmetrical to the center of the first Josephson step. Thus, the BS nonlinear behavior with microwaves, is analogous to Josephson junction and I-V curve has hyperbolical shape in vicinity of the step in accordance to AL theory.

Fig.6a shows the autonomous I-V curve of BS J1.1 and the family of P(V) at f= 1.7 GHz at different power levels of external microwave field at $f_e \approx 21$ GHz. The self-radiation attributed to fluxon flow in BS of this series (#J) at high frequency (7-10 and 18-21 GHz) is very weak and noisy in comparison with other because of strong disordering of vortices motion by granular HT$_c$S material macrostructure probably[14]. The external weak microwave power P_e(70 dB)$\approx$ 100 fW changes the amplitude of self radiation. Fig.6b shows the maximums of P(V) dependence vs external power. The background level P(V) indicates the noise power intensity, which has tendency to decrease for I>I_c. Additional microwave affection as it is seen from Fig.6 could depress this disordering and self-radiation power increases. Note,

----

[*1] Recently very narrow bandwidth of radiation $\Delta F=22$ MHz was observed at lower frequency f=300 MHz for HT$_c$ thin film[20]

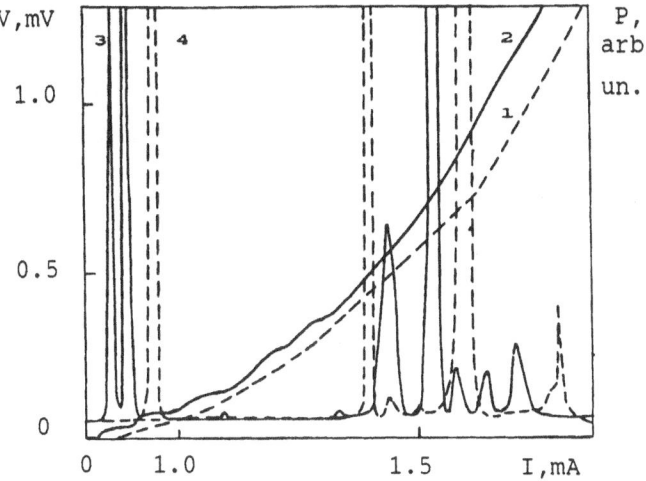

Fig. 5.   I-V curves (1,2) and P(I) (curves 3, 4) of BS in autonomous
          regime (dashed lines) and in presence of microwave field at
          $f_e \simeq 21$ GHz out of bandwidth of receiver (solid lines). The
          "heads"   of the largest peaks are cut out.

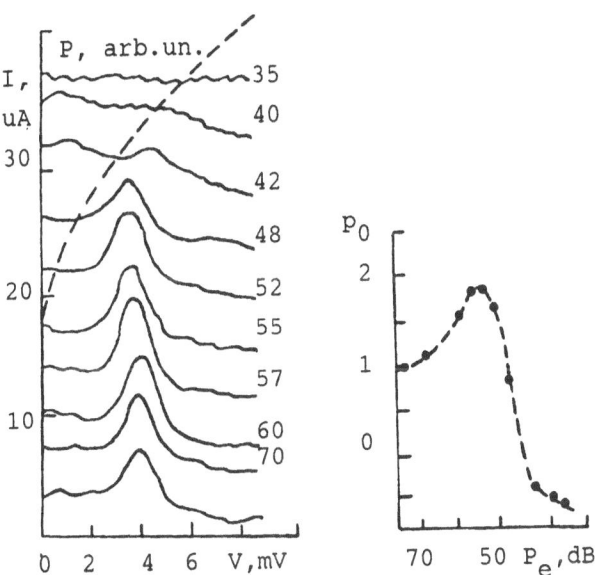

Fig. 6.   a) - The autonomous   I-V curve of the BS (dashed line)
          and a family of microwave radiation power P(V) dependencies
          at $f \approx 1.7$ GHz for different attenuation levels (dB) of
          external microwave power P_e at $f_e \simeq = 21$ GHz at T=4.2K. Zero levels
          of P(V) are shifted parallel to Y axis;b - the maximums P vs
          external power P_e (arbitrary units).

423

that in this case the affecting frequency is more than ten times higher than detected one. Taking into account these results one can suppose that external microwave field could turn in order the vortices motion process .

Fig.7 shows the I-V curve and reflected power δP_e vs current dependence for the BS J1.1. A small $P_e \simeq 100$ fW did not perturb the I-V curve of BS. As it was expected, experimentally obtained $\delta P(I)$ function demonstrated nonmonotonic singularities very similar to those, observed for conventional dc biased superconducting film[14] with the presence of external magnetic field. In our case no magnetic field was applied and only direct current flow is responsible for vortices motion. Such impedance behavior indicates that the vortices motion driven by dc is regular although no radiation was detected. Step-like singularities observed at certain values of external dc electric field in accordance to relation[14]:

$$E_{p,q} \simeq p/q \cdot f_e \cdot (\Phi_0 H/c)^{1/2}, \qquad (9)$$

where H - magnetic field, p,q-integers. The data analysis gives voltage - interval between two impedance jumps, corresponded to p=±1, q=±1 and p=±2, q=±1% $\Delta U = V_{\pm 2,\pm 1} - V_{\pm 1,\pm 1} \simeq 3$ mV which for $f_e \simeq 21$ GHz related to estimated value of H≈0.05 Oe- the field which exist in our sample chamber, shielded by μ-metal screen. The steps at $\Delta U/2$ and $\Delta U/3$ related to harmonics of main frequency - $2f_e$ and $3f_e$. We did not observed any subharmonic singularities with q>1. Asymmetry of $\delta P(I)$ function is inherent to vortex processes and recently it has been observed for I-V curve[5,6].

CONCLUSION

1. The investigated YBaCuO BS with small current densities ($j_c <$ 10^6 A/cm^2) can be considered as a Josephson media with nontunnel conductivity of intergrain links in clean limit.

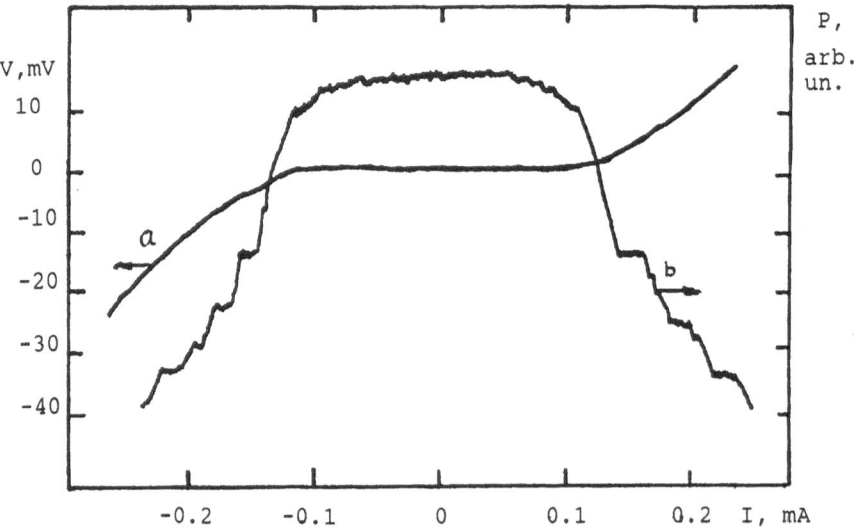

Fig. 7. a) - the experimental I-V curve of BS J1.1 at T=4.2 K, b)-
current dependence of reflected from the BS microwave probe
power $\delta P(I)$ at $f_e \simeq 21$ GHz.

2. The electrophysical and microwave properties of the BS of micrometer dimensions could be explained by vortices motion under dc and rf bias current even when external magnetic field is absent.

3. In BS with dimensions $W, L \sim \lambda_J$ the coherent vortices motion under bias current are realized in case of of small nonhomogenities and coherent effects, like a Shapiro steps on I-V curve, self radiation of electromagnetic waves, are developed.

4. In the case of coherent vortices motion in BS the radiation with line width close to one for Josephson junction is observed experimentally. The maximum registered generation power was $3 \cdot 10^{-11}$ W.

5. The pinning centers presence prevent the vortices coherent motion, but, nevertheless, the regular character of their motion is not disturbed. The external microwave irradiation can rise the vortices motion regularity degree, what is expressed in the self radiation intensity rising, and also in the BS microwave impedance nonmonotonic dependence vs dc electrical field.

ACKNOWLEDGMENTS

We are grateful to T. Claeson and Z. Ivanov for many fruitful discussions and laser ablation sample fabrication. We authors would like to acknowledge A.L. Gudkov, V.N. Laptev and V.I. Makhov for the sample fabrication, J.B. Hansen, R.P. Hubner, K.K. Likharev and A. Ustinov for dicusions.

This work was supported by the USSR scientific Council on the HT_cS problem and performed in the framework of projects No 42 and No 2068 of the USSR State Program "High Temperature Superconductivity"

REFERENCES

1. B. Giovanini and L. Weiss, Critical Current and Order Parameter in Josephon Arrays, Sol. St. Comm. 27:1005 (1978).
2. A. Roboutou, J.Rosenblatt, P. Peyral, Coherence and Disorder in Arrays of Point Contacts, Phys. Rev. Lett. 45:1035 (1980).
3. J.R. Clem, Cranular and Superconducting - Glass Properties of the High- Temperature Superconductors, Physica C153+155:50 (1988).
4. E.B.Sonin and A.K.Tagantsev, Electrodynamics of the Josephson Medium in High-T_c Superconductors, Phys. Lett. 140A:127 (1989).
5. L.E.Amatuni, A.A. Akhumian, K.Y. Constantinian et al., Coherent Vortex Motion in High-T_c Superconducting Bridges, in: Ext Abst. Intern. Conf. Superc. Electronics, Japan, 1989, p.237.
6. L.E.Amatuni, A.A. Akhumyan, R.B. Airapewtyan et al., Intrinnsic Electromagnetic Radiation of Hihg- T_c Superconducting Thin-Film Bridge Structure, JETP Lett. 50: 386 (1989).
7. M.A.M. Gijs, D. Terpstra and H. Rogalla, Discrete Vortex Motion in Thin Film YBaCuO Weak Links, Sol. St. Comm. 71:1881 (1989).
8. V.N.Gubankov, V.P. Koshelets, K.K. Likharev and G.A. Ovsyannikov, Coherent Motion of Vortices in Superconducting Bridges of Large Dimensions, JETP Lett. 18: 171 (1973).
9. L.G.Aslamazov,, A.I.Larkin, Josephon Effect in Wide Superconducting Contacts, Sov. Phys. JETP 41: 381 (1975).

10. K.K.Likharev, Vortex Motion and the Josephson Effect in
    Superconducting Thin Bridges, Sov. Phys. JETP. 34:906 (1972).
11. J. Konopka and G. Jung, Emission of Microwaves from DC Biased YBaCuO
    Thin Films. Europhys. Lett.8:549 (1989).
12. A.F.Volkov, On the Josephson Effect in Granular Superconductors,in:
    Ext Abst. Intern. Conf.Superc. Electronics, Japan, 1989, p.288.
13. W. Xia and P.L. Death, Defects, Vortices, and Critical Current in
    Josephson-Junction Arrays, Phys. Rev. Lett. 63:1428 (1989).
14. A.T. Fiory, Interference Effects in a Superconducting Aluminum
    Film; Vortex Structure and Interactions, Phys. Rev. 7B:1881 -
    (1973).
15. G.A.Ovsyannikov, Z.G. Ivanov, G. Brorsson, T. Claeson, Hyper-
    vortices in Granular Superconducting Thin Film Bridges, Physica
    B165+166:1609 (1990).
16. D.G.Emelyanenkov, Yu.N. Inkin, B.I. Makhov et al., in:Proc. HTSC,
    1,(1989) (in Russian).
17. K.K. Likharev, Superconducting Weak links, Rev. Mod. Phys.. 51:-
    101 1979).
18. V.N.Gubankov, V.P. Koshelets, G.A. Ovsyannikov, Microwave Jose-
    phson Generation in Thin Film Superconducting Bridges, in:
    Proc. of Low Temp. Phys., Helsenki, v.4, p.120 (1976).
19. A.L.Gudkov, K.K. Likharev, M.Yu. Kupriyanov, Properties of
    Josephson Junction with Amorphous Silicon Interlayers, Sov.
    Phys. JETP 68:1478 (1988).
20. G. Jung, J. Konopka and S.Vitae, Synchronization of Radiating
    Intrinsic High-T_c Joseph Junctions, Physica B165+166:106 (1990).

BIFURCATIONS AND NONLINEAR PROPERTIES
OF THE BCS GAP EQUATION

Mads Peter Soerensen

Laboratory of Applied Mathematical Physics
The Technical University of Denmark, Bldg. 303
DK–2800 Lyngby, Denmark

INTRODUCTION

By adopting the BCS strategy in a tight binding model for layered high–temperature superconductors, it is possible to get a remarkable agreement with a number of experimental facts[1]. Accordingly, a more detailed study of the nonlinear properties of the BCS gap equation is of interest. One way to proceed is to investigate the iterative solution of the BCS gap equation defining a nonlinear map which can be studied with the tools developed for nonlinear dynamics[2].

Here we shall study the anisotropic BCS gap equation rather more direct using bifurcation theory on nonlinear algebraic equations. The gap equation is such an equation with the gap parameter as state variable and e.g. the temperature and Fermi energy as bifurcation parameters. For applications it is important to recognize that different equations are equivalent to simpler equations identifying the type of bifurcation which can occur[3]. In the following we shall study in particular pitchfork bifurcations in the BCS gap equation and their implication on physical properties as e.g. the electronic specific heat.

THE BCS GAP EQUATION

In Fourier space the BCS gap equation reads[4]

Nonlinear Superconductive Electronics and Josephson Devices
Edited by G. Costabile *et al.*, Plenum Press, New York, 1991

427

$$\Delta(k) = \frac{1}{N} \sum_{k'} V(k,k') \frac{\Delta(k')}{2E(k')} \tanh(\tfrac{1}{2} \beta E(k')) \ , \tag{1}$$

where $E(k) = \sqrt{\epsilon^2(k) + \Delta^2(k)}$ is the quasiparticle excitation energy, $\epsilon(k)$ is the quasiparticle energy in the normal state measured relative to the Fermi energy μ. Here, k denotes the wavevector and N is the number of lattice sites. $V(k,k')$ and $\Delta(k)$ are the pairing interaction and the gap parameter, respectively. β equals $1/k_B T$, where T is the absolute temperature and k_B is Boltzmann's constant.

The associated pairing Hamiltonian describing the superconducting state is

$$H = \sum_{k,\sigma} \epsilon(k) a^+_{k,\sigma} a_{k,\sigma} - \sum_k \sum_{k'} V(k,k') \, a^+_{k\uparrow} a^+_{-k\downarrow} a_{-k'\downarrow} a_{k'\uparrow} \ . \tag{2}$$

$a^+_{k,\sigma}$ and $a_{k,\sigma}$ being the creation and annihilation operators of an electron with wavevector k and spin σ. Within the tight binding description $\epsilon(k)$ is determined by Fourier transforming the kinetic part $H_{kin} = \sum_{ij} t_{ij} a^+_{i\sigma} a_{j\sigma}$ of the real space Hamiltonian. The t_{ij}'s are the transfer integrals describing the hopping of electrons from site j to site i and $\epsilon(k)$ becomes

$$\epsilon(k) = A[-2(\cos(k_x a) + \cos(k_y a)) + 4B\cos(k_x a) \cos(k_y a)$$

$$-2C \cos(k_z s) - \mu] \ . \tag{3}$$

In Eq. (3) we assume a unit cell with lattice constants $(a_x, a_y, a_z) = (a, a, s)$ and $k_\alpha a_\alpha = n_\alpha \pi/N_\alpha$, $-N_\alpha < n_\alpha \leq N_\alpha$, $\alpha = x, y, z$, corresponding to square sheets stacked in layers separated by the lattice constant s. If t_1, t_2, t_3 denote the hopping probabilities between nearest neighbours, next nearest neighbours, and between the sheets, respectively, then $A = -2t_1$, $AB = t_2$, and $AC = -t_3$. Typical values of A, B, and C are A = 0.153eV, B = 0.45, and C = 0.1 [5,6]. The pairing potential $V(k,k')$ is determined from[7].

$$V(k,k') = \frac{1}{N} \sum_r \sum_{r'} e^{-ik(r-r')} V(r,r') e^{ik'(r-r')} \ . \tag{4}$$

r and r' are the position vectors for the two pairing electrons. Assuming

$$V(r_i, r_j) = \sum_i \sum_j g_{ij} \, \delta(r_i - r_j - \Delta r_{ij}) \tag{5}$$

and restricting the sum over nearest neighbours with the interaction strength $g_{ij} = g_2$, we obtain

$$V(\mathbf{k},\mathbf{k'}) = 2g_2[\cos((k_x - k'_x)a) + \cos((k_y - k'_y)a)]. \tag{6}$$

Equation (6) can easily be extended to the case of onsite, next nearest, and interlayer interaction, respectively.

By taking the periodicity and the crystal symmetry into account the gap parameter can be expressed as

$$\Delta(\mathbf{k}) = 2\Delta_{2x}\cos(k_x a) + 2\Delta_{2y}\cos(k_y a) \ . \tag{7}$$

Introducing the notation

$$F(\mathbf{k}) = \frac{\Delta(\mathbf{k})}{2\,E(\mathbf{k})}\tanh(\tfrac{1}{2}\,\beta E(\mathbf{k})) \ , \tag{8}$$

the gap equation (1) using (7) reads

$$\Delta_{2x} = \frac{1}{N}\,g_2 \sum_{\mathbf{k}} \cos(k_x a)\ F(\mathbf{k}) = f_1(\Delta_{2x},\Delta_{2y}) \tag{9a}$$

$$\Delta_{2y} = \frac{1}{N}\,g_2 \sum_{\mathbf{k}} \cos(k_y a)\ F(\mathbf{k}) = f_2(\Delta_{2x},\Delta_{2y}) \ . \tag{9b}$$

Here, f_1 and f_2 are nonlinear functions of $(\Delta_{2x}, \Delta_{2y})$. The solutions of the gap equation (9) are classified as follows: The cases $\Delta_{2y} = \Delta_{2x}$ and $\Delta_{2y} = -\Delta_{2x}$ are denoted s–wave pairing and d–wave pairing, respectively. All other cases are denoted mixed s– and d–wave pairing.

The gap parameter $\Delta(\mathbf{k})$ can also be found from minimizing the Gibbs free energy[4]

$$G(\Delta_{\mathbf{k}}) = \frac{1}{N}\sum_{\mathbf{k}}\left[\epsilon(\mathbf{k}) - E(\mathbf{k}) + 2\Delta(\mathbf{k})\ F(\mathbf{k}) - \frac{2}{\beta}\ell n(1 + e^{-\beta E(\mathbf{k})})\right]$$

$$-\frac{1}{N^2}\sum_{\mathbf{k}}\sum_{\mathbf{k'}} V(\mathbf{k},\mathbf{k'})\ F(\mathbf{k})\ F(\mathbf{k'}) \tag{10}$$

at fixed chemical potential or Fermi energy μ. The stationary points of the Gibbs energy satisfies the gap equation (1). If more than one stationary point exist, the physical relevant solution is the one with lowest Gibbs free energy. In the case of a gap depending on two parameters such as Δ_{2x} and Δ_{2y} in Eq. (7), it is easy to get a quick overview of all stationary points, their stability and free energy, by drawing numerically a contour plot of the Gibbs energy in (10) as function of $(\Delta_{2x}, \Delta_{2y})$.

BIFURCATIONS IN THE BCS GAP EQUATION

Returning to the gap equation (9) we observe, that according to the implicit function theorem, the gap parameters Δ_{2x} and Δ_{2y} can be written as functions of (T,μ) in a nighbourhood of a solution point, provided the Jacobian

$$
J = \begin{bmatrix} \dfrac{\partial f_1}{\partial \Delta_{2x}} & \dfrac{\partial f_1}{\partial \Delta_{2y}} \\ \dfrac{\partial f_2}{\partial \Delta_{2x}} & \dfrac{\partial f_2}{\partial \Delta_{2y}} \end{bmatrix} \tag{11}
$$

taking at this solution point is regular. As T and μ are varied the Jacobian J may become singular and a bifurcation may occur. Due to the symmetry property that if $(\Delta_{2x}, \Delta_{2y})$ is a solution of (9), then $\pm(\Delta_{2y}, \Delta_{2x})$ is also a solution, we expect the appearance of pitchfork bifurcations as T and μ are varied. In particular the transition from the normal state $(\Delta_{2x} = \Delta_{2y} = 0)$ to the superconducting state $(\Delta_{2x} \neq 0$ and $\Delta_{2y} \neq 0)$ as the temperature T is varied is a pitch fork bifurcation. This is evident from Fig. 1 showing the bifurcation diagram obtained numerically using the parameter values

$$
B = 0.45, \quad C = 0.1, \quad g_2 = 3.0, \quad \text{and} \quad \mu = -2.3 \ . \tag{12}
$$

In solving the gap equation numerically, the parameter A and Boltzmann's constant are set to unity, implying that all energies are measured in units of A and the absolute temperature is measured in units of A/k_B. The transistion temperature is denoted T_c and in Fig. 1 T_c equals 0.541.

In order to identify strictly the pitchfork bifurcation at T_c, one can use the Liapunov–Schmidt reduction procedure to transform the two dimensional gap equation (9) to an one dimensional equation $g(x,T) = 0$ which is equivalent to $x^3 - Tx = 0$ [3]. If the function g satisfies

$$
g = g_x = g_{xx} = g_T = 0 \quad \text{and} \quad g_{xxx} g_{xT} > 0 \tag{13}
$$

430

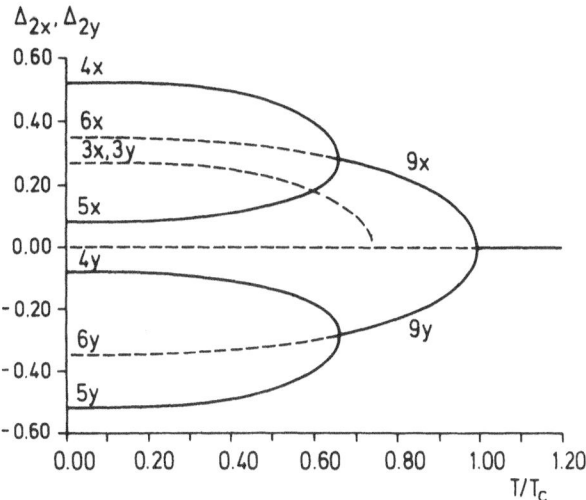

FIG. 1. Bifurcation diagram showing Δ_{2x} and Δ_{2y} versus the reduced temperature T/T_c using the parameter values in Eq. (12). $T_c=0.541$. Solid curves show stable solution points and dashed curves show unstable ones. The curve 3x=3y corresponds to s–wave pairing, 6x=–6y, and 9x=–9y to d–wave pairing, and 4x=–5y and 4y=–5x to mixed s– and d–wave pairing.

at a bifurcation point, the number of solutions of $g(x,T) = 0$ jumps from one to three as T crosses T_c from above. The subscripts x and T denote partial derivatives with respect to the the state variable x and the bifurcation parameter T, respectively. This means that Eq. (13) identifies the pitchfork bifurcation. Using results from Reference 3 the partial derivatives of g can be determined directly from f_1 and f_2. By numerical differentiation the following values of the derivatives in (13) have been found at the bifurcation point $T=T_c$: $g_{xx} = -6.6 \cdot 10^{-12}$, $g_T = -1.5 \cdot 10^{-19}$, $g_{xxx} = 3.60$, and $g_{xT} = 0.81$, showing that within numerical errors Eq. (13) is fulfilled and accordingly the bifurcation at T_c is a pitchfork bifurcation.

At temperatures just below T_c, Δ_{2y} equals $-\Delta_{2x}$, i.e. the solution has d–wave symmetry. Decreasing the temperature further leads to a second pitchfork bifurcation at $T_{c1} = 0.37$ resulting in a transition from d–wave to mixed s– and d–wave symmetry. In Fig. 1 all equilbrium points of the Gibbs energy are shown and their stability can be deduced from Fig. 2 depicting contour plots of the Gibbs energy at the temperatures (a) T = 0.8, (b) T = 0.4, and (c) T = 0. Fig. 2a reveals the stable trivial solution $(\Delta_{2x}, \Delta_{2y}) = (0.0)$ at T = 0.8 above T_c. At the temperature T = 0.4 lying between T_{c1} and T_c the trivial solution $(\Delta_{2x}, \Delta_{2y}) = (0.0)$ has turned into a saddle point and two stable equilibrium points $(\Delta_{2x},\Delta_{2y}) = (\Delta, -\Delta) \neq (0.0)$ and $(\Delta_{2x},\Delta_{2y}) = (-\Delta,\Delta)$ have emerged in addition to the trivial solution (Fig. 2b). Note that due to the symmetry properties of the solutions of the gap equation, only one stable equilibrium point $(\Delta_{2x},\Delta_{2y})$ is shown. The other one is given by $(\Delta_{2x}',\Delta_{2y}') = $

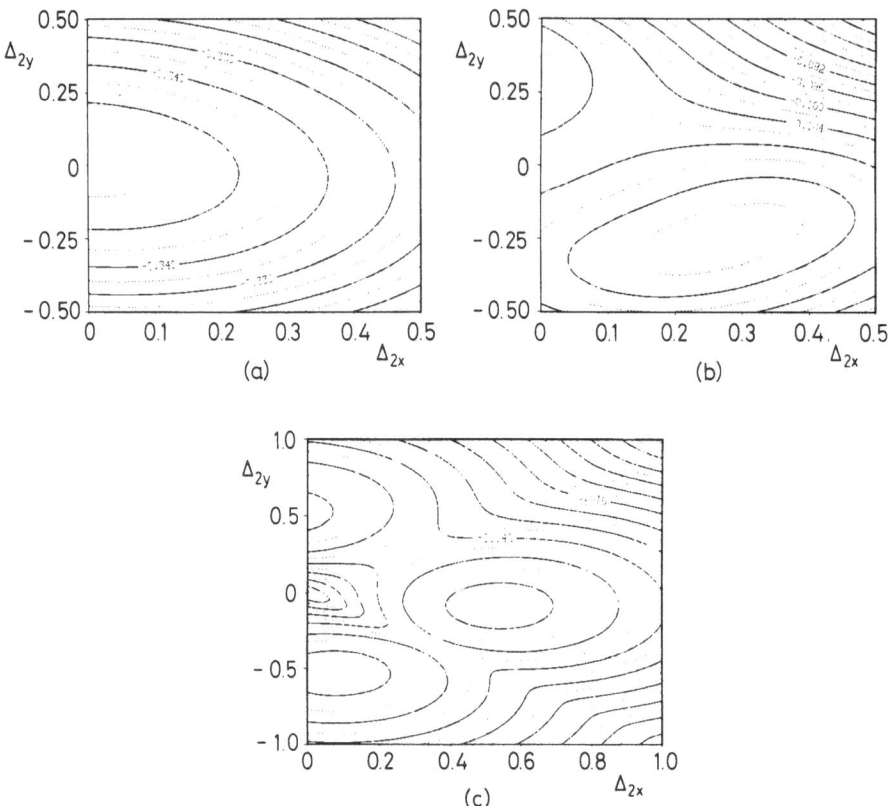

FIG. 2. Contour plots of the Gibbs energy as function of $(\Delta_{2x}, \Delta_{2y})$ for the parameter values in Eq. (12) and at three different temperatures (a) T=0.8 > T_c, (b) T=0.4 < T_c, and T=0.0.

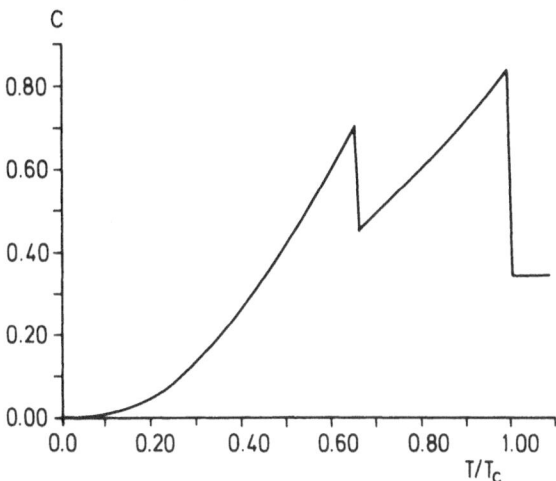

FIG. 3. The temperature dependence of the specific heat using the parameter values in Eq. (12). The discontinuities signal the pitchfork bifurcations from mixed s,d–wave to a d–wave state and from d–wave to the normal state.

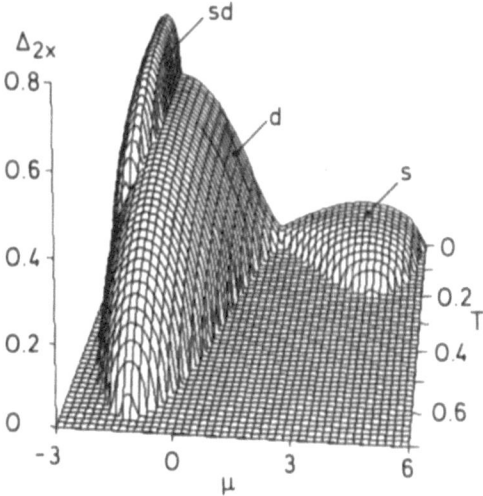

FIG. 4. The gap parameter Δ_{2x} shown as function of T and μ. The lettering s, d, and sd, identifies regions with s–, d– and mixed s– and d–wave symmetry, respectively.

$- (\Delta_{2x}, \Delta_{2y})$. In Fig. 2c the temperature has been decreased to zero and the figure shows 4 stable equilibrium points corresponding to mixed s– and d–wave symmetry. In addition 4 saddle points are present lying along the lines $\Delta_{2y} = \Delta_{2x}$ (s–wave symmetry) and $\Delta_{2y} = -\Delta_{2x}$ (d–wave symmetry), respectively. Finally, we observe that the zero solution $(\Delta_{2x}, \Delta_{2y}) = (0.0)$ has turned into a local maximum. The dependence on the temperature T of all stable as well as unstable equilibrium points has been depicted in Fig. 1. The solid lines show stable equilibrium points and the dashed lines show the unstable ones.

In the electronic specific heat C given by

$$C = T \frac{dS}{dT} = \frac{1}{2} \beta^2 k_B \frac{1}{N} \sum_{\mathbf{k}} [E(\mathbf{k}) + \beta \frac{dE(\mathbf{k})}{d\beta}] \, E(\mathbf{k}) \, \text{sech}^2(\frac{1}{2} \beta E(\mathbf{k})) \qquad (14)$$

and shown in Fig. 3 as function of temperature T, the two pitchfork bifurcations at T $= T_c$ and $T = T_{c1}$ manifest themselves as discontinuities. In Eq. (14) S denotes the entropy.

In reference 2 the bifurcation phenomena have been mapped out for the case of varying the Fermi energy at zero temperature. For $1.7 < \mu < 6.0$ the solution of the gap equation has s–wave symmetry, for $-1.3 < \mu < 1.7$ the solution has d–wave symmetry and for $-3.0 < \mu < -1.3$ it is mixed s– and d–wave. In Fig. 4 the gap parameter $\Delta_{2x}(T,\mu)$ for the solution with lowest Gibbs energy is depicted as function of T and μ. The transition temperature T_c is given by the line in the (T,μ)–plane where Δ_{2x} vanish. In Fig. 4 the solutions with s– , d– , and mixed s– and d–wave symmetry are marked with the lettering s, d, and sd, respectively. The transitions from the normal to the superconducting state and from d–wave to mixed s– and d–wave symmetry result from pitchfork bifurcations. The transition from s– to d–wave symmetry at low temperatures is a little more complicated[2]. For μ–values around 1.7 the s–wave solution coexist with the d–wave solution, and the transition from one type of solution to the other results from an exchange of minimum Gibbs energy.

SUMMARY

The BCS strategy has been applied to a tight binding model of a layered high–temperature superconductor with nearest neighbour pairing interaction in the planes. The BCS gap equation is written as a nonlinear equation determining the gap state variables, with the auxiliary parameters T and μ being the bifurcation parameters. The implicit function theorem then states that if the Jacobian in Eq.

(11) is regular in a solution point, the gap parameters can be written as functions of T and μ. In points where the Jacobian is singular, bifurcations may take place. The solution of the gap equation can be classified according to s–wave, d–wave, and mixed s–and d–wave symmetry. It is evident from the numerical solutions that the transition from the normal state to the superconducting state results from a pitchfork bifurcation as the temperature is decreased below the critical temperature. Similarly the transition from d–wave to mixed s– and d–wave symmetry follows from a pitchfork bifurcation.

The notion of bifurcations taking place in the anisotropic gap equation is important for physical interpretations of the solutions of the gap equation. If the temperature is changed and a bifurcation point is reached, the slope of the gap will be infinite at the bifurcation point due to the Jacobian being singular. In the electronic specific heat the derivative of the gap with respect to T enters and accordingly the specific heat shows discontinuities at the bifurcation points.

Finally, we mention, that breaking the symmetry is expected to lead to a perturbed pitchfork bifurcation at $T=T_{c1}$. Even for small perturbations the pitchfork bifurcation changes dramatically and accordingly a number of physical quantities as the electronic specific heat or NMR relaxation rate may change their temperature dependence dramatically compared to what is usually observed within the framework of the BCS theory.

ACKNOWLEDGMENTS

P. Gross, M. Brøns and H. True are gratefully acknowledged for their advices and stimulating discussions on bifurcation theory.

REFERENCES

1.   T. Schneider and M.P. Soerensen, "Correlation between Hall Coefficient, Penetration Depth, Transition Temperature, Gap Anisotropy and Hole Concentration in Layered High–Temperature Superconductors", submitted to Z. Phys. B.

2.   M.P. Soerensen, T. Schneider and M. Frick, "Nonlinear Properties of the BCS Gap Equation", presented at the NATO A.R.W. on "Microscopic Aspect of Nonlinearity in Condensed Matter", Florence, Italy, June 7–13, 1990.

3. M. Golubitsky and D.G. Schaeffer, "Singularities and Groups in Bifurcation Theory". Springer Verlag, New York Inc., 1985.

4. A.J. Leggett, **Rev. Mod. Phys.**, 47(2):331 (1975).

5. T. Schneider and M. Frick, "Experimental Constraints and Theory of Layered High–Temperature Superconductors", Proceedings of the 16th Course "Earlier and Recent Aspects of Superconductivity" of the International School of Materials Science and Technology, Erice, Italy, July 4–16, 1989 (Springer Verlag).

6. T. Schneider, H. de Raedt, and M. Frick, Z. Phys. B76:3, 1989.

7. J.R. Schrieffer, "Theory of Superconductivity", The Benjamin/Cummings Publishing Company, Inc., Massachusetts, 1964.